MARINE FISHES
OF THE GREAT BARRIER REEF
AND SOUTH-EAST ASIA

Burma
Thailand
Vietnam
Kampuchea
Bay of Bengal
Andaman Is.
Andaman Sea
Nicobar Is.
Gulf of Thailand
South China Sea
Sulu Sea
Celebes Sea
Strait of Malacca
Malaysia
Singapore
Malaysia
Borneo
Sumatra
INDONESIA
Sulawesi
Sunda Islands
Java Sea
Java
Flores Sea
Bali
Lombok
Sumbawa
Sumba
Flores
Christmas I.
Cocos (Keeling) Islands
Indian Ocean
Western Australia
20°
10°
Equator
10°
20°
Tropic of Capricorn
90°
100°
110°
120°

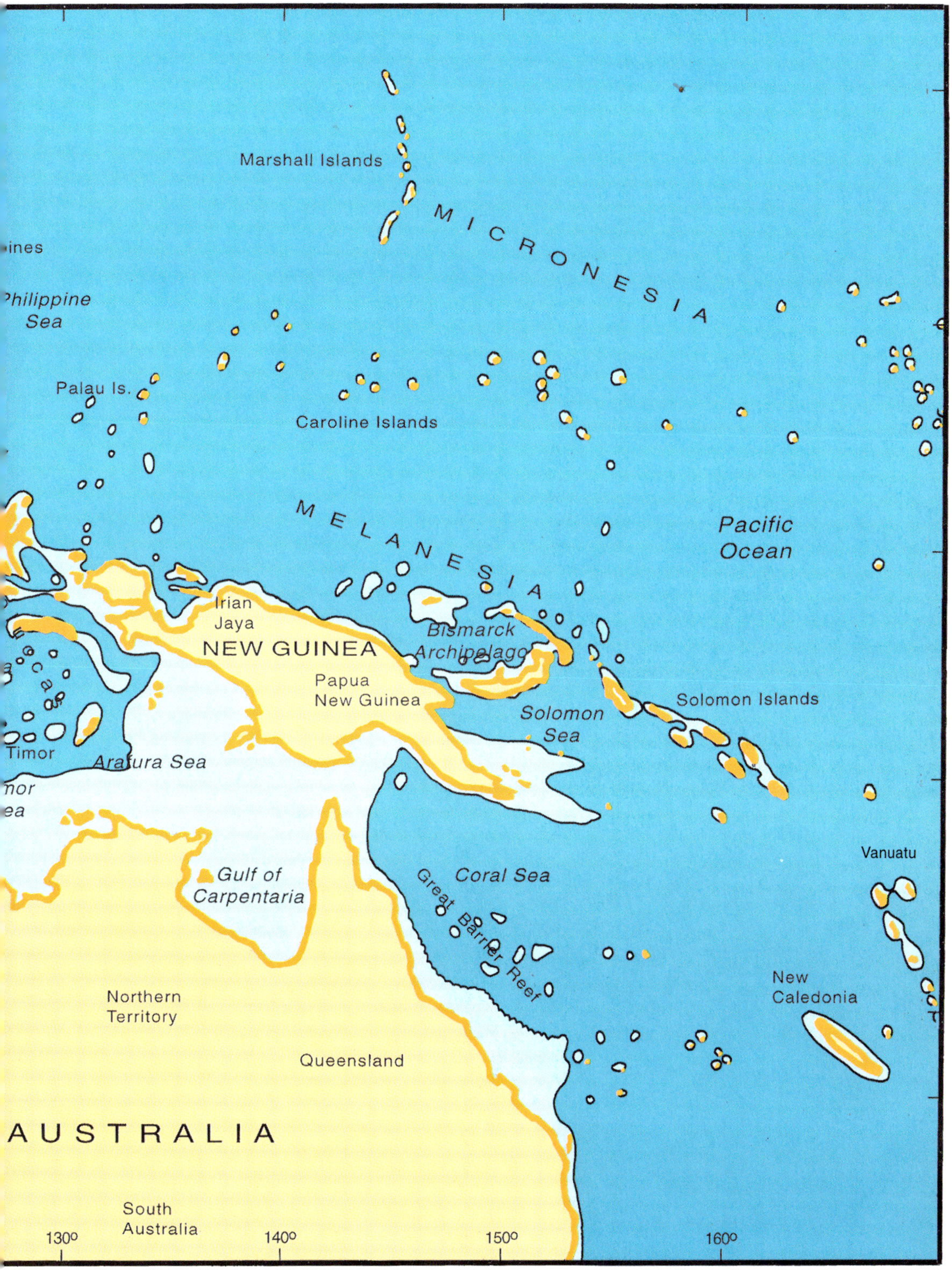

Marshall Islands
MICRONESIA
ines
Philippine Sea
Palau Is.
Caroline Islands
MELANESIA
Pacific Ocean
Irian Jaya
NEW GUINEA
Papua New Guinea
Bismarck Archipelago
Solomon Islands
Solomon Sea
Timor
Arafura Sea
Coral Sea
Great Barrier Reef
Gulf of Carpentaria
Vanuatu
New Caledonia
Northern Territory
Queensland
AUSTRALIA
South Australia
130°
140°
150°
160°

MARINE FISHES

OF THE GREAT BARRIER REEF AND SOUTH-EAST ASIA

BY
GERRY ALLEN

ILLUSTRATIONS BY
ROGER SWAINSTON
AND JILL RUSE

PERIPLUS
EDITIONS

ABOUT THE AUTHORS

DR GERRY ALLEN is the author of nearly 300 scientific articles and 25 books. He has been Senior Curator in the Department of Ichthyology since 1974 and is an internationally recognised authority on the classification of coral reef fishes. He is also an accomplished underwater photographer who has dived extensively throughout the world's tropics. *Photo: Roger Steene*

ROGER SWAINSTON is acclaimed as one of the world's finest fish artists. His illustrations are more than accurate scientific records: his meticulous attention to detail, coupled with years of experience in, and love of angling, diving, and marine life, infuse his paintings with a special flavour. *Photo: Barry Hutchins*

JILL RUSE was born in Western Australia and has nearly 25 years of experiece with wildlife illustration. She has done a variety of freelance jobs for the WA Museum for the past 12 years. This is her first fish assignment, but hopefully not her last. Jill does not dive, but has an excellent eye for colours and natural shapes. Most of her paintings in this volume were based on Dr. Allen's underwater photographs. *Photo: Douglas Elford*

© Western Australian Museum, 1997

The National Library of Australia
Cataloguing-in-Publication entry:

Allen, Gerald R., 1942-.
Marine fishes of the Great Barrier Reef and South-east Asia.

Includes index.
ISBN 0 7309 8751 5.

1. Marine fishes – Australia, Northern. 2. Marine fishes – Queensland – Great Barrier Reef. 3. Marine fishes – Asia, Southeastern. I. Western Australian Museum. II. Title.

597.0994

First Published 1988
Revised Edition 1990
Second Revised Edition 1993
Reprinted 1994
Reprinted 1995
Third Revised Edition 1997

Illustrated by Roger Swainston and Jill Ruse

Cover design by Robyn Mundy

Printed by Kaleidoscope Print & Prepress, Perth, Western Australia

Published by the Western Australian Museum Francis Street, Perth, Western Australia 6000

Dedicated to my father, Rex Robert Allen

FOREWORD

The present volume is an expanded version of *The Marine Fishes of North-Western Australia*, first published by the Western Australian Museum in 1988. The former title enjoyed a remarkable success, selling 17,000 copies. The book has proven to be an indispensible reference, especially for anglers, divers, and aquarists. However, one of the shortcomings of the original edition was its limited coverage - essentially coastal waters, therefore excluding the remote reefs at the edge of Western Australia's continental shelf. The original idea behind the new book was to remedy this situation by including fishes from the major offshore reefs - Hibernia, Cartier, Ashmore, Seringapatam, Scott, and Rowley Shoals. These areas have a coral reef fish fauna that is considerably richer than mainland Western Australia, and which in many respects is similar to that found on Queensland's Great Barrier Reef. After carefully analysing the necessary modifications needed for the expanded coverage I realised that with a little extra effort, the book could be tailored to suit the need's of fish enthusiasts not only in Western Australia, but across the entire northern half of our continent, including the Great Barrier Reef. But why stop there? The addition of relatively few species from the Indonesian- Malaysian Archipelago would make the book an invaluable reference for the entire region.

The key to transforming this idea into reality was finding a talented artist. Roger Swainston, whose excellent paintings were utilised in the original *Marine Fishes of North-Western Austrlalia* and still form the basis of the new title, was living abroad and no longer available. Fortunately I was introduced to Jill Ruse, a freelance artist, who had done various jobs in the past for the Museum's Publications Department. Jill had never painted fishes previously, but she caught on quickly. The thirty-six new plates that adorn this book are proof of her superb skills.

Contents

	Page
Foreword	7
Introduction	11
Area of Coverage	12
Faunal Composition	13
Biology of Reef Fishes	14
Ecology of Reef Fishes	16
Classification of Fishes	18
Preserving Fishes	19
Sending Specimens to the Museum	20
Fish Photography	21
Dangerous Fishes	22
Fish vs. Fishes	23
How to Use this Book	24
Colour Patterns	26
Edibility Ratings	26
Acknowledgements	27
Guide to Families	29
Index	255

Fish Plates (with artist indicated)

		Page
1	Sharks (Roger Swainston)	40
2	Sharks (Roger Swainston)	42
3	Sharks and Rays (Roger Swainston)	44
4	Sharks and Rays (Roger Swainston)	46
5	Rays and Ghost Sharks (Roger Swainston)	48
6	Herrings and Relatives (Roger Swainston)	50
7	Moray Eels (Roger Swainston)	52
8	Moray Eels (Jill Ruse)	54
9	Snake Eels and Conger Eels (Roger Swainston)	56
10	Snake Eels and Garden Eels (Jill Ruse)	58
11	Catfishes and Lizardfishes (Roger Swainston)	60
12	Frogfishes, Anglerfishes, Clingfishes, and Cuskeels (Roger Swainston)	62
13	Flyingfishes, Garfishes, Longtoms, and Hardyheads (Roger Swainston)	64
14	Squirrelfishes (Roger Swainston)	66
15	Squirrelfishes (Jill Ruse)	68
16	Knight Fishes, Dories, Flutemouths, etc. (Roger Swainston)	70
17	Seahorses and Pipefishes (Roger Swainston)	72
18	Pipefishes and Scorpionfishes (Jill Ruse)	74
19	Scorpionfishes (Roger Swainston)	76
20	Scorpionfishes (Roger Swainston)	78
21	Velvetfishes, Gurnards, and Flatheads (Roger Swainston)	80
22	Gropers (Roger Swainston)	82
23	Gropers (Roger Swainston)	84
24	Gropers (Roger Swainston)	86
25	Gropers (Jill Ruse)	88
26	Gropers (Jill Ruse)	90
27	Anthias (Jill Ruse)	92
28	Longfins, Dottybacks, and Relatives (Roger Swainston)	94
29	Dottybacks (Jill Ruse)	96
30	Perchlets, Grunters, and Whitings (Roger Swainston)	98
31	Bigeyes and Cardinalfishes (Roger Swainston)	100
32	Cardinalfishes (Roger Swainston)	102
33	Cardinalfishes (Roger Swainston)	104
34	Cardinalfishes (Jill Ruse)	106
35	Cardinalfishes (Jill Ruse)	108
36	Cardinalfishes (Jill Ruse)	110
37	Trevallies (Roger Swainston)	112
38	Trevallies (Roger Swainston)	114
39	Dolphinfishes and Trevallies (Roger Swainston)	116
40	Trevallies, Ponyfishes, etc. (Roger Swainston)	118

		Page			Page
41	Seaperches (Snappers) (Roger Swainston)	120	74	Wrasses (Roger Swainston)	186
42	Seaperches (Snappers) (Roger Swainston)	122	75	Wrasses (Roger Swainston)	188
43	Seaperches (Snappers) (Jill Ruse)	124	76	Wrasses (Jill Ruse)	190
44	Seaperches (Snappers) and Sweetlips (Jill Ruse)	126	77	Wrasses (Roger Swainston)	192
			78	Wrasses (Roger Swainston)	194
45	Sweetlips, Fusiliers, and Banjofishes (Roger Swainston)	128	79	Wrasses (Jill Ruse)	196
			80	Parrotfishes (Roger Swainston)	198
46	Fusiliers and Tilefishes (Jill Ruse)	130	81	Parrotfishes (Roger Swainston)	200
47	Emporers and Breams (Roger Swainston)	132	82	Parrotfishes (Jill Ruse)	202
48	Emperors (Jill Ruse)	134	83	Grubfishes, Jawfishes, and Stargazers (Roger Swainston)	204
49	Threadfin Breams (Roger Swainston)	136	84	Blennies (Roger Swainston)	206
50	Monocle Breams and Silver Biddies (Roger Swainston)	138	85	Blennies (Roger Swainston)	208
			86	Blennies (Jill Ruse)	210
51	Monocle Breams and Goatfishes (Jill Ruse)	140	87	Threefins and Dragonets (Roger Swainston)	212
52	Goatfishes (Roger Swainston)	142	88	Dragonets and Gobies (Jill Ruse)	214
53	Croakers, Bullseyes, and Drummers (Roger Swainston)	144	89	Gobies (Roger Swainston)	216
			90	Gobies (Jill Ruse)	218
54	Archerfishes and Batfishes (Roger Swainston)	146	91	Gobies and Dartfishes (Jill Ruse)	220
55	Butterflyfishes (Roger Swainston)	148	92	Gobies, Gudgeons, and Spinefeet (Roger Swainston)	222
56	Butterflyfishes (Roger Swainston)	150	93	Surgeonfishes and Spinefeet (Jill Ruse)	224
57	Butterflyfishes (Jill Ruse)	152	94	Surgeonfishes (Roger Swainston)	226
58	Butterflyfishes and Angelfishes (Jill Ruse)	154	95	Surgeonfishes (Jill Ruse)	228
59	Angelfishes (Roger Swainston)	156	96	Billfishes and Tunas (Roger Swainston)	230
60	Angelfishes (Jill Ruse)	158	97	Mackerels and Tunas (Roger Swainston)	232
61	Damselfishes (Roger Swainston)	160	98	Flounders and Soles (Roger Swainston)	234
62	Damselfishes (Jill Ruse)	162			
63	Damselfishes (Jill Ruse)	164	99	Duckbills, Driftfishes, and Tripodfishes (Roger Swainston)	236
64	Damselfishes (Roger Swainston)	166			
65	Damselfishes (Jill Ruse)	168	100	Triggerfishes (Roger Swainston)	238
66	Damselfishes (Roger Swainston)	170	101	Triggerfishes and Leatherjackets (Jill Ruse)	240
67	Damselfishes and Hawkfishes (Jill Ruse)	172	102	Leatherjackets (Roger Swainston)	242
68	Hawkfishes and Eel Blennies (Roger Swainston)	174	103	Boxfishes and Puffers (Roger Swainston)	244
69	Mullets, Threadfins, and Barracudas (Roger Swainston)	176	104	Boxfishes and Puffers (Jill Ruse)	246
			105	Puffers and Porcupinefishes (Roger Swainston)	248
70	Wrasses (Roger Swainston)	178			
71	Wrasses (Jill Ruse)	180	106	Supplementary Plate (Roger Swainston)	250
72	Wrasses (Jill Ruse)	182			
73	Wrasses (Jill Ruse)	184			

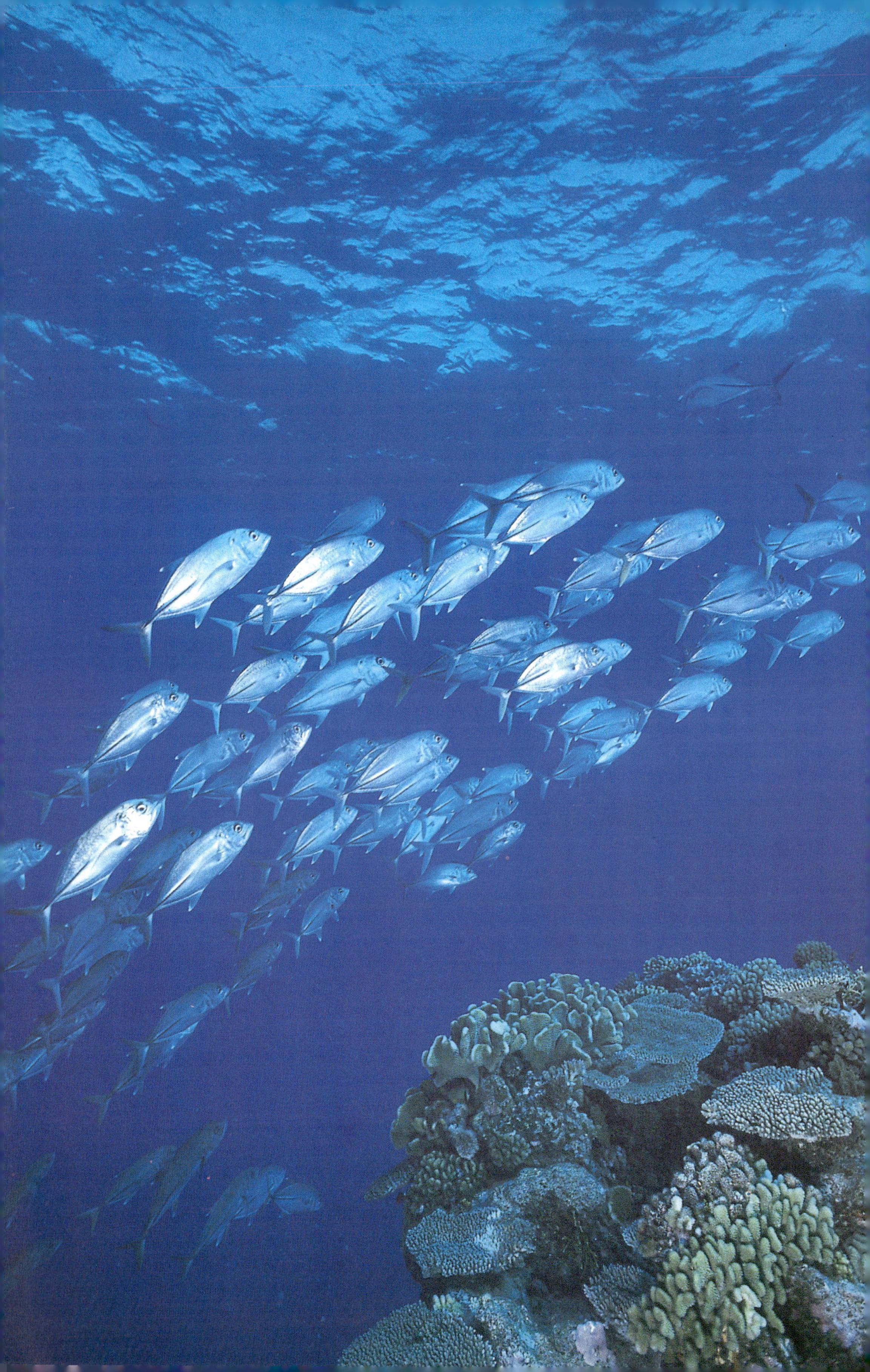

INTRODUCTION

There are approximately 22,000 types of fishes inhabiting our planet, including approximately 13,500 marine species. Seas encompassing northern Australia and the tropics that lie immediately northward are inhabited by the richest fish fauna on earth. Official counts are lacking, but an estimated 4,000 species occur in the region, or about 30 per cent of the world's total marine fishes. For most families of tropical marine organisms there are more species represented in this area than any place on earth. For example, among the reef-dwelling damselfishes, a family containing about 330 species worldwide, about 175 species or over 50 percent of the world's total inhabit the region sometimes referred to as the Indo-Australian Archipelago.

Why is this region populated by so many species? No doubt a combination of several factors are responsible. Certainly among the most important are a lengthy history of favourable climatic conditions (ie., warm water), a diversity of habitat types, and a tumultuous geological and hydrological past. In the latter category events such as sea level changes, ocean current shifts, volcanism, and continental drift have created isolating barriers that greatly enhanced the speciation process.

One of the most significant factors responsible for the region's great plethora of fishes and other marine organisms is its vast tropical shoreline with an array of diverse habitats. Coral reefs are the most complex habitat system. They provide abundant living space and seemingly endless "survival opportunities" for a wealth of creatures. Spires of tabular and branching corals are the reef's equivalent of multi-story tenements. Not only do they house thousands of living polyps, they also serve as a retreat for legions of brightly coloured fishes that swarm above. Ledges, caves and crevices form the inner sanctum of the reef city, which is populated by shy, seldom seen fishes that may only emerge for night-time feeding patrols.

The sand and rubble fringe surrounding individual reef complexes appears devoid of fish life. But first impressions are deceptive. Close inspection reveals an entire community of specially adapted species. Although this habitat contains far fewer fishes than nearby reefs, its inhabitants are equally interesting. Common adaptations include camouflage colouration and burrowing behaviour. Other major habitats in the region include inshore coastal environments - vast stretches of sandy or rocky shores, interspersed with estuaries and coastal mangroves - and open offshore oceanic habitats. All of these zones are populated by distinct fish communities.

The book contains nearly 2,000 hand-painted illustrations featuring 1,635 individual species. The paintings were completed over a 3-year period and are primarily based on photographs or colour transparencies of either live fishes taken underwater or freshly caught specimens. In many cases preserved specimens at the Western Australian Museum have been consulted to ensure accuracy of detail and proportions. The end result is a colourful and highly comprehensive guide to the sea fishes of northern Australia and the adjacent South-east Asian region.

◀ Schools of Bigeye Trevally *(Caranx sexfasciatus)* are common on reefs throughout the region. (P. Kuhn)

AREA OF COVERAGE

This book provides coverage of tropical Western Australia, Northern Territory, Queensland, and the region immediately north of Australia encompassing Malaysia, Indonesia, Philippines, New Guinea, and Solomon Islands. The main emphasis is placed on reef and shore fishes - relatively good coverage is provided except for gobies (Gobiidae) and threefins (Tripterygiidae). Both of these families contain small, cryptic, seldom seen species, whose classification has not been satisfactorily studied - many of the species are difficult to identify, even by trained specialists.

Relatively good coverage is also extended to families containing species of interest to anglers. Foremost in this respect are the trevallies and their relatives (Carangidae), tunas and mackerels (Scombridae), and billfishes (Xiphiidae and Istiophoridae). Less than full coverage is given to families containing cryptic, hence seldom encountered species (Antennariidae for example), or species living in deeper sections of continental shelves. Excluded from coverage are the true deep sea fishes which for the most part, live well offshore, below 200 m depth (although some make daily migrations to the surface).

▲ A diver inspects a colony of anemonefishes (*Amphiprion melanopus*) in the Coral Sea off Australia's north-east coast. (R. Steene)

FAUNAL COMPOSITION

Most of the region's reef and shore fishes can be broadly described as being of Indo-Pacific origin. In other words, they belong to the overall reef fish community that ranges across the vast reaches of the tropical Indian and Pacific oceans. Although individual species, and particularly the mixture of species present, vary greatly from one locality to the next in this huge region, there is a general faunal theme that pervades. Nearly all families and many genera are widely distributed throughout the region. Also the dominant (i.e., most speciose) families tend to be the same regardless of locality. Dominant groups across this region usually include such families as gobies, wrasses, damselfishes, gropers, moray eels, cardinalfishes, and surgeonfishes.

The reason for the region's relative homogeneity in faunal composition is at least partly explained by examining the life cycle of reef fishes. With few exceptions most species have a pelagic larval stage which is transported by ocean currents for variable distances depending on hydrological conditions and duration of the larval period. Until recently the length of larval life for most fishes was an unknown factor, but thanks to otolith aging techniques our knowledge in this area is rapidly expanding. Essentially this technique consists of counting daily microscopic growth rings that appear on the bones of the inner ear (otoliths). We now know that the larval duration is highly variable, ranging from just a few days up to nearly two months, with an average length of about 3-4 weeks. Although the adults of most reef fishes are highly home-ranging or territorial, a homogeneous gene-pool is maintained over a broad area by the dispersal capabilities of the larval stage.

Many Indo-Australian reef fishes are distributed widely across the Indian and West and Central Pacific oceans. Species such as the Racoon Butterflyfish (Plate 55-13) and Pacific Gregory (Pl. 66-12) range from the shores of eastern Africa to the Hawaiian Islands and a few others such as the Moorish Idol (Pl. 92-9) and Longnose Hawkfish (Pl. 67-18) extend even farther, to the coast of the Americas. Indeed, throughout the Indo-Pacific region a significant segment of the fauna consists of similar widespread species. Another important component of the fauna consists of species that have more limited regional distributions. A number of species such as the Honey-Head Damsel (Pl. 61-8) and Rainbow Monocle Bream (Pl. 51-12) are mainly confined to what biogeographers refer to as the Indo-Australian Archipelago, which encompasses the Malay Peninsula, Indonesia, Philippines, northern Australia, and Melanesia. At the bottom end of the scale, a few species such as the Banggai Cardinalfish (Pl. 34-7) have an extremely limited range. It is only found among a small group of islands off central-east Sulawesi. This fish, and others that are similarly restricted, usually lack a pelagic larval stage, which prevents their dispersal.

BIOLOGY OF REEF FISHES

The region's tremendous diversity of inshore fishes is reflected in a wide variety of reproductive habits and life history strategies. The following discussion is intended to give an overview of the most common patterns. More detailed information is available in the scientific literature or semi-popular works such as Thresher's (1984) *Reproduction in Reef Fishes*. The majority of reef fishes are egg layers that employ external fertilisation. Relatively few species bear live young that are prepared to fend for themselves at birth. Included in the latter category are sharks, rays, and cusk eels. Basically two patterns of oviparous or egg-laying reproduction is evident in most reef species. Females of many fishes, including the highly visible wrasses and parrotfishes, scatter relatively large numbers of small, positively buoyant eggs into open water where they are summarily fertilised by the male. The spawning event is typically preceded by nuptial chasing, temporary colour changes, and courtship display in which fins are erected. This behaviour is generally concentrated into a short period, often at sundown or shortly afterwards. This pattern is seen in diverse groups such as lizardfishes, angelfishes, wrasses, parrotfishes, and boxfishes. Typically either pair or group spawning occurs in which the participants make a rapid dash towards the surface, releasing their gonadal products at the apex of the ascent.

The fertilised eggs float near the surface and are dispersed by waves, winds, and currents. Hatching occurs within a few days and the young larvae are similarly at the mercy of the elements. Recent studies of the daily growth rings found on the ear bones (otoliths) of reef fishes indicate that the larval stage generally varies from about 1-8 weeks depending on the species involved. The extended larval period no doubt accounts for the wide dispersal of many reef species. For example, many fishes that occur in our region have geographic ranges that extend from East Africa to Polynesia.

A second reproductive pattern involves species that lay their eggs on the bottom, frequently in rocky crevices, empty shells, sandy depressions, or on the surface of invertebrates such as sponges, corals, or gorgonians. Among the best known fishes in this category are the damselfishes, gobies, and triggerfishes. These fishes often prepare the surface prior to egg deposition by cleaning away detritus and algal growth. Bottom spawners also exhibit elaborate courtship rituals which involve much aggressive chasing and displaying. This behaviour has probably been best studied amongst the damselfishes. In addition, one or both parents may exhibit a certain degree of nest-guarding behaviour in which the eggs are kept free of debris and guarded from potential egg feeders such as wrasses and butterflyfishes. A very specialised mode of parental care is seen in cardinalfishes, in which the male broods the egg mass in its mouth. Similarly, male pipefishes and seahorses brood their eggs on a highly vascularised region of the belly or underside of the tail. As a rule the eggs of benthic nesting fishes are more numerous, larger, have a longer incubation period, and are at a more advanced developmental stage when hatched, compared to the eggs and larvae of pelagic spawning fishes. Hatching may require up to one week (in anemonefishes for example) and the larvae then lead a pelagic existence for up to several weeks before settling on the bottom in a suitable reef habitat.

There is very little information on the longevity of most reef fishes. Perhaps one of the longest life spans is that of the Lemon Shark which may reach 50 years or more. Most of the larger reef sharks probably live at least to an age of 20-30 years. In general the larger reef fishes such as gropers, snappers, and emperors tend to live longer than smaller species. Otolith aging techniques indicate that large gropers may live at least 25 years and some snappers approximately 20 years. Most of our knowledge of smaller reef fishes has resulted from aquarium studies. The values obtained from captive fishes may exceed the natural longevity due to lack of predation and the protective nature of the artificial environment. Batfishes (*Platax*) are known to survive for 20 years and even small species such as damselfishes and angelfishes may reach an age of 10 years or more.

▲ Many-spotted Sweetlips and small reef fishes (mainly Anthias) are typical reef inhabitants throughout the region. (R. Steene)

ECOLOGY OF REEF FISHES

The majority of fishes included in this book are generally considered to be inhabitants of coral reefs. However, reefs are highly complex systems, consisting of numerous microhabitats. In general, coral reef fishes are finely synchronised to their environment.

Each species exhibits very precise habitat preferences that are dictated by a combination of factors including the availability of food and shelter, and various physical parameters such as depth, water clarity, currents, and wave action. The huge number of species found on coral reefs is a direct reflection of the high number of habitat opportunities afforded by this environment.

Coral reef fishes generally exhibit a higher degree of habitat partitioning than do fishes from cooler seas. A good example of the fine scale on which this principle operates is the Urchin Clingfish (Pl. 12-6). It is usually found amongst the spines of *Diadema* sea urchins or nearby branching corals and feeds primarily on the tube feet of its host urchin or on coral-burrowing molluscs. The coral reef offers numerous examples of fishes that have similar narrow habitat and feeding requirements. Water depth is also an important partitioning factor, and again there are numerous examples of coral reef fishes that have well defined depth ranges. In the very broadest sense there are three main depth categories for reef fishes: shallow (0-4 m), intermediate (5-19 m), and deep (20 m+). The depth limits of these zones may locally vary depending largely on the degree of shelter and sea conditions. The shallow environment is typified by wave action which in highly protected areas such as coastal bays or lagoons may exert its effect down to only a few cm. On the contrary in exposed outer reef structures the effect of surface waves may sometimes be felt below 10 m. The intermediate zone harbours the greatest abundance of fishes and live corals. Here wave action is minimal, although currents are often strong, and sunlight is optimal for reef-building corals. The deep outer reef slope is characterised by reduced light levels, hence fewer corals and fishes. Although species numbers are reduced the species that occur in this habitat are among the most interesting of coral reef fishes. A high percentage of the new fishes that have been discovered on coral reefs in the past three decades were collected on deep reefs by SCUBA diving scientists.

The region's reef environments can be broadly classified into two major categories: sheltered inshore reefs or lagoons and outer reefs. Under optimum conditions both of these environments can support extensive beds of nearly 100 per cent coral cover. Inshore or coastal reefs may be strongly influenced by freshwater runoff and resultant siltation. Underwater visibility on these reefs is often greatly reduced, particularly during the wet season when rivers are flowing at their maximum. Coastal reefs and lagoons are further characterised by extensive sand or silt bottom areas that may support broad seagrass beds. In most coastal reef or lagoon situations the maximum depth seldom exceeds 25 m and due to heavy siltation coral growth is usually sparse below 15 m depth.

Outer reefs often have a classical reef structure consisting of a broad shallow reef flat, a raised algal ridge, reef front zone of surge channels, and a steep outer slope. But on some islands the bottom plunges into the depths directly from the rocky shore. The clearest waters are found on outer reef slopes and underwater visibility may sometimes exceed 30 m. Coral growth is most abundant between about 5 and 15 m depth, although in some areas appreciable growth may extend well below this limit. In shallower water corals are inhibited by the pounding surge, and in deeper water by the much reduced penetration of light. Although most reef-building corals do not thrive below 30-40 m, certain reef fishes may penetrate well below these depths. Observations made in research submarines at Hawaii and Enewetak Atoll indicate that reef species, including some damselfishes, butterflyfishes, and squirrelfishes, may occur to depths approaching 200 m.

▲ A perfect day on Australia's Great Barrier Reef. (R. Steene)

CLASSIFICATION OF FISHES

Although the fundamentals of biological nomenclature and classification are common knowledge to many, it is my experience that the average non-biologist frequently has little idea of the basis of scientific names or how fishes are classified. It therefore seems worthwhile to include a brief section on the rudiments of this subject.

Every described organism, be it a single-celled amoeba, crab, bird, fish or mammal has a scientific or Latin name. It is composed of two parts and is generally italicised. The first part is the genus or generic name and the second is the species or specific name. For example the Five-lined Seaperch is *Lutjanus quinquelineatus*. The generic name *Lutjanus* pertains to a group of closely related species which share a number of common features related to general shape, scalation, type of teeth, fin-ray counts, etc. The specific name *quinquelineatus* applies only to a single entity that is distinguished from its relatives by a unique set of characteristics, often including colour pattern. Related genera (plural of genus) are grouped together in a family, whose spelling always ends in - idae. An illustrated list of families is presented on pages 9-18. Worldwide there are about 450 families - more than 300 are represented in Australia and surrounding regions. A group of similar families is placed in one of the 35 orders of fishes whose spelling always ends in -iformes. The highest rungs on the 'ladder' of classification pertain to class and phylum. The class Myxini contains the jawless hagfishes and lampreys (no species included in this book); Chondrichthyes contains sharks and rays; and the third class Osteichthyes contains the majority of fishes. All fishes, as do other higher animals including amphibians, reptiles, birds and mammals, belong to the phylum chordata. Therefore, in summary the classification of the Five-lined Seaperch can be represented as follows.

Phylum – Chordata (all animals with notochord)
Class – Osteichthyes (all bony fishes)
Order – Perciformes (most reef fishes)
Family – Lutjanidae (seaperches and relatives)
Genus – *Lutjanus* (closely related seaperches)
Species – *quinquelineatus* (5-lined seaperch)

Characters that are most often used to separate species, and often genera, include external features such as the number of fin rays, size and number of scales, ratio of various body proportions, and colour pattern. For higher classification at levels above genus internal structure, particularly those pertaining to skeletal elements, are often indicative of relationships.

Many species previously unknown to science have been found in our region over the past few decades. When a new fish is discovered it is given a scientific name by the researcher who formally publishes a detailed description in a recognised scientific journal. Scientific names are frequently descriptive. For example *quinquelineatus is* Latin for five lines and is therefore appropriate for the Fivelined Seaperch (see Pl. 42-7 in species section). New fishes are sometimes named after the locality from where they are collected, for example *japonicus* (Japan) or *novaeguineae* (New Guinea). A third category of specific names are based on the names of people, often the person who first discovers the fish (respectable reseachers never name fishes after themselves). Fishes named after a male end in - i, those after females in - ae.

PRESERVING FISHES

It is sometimes desirable to preserve specimens, particularly if a positive identification by museum authorities is required. Also small, unusual or rare fishes can be kept as curios or as teaching aids for children. The recommended method of preservation in any case is exactly the same one that is employed by fish biologists in museums. The basic ingredient is full strength formalin which can be obtained from a pharmacy. The preserving solution is made by diluting one part of formalin with nine parts of water. The fish should be fully immersed in the solution. If larger than about 15-20 cm a slit along the side of the belly will facilitate preservation of the internal organs. For long term storage it is desirable to transfer the specimen to a 70 per cent ethyl alcohol solution (70 per cent ethanol, 30 per cent water) after the fish is fully fixed in formalin (i.e. after several weeks). However, the fish may be held in the initial formalin solution for several years without deleterious effects.

Unfortunately colours fade rapidly in preservative. Therefore photography (see below) is a valuable method of accurately recording the colour pattern.

SENDING SPECIMENS TO THE MUSEUM

Although most of the specimens in the reference collections of the various museums around Australia and South-east Asia are collected on special expeditions by museum staff, occasionally valuable fishes are donated by the public. Also it may be desirable for people living far from their local museum to send specimens in for identification, particularly if the fish in question is suspected to represent a new record for the area or perhaps is very rare. Also members of the public may have the opportunity to collect fishes in remote areas that are not easily reached by museum scientists. For example, several years ago, a medical officer aboard an experimental offshore drilling platform obtained a valuable collection of deep reef fishes on Australia's North West Shelf that were accidentally captured when the drill was brought up from 120 m depth. In this case none of the fishes were recognised by the crew which included several anglers. The specimens were wisely preserved and sent to the Western Australian Museum. Several species from this collection proved to be previously unknown to science.

Specimens can easily be sent to museums via parcel post if first properly preserved (see above section). They should be removed from the preserving solution, rinsed, and wrapped in moist cloth (cheesecloth is ideal) or newspaper, then sealed in several layers of plastic bags. The bags can then be posted in a well padded cardboard box. In Australia fishes can be sent to any of the following institutions depending on their state of origin: (1) Department of Ichthyology, Western Australian Museum, Francis Street, Perth, W.A. 6000, (2) Department of Ichthyology, Northern Territory Museum, P.O. Box 4646, Darwin, N.T.0801, and (3) Department of Ichthyology, Queensland Museum, P.O. Box 3300, South Brisbane, Qld. 4101.

Diver at work, Madang, Papua New Guinea. Underwater photography is the best way to record the fantastic range of colours shown by coral reef fishes. (R. Steene)

FISH PHOTOGRAPHY

Nearly everyone carries a camera on fishing and diving expeditions these days. Good photographs can be valuable in determining the identification of a questionable fish, particularly if the catch has already been eaten. Anglers frequently rely on hastily taken snapshots in order to later identify their catch. Their usefulness is sometimes diminished because little care was taken in preparing the fish. The following steps will ensure the photos are of good diagnostic quality: (1) The specimen should be photographed when fresh as live colours fade rapidly after death. (2) An attempt should be made to spread out the fins. With small fishes you can hold the fins erect with sewing pins on a piece of flat styrofoam or cardboard. (3) Wet fish should be blotted dry with a cloth or paper to prevent harsh glare when photographed. (4) The specimen should be placed on a suitable contrasting background and photographed as close as the lens will allow for sharp focus, attempting to fill the frame. (5) It is helpful if a ruler or some other object of known length can be placed besides the fish when it is photographed in order to determine its length later on.

Underwater photography is a fascinating hobby and will add a new dimension to your diving activities. Fish photography, if done on a regular basis, is an excellent method of learning the fishes of an area. Most beginners start out with a Nikonos or one of the relatively inexpensive automatics in a perspex housing. However, to obtain high quality fish portraits it is advisable to use an SLR camera housed in a special case made of perspex or aluminum alloy. In addition, strobe lighting is a must. The cost of the basic outfit ranges from about $3,000- 5,000, so only the serious photographer will consider this alternative. Even for accomplished divers it requires much practice and patience before good results are obtained. The combination of a moving subject on variable backgrounds present a great challenge.

DANGEROUS FISHES

The region's seas are generally safe for normal swimming and wading activities, but there are a number of fishes potentially capable of causing injury. They can be divided into several broad categories including species that bite, sting, or which may cause poisoning if consumed.

Biters – first and foremost in this category are the whaler sharks and their relatives (Plates 1 to 3). In addition there are a number of smaller reef fishes which, although they pose no threat to swimmers, can inflict painful bites if handled carelessly by anglers. For example, barracudas (Plate 69), razorfishes (Plates 78 and 79), and triggerfishes (Plates 100 and 101) are notorious in this respect. As a rule of thumb any fish with large, obvious teeth should be handled with care.

Stingers – virtually any fish which possesses rigid fin spines is capable of inflicting wounds if handled carelessly. Most are non-venomous and can be treated in the same manner as any puncture wound. Surgeonfishes (Plates 93 to 95) are equipped with scalpel-like spines that are either fixed in an erect position or fold into a groove along the base of the tail. Spearfishermen in particular, need to exercise special care when removing these fish from spears as large specimens can sever a finger. The most dangerous category of stingers includes fishes which have venomous spines. The best known of these are stingrays (Plates 4 and 5), catfish (Plate 11), scorpionfishes (Plates 18 to 20), and spinefeet (Plates 92 and 93). For all of these fishes the recommended first-aid procedure is to immerse the injured area in hot water (as hot as bearable), repeating until the pain subsides. Apparently the protein base of the toxin is denatured by heat and relief is sometimes immediate. In cases where the victim is stung by several spines, or if the wound is deep, medical assistance should be obtained. Firefishes, Lionfishes and stonefishes (Plates 18 and 19) have very potent venom in all fin spines. Several deaths have occurred as a result of people who failed to receive immediate first aid after treading on stonefishes.

Poisonous fishes – there are two main types of fishes in this category. The first includes species that have naturally occurring poisons either in their external mucus or in some internal organs, frequently the viscera or gonads. The best known examples are pufferfishes, porcupinefishes, and boxfishes (Plates 103 to 105). Although these fishes are eaten by the Japanese when specially prepared by licensed chefs, they are considered extremely dangerous and specimens from local waters should never be eaten. The symbol P is used in the species accounts to indicate fishes that are naturally poisonous. The second group of poisonous fishes includes species that acquire toxic properties during their life cycle by accumulating a dinoflagellate that lives on dead coral or among algae and is first consumed by herbivorous fishes which are eventually eaten by larger predatory fishes. The toxin known as ciguatera is accumulative and large fishes such as the Red Bass (Pl. 41-12) and Barracuda (Pl. 69-12) are potentially the most dangerous. The symptoms from eating ciguatoxic fish appear from one to 10 hours later and range from mild dizziness, diarrhoea, and a numb sensation of the lips, hands, and fingers, to extreme nausea, coma, and total respiratory failure. The degree of poisoning depends on the amount of fish that is consumed and the concentration of toxin it contains. As a matter of safe practice it would be wise to avoid eating large barracuda, Red Bass, or extraordinary large gropers, all of which have been implicated in ciguatera poisonings in other regions.

FISH VS. FISHES

Confusion is frequently expressed over the use of the words fish and fishes. The term "fish" in particular, is often used inappropriately. It is grammatically correct to use fish when referring to a single individual or more than one individual if only a single species is involved. For example, one might say "there were 100 fish in that school of Spanish Mackerel." The term fishes is a plural form that is used when referring to two or more different species. For example, "we saw hundreds of fishes while diving on the reef".

HOW TO USE THIS BOOK

This book is designed as a pictorial guide that relies on visual comparison between the painted illustrations and actual specimens, photographs, or underwater observations. Distinguishing features are highlighted in the text accompanying each plate, in most cases referring to colour pattern or the shape of the body or fins. These features are useful for differentiating the species in question from its close relatives, or species it is likely to be confused with. A guide to families based on outline drawings precedes the species section. When attempting to place a fish in the proper family particular attention should be given to the head and body shape, number of dorsal fins, placement of fins and their positions relative to one another, and presence or absence of spiny elements in dorsal and anal fins in particular. The use of technical scientific words is deliberately avoided, but a few terms relating to the external features of fishes are useful for identification and are illustrated below.

Each species account appearing on the page opposite the corresponding plate includes the common name and scientific name followed by the name of the person who first described it and the date of description. If the person's name appears in parentheses it indicates that the species was originally placed in a genus different from its presently recognised one.

Common names are invariably contentious in that species that range widely often have several common names according to locality. This problem is greatly compounded in South-east Asia because of the huge number of languages and dialects spoken. Because of this problem, it is unfortunately not possible to use local names. Therefore Australian common names are utilised. The basis for many of these is Munro's *Fishes of New Guinea*, Marshall's *Fishes of the Great Barrier Reef*, Allen and Swainston's *Marine Fishes of North-Western Australia*, and Randall, Allen, and Steene's *Fishes of the Great Barrier Reef and Coral Sea*.

The text for each species contains general information on habitat, feeding habits, distinguishing features, and geographic distribution. Several of the distributional terms need to be explained in more detail. Indo-E. Pacific refers to a distribution that extends from East Africa to the Americas; E. Indian Ocean and W. Pacific is generally from the Maldives eastward to the western fringe of the Pacific, including Micronesia and Melanesia; W. and C. Pacific refers to the area encompassing the western fringe of the Pacific from Japan to Australia and extending eastward to embrace much of Oceania, often to Samoa, Tuamotus, or Society Islands south of the equator, and the Line Islands (and sometimes Hawaii) north of the equator; Indo-Australian Archipelago embraces the region including the Malay Peninsula, Indonesia, Philippines, northern Australia, and the islands of Melanesia.

The maximum known total length, measured from snout tip to the end of the tail is given at the end of each species account and for a few exceptionally large fishes the maximum recorded weight is also given. I have purposely omitted information on relative abundance (i.e., common, rare, etc.) as this parameter is subject to considerable local variation depending on availability of suitable habitat and in the case of migratory species, the time of year.

In addition to the individual species accounts, 'boxes' of text are included for most plates which contain general information for families, pertaining to such topics as number of species worldwide, ecology including food habits, and any noteworthy behavioural or morphological characteristics.

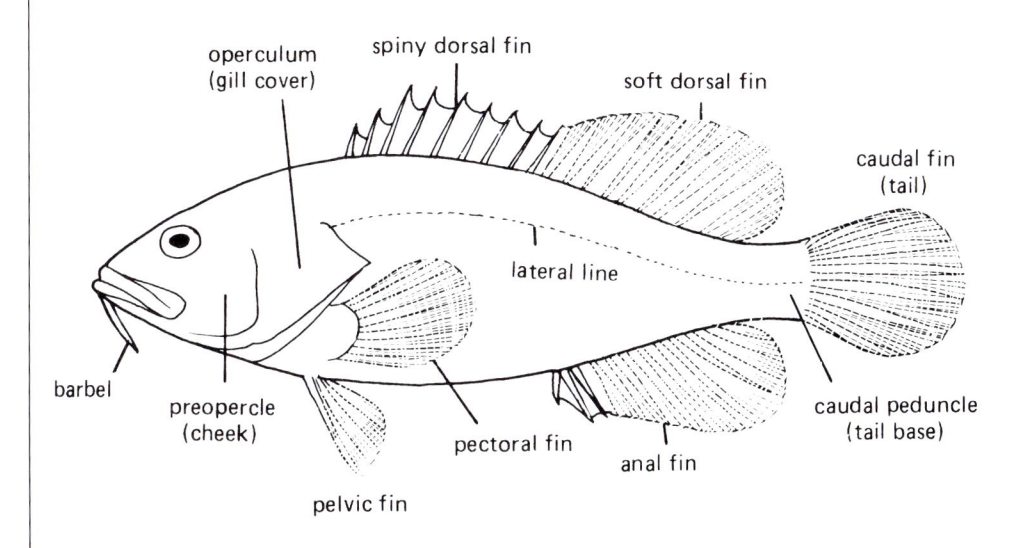

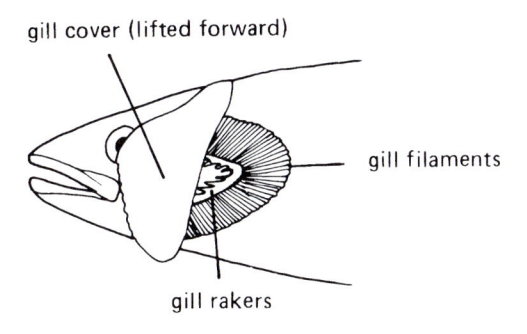

Figure 2. Diagram of a 'typical' fish showing external features.

COLOUR PATTERNS

A major shortcoming of any field guide to fishes is that it is virtually impossible to illustrate all of the variations in colour that commonly occur within a single species. With a few exceptions the colours shown here are the "normal" or average ones displayed by live fish in their natural habitat. Anglers especially will be well aware that many fishes can drastically alter their coloration after being caught. Variation in colour pattern within individual species may also be related to age, sex, environmental conditions, or geography. Angelfishes (Plates 21-22), damselfishes (Plates 23-28), wrasses (Plates 28-32), and parrotfishes (Plates 33-34) are particularly notorious for dramatic changes in livery between the juvenile and adult stages. Wrasses and parrotfishes are also well known for their often different male and female patterns. Mainly due to budgetary restrictions it was not possible to illustrate all the variations related to sex and age, but they are included for a number of the more common species.

EDIBILITY RATINGS

Star symbols that give an indication of the eating qualities of a particular fish appear at the bottom right of each species account unless the fish is too small for human consumption or if there is no available information. Symbols are as follows: ★ = poor eating; ★★ = fair eating; ★★★ = good eating; ★★★★ = excellent eating; P = poisonous. The star symbols are intended as approximate guides only. Wide variation in the edibility of a given species may be caused by a number of factors of which, degree of freshness and method of preparation are particularly important.

ACKNOWLEDGEMENTS

I thank the Chief Executive Officer and Board of Trustees of the Western Australian Museum for their support of this project. Details of planning, typesetting, and technical layout were facilitated by Museum Publications Department staff members Ann Ousey, Greg Jackson, Vince McInerney, and Malcolm Parker. I am also indebted to Sue Morrison, Technical Officer in the Fish Section of the Museum's Department of Aquatic Zoology for her assistance with the preparation of this volume.

Field work in north-western Australia was greatly assisted by the following people: Tony and Avril Ayling, John Braun, Norrie Cross, Eve and Bill Curry, Craig Howson, Hugh Morrison, Ian Parker, Neil Sarti, and Barry Wilson. I am particularly grateful for the assistance and companionship of the Museum's marine biological group (past and present) including Paddy Berry, Clay Bryce, Ray George, Barry Hutchins, Diane Jones, Louisette Marsh, Gary Morgan, Shirley Slack-Smith, and Fred Wells.

Roger Steene of Cairns, Queensland, an Honoray Associate of the W.A. Museum's Department of Aquatic Zoology, accompanied me on numerous field trips throughout the region and was particularly helpful in providing collecting assistance, and photographic coverage of numerous fishes featured in this book.

Walter Starck, former owner of the research vessel "El Torito", graciously offered the use of his ship for fieldwork at Papua New Guinea, Solomon Islands, and on the Great Barrier Reef. Walter first introduced me to the underwater realm of this fascinating region 25 years ago.

I am particularly grateful to the following owners and dive managers of various resorts in the South-east Asian region for providing accommodation, diving facilites, and boat transport: Ron Holland, Jenny Majalup, Graham and Donna Taylor, (Borneo Divers, Sabah), Alan Raabe, Max and Cecilie Benjamen (Walindi Plantation Resort, New Britain), Anton Saksono (Pulau Putri Island Resort, Java Sea), Kal Muller (Komodo Tour and Travel), Hanny and Inneke Batuna (Manado Murex Resort, Sulawesi), Frans Seda (Sao Wisata Resort, Maumere, Flores), and Mark Eckenbarger (Kungkungan Bay Resort, N. Sulawesi). Dive guide, Wally Sagian of Denpasar, Bali assisted with collecting and photography on Bali, Komodo, and Flores. Rudie Kuiter and Roger Steene also provided diving companionship and assistance at several Indonesian sites. Phil Munday offered similar assistance at Kimbe Bay, New Britain.

Excellent facilities for extensive field work and research on coral reef fishes was provided by the Christensen Research Institute, Madang, Papua New Guinea under the directorships of Matthew Jebb and Larry Orsak. I am also grateful to Diane Christensen and the Board of Directors of CRI for providing funding and research opportunities. My son, Mark Allen, who is a keen diver and naturalist, capably assisted with field work at Madang on three occasions.

Numerous scientists in Australia and overseas have contributed taxonomic knowledge resulting in a better understanding of the region's fish fauna, either through publications or by direct assistance with problematical identifications.

Those particularly helpful in this regard included Doug Hoese, Jeff Leis, and John Paxton (Australian Museum), Rudie Kuiter and Martin Gomon (Museum of Victoria), Barry Russell and Helen Larson (Northern Territory Museum), Rolly McKay (Queensland Museum), Peter Last (CSIRO Fisheries, Hobart, Tasmania), Tony Gill (Natural History Museum, London), Ronald Fricke (Natural History Museum, Stuttgart), William Eschmeyer and John McCosker (California Academy of Sciences), Ed Murdy, Jeff Williams, and Victor Springer (Smithsonian Institution), Theodore Pietsch (University of Washington), Stuart Poss (Gulf Coast Research Lab, U.S.A.), Jack Randall, Richard Pyle, and Arnold Suzimoto (Bishop Museum, Honolulu), Bill Smith-Vaniz (National Biological Science Centre, Gainesville, Florida) and Richard Winterbottom (Royal Ontario Museum).

Finally, this guide to the region's fishes would not have been possible without the wonderful artwork of Jill Ruse and Roger Swainston. It has been a great pleasure working with both of these highly talented artists.

GUIDE TO FAMILIES

The following pages contain outline drawings of typical members of the families contained in the book. Scientific family names are indicated below each drawing and the Plate number is given in parentheses

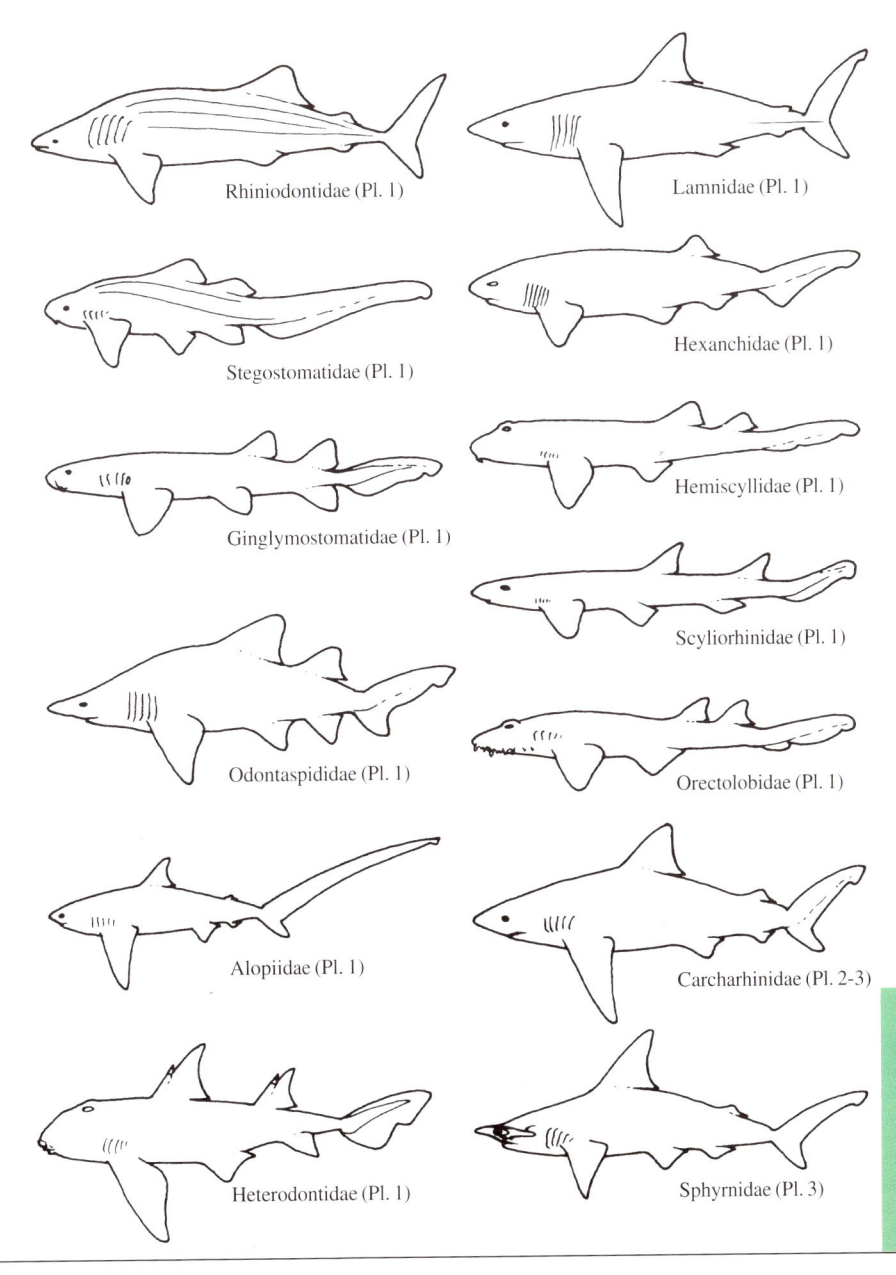

Rhiniodontidae (Pl. 1)

Lamnidae (Pl. 1)

Stegostomatidae (Pl. 1)

Hexanchidae (Pl. 1)

Ginglymostomatidae (Pl. 1)

Hemiscyllidae (Pl. 1)

Scyliorhinidae (Pl. 1)

Odontaspididae (Pl. 1)

Orectolobidae (Pl. 1)

Alopiidae (Pl. 1)

Carcharhinidae (Pl. 2-3)

Heterodontidae (Pl. 1)

Sphyrnidae (Pl. 3)

Pristidae (Pl. 3)

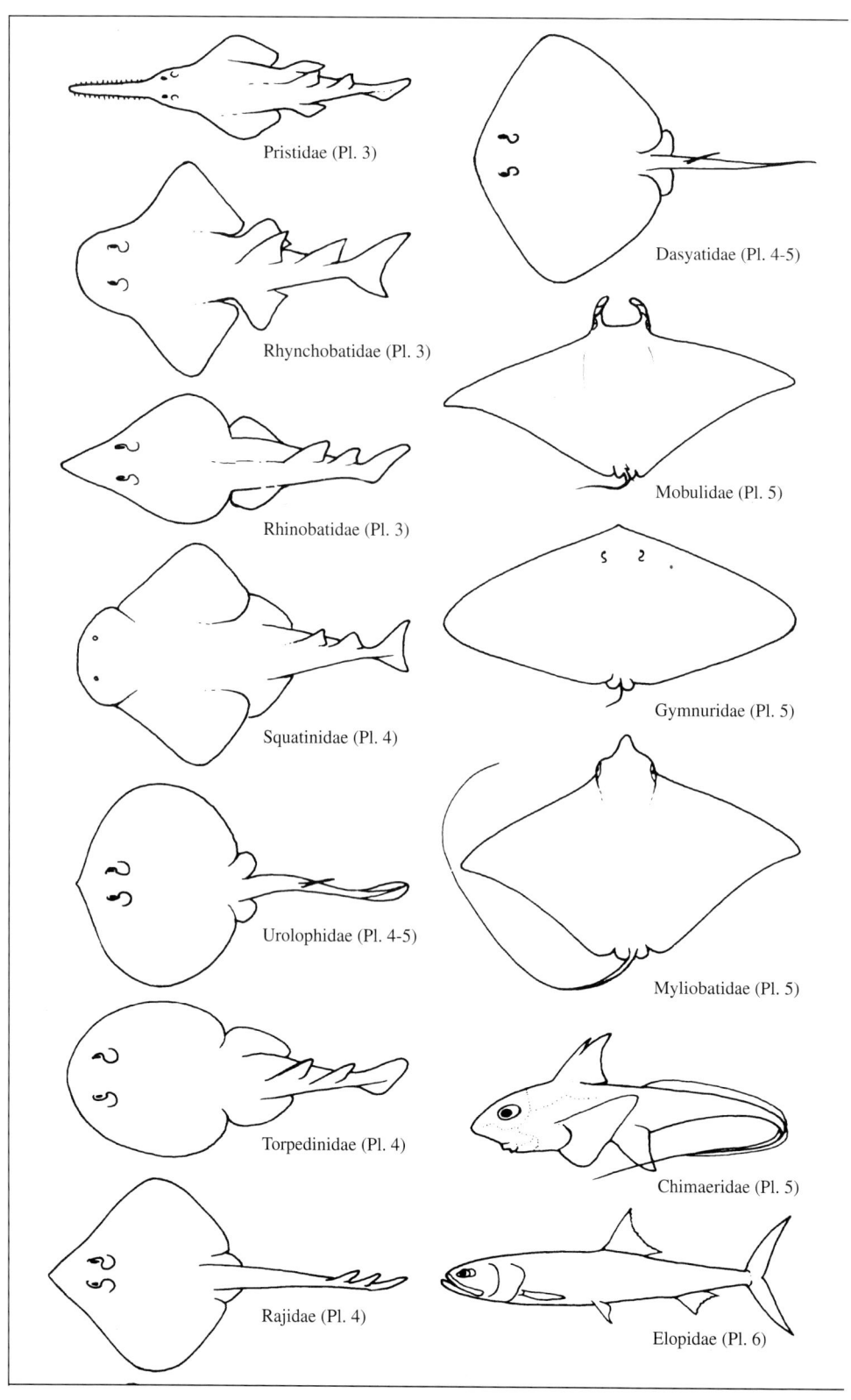

Dasyatidae (Pl. 4-5)

Rhynchobatidae (Pl. 3)

Mobulidae (Pl. 5)

Rhinobatidae (Pl. 3)

Gymnuridae (Pl. 5)

Squatinidae (Pl. 4)

Urolophidae (Pl. 4-5)

Myliobatidae (Pl. 5)

Torpedinidae (Pl. 4)

Chimaeridae (Pl. 5)

Rajidae (Pl. 4)

Elopidae (Pl. 6)

Albulidae (Pl. 6)

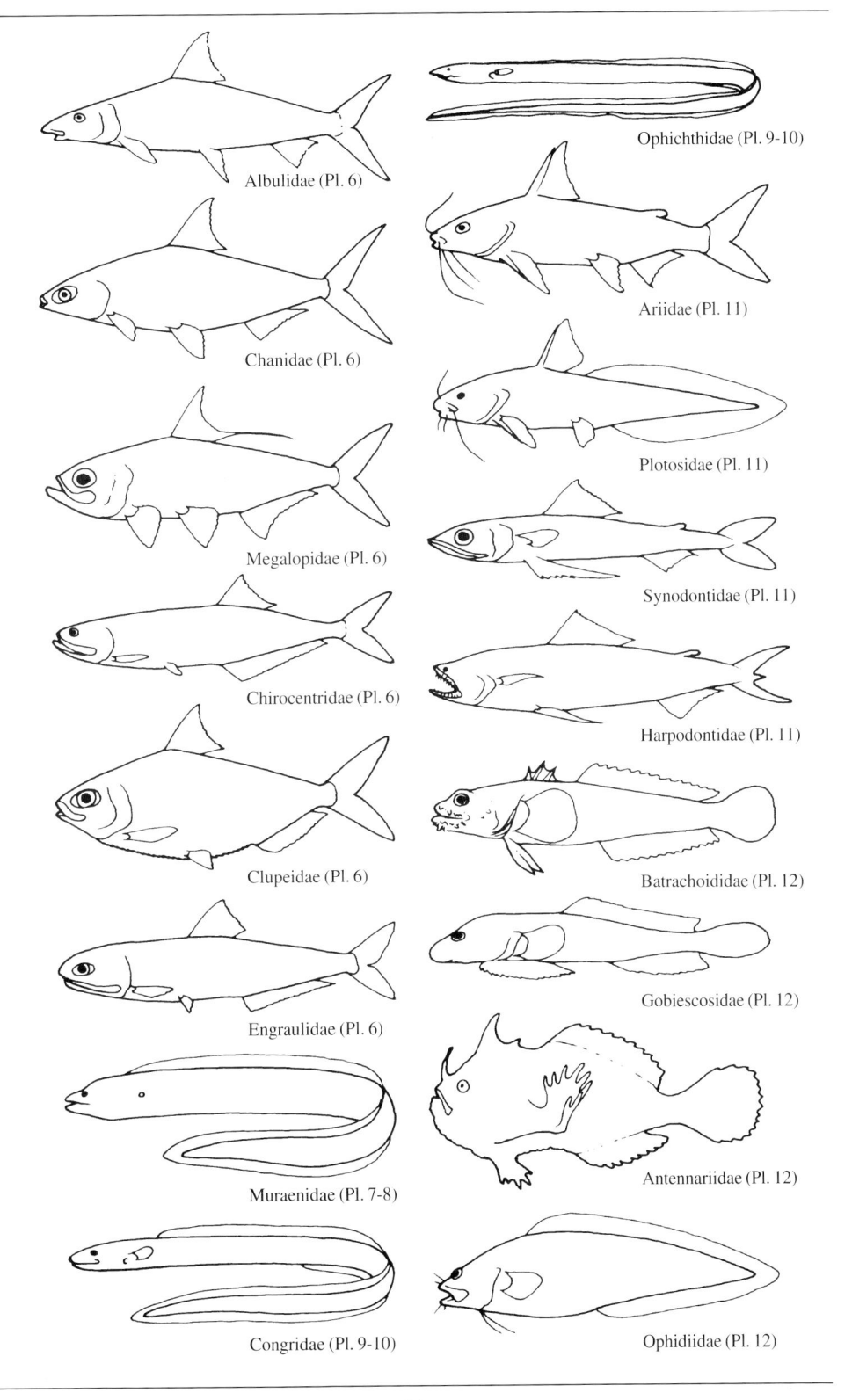

Chanidae (Pl. 6)

Megalopidae (Pl. 6)

Chirocentridae (Pl. 6)

Clupeidae (Pl. 6)

Engraulidae (Pl. 6)

Muraenidae (Pl. 7-8)

Congridae (Pl. 9-10)

Ophichthidae (Pl. 9-10)

Ariidae (Pl. 11)

Plotosidae (Pl. 11)

Synodontidae (Pl. 11)

Harpodontidae (Pl. 11)

Batrachoididae (Pl. 12)

Gobiescosidae (Pl. 12)

Antennariidae (Pl. 12)

Ophidiidae (Pl. 12)

Carapidae (Pl. 12)

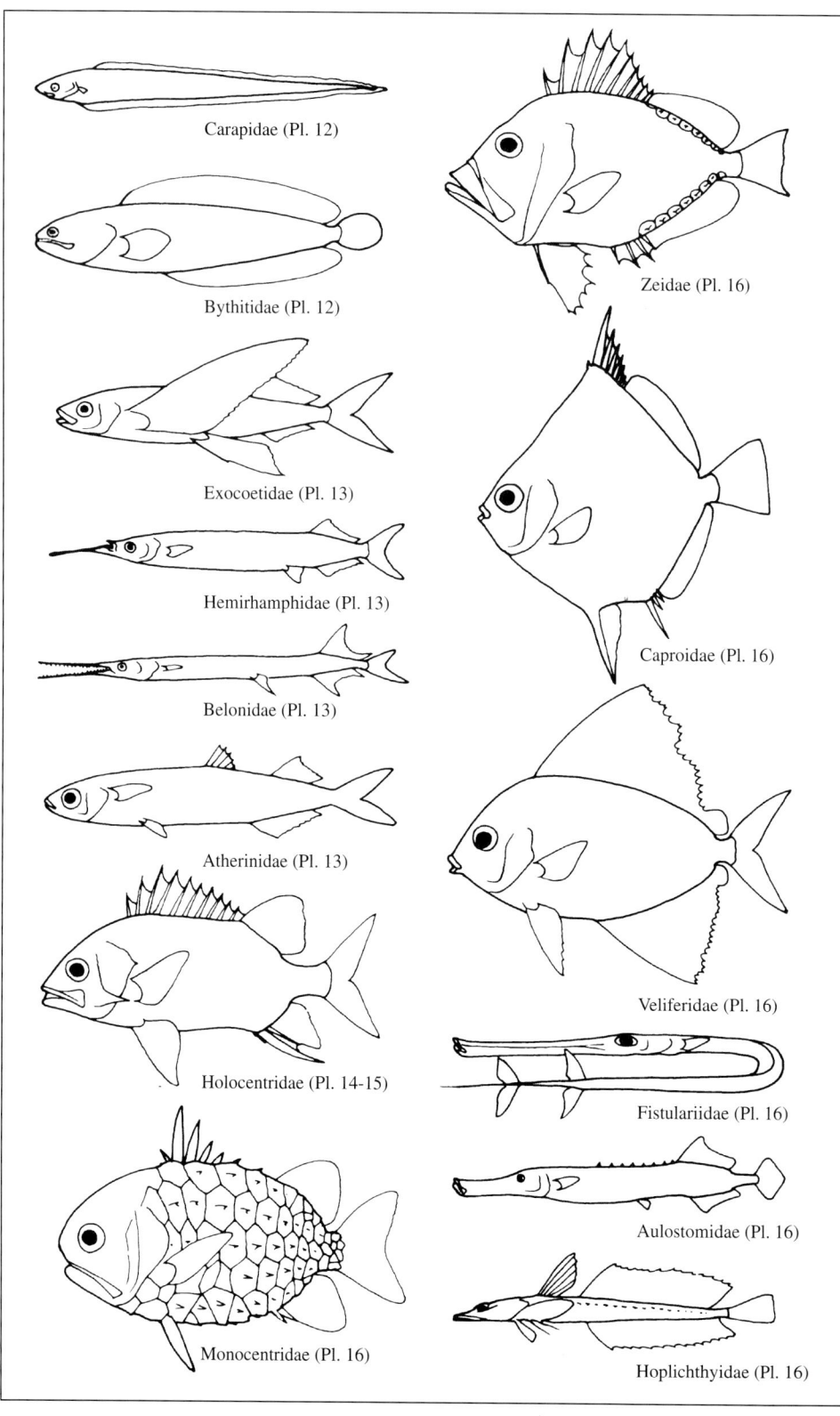

Bythitidae (Pl. 12)

Exocoetidae (Pl. 13)

Hemirhamphidae (Pl. 13)

Belonidae (Pl. 13)

Atherinidae (Pl. 13)

Holocentridae (Pl. 14-15)

Monocentridae (Pl. 16)

Zeidae (Pl. 16)

Caproidae (Pl. 16)

Veliferidae (Pl. 16)

Fistulariidae (Pl. 16)

Aulostomidae (Pl. 16)

Hoplichthyidae (Pl. 16)

Dactylopteridae (Pl. 16)

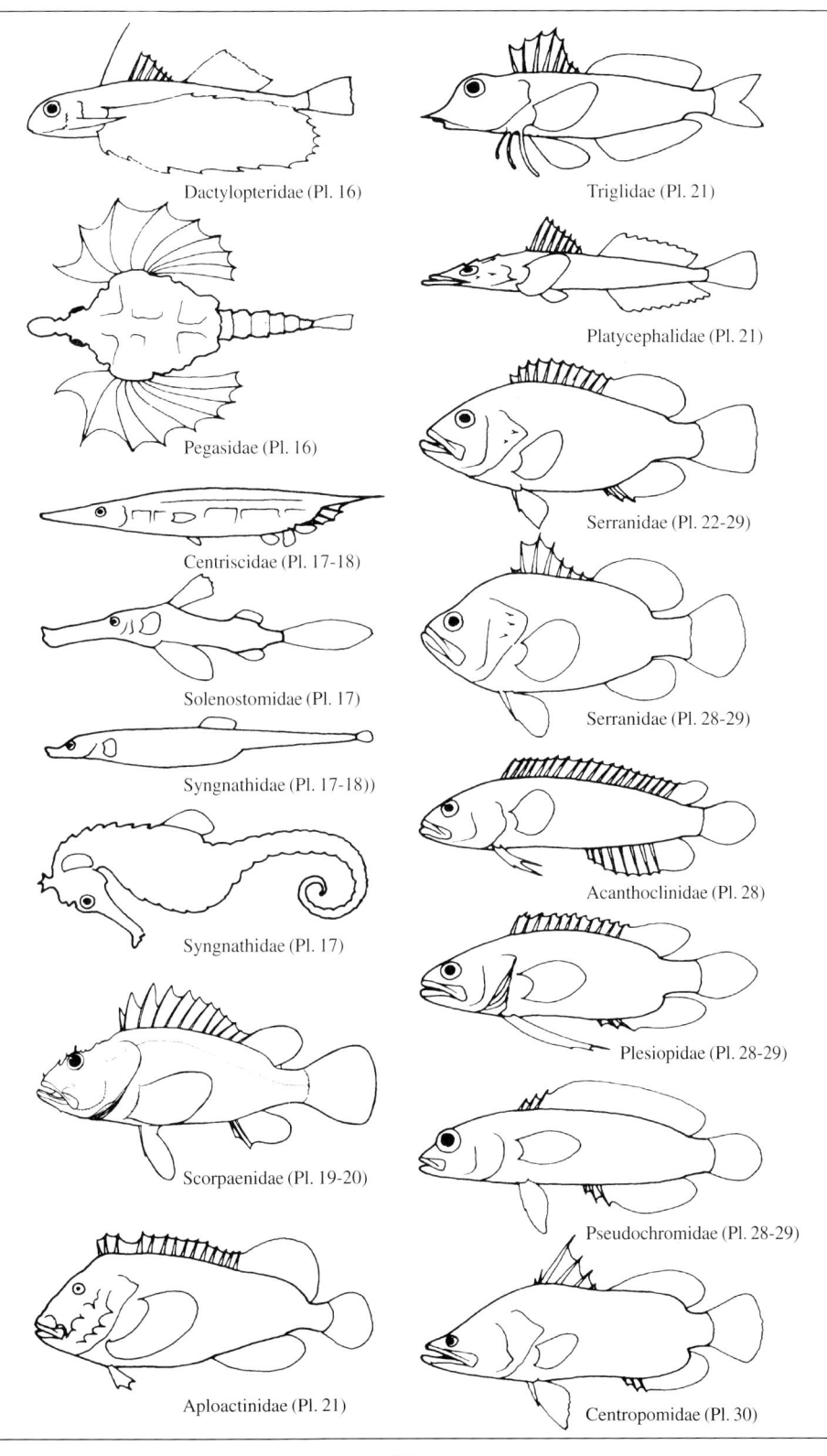

Triglidae (Pl. 21)

Pegasidae (Pl. 16)

Platycephalidae (Pl. 21)

Serranidae (Pl. 22-29)

Centriscidae (Pl. 17-18)

Solenostomidae (Pl. 17)

Serranidae (Pl. 28-29)

Syngnathidae (Pl. 17-18)

Acanthoclinidae (Pl. 28)

Syngnathidae (Pl. 17)

Plesiopidae (Pl. 28-29)

Scorpaenidae (Pl. 19-20)

Pseudochromidae (Pl. 28-29)

Aploactinidae (Pl. 21)

Centropomidae (Pl. 30)

33

Ambassidae (Pl. 30)

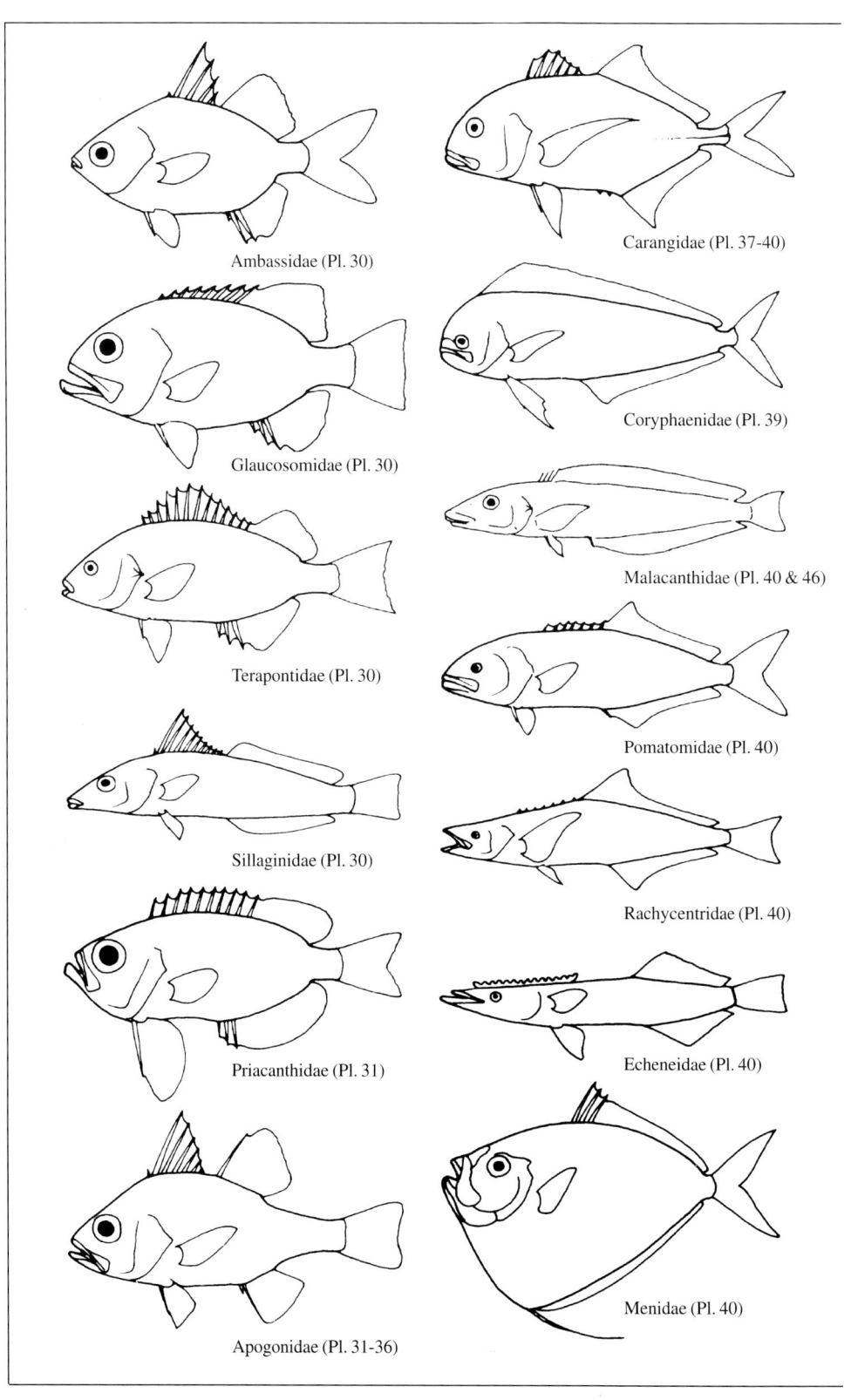

Carangidae (Pl. 37-40)

Glaucosomidae (Pl. 30)

Coryphaenidae (Pl. 39)

Terapontidae (Pl. 30)

Malacanthidae (Pl. 40 & 46)

Sillaginidae (Pl. 30)

Pomatomidae (Pl. 40)

Rachycentridae (Pl. 40)

Priacanthidae (Pl. 31)

Echeneidae (Pl. 40)

Apogonidae (Pl. 31-36)

Menidae (Pl. 40)

Leiognathidae (Pl. 40)

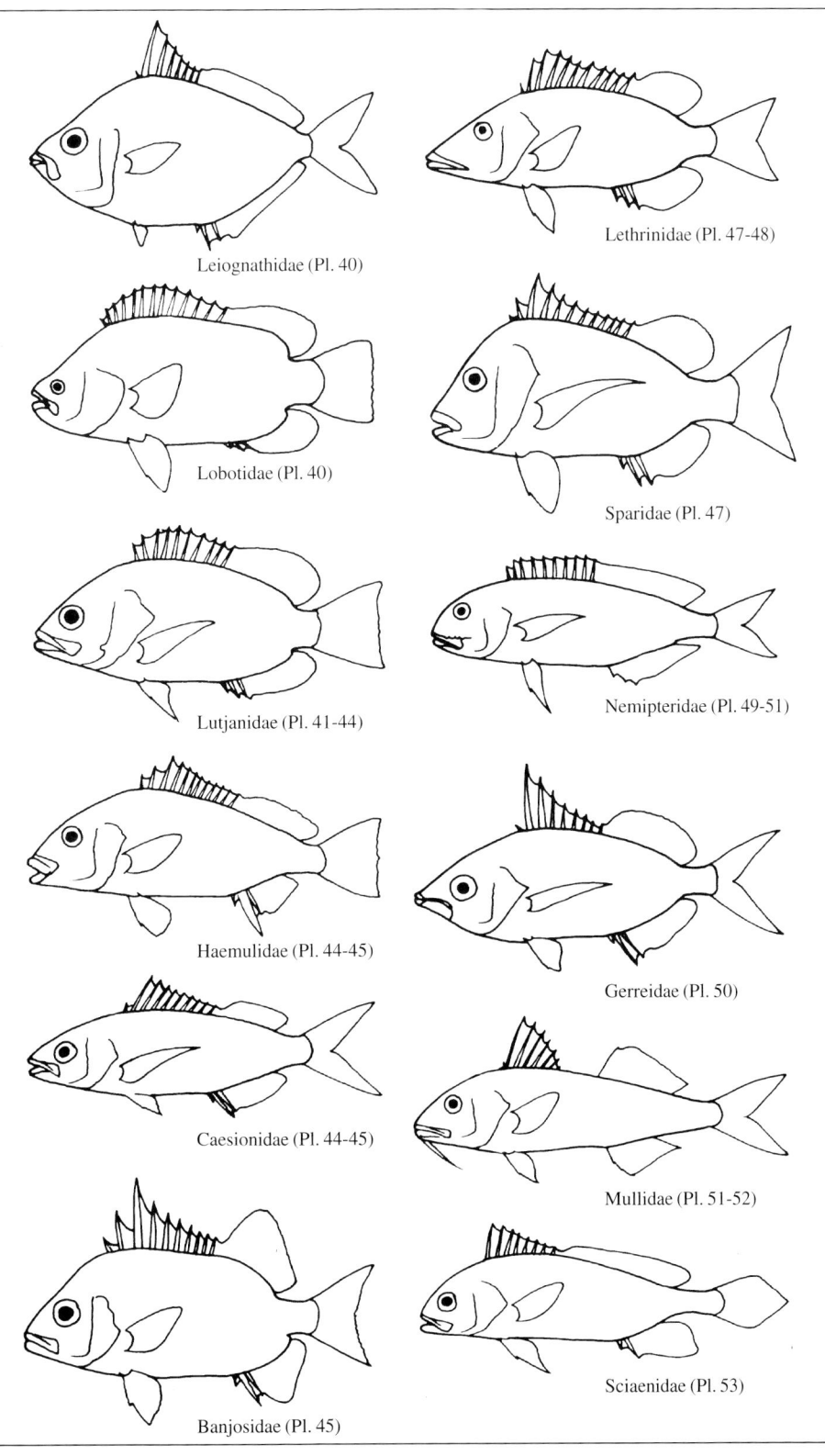

Lethrinidae (Pl. 47-48)

Lobotidae (Pl. 40)

Sparidae (Pl. 47)

Lutjanidae (Pl. 41-44)

Nemipteridae (Pl. 49-51)

Haemulidae (Pl. 44-45)

Gerreidae (Pl. 50)

Caesionidae (Pl. 44-45)

Mullidae (Pl. 51-52)

Banjosidae (Pl. 45)

Sciaenidae (Pl. 53)

Monodactylidae (Pl. 53)

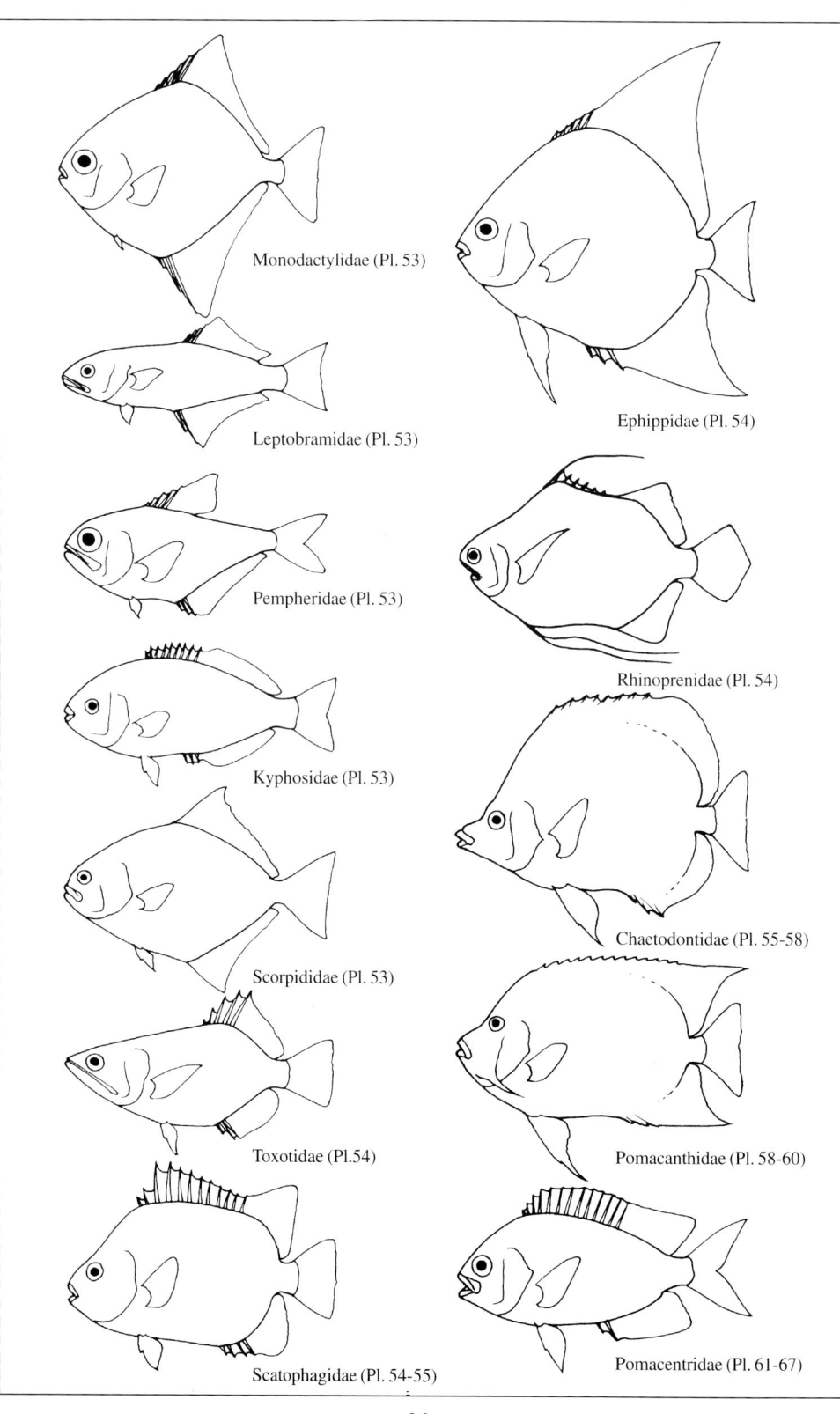

Leptobramidae (Pl. 53)

Pempheridae (Pl. 53)

Kyphosidae (Pl. 53)

Scorpididae (Pl. 53)

Toxotidae (Pl.54)

Scatophagidae (Pl. 54-55)

Ephippidae (Pl. 54)

Rhinoprenidae (Pl. 54)

Chaetodontidae (Pl. 55-58)

Pomacanthidae (Pl. 58-60)

Pomacentridae (Pl. 61-67)

Pentacerotidae (Pl. 68)

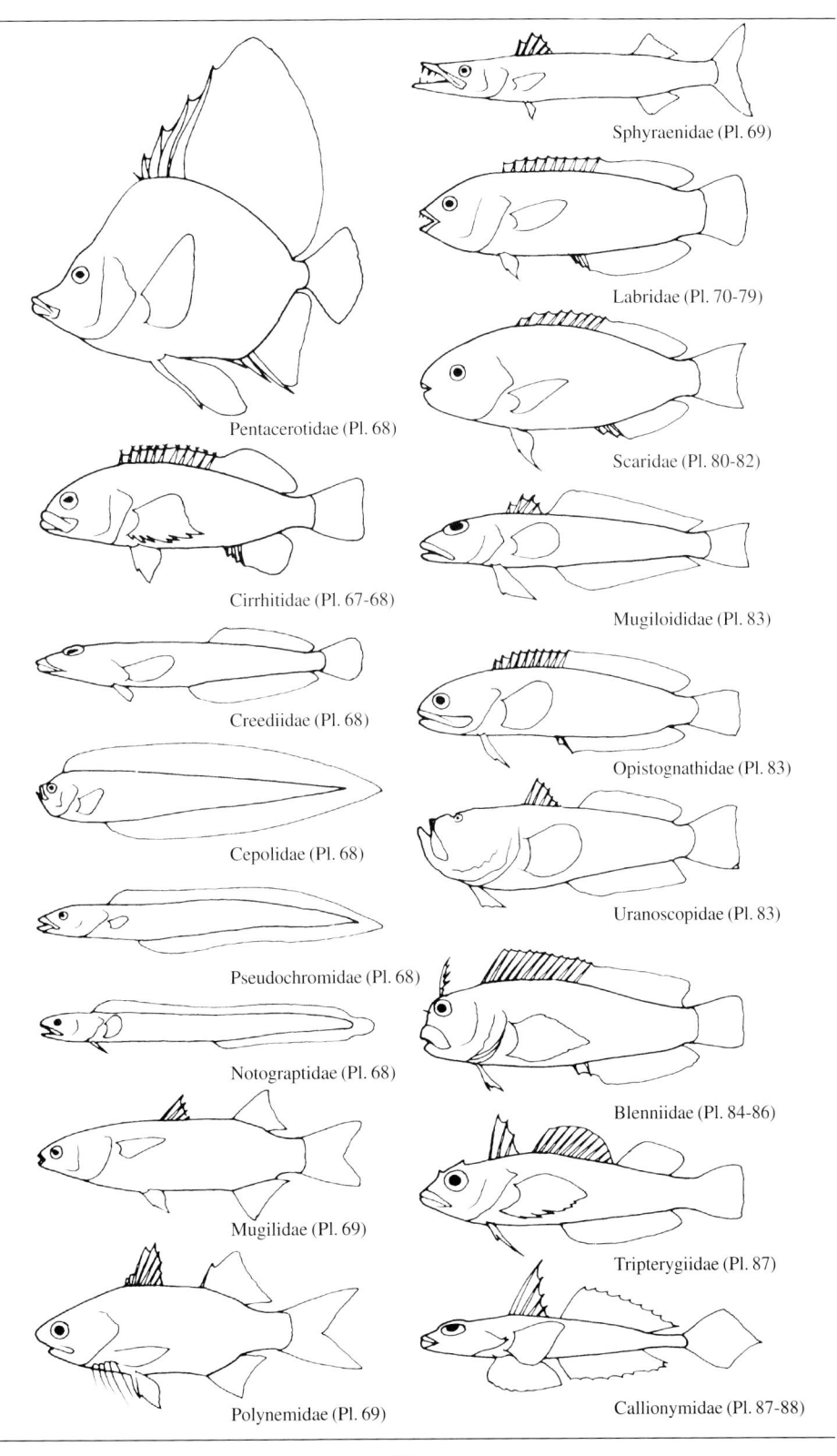

Cirrhitidae (Pl. 67-68)

Creediidae (Pl. 68)

Cepolidae (Pl. 68)

Pseudochromidae (Pl. 68)

Notograptidae (Pl. 68)

Mugilidae (Pl. 69)

Polynemidae (Pl. 69)

Sphyraenidae (Pl. 69)

Labridae (Pl. 70-79)

Scaridae (Pl. 80-82)

Mugiloididae (Pl. 83)

Opistognathidae (Pl. 83)

Uranoscopidae (Pl. 83)

Blenniidae (Pl. 84-86)

Tripterygiidae (Pl. 87)

Callionymidae (Pl. 87-88)

37

Gobiidae (Pl. 88-92)

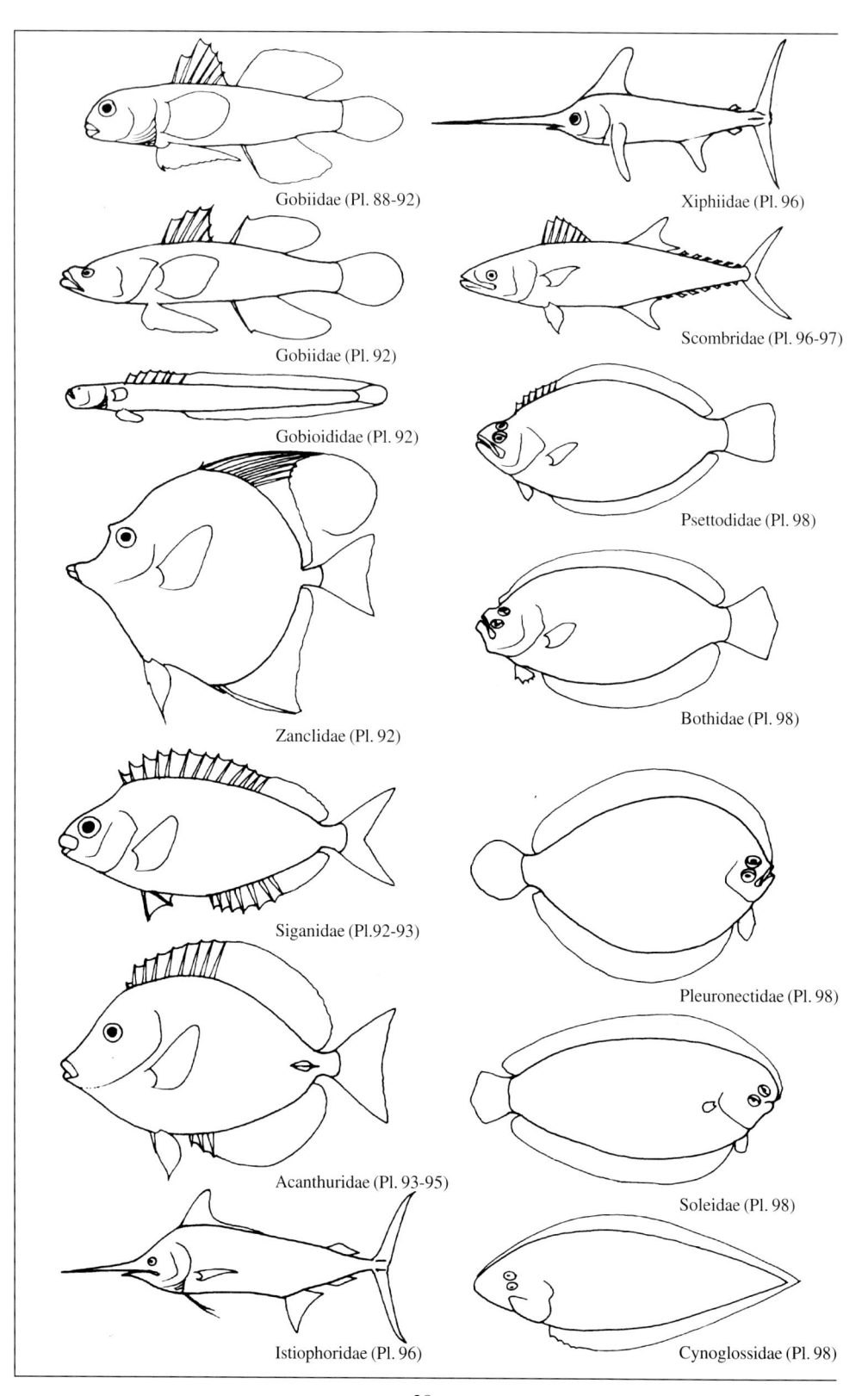

Gobiidae (Pl. 92)

Gobioididae (Pl. 92)

Zanclidae (Pl. 92)

Siganidae (Pl.92-93)

Acanthuridae (Pl. 93-95)

Istiophoridae (Pl. 96)

Xiphiidae (Pl. 96)

Scombridae (Pl. 96-97)

Psettodidae (Pl. 98)

Bothidae (Pl. 98)

Pleuronectidae (Pl. 98)

Soleidae (Pl. 98)

Cynoglossidae (Pl. 98)

Percophididae (Pl. 99)

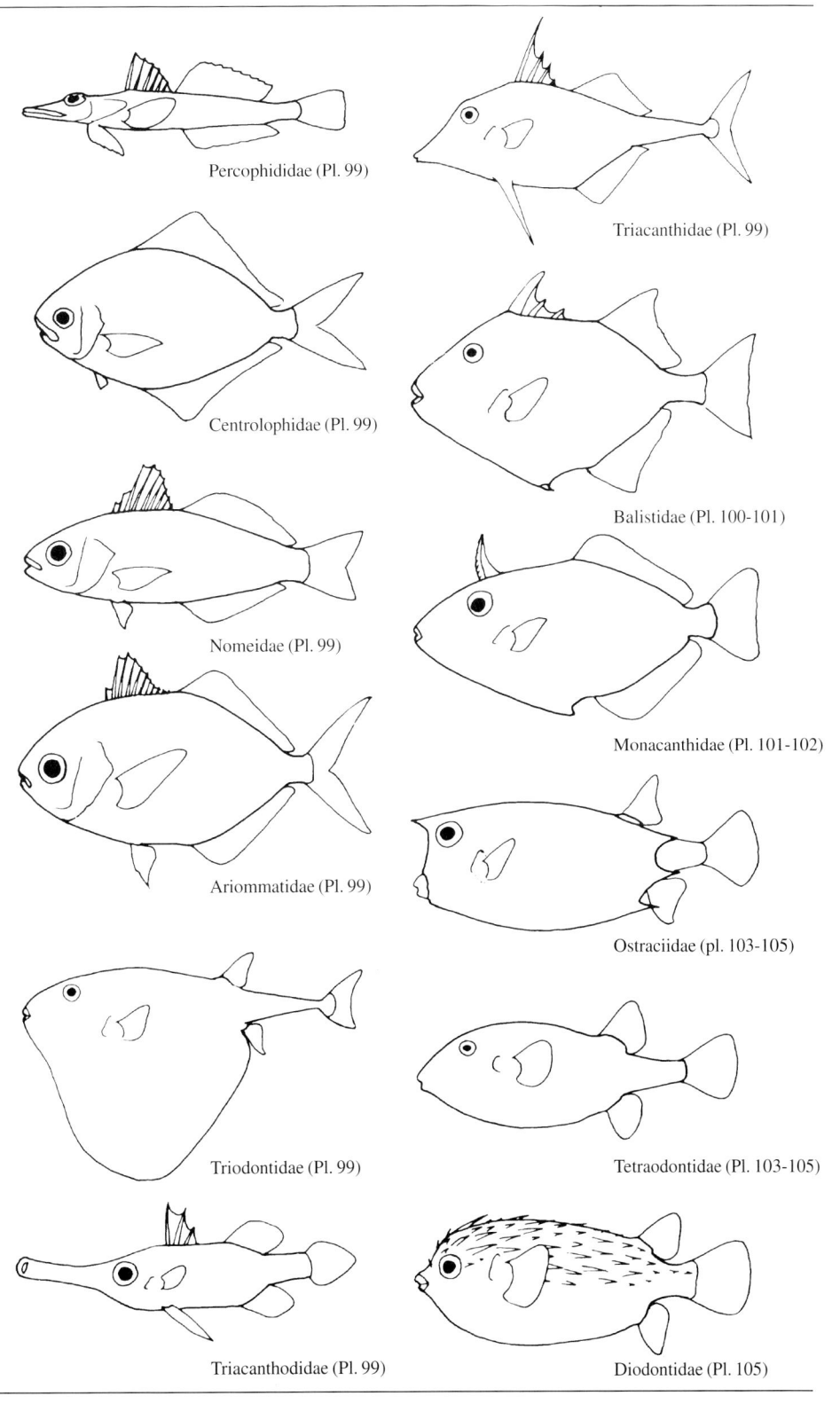

Triacanthidae (Pl. 99)

Centrolophidae (Pl. 99)

Balistidae (Pl. 100-101)

Nomeidae (Pl. 99)

Monacanthidae (Pl. 101-102)

Ariommatidae (Pl. 99)

Ostraciidae (pl. 103-105)

Triodontidae (Pl. 99)

Tetraodontidae (Pl. 103-105)

Triacanthodidae (Pl. 99)

Diodontidae (Pl. 105)

PLATE 1: SHARKS

1 WHALE SHARK
Rhincodon typus Smith, 1828
Inhabits coastal waters, also occurs well offhore; distinguished by huge size and pattern of white spots; world's largest fish, but harmless plankton feeder; rarely seen but sightings off North West Cape, Western Australia during March-April are a regular occurrence; found throughout the region; worldwide temperate and tropical seas; possibly to 18 m; but seldom above 12 m. (RHINCODONTIDAE)

2 LEOPARD SHARK
Stegostoma fasciatum (Hermann,1783)
Inhabits coastal waters and offshore areas in the vicinity of coral reefs, may be seen resting on the bottom; distinguished by large tail, dark spots, and ridges on side; also known as Zebra Shark; harmless; found throughout the region; Indo-W. Pacific; to 350 cm. (STEGASTOMATIDAE)

3 TAWNY NURSE SHARK
Nebrius ferrugineus (Lesson, 1830)
Inhabits shallow reefs; distinguished by brown colour; equal-sized dorsal fins and moderately long barbels on snout; harmless; found throughout the region; Indo-W. Pacific; to 320 cm. (GINGLYMOSTOMATIDAE)

4 GREY NURSE SHARK
Carcharias taurus Rafinesque, 1810
Inhabits coastal waters, often occurs near the bottom in small schools, distinguished by pair of dorsal and anal fins nearly of equal size, long curved fang-like teeth, and lack of barbels on snout; usually harmless, but will attack if provoked; subtropical Australian seas; Atlantic and Indo-W. Pacific; to 360 cm. (ODONTASPIDAE)

5 SMALL TOOTH THRESHER SHARK
Alopias pelagicus Nakamura, 1935
Inhabits oceanic waters, but occasionally caught near shore; distinguished by very long upper tail lobe (used to stun schools of fish); harmless; found throughout the region; Indo-E. Pacific; to 330 cm. (ALOPIIDAE)

6 BULLHEAD SHARK
Heterodontus zebra (Gray, 1831)
Inhabits flat bottoms on the continental shelf to at least 50 m depth; distinguished by barred pattern and sharp spine at front of both dorsal fins; harmless except dorsal spines can cause painful wound; found throughout the region; mainly W. Pacific; to 122 cm. (HETERODONTIDAE)

7 SHORTFIN MAKO
Isurus oxyrinchus Rafinesque, 1809
Inhabits oceanic waters usually well offshore, but sometimes visits coastal areas; distinguished by slender shape, equal-sized tail fin lobes, and slender dagger-like teeth; also known as Blue pointer; dangerous; found throughout the region; worldwide temperate and tropical seas; to 400 cm. (LAMNIDAE)

8 BLUNTNOSE SIXGILL SHARK
Hexanchus griseus (Bonnaterre, 1788)
Inhabits coastal waters, also occurs well offshore in deeper waters of the continental shelf; distinguished by absence of second dorsal fin; found throughout the region; worldwide tropical seas; to 180 cm. (HEXANCHIDAE)

9 SPECKLED CATSHARK
Hemiscyllium trispeculare Richardson, 1843
Inhabits shallow coral reefs; distinguished by pale edged black spot partially surrounded by smaller black spots just behind gill slits; N. Australia only; to 65 cm. (HEMISCYLLIDAE)

10 EPAULETTE SHARK
Hemiscyllium ocellatum (Bonaterre, 1788)
Inhabits shallow coral reefs; similar to 9, but lacks smaller black spots adjacent to large spot behind gill slits; harmless; N. Australia and New Guinea; to 107 cm. (HEMISCYLLIDAE)

11 BROWN-BANDED CATSHARK
Chiloscyllium punctatum Müller & Henle, 1838
Inhabits shallow coral reefs; distinguished by strongly-barred pattern and barbels on snout; harmless; found throughout the region; E. Indian Ocean and W. Pacific; to 104 cm. (HEMISCYLLIDAE)

12 MARBLED CATSHARK
Atelomycterus macleayi Whitley, 1939
Inhabits coastal waters on sand or rocky bottoms; distinguished by small size, slender shape, no barbels on snout, and pattern of black spots and faint broad dark bars; N. Australia only; to 60 cm. (SCYLIORHINIDAE)

13 RETICULATED SWELLSHARK
Cephaloscyllium fasciatum Chen, 1966
Inhabits deeper waters of the continental shelf; distinguished by rounded, inflatable stomach, blunt snout, narrow eye-slits and pattern of spots and lines; harmless; found throughout the region; N. Australia and S.E. Asia; to 80 cm. (SCYLIORHINIDAE)

14 BANDED WOBBEGONG
Orectolobus ornatus (de Vis, 1883)
Inhabits shallow coastal reefs frequently on sand or weed bottoms; distinguished by ornate colour pattern and numerous skin flaps on mouth and lower part of head; harmless, but will bite if accidentally trod on; entire Australian coastline and New Guinea; to 300 cm. (ORECTOLOBIDAE)

15 NORTHERN WOBBEGONG
Orectolobus wardi Whitley, 1939
Inhabits coastal waters; distinguished by pale-edged dark saddles and bands, frequently has black spots on edge of dorsal fins and tail, skin flaps on head not as well developed as in 14; harmless; N. Australia only; to 100 cm. (ORECTOLOBIDAE)

16 TASSELLED WOBBEGONG
Euchrossorhinus dasypogon (Bleeker,1867)
Inhabits coral reefs; distinguished by numerous branched skin flaps on both chin and side of head (absent on chin in 14 and 15) and very broad, rounded head; harmless; N. Australia and New Guinea; to 350 cm. (ORECTOLOBIDAE)

SHARK TEETH

Sharks typically have an outer row of well developed, upright teeth and several inner rows of teeth in various stages of development which are folded downward. Teeth are continuously produced throughout the life of the shark and each row moves forward to replace the next row every few weeks. The teeth are a valuable means of indentifying species, particulary among the whalers (Plate 2). Typical examples from the upper and lower jaw of a number of sharks are included on plates 1 and 2.

1-4
1-3
1-7
2-1
2-2
2-3
2-4
2-5
2-6
1
juv.
2
3
4
5
6
7
9
8
10
14
11
16
12
13
15
2-9
2-8
2-7
2-10
2-12
2-11
2-13
2-14
2-15
2-16
3-1

PLATE 2: SHARKS (FAMILY CARCHARHINIDAE)

1 SILVERTIP SHARK
Carcharhinus albimarginatus (Rüppell, 1837)
Inhabits offshore coral reefs, usually below 20 m depth on outer edge of reefs; distinguished by white tips on dorsal, tail, and pectoral fins; dangerous; found throughout the region; Indo-E Pacific; to 300 cm.

2 BIGNOSE SHARK
Carcharhinus altimus (Springer, 1950)
Inhabits coastal waters; distinguished by long rounded or bluntly pointed snout when viewed from above, no conspicuous fin markings, and skin ridge between dorsal fins; potentially dangerous; found throughout the region; worldwide tropical seas, to 300 cm

3 GREY REEF SHARK
Carcharhinus amblyrhynchos (Bleeker, 1856)
Inhabits inshore and offshore coral reefs; usually seen adjacent to dropoffs on the outer edge of reefs; distinguished by black margin on tail and lacks skin ridge between dorsal fins; dangerous; found throughout the region; Indo-W. Pacific; to 255 cm. Cairns 10/98

4 PIGEYE SHARK
Carcharhinus amboinensis (Müller & Henle, 1839)
Inhabits coastal waters, sometimes entering estuaries and rivers; a large stout grey shark without distinguishing marks, has a large dorsal fin and lacks a skin ridge between the dorsal fins; dangerous; found throughout the region; Indo-W. Pacific; to 280 cm.

5 BRONZE WHALER
Carcharhinus brachyurus (Günther, 1870)
Inhabits coastal waters; often confused with **13**, but has narrower upper teeth and no skin ridge between dorsal fins; dangerous; subtropical and temperate Australian seas; worldwide temperate and tropical seas; to 325 cm.

6 LONG-NOSED GREY SHARK
Carcharhinus brevipinna (Müller & Henle, 1839)
Inhabits coastal waters, often occurs in schools; distinguished by black tips on most fins, but lacks white margins around black areas as found in **11**; potentially dangerous; also known as Smooth-fanged or Inkytail Shark; found throughout the region; Atlantic and Indo-W. Pacific; to 300 cm.

7 BULL SHARK
Carcharhinus leucas (Valenciennes, 1839)
Inhabits coastal waters, enters estuaries and rivers; land-locked freshwater populations occur in some areas outside Australia; a large stocky shark with short blunt snout when viewed from above, broad triangular teeth, and lacks skin ridge between dorsal fins; dangerous; also known as River Shark and Estuary Whaler; found throughout the region; worldwide temperate and tropical seas; to 340 cm.

8 SILKY SHARK
Carcharhinus falciformes (Bibron, 1839)
Inhabits oceanic waters, usually well offshore; a large slender grey shark with a moderately long rounded snout, short first dorsal fin, and elongate tips on anal and second dorsal fins; potentially dangerous; found throughout the region; worldwide temperate and tropical seas; to 330 cm.

9 WHITE CHEEK SHARK
Carcharhinus dussumieri Valenciennes, 1839
Inhabits coastal waters; see remarks for **15** below; harmless; N.W. Australia and S.E. Asia; N.Indian Ocean and W.Pacific; to 100 cm.

10 BLACKTIP SHARK
Carharhinus limbatus (Valenciennes, 1839)
Inhabits coastal waters; similar to **6**, but more black on fin tips and has black spot on pelvic fin; potentially dangerous; found throughout the region; worldwide temperate and tropical seas; to 255 cm.

11 BLACKTIP REEF SHARK
Carcharhinus melanopterus (Quoy & Gaimard, 1824)
Inhabits reef flats and coral reef lagoons; distingusished from other 'black-tipped' sharks by white margin around black areas, especially noticeable on first dorsal fin; usually not dangerous unless cornered; found throughout the region; Indo-W .Pacific; to 180 cm.

12 OCEANIC WHITETIP SHARK
Carcharhinus longimanus (Poey, 1861)
Inhabits oceanic waters, ususally well offshore; distinushed by over-sized pectoral fin and broad rounded dorsal fin, both of these fins broadly white tipped; dangerous; found throughout the region; worldwide temperate and tropical seas; to 396 cm.

13 BLACK WHALER
Carcharhinus obscurus (Le Sueur, 1818)
Inhabits coastal waters, also found well offshore; similar to **5**, but has wider, more triangler teeth in upper jaw and has a low skin ridge between dorsal fins; dangerous; also known as Dusky Shark; subtropical and tropical Australia; worldwide temperate and tropical seas; to 362 cm.

14 SANDBAR SHARK
Carcharhinus plumbeus (Nardo, 1827)
Inhabits coastal waters; distinguished by very tall first dorsal fin that arises above rear base of pectoral fin; dangerous; also known as Sand or Thickskin Shark and Northern Whaler; found throughout the region; worldwide temperate and tropical seas; to 300 cm.

15 BLACKSPOT SHARK
Carcharhinus sealei (Pietschmann, 1916)
inhabits coastal waters; similar to **9**, but has falcate rather than triangular first dorsal fin; harmless; found throughout the region; Indo-W. Pacific; to 95 cm.

16 SPOT-TAIL SHARK
Carcharhinus sorrah (Valenciennes, 1839)
Inhabits coastal waters in the vicinity of coral reefs; distinguished by conspicuous black tips on pectoral and second dorsal fins, and lower lobe of tail; dangerous; found throughout the region; Indo-W. Pacific; to 160 cm

---

SHARK ATTACK!

The sharks illustrated in Plate 2 are members of the family Carcharhinidae, commonly known as whalers. Although many of the species have never been implicated as far as attacks on humans are concerned, the familily contains several which have a bad reputation. All should be handled with respect when removing hooks and none should be deliberately provoked by spearing fishes or offering food when diving in their company.

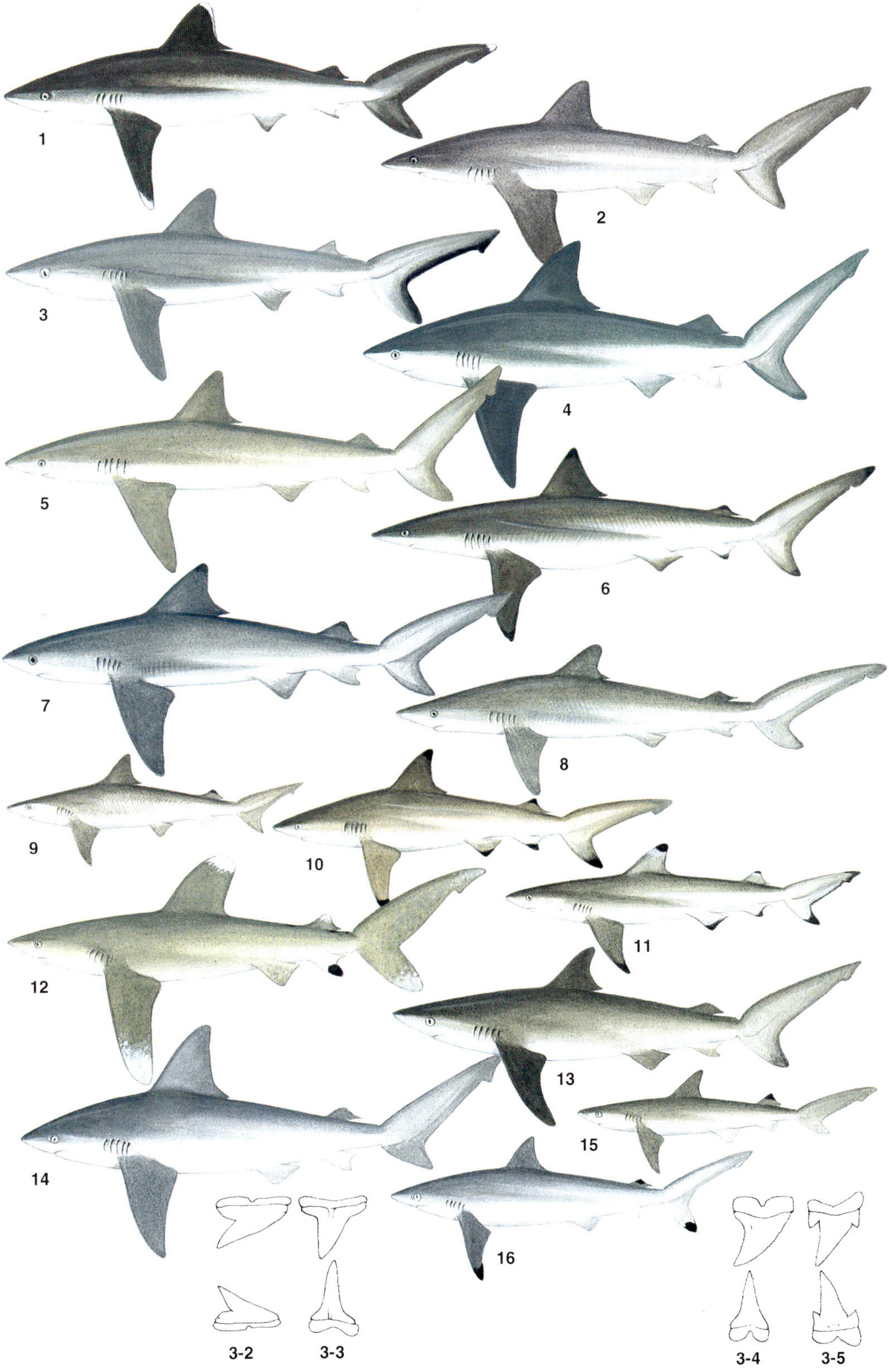
1
2
3
4
5
6
7
8
9
10
11
12
13
14
15
16
3-2
3-3
3-4
3-5

PLATE 3: SHARKS AND RAYS

1 TIGER SHARK
Galeocerdo cuvier (Peron & Le Sueur, 1822)
Inhabits deeper offshore areas, frequently near reefs, distinguished by blunt (when viewed from above) snout, stripes on side (faint or absent in large adults), keel on side of tail base, and strongly curved teeth; very dangerous; found throughout the region; worldwide temperate and tropical seas; to 650 cm and 520 kg. (CARCHARHINIDAE) ★★★

2 SLITEYE SHARK
Loxodon macrorhinus Müller & Henle, 1839
Inhabits continental shelf waters between 7-80 m depth; distinguished by notch or slit on rear edge of eye socket, long slender snout, and large eye; harmless; found throughout the region; Indo-W. Pacific; to 91 cm. (CARCHARHINIDAE) ★★★

3 LEMON SHARK
Negaprion acutidens (Rüppell, 1837)
Inhabits inshore waters, in bays, estuaries and in coral reef lagoons; distinguished by yellow-brown colour, short snout and stocky body with 2 dorsal fins about equal sized; generally harmless to divers but potentially dangerous; found throughout the region; Indo-W. Pacific; to 335 cm and 91 kg. (CARCHARHINIDAE) ★★★

4 BLUE SHARK
Prionace glauca (Linnaeus, 1758)
Inhabits surface waters, usually well offshore; similar to Mako (**1-7**) at least in colour, but has smaller gill slits; longer pectoral fins and serrated teeth; also called Blue whaler; dangerous; found throughout the region; worldwide temperate and tropical seas; to 380 cm and 140 kg. (CARCHARHINIDAE) ★★★

5 WHITETIP SHARK
Triaenodon obesus (Rüppell, 1837) Cairns 10/48
Inhabits coral reefs, frequently seen resting on the bottom in caves or in the open; distinguished by slender shape and white tips on first dorsal and caudal fin; usually harmless, but has attacked humans; found throughout the region; Indo-E. Pacific; to 215 cm. (HEMIGALEIDAE) ★★

6 MILK SHARK
Rhizoprionodon acutus (Rüppell, 1837)
Inhabits coastal bays and off sandy beaches, also offshore areas to 200 m depth; similar to **2** above, but lacks notch on rear edge of eye socket; harmless; found throughout the region; E. Atlantic and Indo-W. Pacific; to 178 cm. (CARCHARHINIDAE) ★★

7 SCALLOPED HAMMERHEAD
Sphyrna lewini (Griffith & Smith, 1834)
Inhabits coastal waters and also encountered well offshore, frequently near the surface; distinguished from other hammerheads in the region by an indentation in the middle of the front edge of the head; dangerous; found throughout the region; worldwide temperate and tropical seas; to 420 cm and 76 kg. (SPHYRNIDAE) ★★

8 WINGHEAD SHARK
Eusphyra blochii (Cuvier, 1817)
Inhabits shallow coastal waters; distinguished by huge wing-shaped head (nearly 1/2 length of body); considered harmless; found throughout the region; N. Indian Ocean and Indo-Australian Archipelago; to 152 cm. (SPHYRNIDAE) ★★

9 GREAT HAMMERHEAD
Sphyrna mokarran (Rüppell, 1837)
Inhabits coastal waters and also found well offshore; distinguished by very flat front edge of head and tall, sail-like first dorsal fin with pointed tip, otherwise body shape similar to **7** above; dangerous; found throughout the region; worldwide temperate and tropical seas; to 610 cm. (SPHYRNIDAE) ★★

10 SMOOTH HAMMERHEAD
Sphyrna zygaena (Linnaeus, 1758)
Inhabits coastal waters and also found well offshore; distinguished from other hammerheads in the region by smooth front edge of head; dangerous; found throughout the region; worldwide temperate and tropical seas; to 400 cm; 96 kg. (SPHYRNIDAE) ★★

11 GREEN SAWFISH
Pristis zijsron Bleeker, 1851
Inhabits mud bottoms, entering estuaries; distinguished by relatively long saw-like snout; similar to **12** below, but has 24-28 pair of teeth extending along entire edge of snout; not dangerous unless cornered; found throughout the region; Indo-W. Pacific; usually to 600 cm, but reported to attain 730 cm. (PRISTIDAE) ★★

12 NARROW SAWFISH
Anoxypristis cuspidatus (Latham, 1794)
Inhabits mud bottoms inshore to about 40 m depth; similar in general appearance to **11** above, but body not as stout and has very long, narrow snout with 18-22 pairs of lateral teeth, but teeth absent on rear part; found throughout the region; N. Indian Ocean and Indo-Australian Archipelago; to 350 cm. (PRISTIDAE) ★★

13 WIDE SAWFISH
Pristis microdon Latham, 1794
Inhabits mud bottoms of bays and estuaries, also enters large rivers and goes well upstream; similar to **11** above, but has shorter, broader snout with 18-23 (usually 20-22) pairs of lateral teeth; also known as Freshwater Sawfish; found throughout the region; Indo-W. Pacific; usually to about 200 cm, but reputed to reach 700 cm. (PRISTIDAE) ★★

14 SHARK RAY
Rhina ancylostoma (Bloch & Schneider, 1801)
Inhabits coastal waters, on mud or sand bottoms; distinguished from **15** below by rounded head and granular patches or ridges above each eye and on middle of forehead; found throughout the region; Indo-W. Pacific; to 260 cm. (RHYNCHOBATIDAE) ★★

15 WHITE-SPOTTED SHOVELNOSE RAY
Rhynchobatus djiddensis (Forsskål, 1775)
Inhabits sandy areas, sometimes seen resting motionless; distinguished from **14** above by pointed head and black spot above pectoral fin base; sometimes incorrectly referred to as Shovelnose Shark; found throughout the region; Indo-W Pacific; to 300 cm and 75 kg. (RHYNCHOBATIDAE) ★★★

16 SPOTTED SHOVELNOSE RAY
Aptychotrema sp.
Inhabits coastal waters on sand bottoms; similar to **17** below, but has broad brown margin around white spots; an undescribed species known only from the Timor Sea, off Melville I., N. Territory; to 120 cm. (RHINOBATIDAE) ★★

17 GIANT SHOVELNOSE RAY
Rhinobatos typus Bennett, 1830
Inhabits coastal waters; similar to **16** above, but lack spots and snout is broader with more rounded tip; found throughout the region; E. Indian Ocean and Indo-Australian Archipelago; to at least 270 cm. (RHINOBATIDAE) ★★

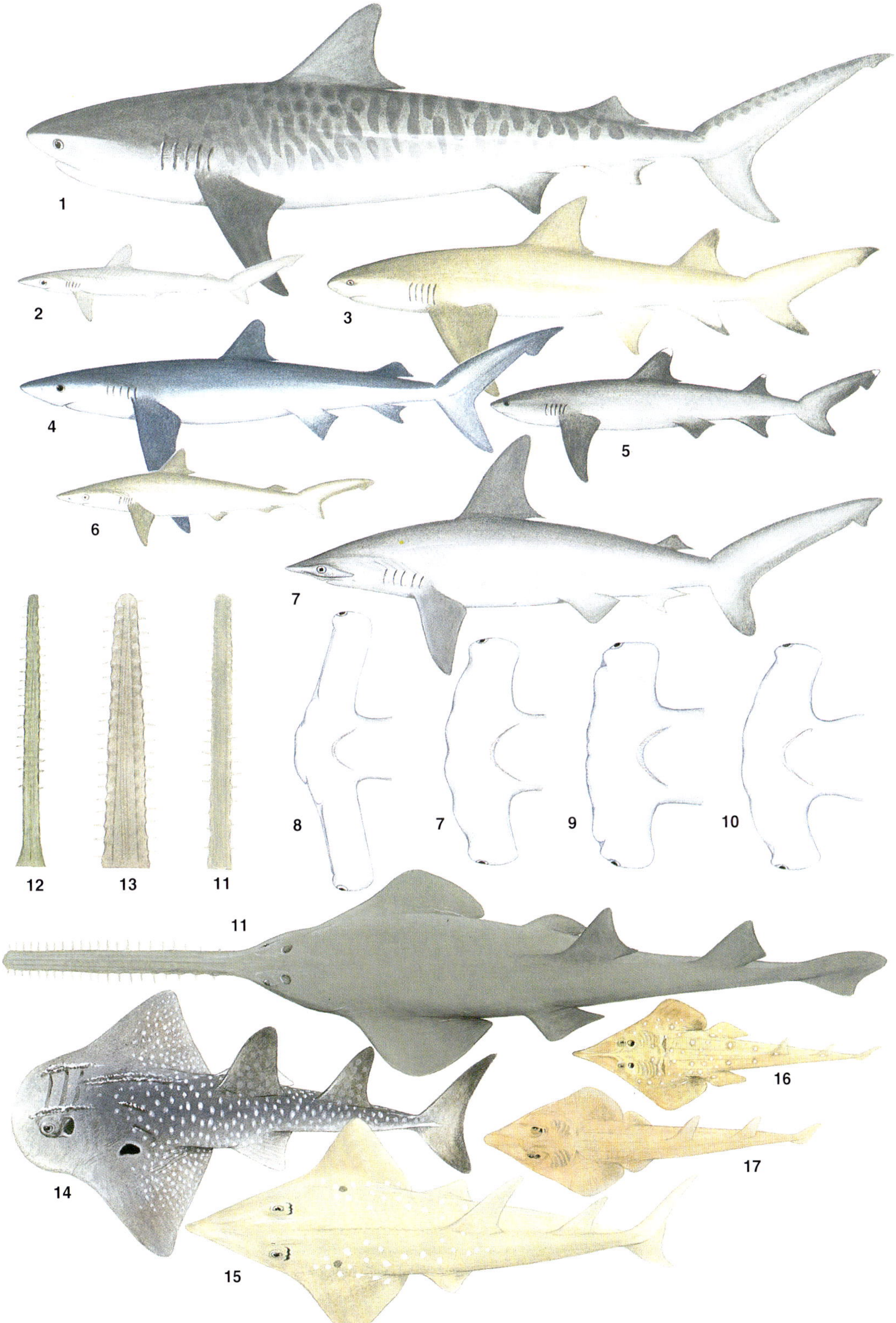
1
2
3
4
5
6
7
12
13
11
8
7
9
10
11
14
16
17
15

PLATE 4: SHARKS AND RAYS

1 WESTERN ANGEL SHARK
Squatina sp.
Inhabits deep (150-310 m) trawling grounds; a ray-like shark with distinctive shape, has greatly enlarged pectoral fins that are not entirely fused to the head and body as in rays, also has bilobed tail and 2 small dorsal fins; W. Australia between Shark Bay and Broome; to at least 64 cm. (SQUATINIDAE) ★★

2 YELLOW SHOVELNOSE-RAY
Aptychotremata sp.
Inhabits sand bottoms; similar to **16** on previous page, but lacks spots and snout is blunter; continental shelf of N.W.Australia; to at least 65 cm. (RHINOBATIDAE) ★★

3 BROWN STINGAREE
Urolophus westraliensis Last & Gomon, 1987
Inhabits sand bottoms in depths of 150-210 m; distinguished by sharply pointed snout tip, at least one serrated spine on tail, similar to **4** below, but is plain brown without markings or has 3 indistinct bars on disc; N.W. Australia between Dampier and Buccaneer Archipelago; to at least 36 cm. (UROLOPHIDAE) ★★

4 BLOTCHED STINGAREE
Urolophus mitosis Last & Gomon, 1987
Inhabits sand bottoms to depths of 200 m; similar to **3** above and **8** (next page) but has pale elongate blotches and lines that surround clusters of dark spots; N.W. Australia, known thus far only from off Port Hedland; to at least 29 cm. (UROLOPHIDAE) ★★

5a BANDED NUMBFISH
Narcine westraliensis McKay, 1966
Inhabits sand bottoms; distinguished by flattened "tadpole" shape, large round head, and 2 small dorsal fins; colour ranges from plain to spotted; capable of producing mild electrical shock; N.W. Australia only; to 28 cm.

5b ORNATE NUMBFISH
Narcine sp.
This small species has often been confused with Banded Numbfish. It only occurs in the Gulf of Carpentaria; to 17cm. (NARCINIDAE)

6 NUMBFISH
Hypnos monopterygium (Shaw & Nodder, 1795)
Inhabits sand-weed areas; distinguished by round body with smaller rounded pelvic lobe at rear which bears the tail and 2 small dorsal fins; colour varies from light brown to blackish; can produce a strong electrical shock if handled or accidentally trod on; also known as Coffin Ray; St. Vincents Gulf, S. Austraila to Broome, W. Australia and S.Queensland to S. New S. Wales; Australia only; to 69 cm. (HYPNIDAE)

7 EYED SKATE
Raja sp.
Inhabits sand bottoms of continental shelf in 60-200 m depth; distinguished by pointed snout, bilobed pelvic fins, single row of thorns or small spines down middle of tail, and 2 small dorsal fins near end of tail, also has fragmented ocellus-type markings on each side of back; also known as False Argus Skate; continental shelf of N.W. Australia; to at least 20 cm width. (RAJIDAE)

8 WESTERN ROUND SKATE
Irolita sp.
Inhabits sand bottoms of continental shelf in 150-200 m depth; distinguished by round shape, bilobed pelvic fins, short spines or thorns in 1 or more rows on tail, and 2 small dorsal fins near end of tail, also has blue-grey spots or blotches scattered on back; N.W. Australia from Shark Bay to Port Hedland; to 42 cm. (RAJIDAE)

9 BROWN STINGRAY
Dasyatis annotatus Last, 1987
Inhabits sand bottoms in 40-65 m depth; distinguished by kite-shape and 2 long serrated spines on tapering tail; similar to **10** below, but is plain grey-brown without markings and has a more pointed snout; Timor and Arafura seas off N. Australia; to at least 24 cm width and 45 cm length. (DASYATIDAE) ★★

10 BROWN RETICULATED STINGRAY
Dasyatis leylandi Last, 1987
Inhabits sand bottoms; two distinct colour forms are known, one with numerous irregular pale blotches and another with pepper-like spotting, both have broad dark band or "mask" between eyes; also known as Painted Maskray; N. Australia and S. New Guinea; to at least 25 cm width. (DASYATIDAE) ★★

11 BLUE-SPOTTED STINGRAY
Dasyatis kuhlii (Müller & Henle, 1841)
Inhabits sand bottoms, frequently in the vicinity of coral reefs; sometimes buries itself with only the eyes protruding above the sand; distinguished by blue spots and frequently has scattered black spots on disc, found throughout the region; Indo-W. Pacific; maximum width about 45 cm. length to 70 cm. (DASYATIDAE) ★★

SHARK FACTS

Sharks and their cousins, the rays, are very specialised fishes that represent a primitive stage of evolutionary development. The sharks represent an extremely ancient lineage. They were known in Devonian seas, over 350 million years ago. Many of the present genera of sharks, skates, and rays date back more than 100 million years. Approximately 350 species of sharks are currently known. Only a small number of these are considered dangerous. Sharks represent an extremely diverse assemblage, occurring in all seas, and in a variety of depths ranging from shallow intertidal pools to deep oceanic trenches, miles below the surface.

RAYS AND RELATIVES

Rays and their relatives are classified in the Order Rajiformes. Both sharks and rays are characterised by a cartilagenous skeleton. Other typical features of rays include a greatly flattened body that is often disc-shaped, and the presence of five, ventrally located gill openings. Most species bear their young alive except the Rajidae, which deposit egg cases. Many rays have venomous spikes or spines on the tail base that are capable of inflicting painful wounds. The members of the family Torpedinidae possess powerful electric organs situated in the head region. Rays dwell in a variety of habitats, ranging from oceanic depths to shallow reefs, estuaries, and even freshwater streams. They range in size from about 30-40 cm disc-width (some skates) to more than 4 m (manta rays).

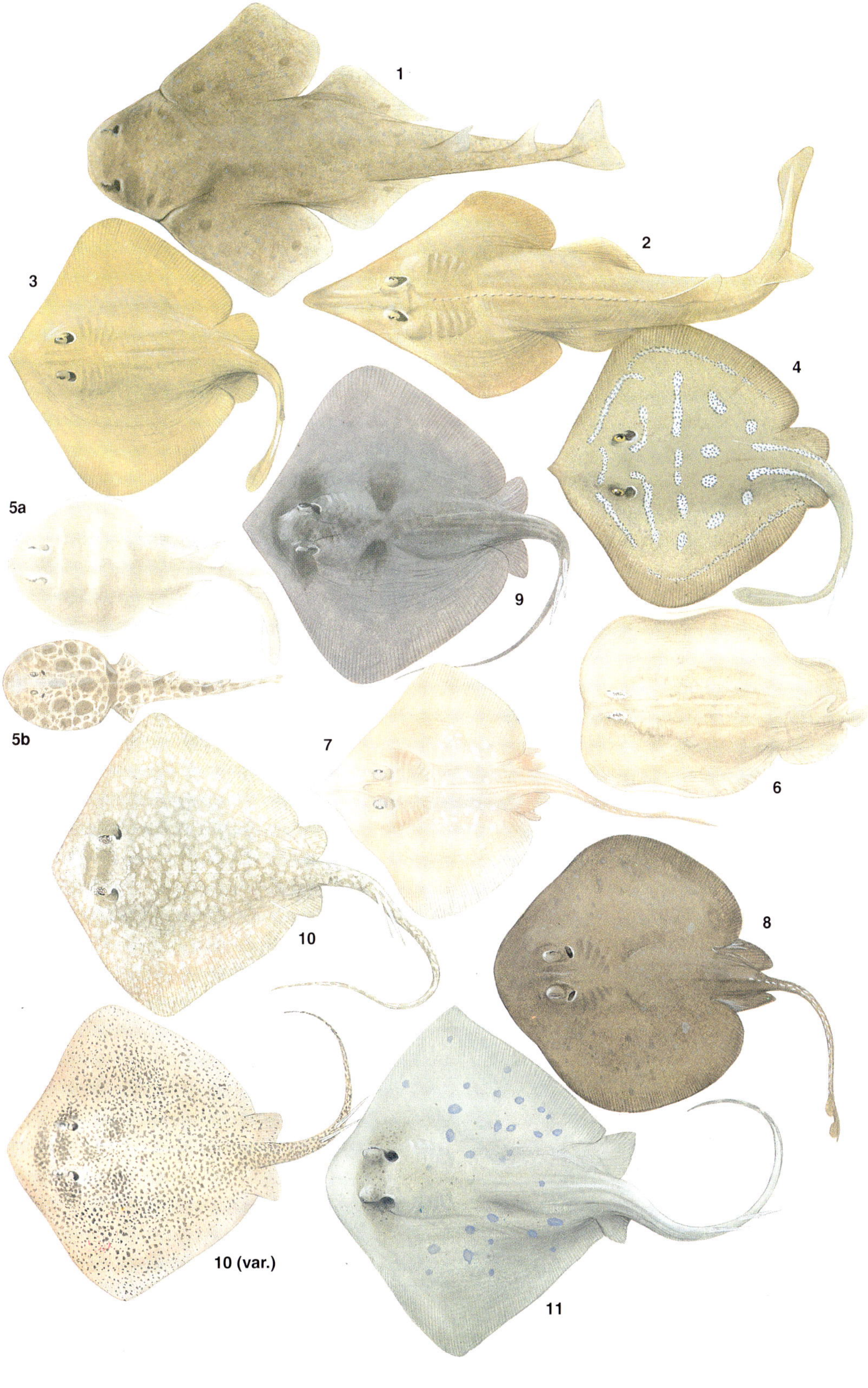
1
2
3
4
5a
9
5b
7
6
10
8
10 (var.)
11

PLATE 5: RAYS AND GHOST SHARKS

1 BLACK STINGRAY
Dasyatis thetidis Ogilby, 1899
Inhabits coastal waters off beaches and over sand or mud bottoms to at least 300 m depth; distinguished by blue-grey to blackish colour and short tubercles on top of head and over middle of back; has a pair of spines on tail that can inflict serious wounds; temperate and subtropical Australia northward to Shark Bay in the west and Coff's Harbour in the east; also New Zealand, and South Africa; to 180 cm disc width, 400 cm total length and 65 kg. (DASYATIDAE) Cairns 10/98 ★★

2 COWTAIL STINGRAY
Pastinachus sephen (Forsskål, 1775)
Inhabits flat sand or mud bottoms near shore, also common in brackish mangrove estuaries and in the lower reaches of rivers; distinguished by grey-brown to blackish colour and broad flap of skin on lower edge of tail, large specimens may have tubercles on back similar to **1** above; has 2 dangerous spines on tail; found throughout the region; Indo-W. Pacific to 180 cm disc width. (DASYATIDAE) ★★

3 BLACK-BLOTCHED STINGRAY
Taeniura meyeni Müller & Henle, 1841
Inhabits sandy bottoms in the vicinity of coral reefs; distinguished by round shape and dense pattern of black spots; has pair of dangerous spines on tail; found throughout the region; Indo-W. Pacific; to 180cm disc width and 330 cm total length. (DASYATIDAE) ★★

4 BLUE-SPOTTED FANTAIL STINGRAY
Taeniura lymma (Forsskål, 1775)
Inhabits flat sand bottoms in the vicinity of coral reefs; distinguished by kite shape and pattern of bright blue spots; has 1-2 dangerous spines on tail; found throughout the region; Indo-W. Pacific; to at least 30 cm disc width and 70 cm total length. (DASYATIDAE)

5 BLACK-SPOTTED STINGRAY
Himantura toshi Whitley, 1939
Inhabits sandy beaches, sand flats near reefs, or shallow mangrove estuaries; distinguished by very long, whip-like tail and pattern of small black spots; has pair of dangerous spine on tail (sometimes missing); N. Australia and S. New Guinea; to at least 70 cm disc width and 180 cm total length. (DASYATIDAE) ★★

6 LEOPARD WHIPRAY
Himantura undulata (Bleeker, 1852)
Inhabits sandy beaches, sand flats near reefs, or shallow mangrove estuaries; distinguished by very long, whip-like tail and leopard-like spot pattern; has dangerous spine on tail; a similar species, *H. uarnak* (not shown), also found throughout the region - it has a reticulate, maze-like pattern; N. Australia and S. New Guinea; to at least 70 cm disc width and 180 cm total length. (DASYATIDAE) ★★

7 MANTA RAY
Manta birostris (Donndorff, 1798)
Inhabits coastal waters and the vicinity of offshore reefs; distinguished by large size, pair of protruding flaps at front of head, and short tail; one of the largest of all fishes, it is a harmless plankton feeder well known for its ability to make spectacular leaps above the water surface; found throughout the region; worldwide circumtropical; to 670 cm disc width and over 2 tons in weight. (MOBULIDAE) ★★★

8 PATCHWORK STINGAREE
Urolophus flavomosaicus Last & Gomon, 1987
Inhabits coastal waters on flat sand bottoms; similar to **3** and **4** on previous page, but has ornate pattern of white spots and reticulated white and brown lines; found throughout the region; central W. Australia and southern half of Queensland; to 60 cm. (DASYATIDAE) ★★

9 RAT-TAILED RAY
Gymnura australis (Ramsay & Ogilby, 1886)
Inhabits shallow coastal waters; distinguished by broad triangular "wings" and very short rat-like tail; also known as Butterfly Ray; northern half of Australia and S. New Guinea; to at least 80 cm disc width. (GYMNURIDAE)

10 BARBLESS EAGLE RAY
Aetomyleus nichofii (Schneider, 1801)
Inhabits coastal waters in the vicinity of reefs; distinguished by bulging head and protruding snout similar to **11** below, but lacks white spots and has pale blue cross bands on back; found throughout the region; C. and E. Indian Ocean to W. Pacific; to 58 cm disc width and about 100 cm total length. (MYLIOBATIDAE) ★★

11 SPOTTED EAGLE RAY
Aetobatus narinari (Euphrasen, 1790)
Inhabits coastal waters in the vicinity of reefs; similar to **10** above, but has white spots on back and 2-6 barbed spines on base of tail; found throughout the region; worldwide circumtropical; to at least 300 cm disc width and total length of 880 cm. (MYLIOBATIDAE) ★★

12 BLACKFIN GHOST SHARK
Hydrolagus lemures (Whitley, 1939)
Inhabits deeper offshore waters of continental shelf in 200-500 m depth; distinguished by 'rodent-like' head with small mouth below eye, large pectoral fins, long tapering tail; several similar species found in region, but lack dark edge on first dorsal fin; widespread along most of the Australian continental shelf; to 30 cm. (CHIMAERIDAE)

BEWARE OF SPINES!

Many of the rays illustrated on Plates 4 and 5 are characterised by one or more venomous spines on the tail. Stings inflicted by these spines are extremely painful and fatalities have occurred when either heart, abdomen, or lungs were badly perforated. Caution should be exercised when wading on sandy bottoms. It is advisable to use a walking stick to probe just ahead or at least walk with a shuffling gait rather than in normal strides. This sort of movement will often prevent treading directly on the back of a partially buried ray. If a ray is stepped on it has the ability to thrust its tail upward and forward, impaling the victim with remarkable speed. Pain is immediate and intense, and may persist for several days. Immersion in hot water (about 50°C) for 30-90 minutes may dramatically relieve the pain as the venom is a protein which is heat labile. Medical assistance should always be obtained as the wound may be far more serious than it appears.

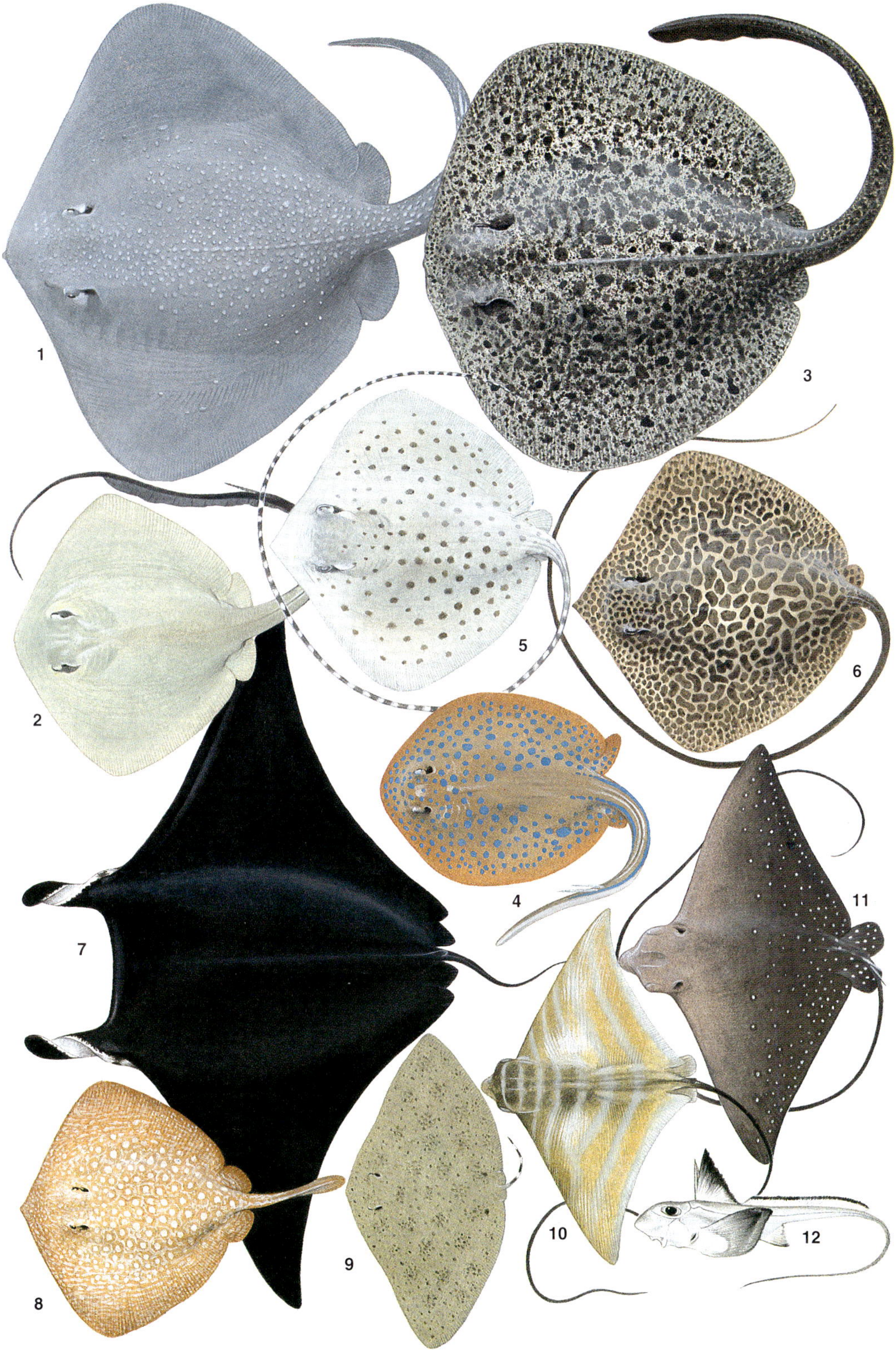
1
3
2
5
6
4
11
7
10
12
9
8

PLATE 6: HERRINGS AND RELATIVES

1 GIANT HERRING
Elops hawaiiensis Regan, 1909
Inhabits coastal waters and mangrove areas; distinguished by slender body and relatively large mouth; found throughout the region; Indo-C. Pacific; to 120 cm. (ELOPIDAE) ★

2 BONEFISH
Albula neoguinaica Valenciennes, 1847
Inhabits estuaries and mudflats; distinguished by protruding snout; found throughout the region; Indo-W. Pacific; to 110 cm. (ALBULIDAE) ★

3 MILKFISH
Chanos chanos (Forsskål, 1775)
Inhabits coastal waters near reefs; distinguished by small mouth and scissor-like tail; found throughout the region; Indo-W. Pacific; to 120 cm, 10.6 kg. (CHANIDAE) ★

4 OXEYE HERRING
Megalops cyprinoides (Broussonet, 1782)
Inhabits coastal waters and mangrove areas; distinguished by large eye and mouth, and filament at end of dorsal fin; found throughout the region; Indo-W. Pacific; to 150 cm, 2.3 kg. (ELOPIDAE) ★

5 WOLF HERRING
Chirocentrus dorab Forsskål, 1775
Inhabits coastal waters; distinguished by large fangs; found throughout the region; Indo-W. Pacific; to 140 cm. (CHIROCENTRIDAE) ★

6 SMOOTH-BELLY SARDINE
Amblygaster leiogaster (Valenciennes, 1847)
Inhabits coastal waters in large schools; similar to **8** below, but pelvic and anal fins farther apart; N.W. Australia and Indo-Malay Archipelago; Indo-W. Pacific; to 25 cm. (CLUPEIDAE) ★★

7 NORTHERN PILCHARD
Amblygaster sirm (Walbaum, 1792)
Inhabits coastal waters; distinguished by row of spots on side; N.W. Australia to Gulf of Carpentaria and throughout S.E. Asia; Indo-W. Pacific; to 22 cm. (CLUPEIDAE) ★★

8 SLENDER SARDINE
Dussumieria elopsoides Bleeker, 1849
Inhabits coastal waters; similar to **6** above, but pelvic and anal fins closer together; N.W. Australia to Gulf of Carpentaria and throughout S.E. Asia; Indo-W. Pacific; to 25 cm. (CLUPEIDAE) ★★

9 BLUESTRIPE HERRING
Herklotsichthys quadrimaculatus (Rüppell, 1837)
Inhabits coastal waters; distinguished from all other plain coloured herrings on this page by a pair of fleshy outgrowths on margin of gill cover; found throughout the region; Indo-W. Pacific; to 16 cm. (CLUPEIDAE) ★★

10 KONINGSBERGER'S HERRING
Herklotsichthys koningsbergeri (Weber & de Beaufort, 1912)
Inhabits beaches and inlets; distinguished by double row of spots on side; found throughout the region; N.W. Australia to Gulf of Carpentaria and S. New Guinea; to 15 cm. (CLUPEIDAE) ★★

11 GIZZARD SHAD
Anodontostoma chacunda (Hamilton, 1822)
Inhabits coastal waters and mangrove areas; similar to **14** below, but lacks filament at rear of dorsal fin; found throughout the region; N.W. Australia to Gulf of Carpentaria and throughout S.E. Asia; Indo-Australian Archipelago and N. Indian Ocean; to 20 cm. (CLUPEIDAE) ★★

12 BANDED ILISHA
Ilisha striatula Wongratana, 1983
Inhabits coastal waters; similar to **13** below, but has faint stripe along middle of side (not shown) and lacks dark spot behind gill cover; N.W. Australia and Indonesia; mainly N. Indian Ocean; to 22 cm. (PRISTIGASTERIDAE) ★★

13 DITCHELEE
Pellona ditchela Valenciennes, 1847
Inhabits coastal bays and estuaries; similar to **12** above, but lacks stripe on sides and has dark spot behind gill cover; found throughout the region; Indo-W. Pacific; to 18 cm. (PRISTIGASTERIDAE) ★★

14 HAIRBACK HERRING
Nematalosa come (Richardson, 1846)
Inhabits coastal bays and estuaries; similar to **11** above but has filament at end of dorsal fin; found throughout the region; mainly Indo-Australian Archipelago north to E. China Sea; to 23 cm. (CLUPEIDAE) ★★

15 SLENDER SPRAT
Spratelloides gracilis (Temminck & Schlegel, 1846)
Inhabits coastal waters; distinguished by slender shape and silvery stripe on sides; found throughout the region; Indo-C. Pacific; to 11 cm. (CLUPEIDAE) ★★

16 BLUE SPRAT
Spratelloides robustus Ogilby, 1897
Inhabits coastal waters and estuaries; distinguished by bluish back and lack of silver stripe on sides; Australia only (W.A. and New. S.Wales); to 9 cm. (CLUPEIDAE) ★★

17 GOLDSTRIPE SARDINE
Sardinella gibbosa (Bleeker, 1849)
Inhabits coastal waters, distinguished by thin gold-coloured stripe on sides; N.W. Australia and throughout S.E. Asia; Indo-W. Pacific; to 19 cm. (CLUPEIDAE) ★★★

18 BAREBACK ANCHOVY
Papuengraulis micropinna Munro, 1964
Inhabits coastal bays and estuaries; distinguished by thread-like dorsal fin; N. Australia and S. New Guinea; to 15 cm. (ENGRAULIDAE) ★★

19 INDIAN ANCHOVY
Stolephorus indicus (van Hasselt, 1823)
Inhabits coastal waters; distinguished by rounded snout and broad silvery stripe on sides; found throughout the region; Indo-W. Pacific; to 16 cm. (ENGRAULIDAE) ★★

20 LONGFIN ANCHOVY
Setipinna tenuifilis (Valenciennes, 1848)
Inhabits coastal waters; similar to **18** above, but has normal dorsal fin; N.W. Australia and S. New Guinea; E. Indian Ocean and W. Pacific; to 20 cm. (ENGRAULIDAE) ★★

21 HAMILTON'S ANCHOVY
Thryssa hamiltonii (Gray, 1835)
Inhabits estuaries and mudflats; distinguished by rounded snout, large mouth, and spot behind gill cover; *Thryssa setirostris* (not shown) similar, but with extremely long posterior extension of upper jaw; found throughout the region except E. Queensland; N. Indian Ocean and W. Pacific; to 25 cm. (ENGRAULIDAE) ★★

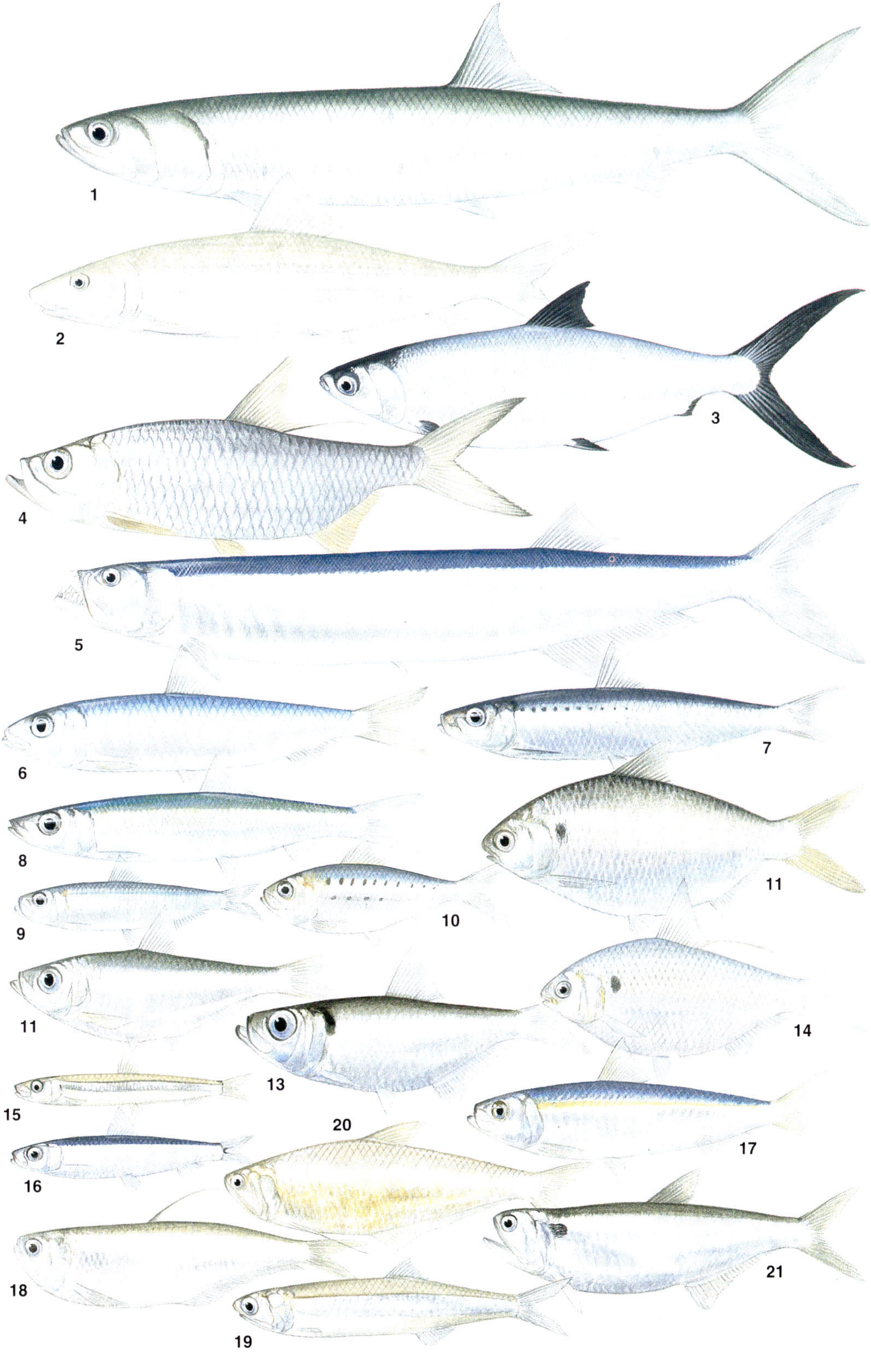
1
2
3
4
5
6
7
8
11
9
10
11
14
13
15
20
17
16
18
21
19

PLATE 7: MORAY EELS (MURAENIDAE)

1 STARRY EEL
Echidna nebulosa (Thünberg, 1789)
Inhabits shallow coral reefs; distinguished by whitish body with 2 longitudinal rows of darkish pale-centred blotches and lacks sharp fangs, also known as Clouded reef-eel; found throughout the region; Indo-E. Pacific; to 70 cm.

2 GIRDLED REEF EEL
Echidna polyzona (Richardson, 1845)
Inhabits shallow coral reefs, often exposed to surge; distinguished by alternating light and dark bars of approximately equal width and lacks sharp fangs; N.W. Australia, E. Queensland, and throughout S.E. Asia; Indo-W. Pacific; to 60 cm.

3 ZEBRA EEL
Echidna zebra (Shaw and Nodder, 1797)
Inhabits shallow coral reefs, often exposed to surge; distinguished by narrow pale bands encircling head and body, and lacks sharp fangs; N.W. Australia, E. Queensland, and throughout S.E. Asia; Indo-E. Pacific; to 150 cm.

4 LATTICE-TAIL MORAY
Gymnothorax buroensis (Bleeker, 1857)
Inhabits offshore coral reefs; distinguished by brown colour on front of body and blackish colour on posterior part with pale spotting; N.W. Australia, E. Queensland, and throughout S.E. Asia; Indo-E. Pacific; to 31 cm.

5 SPOTTED MORAY
Gymnothorax eurostus (Abbot, 1861)
Inhabits coral reef crevices; distinguished by numerous small yellowish spots becoming larger on rear part of body; also dark spots or blotches evident mainly on front half; N.W. Australia, E. Queensland, and throughout S.E. Asia; Indo-E. Pacific; to 40 cm 0.23 kg.

6 SIEVE-PATTERNED MORAY
Gymnothorax criboris Whitley, 1932
Inhabits coral reef crevices; distinguished by several dark spots behind eye; network of fine interconnected lines on front half of body, and network of darker brown surrounding pale blotches on posterior half; also known as Brown-flecked reef eel; N. Australia southward to Sydney; to 75 cm.

7 BLACK-BLOTCHED MORAY
Gymnothorax favagineus Bloch & Schneider, 1801
Inhabits coral reef crevices; distinguished by bold spot pattern; one of the largest of moray eels, but usually harmless unless provoked; its sharp fangs can cause serious injury; also known as Tesselated moray and Giraffe eel; found throughout the region; Indo-W. Pacific; to 300 cm.

8 YELLOW-EDGED MORAY
Gymnothorax flavimarginatus (Rüppell, 1830)
Inhabits coral reef crevices; generally yellow-brown in colour with fine dark spotting on head and body and black patch at gill opening; juveniles are dark brown with fine yellow-green margin on dorsal and anal fins; also known as Leopard eel; N.W. Australia, E. Queensland, and throughout S.E. Asia; Indo-E. Pacific; to 50 cm.

9 FIMBRIATED MORAY
Gymnothorax fimbriatus (Bennett, 1832)
Inhabits coral reef crevices; distinguished by tan or light brown colour with loose network of branched dark bands; N.W. Australia, E. Queensland, and throughout S.E. Asia; Indo-W. Pacific; to 80 cm.

10 GIANT MORAY
Gymnothorax javanicus (Bleeker, 1859)
Inhabits offshore coral reefs; distinguished by yellow-brown head with small dark spots and large dark patch at gill opening; adults have leopard-like spotting on body; a large eel that can be dangerous if provoked; several attacks have been reported; found throughout the region; Indo-W. Pacific; to 250 cm.

11 PEARLY MORAY
Gymnothorax margaritophorus Bleeker, 1865
Inhabits coral reef crevices; distinguished by series of dark blotches just behind eye, pale chin and breast, and "lattice" pattern on rear part of body; N.W. Australia, E. Queensland, and throughout S.E. Asia; Indo-W. Pacific; to 40 cm.

12 MOTTLED MORAY
Gymnothorax undulatus (Lacepède, 1803)
Inhabits coral reef crevices, distinguished by "chain-link" pattern of narrow pale bands; juvenile with diffuse vertical bars most noticeable towards tail; found throughout the region; Indo-E. Pacific; to 150 cm.

13 BARTAIL MORAY
Gymnothorax zonipectus Seale, 1906
Inhabits coral reef crevices; distinguished by white spots on upper and lower jaw; dark spots on body and distinct dark bars on rear portion of dorsal and anal fins; N.W. Australia, E. Queensland, and throughout S.E. Asia; IndoW. Pacific; to 32 cm.

14 PAINTED MORAY
Siderea picta (Ahl, 1789)
Inhabits shallow reef flats and tide pools; sometimes seen entirely out of water at low tide; distinguished by white colouration with numerous small dark spots; also known as Peppered moray; found throughout the region; Indo-E. Pacific; to 68 cm.

15 FRECKLED MORAY
Siderea thrysoidea (Richardson, 1845)
Inhabits shallow coral reefs; distinguished by light brown or tan coloured body (with faint mottlings), white to bluish snout, and silvery eyes; found throughout the region; Indo-W. Pacific; to 35 cm.

MORAY EELS

The eels featured on this plate are all members of the family Muraenidae, commonly known as morays. Most are equipped with needle-sharp teeth which has given them a largely undeserved reputation of being dangerous. While it is true that some larger eels, for example no. 10, have attacked humans, in most cases the eel had been provoked in some manner. Large eels should definitely not be teased with offerings of dead or struggling fish either hand held or on the end of a spear. Exceptions are morays that hang out at popular tourist dive sites and are relatively tame. In this case trust your local guide for advice, but always be cautious! Species in the genus *Echidna* have blunt teeth in contrast to most other eels. This is an adaptation for feeding on shelled molluscs and crustaceans. They exhibit striking colour patterns and are sometimes kept as aquarium pets. The Painted Moray (14) sometimes frightens beachcombers. It occurs in very shallow pools at low tide or is occasionally found high and dry under rocks.

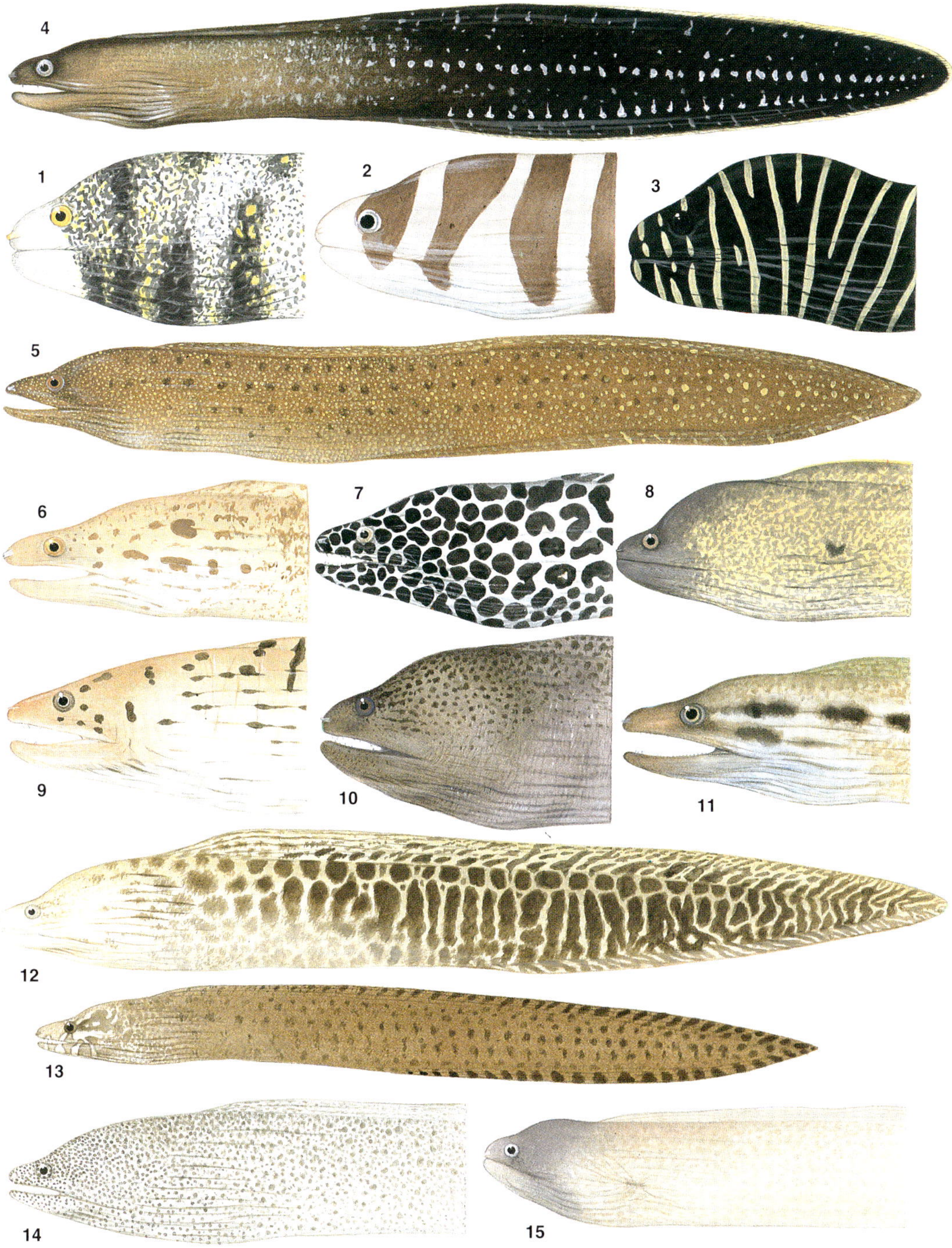
4
1
2
3
5
6
7
8
9
10
11
12
13
14
15

PLATE 8: MORAY EELS (FAMILY MURAENIDAE)

1 BLACK-SPOTTED MORAY
Gymnothorax melanospilus (Bleeker, 1855)
Inhabits coral reef crevices; distinguished by bold black spots on white background, with many of the spots u-shaped; throughout S.E. Asia; E. Indian Ocean and W. Pacific; to 180 cm.

2 LONGFANG MORAY
Enchelynassa canina (Quoy & Gaimard, 1824)
Inhabits coral reef crevices; distinguished by bilobed flap on front nostril, hooked jaws, wrinkled appearance of head, extremely long canine teeth at front of mouth, and several white spots on lower jaw; Great Barrier Reef and throughout S.E. Asia; Indo-E. Pacific; to 150 cm.

3 WHITE-MARGINED MORAY
Enchelycore schismatorhynchus (Bleeker, 1853)
Inhabits coral reef crevices; distinguished by hooked jaws, large fangs, and white margin on dorsal fin; throughout S.E. Asia; Indo-C. Pacific; to 120 cm.

4 HOOKJAW MORAY
Enchelycore bayeri (Schultz, 1953)
Inhabits coral reef crevices; distinguished by plain brown colour, large fangs, and hooked jaws; a relatively small harmless species; Great Barrier Reef, offshore reefs of W.A., and throughout S.E. Asia; Indo-C. Pacific; to 70 cm.

5 DRAGON MORAY
Enchelycore pardalis (Temminck, 1847)
Inhabits coral reef crevices; distinguished by highly ornate pattern of bars and spots, and long tube-like rear nostrlis above front part of eyes; throughout S.E. Asia; Indo-C. Pacific; to 80 cm.

6 GUINEAFOWL MORAY
Gymnothorax meleagris (Shaw & Nodder, 1795)
Inhabits coral reef crevices; distinguished by network of small white spots on dark ground colour, also inside of mouth is white; Great Barrier Reef and throughout S.E. Asia; Indo-C. Pacific; to 100 cm.

7 MOLUCCAN MORAY
Gymnothorax moluccensis (Bleeker, 1865)
Inhabits coral reef crevices; plain brown colour without distinguishing marks; throughout S.E. Asia; W. Pacific; to at least 50 cm.

8 YELLOWMOUTH MORAY
Gymnothorax nudivomer (Playfair & Günther, 1867)
Inhabits coral reef crevices; distinguished by yellow colour on inside of mouth, network of fine whites spots on head and much larger white spots over much of body; Great Barrier Reef and throughout S.E. Asia; Indo-C. Pacific; to 180 cm.

9 TIGER MORAY
Gymnothorax enigmaticus McCosker & Randall, 1982
Inhabits coral reef crevices; distinguished by black bars over entire length of body; similar to **15** below, but bars on head completely encirlcle body and no yellow present on head; Great Barrier Reef, offshore reefs of W.A. and throughout S.E. Asia; Indo-C. Pacific; to 58 cm.

10 HIGHFIN MORAY
Gymnothorax pseudothrysoideus (Bleeker, 1852)
Inhabits coral reef crevices; distinguished by spotted pattern and relatively well developed dorsal fin; Great Barrier Reef and throughout S.E. Asia; W. Pacific; to 80 cm.

11 RICHARDSON'S MORAY
Gymnothorax richardsoni (Bleeker, 1852)
Inhabits coral reef crevices; a small speckled eel that is frequently found under rocks on shallow reef flats, often in weedy areas; throughout S.E. Asia; W. Pacific; to at least 30 cm.

12 SLENDERTAIL MORAY
Gymnothorax gracilicaudus Jenkins, 1903
Inhabits coral reef crevices; distinguished by vertically elongate, branching dark blotches, forming definite bars on anterior half of body, but in several interconnected rows on posterior half; offshore reefs of W.A., Great Barrier Reef, and throughout Oceania; to 32 cm.

13 DWARF MORAY
Gymnothorax melatremus Schultz, 1953
Inhabits coral reef crevices; distinguished by small size, black rim around eye and prominent black mark around gill opening; general colour ranges from brown to bright yellow, sometimes with network of dark markings posteriorly; Great Barrier Reef, offshore reefs of W.A. and throughout S.E. Asia; Indo-W. and C. Pacific; to 26 cm.

14 MARBLED MORAY
Uropterygius marmoratus (Lacepède, 1803)
Inhabits coral reef crevices; distinguished by marbled colour pattern and complete lack of dorsal and anal fins; Great Barrier Reef, offshore reefs of W.A. and throughout S.E. Asia; W. and C. Pacific; to 50 cm.

15 BANDED MORAY
Gymnothorax rueppelliae (McClelland, 1845)
Inhabits coral reef crevices; similar to **9** above, but distinguished by yellowish head and dark bars on head do not encircle the body; Great Barrier Reef, offshore reefs of W.A. and throughout S.E. Asia; Indo-C. Pacific; to 80 cm.

16 RIBBON EEL
Rhinomuraena quaesita Garman, 1888
Inhabits sand or rubble patches on the edge of coral reefs; feeds on fishes and crustaceans; males are bright blue and yellow with elaborate nostril flaps, females yellow except for black anal fin, juveniles and subadults are largely black; harmless; prized as aquarium pets; Great Barrier Reef and throughout S.E. Asia; Indo-C. Pacific; to 120 cm.

17 WHITE RIBBON EEL
Pseudechidna brummeri (Bleeker, 1858)
Inhabits reef flat, sheltered coastal reefs, and lagoons, usuallly on sand-rubble bottoms with rocky outcrops; distinguished by compressed ribbon-like body and overall pale colouration; throughout S.E. Asia; Indo-W. Pacific; to 105 cm.

RIBBON EEL

The bright-coloured ribbon eel (**16**) is aptly named. Although the head is roughly cylindrical, its body is thin and ribbon-like. Unlike most morays, it lives in sandy burrows. It is usually seen protruding its head and up to about one-third of the body length outside the burrow. Aside from coloration and shape, the most distinguishing feature is the enormously expanded nostrils, which form a membranous scoop-like structure. If threatened, for example when closely approached by a diver, the eel swiftly retreats into its burrow, waiting several minutes before emerging.

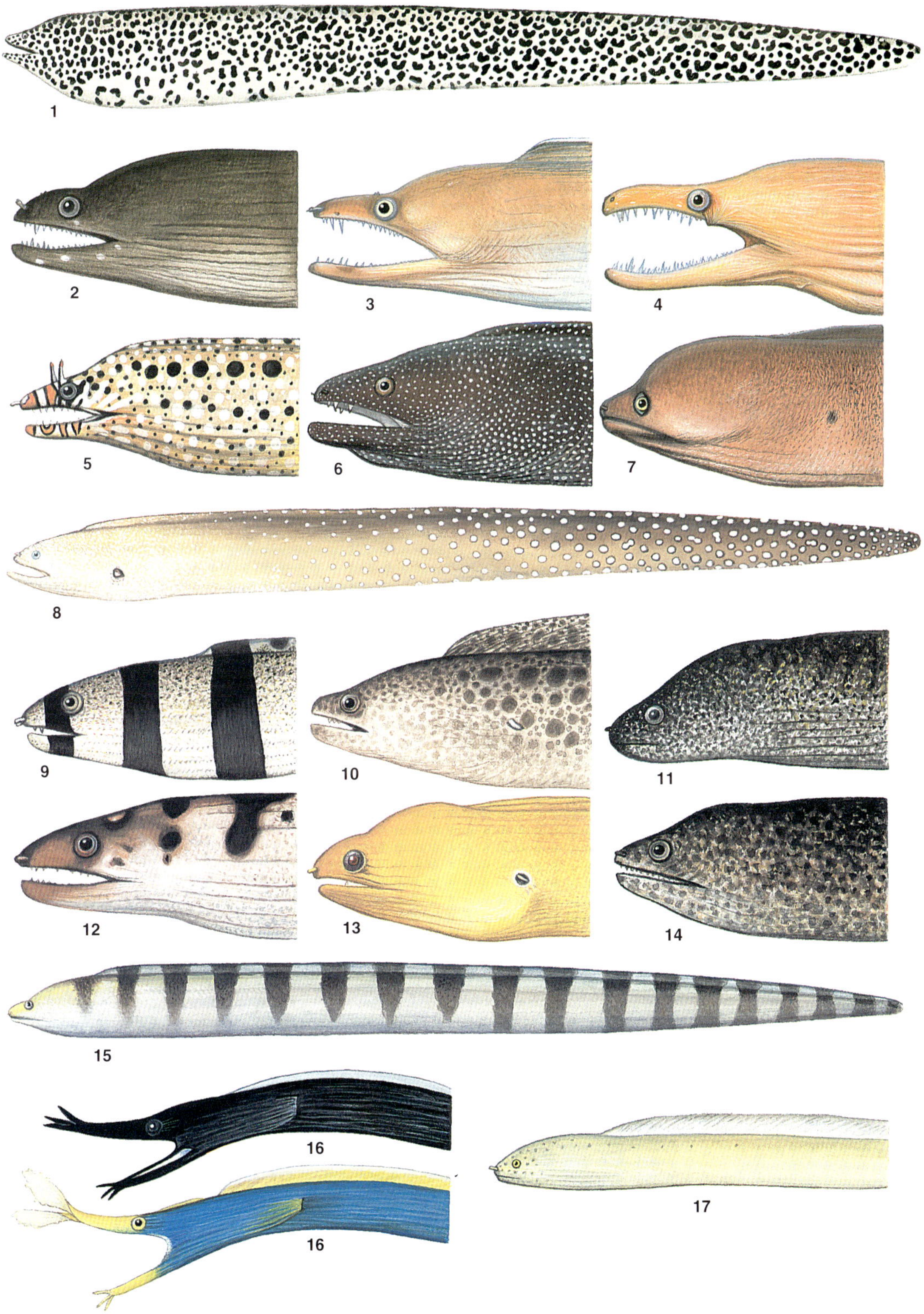
1
2
3
4
5
6
7
8
9
10
11
12
13
14
15
16
16
17

PLATE 9: SNAKE EELS AND CONGER EELS

1 BLACK-EDGED CONGER
Conger cinereus Rüppell, 1830
Inhabits coral reef crevices; distinguished by well developed pectoral fins, relatively tall, black-edged dorsal and anal fins, and diagonal dark band behind mouth; found throughout the region; Indo-C. Pacific; to 103 cm. (CONGRIDAE)

2 MARBLED SNAKE-EEL
Callechelys marmoratus (Bleeker, 1853)
Inhabits sand bottoms near coral reefs; distinguished by dense pattern of irregular black spots; N.W. Australia, E. Queensland, and throughout S.E. Asia; Indo-C. Pacific; to 57 cm. (OPHICHTHIDAE)

3 STARGAZER SNAKE-EEL
Brachysomophis cirrocheilos (Bleeker, 1857)
Inhabits sand bottoms, often with only eyes protruding above surface; distinguished by upward directed eyes near tip of snout, fringe of skin tentacles on lips, fang-like teeth in jaws and roof of mouth, and overall pale colour; N.W. Australia, E. Queensland, and throughout S.E. Asia; Indo-W. Pacific; to 125 cm. (OPHICHTHIDAE)

4 SLENDER WORM-EEL
Muraenichthys gymnotus Bleeker, 1857
Inhabits sand bottoms near coral reefs; distinguished by small worm-like body, olive coloured back, pale belly and lack of pectoral fins; found throughout the region; Indo-C. Pacific; to 17 cm. (OPHICHTHIDAE)

5 FRINGE-LIPPED SNAKE-EEL
Cirrhimuraena calamus (Günther, 1870)
Inhabits sand bottoms; distinguished by fringe of skin tentacles on upper lip, small pectoral fin, brownish colour of back, and abrupt transition to pale on lower half; W. Australia only from Geographe Bay northwards; to 62 cm. (OPHICHTHIDAE)

6 CULVERIN
Leiuranus semicinctus (Lay & Bennett, 1839)
Inhabits sand bottoms; distinguished by series of black saddles on upper two-thirds of body; found throughout the region; Indo-C. Pacific; to 60 cm. (OPHICHTHIDAE)

7 HARLEQUIN SNAKE-EEL
Myrichthys colubrinus (Boddaert, 1781)
Inhabits sand bottoms; similar to 6 above, but black bars completely or nearly encircle body; N.W. Australia, E. Queensland, and throughout S.E. Asia; Indo-C. Pacific; to 88 cm. (OPHICHTHIDAE)

8 FLAPPY SNAKE-EEL
Phyllophichthus xenodontus Gosline, 1951
Inhabits sand bottoms near reefs; distinguished by long pointed snout, leaf-like skin flap at each anterior nostril (near snout tip), and small pectoral fins; found throughout the region; N.W. Australia, E. Queensland, and throughout S.E. Asia; Indo-C. Pacific; to 42 cm. (OPHICHTHIDAE)

9 ONE-BANDED SNAKE-EEL
Ophichthus cephalozona Bleeker, 1864
Inhabits sand bottoms; distinguished by white-edged black saddle on middle of head; N.W. Australia, E. Queensland, and throughout S.E. Asia; mainly W. Pacific; to 80 cm. (OPHICHTHIDAE)

10 OLIVE SNAKE-EEL
Ophichthus rutiodermatoides (Bleeker, 1853)
Inhabits sand bottoms; distinguished by non-descript pattern, pointed snout, pointed teeth, and pectoral fin base on upper half of gill opening; N.W. Australia and Indonesia; mainly Indo-Australian Archipelago; to 68 cm. (OPHICHTHIDAE)

11 BLACK-FINNED SNAKE-EEL
Ophichthus melanochir Bleeker, 1865
Inhabits sand bottoms near coral reefs; distinguished by black edge on dorsal fin, also pectoral fins sometimes entirely or partly black; N.W. Australia and throughout S.E. Asia; Indo-C. Pacific; to 80 cm. (OPHICHTHIDAE)

12 ESTUARY SNAKE-EEL
Pisodonophis boro (Hamilton, 1822)
Inhabits sand or mud bottoms, often in estuaries or freshwater streams; distinguished by non-descript pattern, granular teeth, pectoral fin broad-based (not restricted to upper half of gill opening), and dorsal fin begins behind end of pectoral fins; N.W. Australia and throughout S.E. Asia; Indo-W. Pacific; to 100 cm.

13 BURROWING SNAKE-EEL
Pisodonophis cancrivorous (Richardson, 1848)
Inhabits sand bottoms distinguished by blunt snout (jaws equal in length) granular teeth, pectoral fin broad-based, and dorsal fin begins above pectoral fins; N.W. Australia, E. Queensland, and throughout S.E. Asia; Indo-C. Pacific; to 75 cm. (OPHICHTHIDAE)

14 CHINGILT
Yirrkala lumbricoides (Bleeker, 1864)
Inhabits sand bottoms near coral reefs; distinguished by slender worm-like body; moderately long pointed snout; dorsal fin begins above gill openings, no pectoral fins, and anus about midway between snout and tip of tail; found throughout the region; Indo-W. Pacific; to 44 cm. (OPHICHTHIDAE)

15 VULTURE EEL
Icththyapus vulturis Weber & de Beaufort, 1916
Inhabits sand bottoms near coral reefs; distinguished by general pale colouration, long pointed snout, very small eye and no pectoral fins; N.W. Australia and throughout S.E. Asia; Indo-C. Pacific; to 50 cm. (OPHICHTHIDAE)

SNAKE EELS

All of the species on this plate, except no.1, are members of the family Ophichthidae known as snake eels. Although they are very common, most people, including keen anglers, are unaware of their presence. This is because they spend most of the time buried in the sand. Most of the species have a pointed snout to aid in burrowing. In addition, many have a bony, sharp tail and are equally adept at burrowing forward or backward. The diet of most snake eels consists of small fishes, crabs, and prawns.

A few species particularly, those with banded patterns, are sometimes mistaken for sea snakes, but they are easily distinguished by the lack of scales and possession of a pointed tail (paddle-like in snakes).

The Black-edged Conger Eel (1) belongs to the family Congridae. It is found in rocky areas and amongst coral reef crevices. In some parts of the Indo-Pacfic region its flesh is considered a delicacy. Neither conger or snake eels are dangerous.

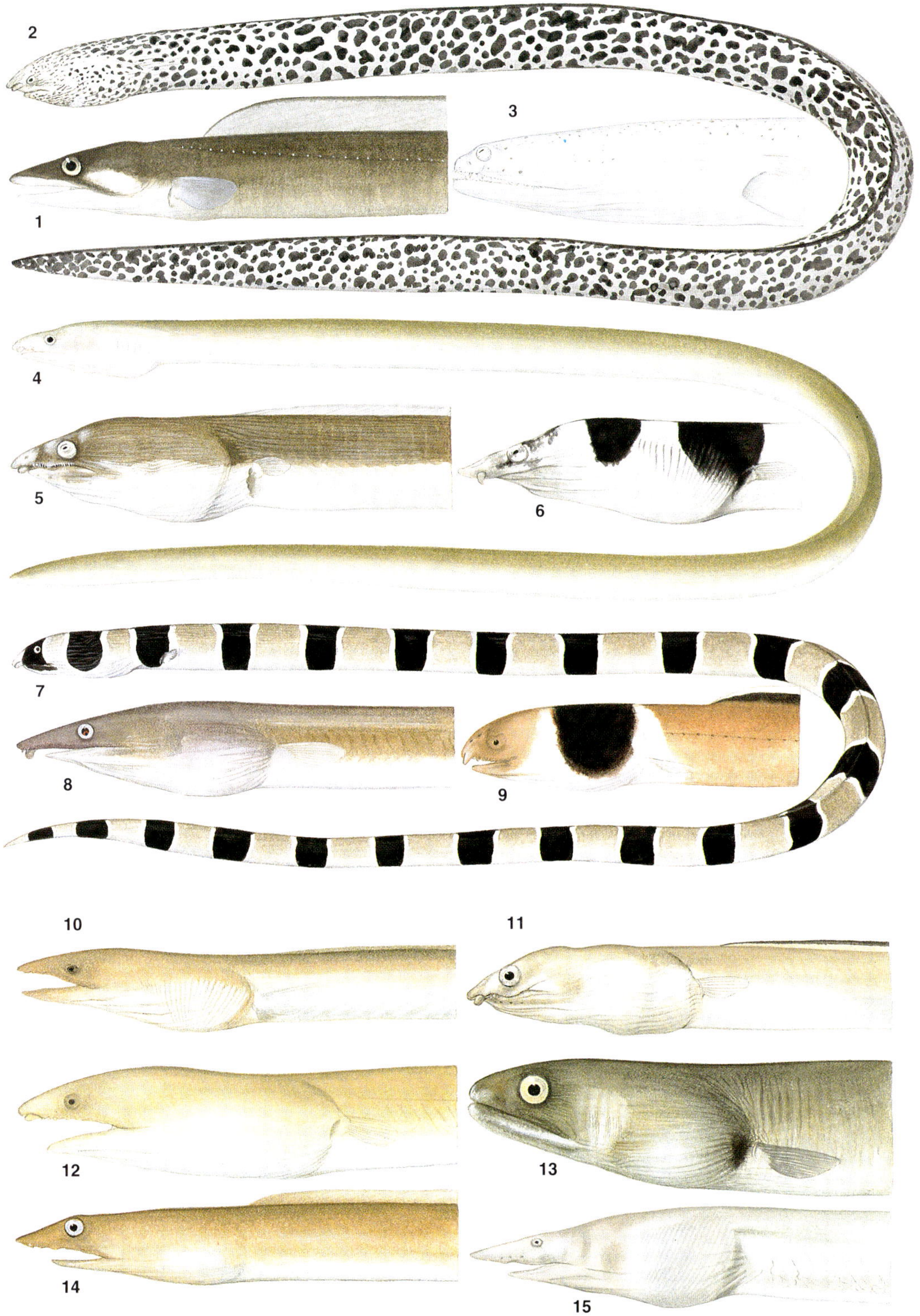
2
3
1
4
5
6
7
8
9
10
11
12
13
14
15

PLATE 10: SNAKE EELS AND GARDEN EELS

1 SHARPSNOUT SNAKE EEL
Apterichtus klazingai (Weber, 1913)
Inhabits sand bottoms near coral reefs, buries in sand and seldom seen; distinguished by pointed snout and by white colour and numerous small light brown spots; Great Barrier Reef and throughout S.E.Asia; Indo-W. Pacific; to 40 cm. (OPHICHTHIDAE)

2 CLOWN SNAKE EEL
Ophichthus bonaparti (Kaup, 1856)
Inhabits sand bottoms near coral reefs, buries in sand - sometimes seen with head protruding; Indonesia and Philippines; to 60 cm. (OPHICHTHIDAE)

3 CROCODILE SNAKE EEL
Brachysomophis crocodilinus (Bennett, 1833)
Inhabits sand bottoms near coral reefs, buries in sand - sometimes seen with head protruding; distinguished by numerous skin flaps lining mouth; similar to *B. cirrocheilos* (Pl. 9-3), but has smaller pectoral fins (their length fits about 10-12 times into head length instead of only 4-5 times); Great Barrier Reef and throughout S.E. Asia; Indo-E. Pacific; to 110 cm. (OPHICHTHIDAE)

4 BLACK-STRIPED SNAKE EEL
Callechelys catostomus (Bloch & Schneider, 1801)
Inhabits sand bottoms near coral reefs, buries in sand and seldom seen; distinguished by broad black stripe; *C. melanotaenia* is a synonym; Great Barrier Reef, offshore reefs of N.W. Australia, and throughout S.E. Asia; to 60 cm. (OPHICHTHIDAE)

5 SPOTTED SNAKE EEL
Myrichthys maculosus (Cuvier, 1817)
Inhabits sand bottoms near coral reefs, buries in sand, but sometimes seen entirely exposed; distinguished by row of large dark spots; Great Barrier Reef, offshore reefs of N.W. Australia, and throughout S.E. Asia; to 50 cm. (OPHICHTHIDAE)

6 COBRA GARDEN EEL
Heteroconger cobra Böhlke and Randall, 1981
Inhabits sand bottoms near coral reefs; mainly covered with brown spots, but distinguished by large dark blotch surrounding u-shaped white marking on head; New Guinea and Solomon Islands; to 40 cm. (CONGRIDAE)

7 MANY-TOOTHED GARDEN EEL
Heteroconger perissodon Böhlke & Randall, 1981
Inhabits sand bottoms near coral reefs; distinguished by white blotch on side of head; Philippines and Indonesia; to 60 cm. (CONGRIDAE)

8 SPLENDID GARDEN EEL
Heteroconger preclara Böhlke & Randall, 1981
Inhabits sand bottoms near coral reefs; distinguished by bold brown and white pattern of bars; Philippines and Indonesia, also Maldive Islands; to 60 cm. (CONGRIDAE)

9 SPOTTED GARDEN EEL
Heteroconger hassi (Klausewitz & Eibl-Eibesfeldt, 1959)
Inhabits sand bottoms near coral reefs; distinguished by fine spotting and pair of large black spots on anterior third of body; Great Barrier Reef, offshore reefs of N.W. Australia, and throughout S.E. Asia; Indo-C. Pacific; to 60 cm. (CONGRIDAE)

10 TAYLOR'S GARDEN EEL
Heteroconger taylori Castle & Randall, 1995
Inhabits sand bottoms near coral reefs; distinguished by leopard-like black spots; Indonesia and New Guinea; to 40 cm. (CONGRIDAE)

11 BARNES'S GARDEN EEL
Gorgasia barnesi Robison & Lancraft, 1984
Inhabits sand bottoms near coral reefs; distinguished by fine brown speckling and dark lips; Indonesia (Flores to Banda); to 100 cm. (CONGRIDAE)

12 SPECKLED GARDEN EEL
Gorgasia sp.
Inhabits sand bottoms near coral reefs; distinguished by fine brownish yellow speckling; an undescribed species still lacking a scientific name; Great Barrier Reef and Coral Sea; W. Pacific; to 46 cm. (CONGRIDAE)

EEL GARDENS

Garden eels (**6-12**) are easy to miss, even though they are frequently abundant. The reason for their apparent scarcity is that they live on flat or sloping sand bottoms - boring terrain for most divers or snorklers, who generally prefer the excitement of coral reefs. These unusual animals do not take a baited hook, and were unknown to the scientific world, until the advent of scuba diving a few decades ago. However, their unusual and interesting habits offer great rewards for the observant diver and underwater photographer.

Although small groups are occasionally encountered, they usually reside in colonies composed of many individuals, sometimes hundreds or even thousands! The eels live in sandy burrows, which they construct. They seldom leave the burrow, but while feeding in the passing current they rise out of their retreat - exposing up to two-thirds of the body length. Zooplankton is their primary food source. The eels even stay in their burrows when spawning - by stretching over to their mates adjacent burrow and entwining bodies.

When disturbed the eels retreat backward into the opening of their lair. If a diver swims through the colony a wave "effect" is created - directly ahead the eels gradually disappear into the sand, while those in the divers wake gradually reappear. The common name of these creatures is derived from the appearance of the colonies when all the members are feeding in an extended position - from a distance they resemble a bed of plant stalks.

These animals offer a real challenge to underwater photographers. A telephoto lens is required, either a 105 or 210 mm. In spite of the magnification advantage of these lenses, it is still necessary to have lots of patience. The best shots result when one can hide behind adjacent reef and hold breaths for long periods - encouraging the eel to rise high into the water column.

1
2
3
4
5
6
7
8
9
10
11
12

PLATE 11: CATFISHES AND LIZARDFISHES

1 GIANT SALMON CATFISH
Arius thalassinus (Rüppell, 1837)
Inhabits coastal waters; there are about 8 species of salmon catfishes in northern waters which are difficult to identify even for experts; *A. thalassinus* is the largest species and has 6 patches of teeth on roof (palate) of mouth (versus 2-4 in others); found throughout the region; Indo-C. Pacific; to 185 cm. (ARIIDAE) ★★★

2 SMALLER SALMON CATFISH
Arius graefei Kner & Steindachner, 1866
Inhabits coastal waters, estuaries, and freshwater streams; difficult to identify, but has gill raker-like processes on back of all gill arches, 4 oval patches of teeth on roof of mouth, and barbel on upper jaw not reaching farther than beginning of dorsal fin; N. Australia and S. New Guinea only; to 50 cm. (ARIIDAE) ★★★

3 NAKED-HEADED CATFISH
Euristhmus nudiceps (Günther, 1880)
Inhabits coastal waters; similar to 4 below, but is more slender with greatest depth of body (i.e. height of body below first dorsal fin) fitting about 10 to 13 times in total length; N. Australia and New Guinea only; to 33 cm.

4 LONG-TAILED CATFISH
Euristhmus lepturus (Günther, 1864)
Inhabits coastal waters; similar to 3 above, but not as slender with greatest depth of body (i.e. height of body below first dorsal fin) fitting about 8 or 9 times in length; N. Australia and S. New Guinea only; to 46 cm. (PLOTOSIDAE)

5 STRIPED CATFISH
Plotosus lineatus (Thünberg, 1791)
Inhabits coastal waters, frequently in the vicinity of coral reefs; juveniles may form tightly packed aggregations containing up to several hundred fish; found throughout the region; Indo-C. Pacific; to 32 cm. (PLOTOSIDAE)

6 WHITE-LIPPED CATFISH
Paraplotosus albilabris (Valenciennes, 1840)
Inhabits coastal reefs, frequently found amongst weed-distinguished from 7 below, by lighter colour and much shorter dorsal fin; found throughout the region; mainly Indo-Australian Archipelago; to 134 cm. (PLOTOSIDAE) ★★★

7 SAILFIN CATFISH
Paraplotosus sp.
Inhabits coastal reefs, usually in the vicinity of coral reefs; distinguished by black colour and tall dorsal fin; N.W. Australia south to Point Quobba; to 30 cm. (PLOTOSIDAE)

8 INDIAN LIZARDFISH
Synodus indicus (Day, 1873)
Inhabits trawling grounds; distinguished by overall light colour with faint stripes on back and 2 dark streaks on upper corner of gill cover; N.W. Australia and throughout S.E. Asia; Indian Ocean and Indo-Australian Archipelago; to 21 cm. (SYNODONTIDAE)

9 TAILSPOT LIZARDFISH
Synodus jaculum Russell & Cressey, 1979
Inhabits the vicinity of coral reefs- distinguished by black spot at base of tail; Ningaloo Reef northwards; N.W. Australia, E. Queensland, and throughout S.E. Asia; Indo-C. Pacific; to 13 cm. (SYNODONTIDAE)

10 BLACK LIZARDFISH
Synodus kaianus (Günther, 1880)
Inhabits trawling grounds; distinguished by overall dark colouration; N.W. Australia and Indonesia; mainly W. Pacific; to 22 cm. (SYNODONTIDAE)

11 BIG-EYED LIZARDFISH
Synodus macrops Tanaka, 1917
Inhabits trawling grounds; distinguished by large size of eye and 3 large dark blotches on side; N.W. Australia and Indonesia; Andaman Sea and W. Pacific; to 20 cm. (SYNODONTIDAE)

12 BLACK-SHOULDERED LIZARDFISH
Synodus hoshinonis Tanaka, 1917
Inhabits trawling grounds; distinguished by prominent black area on upper edge of gill cover; N.W. Australia, E. Queensland, and throughout S.E. Asia; Indo-W. Pacific; to 22 cm. (SYNODONTIDAE)

13 NETTED LIZARDFISH
Synodus sageneus Waite, 1905
Inhabits trawling grounds; distinguished from other lizardfishes by the absence or reduced size of the adipose fin (small fin on back between dorsal fin and tail)- also known as Fishnet lizardfish; N. Australia and S. New Guinea; to 26 cm. (SYNODONTIDAE)

14 VARIEGATED LIZARDFISH
Synodus variegatus (Lacepède, 1803)
Inhabits sand rubble areas in the vicinity of coral reefs; distinguished by mottled appearance with series of dark bars on side; N.W. Australia, E. Queensland, and throughout S.E. Asia; Indo-C. Pacific; to 25 cm. (SYNODONTIDAE)

15 PAINTED GRINNER
Trachinocephalus myops (Schneider, 1801)
Inhabits coastal waters and trawling grounds; distinguished by pug-headed appearance, yellowish colour and bluish stripes on side; found throughout the region; Indo-C. Pacific; to 66 cm. (SYNODONTIDAE) ★★

16 SLENDER GRINNER
Saurida gracilis (Quoy & Gaimard, 1824)
Inhabits sandy areas, frequently near coral reefs; similar to 13 above, but teeth not covered by lips when mouth is closed; N.W. Australia, E. Queensland, and throughout S.E. Asia; Indo-C. Pacific; to 28 cm. (HARPODONTIDAE) ★★

17 LARGE-SCALED GRINNER
Saurida undosquamis (Richardson, 1848)
Inhabits trawling grounds; distinguished by small black spots along upper edge of tail, also known as Checkered lizardfish; found throughout the region; Indo-W. Pacific; to 45 cm. (HARPODONTIDAE) ★★

18 COMMON GRINNER
Saurida tumbil (Bloch, 1795)
Inhabits trawling grounds; distinguished by lack of markings and dark lower lobe of tail; found throughout the region; Indo-W. Pacific; to 43 cm. (HARPODONTIDAE) ★★

19 GLASSY BOMBAY DUCK
Harpodon translucens Saville-Kent, 1889
Inhabits bays and estuaries- distinguished by large curved teeth and flaccid semi-transparent appearance; N. Australia and S. New Guinea only; to 70 cm. (HARPODONTIDAE) ★

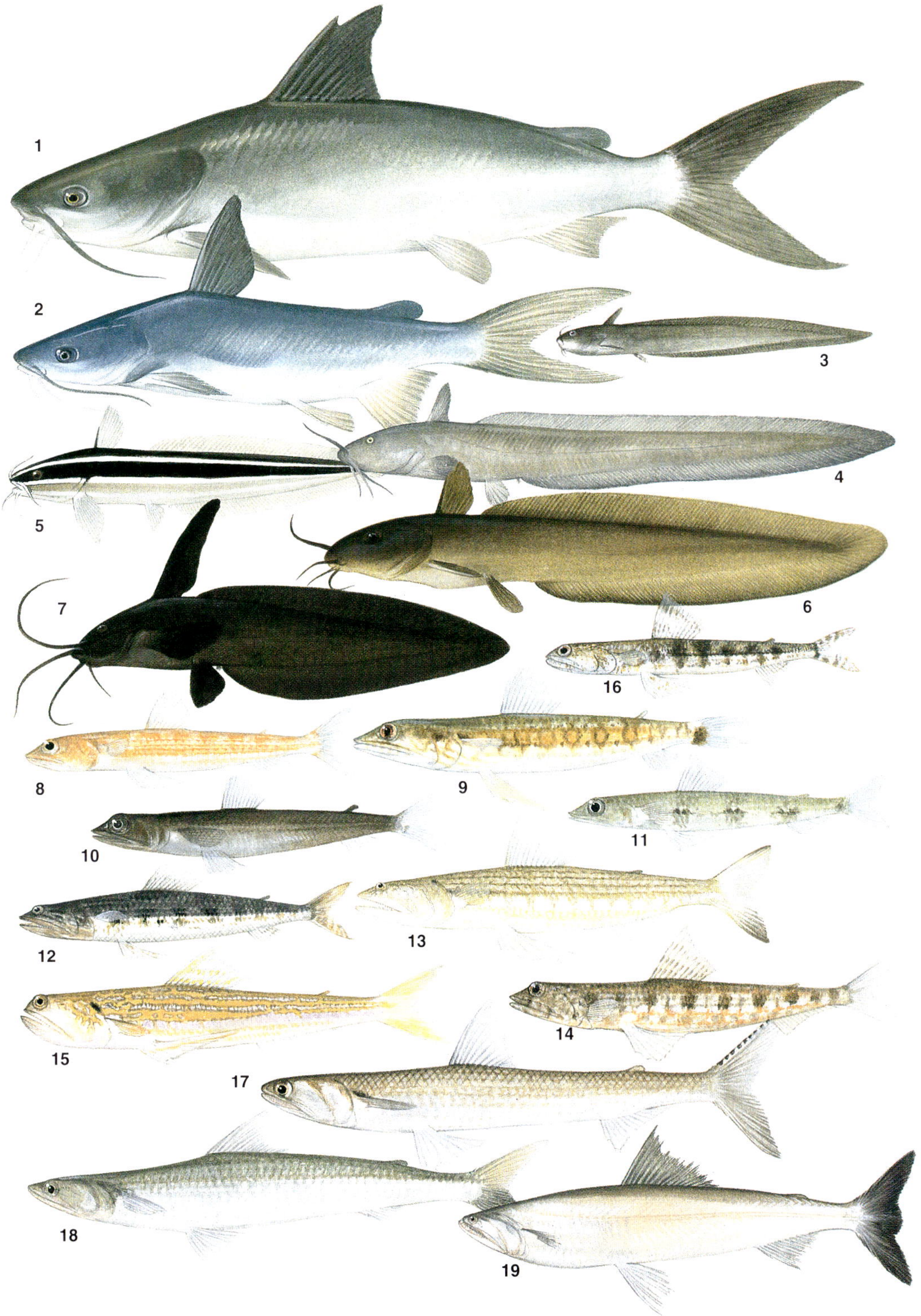
1
2
3
4
5
6
7
8
9
10
11
12
13
14
15
16
17
18
19

PLATE 12: FROGFISHES, ANGLERFISHES, CLINGFISHES AND CUSKEELS

1 DAHL'S FROGFISH
Batrachomoeus dahli (Rendahl, 1922)
Inhabits shallow reefs; distinguished by gill slit extending along entire pectoral fin base and marbled pattern without distinctive cross bars; N.W. Australia only; to 20 cm. (BATRACHOIDIDAE)

2 WESTERN FROGFISH
Batrachomoeus occidentalis Hutchins, 1976
Inhabits offshore trawling grounds; distinguished by gill slit extending along entire pectoral base and distinct cross bars on side; Rottnest Island to Exmouth Gulf, W. Australia; to 20 cm. (BATRACHOIDIDAE)

3 THREE-SPINED FROGFISH
Batrachomoeus trispinosus (Günther, 1861)
Inhabits coastal reefs; similar to 1 above, but has relatively distinct cross bars on side and well contrasted markings on dorsal surface of head; found throughout the region; N. Australia and New Guinea; to 30 cm. (BATRACHOIDIDAE)

4 BANDED FROGFISH
Halophryne diemensis (Lesueur, 1824)
Inhabits reef crevices; distinguished by gill slit extending only half to two-thirds of pectoral fin base and lacks small dark-edged white spots on side; found throughout the region; Indo-Australian Archipelago; to 26 cm. (BATRACHOIDIDAE)

5 OCELLATED FROGFISH
Halophryne ocellatus Hutchins, 1974
Inhabits offshore trawling grounds, may enter craypots; distinguished by restricted gill slit as in 4 above, but has small dark-edged white spots on side; Fremantle to Broome, W.Australia; to 26 cm. (BATRACHOIDIDAE)

6 URCHIN CLINGFISH
Diademichthys lineatus (Sauvage, 1883)
Inhabits coral reefs amongst spines of sea urchins or in branching corals; distinguished by peculiar shape with pale stripe on middle of side; found throughout the region; Indo-W. Pacfic; to 5 cm. (GOBIESOCIDAE)

7 SHARK BAY CLINGFISH
Lepadichthys frenatus Waite, 1904
Inhabits trawling grounds; distinguished by red to brown colour, dark stripe behind eye, and peculiar shape; N. Australia; W. Pacific; to 5 cm. (GOBIESOCIDAE)

8 HUMPBACK ANGLERFISH
Tetrabrachium ocellatum (Günther, 1880)
Inhabits offshore trawling grounds; distinguished by amorphous appearance, low dorsal and anal fins, small eyes and mouth, and numerous white spots; N. Australia and New Guinea; to 8.5 cm. (ANTENNARIIDAE)

9 STRIPED ANGLERFISH
Antennarius striatus (Shaw, 1794)
Inhabits inshore reefs, often in weeds; distinguished by narrow dark streaks and elongate blotches on body and fins; also entirely black variety; found throughout the region; Indo-W. Pacific; to 22 cm. (ANTENNARIIDAE)

10 WHITE-FINGER ANGLERFISH
Antennarius nummifer (Cuvier, 1817)
Inhabits inshore reefs; similar to 11 below, but has distinct tail base and usually a dark spot at base of dorsal fin; found throughout the region; Indo-W. Pacific; to 13 cm. (ANTENNARIIDAE)

11 FRECKLED ANGLERFISH
Antennarius coccineus (Lesson, 1830)
Inhabits offshore coral reefs; similar to 10 above, but lacks distinct tail base and black spot on dorsal fin base; found throughout the region; Indo-E. Pacific; to 13 cm. (ANTENNARIIDAE)

12 SHAGGY ANGLERFISH
Antennarius hispidus (Schneider, 1801)
Inhabits coastal reefs; distinguished by scattered dark blotches on side and fins, and diagonal, elongate streaks and blotches on dorsal fin; found throughout the region; Indo-W. Pacific; to 20 cm. (ANTENNARIIDAE)

13 PAINTED ANGLERFISH
Antennarius pictus (Shaw & Nodder, 1794)
Inhabits coastal reefs; distinguished by scattered dark spots (some pale-edged) and with light patches on nape, cheek, and pectoral regions; found throughout the region; Indo-W. Pacific; to 24 cm. (ANTENNARIIDAE)

14 SPOTTED-TAIL ANGLERFISH
Lophiocharon trisignatus (Richardson, 1844)
Inhabits coastal reefs, also found under wharves; distinguished by dark-edged pale spots on tail; Indo-Australian Archipelago; to 18 cm. (ANTENNARIIDAE)

15 SARGASSUM FISH
Histrio histrio (Linnaeus, 1758)
Usually found in clumps of floating sargassum weed; distinguished by smooth skin (most other anglerfishes have prickles), often with skin flaps and filaments on head and body; found throughout the region; worldwide temperate and tropical seas; to 16 cm. (ANTENNARIIDAE)

16 RED CUSKEEL
Ogilbia sp.
Inhabits coral reef crevices; distinguished by yellow to orange colour; elongate, tapering shape, blunt snout and tail separated from dorsal and anal fins; found throughout the region;W. Pacific; to 9 cm. (BYTHITIDAE)

17 PEARLFISH
Onuxodon margaritiferae (Rendahl, 1921)
Inhabits coastal waters, lives in the mantle cavity of oysters; other similar species found inside sea cucumbers and cushion starfish; distinguished by long slender, transparent body; found throughout the region; Indo-W. Pacific; to 9 cm. (CARAPIDAE)

18 BLACK-EDGED CUSKEEL
Ophidion muraenolepis (Günther, 1880)
Inhabits continental shelf and slope; distinguished by long tapering body, low, dark edged dorsal and anal fins, no barbels, but thread-like pelvic fins just behind chin; N.W. Australia and S. Indonesia; mainly eastern Indian Ocean; to 20 cm. (OPHIDIIDAE)

19 BEARDED CUSKEEL
Brotula multibarbata Temminck & Schlegel, 1847
Inhabits coastal reefs; distinguished by elongate body, barbels around mouth, and thread-like pelvic fins; found throughout the region; Indo-C. Pacific; to 90 cm. (OPHIDIIDAE) ★★

20 GOLDEN CUSKEEL
Sirembo imberis (Temminck & Schlegel, 1847)
Inhabits trawling grounds; similar to 18 above, but with broken stripes and spots on side; found throughout the region; N. Indian Ocean and W. edge of Pacific; to 20 cm. (OPHIDIIDAE)

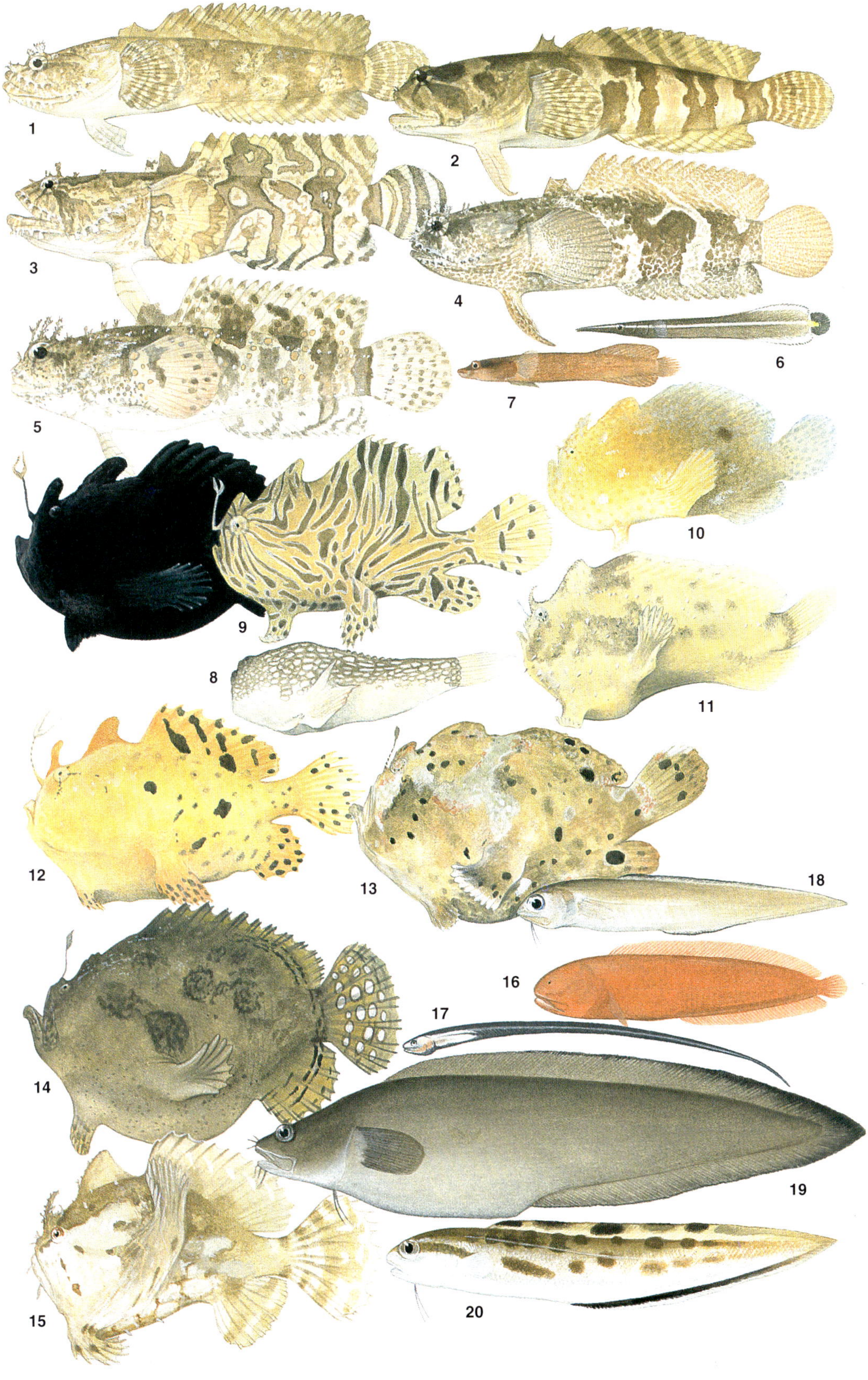
1
2
3
4
5
6
7
8
9
10
11
12
13
18
16
17
14
19
15
20

PLATE 13: FLYING FLSHES, GARFISHES, LONGTOMS AND HARDYHEADS

1 FLYINGFISH
Cypselurus sp.
Inhabits oceanic waters, frequently well offshore; several species of flyingfishes are common in the region, but only one is shown here as they are seldom caught by anglers and identification at the species level is frequently difficult; flyingfishes are distinguished by the wing-like pectoral fins and elongated lower tail lobe which facilitate long gliding flights over the sea surface; found throughout the region; to 27 cm. (EXOCOETIDAE) ★★

2 SNUB-NOSED GARFISH
Arrhamphus sclerolepis Günther, 1866
Inhabits coastal waters, sometimes entering brackish estuaries and the lower reaches of freshwater streams; distinguished by its short snout and lack of elongated lower jaw (as in other garfishes); N. Australia and New Guinea; to 38 cm. (HEMIRAMPHIDAE) ★★★

3 BUFFON'S GARFISH
Zenarchopterus buffonis (Valenciennes, 1847)
Inhabits coastal waters, sometimes entering estuaries; distinguished by truncate tail (forked in other genera of garfishes) and prominent dark brown stripe along midline of snout; male (shown here) has elongate and thickened anal fin; several similar species in region, usually in brackish or fresh waters; found throughout the region; Andaman Sea and Indo-Australian Archipelago: to 13 cm. (HEMIRAMPHIDAE) ★★

4 BARRED GARFISH
Hemiramphus far (Forsskål, 1775)
Inhabits coastal waters, frequently in schools near reefs; distinguished by 4-6 prominent dark bars on side; found throughout the region; Indo-W. Pacific; to 35 cm. (HEMIRAMPHIDAE) ★★★

5 ROBUST GARFISH
Hemiramphus robustus Günther, 1866
Inhabits coastal waters; similar to **4** above, but lacks dark bars on side (may have a single dark blotch below dorsal fin); both these species differ from **6** and **7** below in lacking scales on the triangular-shaped surface of the upper jaw; also known as Three by Two garfish; tropical and temperate Australia only; to 48 cm. (HEMIRAMPHIDAE) ★★★

6 TROPICAL GARFISH
Hyporhamphus affinis (Günther, 1866)
Inhabits coastal waters, occurring in schools; similar to **4** and **5** above, but has scales (versus no scales) on the triangular-shaped surface of the upper jaw, is more slender, and lacks bars or blotches on side; found throughout the region; Indo-C. Pacific; to 25 cm. (HEMIRAMPHIDAE) ★★

7 QUOY'S GARFISH
Hyporhamphus quoyi (Valenciennes, 1847)
Inhabits coastal waters, occurring in schools; distinguished by the relatively short lower jaw; found throughout the region; E. Indian Ocean and W. Pacific; to 34 cm. (HEMIRAMPHIDAE) ★★★

8 LONG-FINNED GARFISH
Euleptorhamphus viridis (van Hasselt, 1823)
Inhabits coastal waters, but sometimes encountered well offshore; distinguished by strongly compressed ribbon-like body, very long lower jaw, and wing-like pectoral fins; exhibits gliding behaviour similar to flyingfishes; found throughout the region; Indo-E. Pacific; to 60 cm. (HEMIRAMPHIDAE) ★★★

9 BARRED LONGTOM
Ablennes hians (Valenciennes, 1846)
Inhabits oceanic waters, often well offshore; distinguished by bars on rear part of body below dorsal fin; found throughout the region; worldwide tropical and subtropical seas; to 120 cm. (BELONIDAE) ★★

10 FLAT-TAILED LONGTOM
Platybelone platyura (Bennett, 1837)
Inhabits offshore waters; distinguished by flattened (dorsoventrally) tail base; found throughout the region; Indo-C. Pacific; to 40 cm. (BELONIDAE) ★★

11 SLENDER LONGTOM
Strongylura leiura (Bleeker, 1851)
Inhabits coastal waters, frequently in bays and estuaries; distinguished by elongate jaws, slender shape, and black bar or streak across base of gill cover (not shown); found throughout the region; Indo-W. Pacific; to 110 cm. (BELONIDAE) ★★

12 CROCODILIAN LONGTOM
Tylosurus crocodilus (Péron & Lesueur, 1821)
Inhabits coastal waters; distinguished by dark fleshy ridge on side of tail base (not shown), found throughout the region; Atlantic and Indo-C. Pacific; to 150 cm. (BELONIDAE) ★★

13 STOUT LONGTOM
Tylosurus gavialoides (Castelnau, 1873)
Inhabits coastal waters; similar to **12** above, but lacks fleshy ridge on side of tail base also has more rounded snout tip (when viewed from above); found throughout the region; E. Indian Ocean and W. Pacific; to 130 cm. (BELONIDAE) ★★

14 SPOTTED HARDYHEAD
Allanetta mugiloides (McCulloch, 1912)
Inhabits coastal waters, frequently in schools off beaches; distinguished by small dark spot below pectoral fin; N. Australia only; to 6 cm. (ATHERINIDAE)

15 FEW-RAYED HARDYHEAD
Craterocephalus pauciradiatus (Günther, 1861)
Inhabits coastal waters, occurring in schools; a non-descript silvery fish which has the anus positioned close to the pelvic fin base, and lacks a notch on the lower cheek margin; N.W. Australia only; to 6 cm. (ATHERINIDAE)

16 SAMOAN HARDYHEAD
Hypoatherina temminckii (Bleeker, 1853)
Inhabits coastal waters; distinguished by slender shape, anus placed far behind pelvic fin base, and prominent silvery midlateral stripe; found throughout the region; Indo-C. Pacific; to 10 cm. (ATHERINIDAE)

17 ENDRACHT HARDYHEAD
Atherinomorus endrachtensis (Quoy & Gaimard, 1824)
Inhabits coastal waters; similar to **14** above, but no spot below pectoral fin and rear part of lower jaw only slightly elevated (versus prominently elevated - mouth must be opened widely to detect this feature); N. Australia and New Guinea; to 10 cm. (ATHERINIDAE)

18 OGILBY'S HARDYHEAD
Atherinomorus ogilbyi (Whitley, 1930)
Inhabits shallow coastal waters, including bays and estuaries, usually in schools; distinguished by dark blotch at tip of pectoral fins; found throughout the region; Australia only - W. Australia, Queensland, and New S. Wales; to 17 cm. (ATHERINIDAE)

juv.

1

2

3

4

5

6

7

8

9

10

11

12

13

14

15

16

17

18

PLATE 14: SQUIRRELFISHES (HOLOCENTRIDAE)

1 BLACKFIN SOLDIERFISH
Myripristis adusta Bleeker, 1853
Inhabits caves and crevices of coral reefs; distinguished by black area on outer half of dorsal and anal fins; found throughout the region; W. Australia, Great Barrier Reef, and throughout S.E. Asia; Indo-W. Pacific; to 30 cm. ★★★

2 DOUBLETOOTH SOLDIERFISH
Myripristis hexagonatus (Lacepède, 1802)
Inhabits caves and crevices of coral reefs; similar to **4** below (both have 2 pairs of tooth patches on front of lower jaw, just outside of the mouth), but differs in having small scales on the inside base ("armpit") of pectoral fin; found throughout the region; Indo-W. Pacific; to 20 cm. ★★★

3 KUNTEE SOLDIERFISH
Myripristis kuntee Valenciennes, 1831
Inhabits caves and crevices of coral reefs; distinguished by broad dark behind head to pectoral fin base and 37-44 scales in lateral line; found throughout the region; Indo-C. Pacific; to 20 cm. ★★★

4 PALE SOLDIERFISH
Myripristis melanostictus Bleeker, 1863
Inhabits caves and crevices of coral reefs; similar to **2** above, but differs in lacking small scales on the inside base ("armpit") of pectoral fin; found throughout the region; Indo-W. Pacific; to 30 cm. ★★★

5 CRIMSON SOLDIERFISH
Myripristis murdjan (Forsskål, 1775)
Inhabits caves and crevices of coral reefs; distinguished by dark margin on upper part of gill cover and 27-32 scales in lateral-line; found throughout the region; Indo-C. Pacific; to 30 cm. ★★★

6 SPOTFIN SQUIRRELFISH
Neoniphon sammara (Forsskål, 1775)
Inhabits patch reefs in lagoons amongst branching corals; distinguished by slender shape, silvery or pale colouration, and black spot at front of dorsal fin; found throughout the region; Indo-C. Pacific; to 32 cm. ★★

7 DEEPWATER SQUIRRELFISH
Ostichthys kaianus (Günther, 1880)
Inhabits offshore trawling ground between about 300-650 m depth; distinguished by broad V-shaped groove at middle of snout and 12 dorsal spines; N.W. Australia and Indonesia; Indo-W. Pacific; to 30 cm. ★★

8 ROUGH SQUIRRELFISH
Pristilepis oligolepis (Whitley, 1941)
Inhabits offshore trawling grounds; distinguished by narrow, elongate groove at middle of snout and 12 dorsal spines; W. Australia and scattered, mainly W. Pacific localities; to 20 cm. ★★

9 CROWNED SQUIRRELFISH
Sargocentron diadema (Lacepède, 1802)
Inhabits coral reefs to 30 m depth; distinguished by broad black margin on dorsal fin; found throughout the region; N.W. Australia, Great Barrier Reef, and throughout S.E. Asia; Indo-C. Pacific; to 16 cm. ★★

10 SPECKLED SQUIRRELFISH
Sargocentron punctatissimum (Cuvier, 1829)
Inhabits rocky shores and reefs exposed to wave action; distinguished by pepper-like spotting on sides; N.W. Australia, Great Barrier Reef, and throughout S.E. Asia; Indo-C. Pacific; to 15 cm. ★★

11 RED SQUIRRELFISH
Sargocentron rubrum (Forsskål, 1775)
Inhabits live and dead coral reefs, usually in protected lagoons; distinguished by red and white stripes of about equal width; found throughout the region; Indo-W. Pacific; to 28 cm. ★★

12 SPINY SQUIRRELFISH
Sargocentron spiniferum (Forsskål, 1775)
Inhabits caves and crevices of coral reefs, distinguished by overall red colouration and very long spine at lower margin of cheek; N.W. Australia, Great Barrier Reef, and throughout S.E. Asia; Indo-C. Pacific; to 40 cm. ★★★

13 BLUESTRIPE SQUIRRELFISH
Sargocentron tiere (Cuvier, 1829)
Inhabits mainly outer exposed reefs to 20 m depth; distinguished by brilliant red colour, and iridescent blue stripes on lower sides; N.W. Australia, Great Barrier Reef, and throughout S.E. Asia; Indo-C. Pacific; to 34 cm. ★★★

14 VIOLET SQUIRRELFISH
Sargocentron violaceum (Bleeker, 1853)
Inhabits caves and crevices of coral reefs; distinguished by overall dusky appearance; found throughout the region; Indo-W. Pacific; to 23 cm. ★★

PREDATORS OF THE NIGHT

The fishes featured on this plate are members of the family Holocentridae, commonly known as squirrelfishes or soldierfishes. They occur in all tropical seas. Most of the approximately 70 species inhabit the Indo-Pacific region. They are characterised by rough scales, prominent fin spines, a large eye, and red coloration. Another remarkable feature is their ability to produce clearly audible "clicking" sounds, believed to function as a form of communication between members of the school. Squirrelfishes differ from soldierfishes in possessing a sharp spine at the back of each cheek, which can inflict a painful wound if handled carelessly. Therefore caution must be exercised when removing them from a hook. Although most of the species are small, the flesh is considered good eating.

Although very abundant on coral reefs, snorkelers seldom see these fishes. During the day they remain hidden deep in the shadows of caves, cracks, and crevices. They begin to appear in the open shortly after sunset. Because most of the reef's fish occupants are active during the day, many invertebrates, particularly crustaceans and echinoderms, have evolved a strategy to avoid them by coming out to feed at night. Squirrelfishes have adapted night-time feeding habits to take advantage of this nocturnal food supply. They feed mainly on crustaceans, particularly small crabs and shrimps.

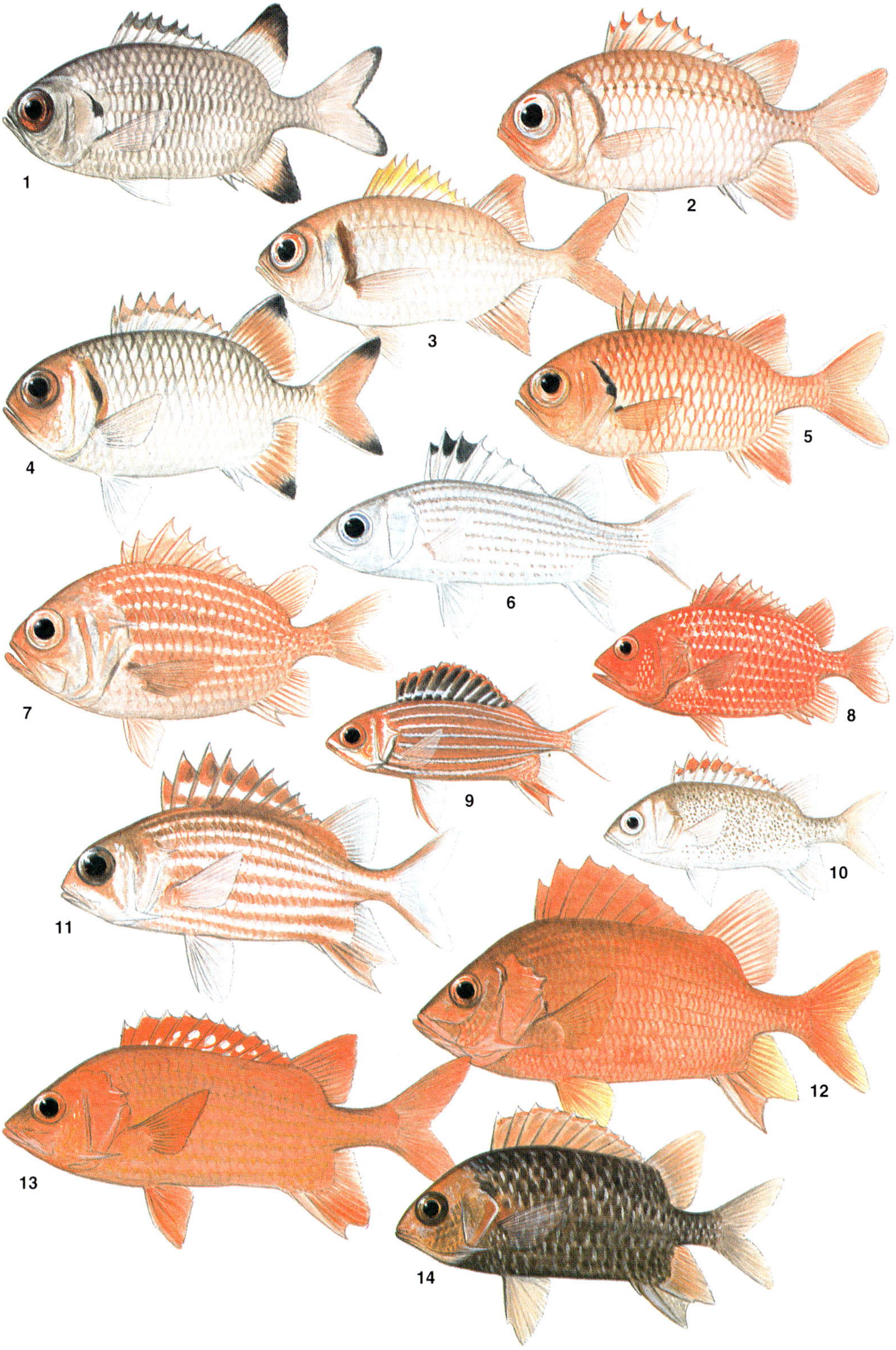
1
2
3
4
5
6
7
8
9
10
11
12
13
14

PLATE 15: SQUIRRELFISHES (FAMILY HOLOCENTRIDAE)

1 BIGSCALE SOLDIERFISH
Myripristis berndti (Jordan & Evermann, 1903)
Inhabits coral reef caves and ledges; all *Myripristis* lack a prominent spine on the lower edge of the cheek that is found in other squirrelfishes; distinguished by 28-31 scales along lateral line, pale body with dark scale edges, yellow dorsal fin, and dark edge on gill cover; Great Barrier Reef, offshore reefs of W.A., and throughout S.E. Asia; Indo-E. Pacific; to 19 cm. ★★★

2 YELLOWFIN SOLDIERFISH
Myripristis chryseres (Jordan & Evermann, 1903)
Inhabits coral reef caves and ledges, usually below 30 m depth; distinguished by bright yellow fins; Great Barrier Reef and Indonesia; widely scattered localities in the Indo-C. Pacific; to 25 cm. ★★★

3 SCARLET SOLDIERFISH
Myripristis pralinia Cuvier, 1829
Inhabits coral reef caves and ledges; distinguished by dark mark that ends abruptly midway on rear edge of gill cover; Great Barrier Reef, offshore reefs of W.A., and throughout S.E. Asia; Indo-C. Pacific; to 20 cm. ★★★

4 LATTICE SOLDIERFISH
Myripristis violacea Bleeker, 1851
Inhabits coral reef caves and ledges; distinguished by bluish-silver colour and broad dark scale edges that give dusky appearance to upper sides; Great Barrier Reef, offshore reefs of W.A., and throughout S.E. Asia; Indo-C. Pacific; to 20 cm. ★★★

5 RED SOLDIERFISH
Myripristis vittata Valenciennes, 1831
Inhabits coral reef caves and ledges, usually on outer slopes below about 20 m; distinguished by bright red-orange colour and white tips on dorsal spines; Great Barrier Reef, offshore reefs of W.A., and throughout S.E. Asia; Indo-C. Pacific; to 20 cm. ★★★

6 SMOOTH SQUIRRELFISH
Neoniphon argenteus (Valenciennes, 1831)
Inhabits coral reefs, frequently amongst branching corals; distinguished by mainly silver colour with faint spots forming longitudinal lines on side, and plain dorsal fin; offshore reefs of W.A. and throughout S.E. Asia; Indo-C. Pacific; to 25 cm. ★★

7 YELLOW-STRIPED SQUIRRELFISH
Neoniphon aurolineatus (Liénard, 1839)
Inhabits coral reefs, under ledges and amongst corals, usually below 40 m depth; distinguished by prominent yellow stripes on side; Great Barrier Reef; scattered localities in Indo-C. Pacific; to 22 cm. ★★

8 BLACK-FINNED SQUIRRELFISH
Neoniphon opercularis (Valenciennes, 1831)
Inhabits coral reefs, under ledges and amongst corals; distinguished by broad black band through anterior dorsal fin; Great Barrier Reef, offshore reefs of W.A., and throughout S.E. Asia; Indo-C. Pacific; to 24 cm. ★★

9 ROUGH-SCALED SOLDIERFISH
Plectrypops lima (Valenciennes, 1831)
Inhabits coral reefs, a cryptic species that hides in deep recesses during the day and rarely ventures far from caves at night; differs from other soldierfishes in having 12 instead of 11 dorsal spines; has stocky body shape similar to **14** below, but lacks spine at lower corner of cheek; Great Barrier Reef and throughout S.E. Asia; Indo-C. Pacific; to 16 cm. ★★

10 TAILSPOT SQUIRRELFISH
Sargocentron caudimaculatum (Ruppell, 1838)
Inhabits coral reefs, frequently in caves and under ledges, but often seen in the open; distinguished by silvery-white spot behind dorsal fin base or entire rear part of fish silvery white; Great Barrier Reef, offshore reefs of W.A., and throughout S.E. Asia; Indo-C. Pacific; to 21 cm. ★★

11 THREESPOT SQUIRRELFISH
Sargocentron cornutum (Bleeker, 1853)
Inhabits coral reef caves and ledges; similar black marks at base of caudal, dorsal, and anal fins as **15** below, but lacks yellow colour on body and has black submarginal band on dorsal fin; Great Barrier Reef, offshore reefs of W.A., Indonesia, Philippines, New Guinea, and Solomon Islands; to 17 cm. ★★

12 SAMURAI SQUIRRELFISH
Sargocentron ittodai (Jordan & Fowler, 1903)
Inhabits coral reef caves and ledges in 5-70 m depth; a red-striped squirrelfish lacking distinguishing marks, but dorsal fin is largely red with white "windows" across the middle; Great Barrier Reef and widely scattered localities in the Indo-W. Pacific; to 17 cm. ★★

13 PINK SQUIRRELFISH
Sargocentron tiereoides (Bleeker, 1853)
Inhabits coral reef caves and ledges; distinguished by silvery-pink stripes between red stripes on side of body, dorsal fin red with white tips; Great Barrier Reef and throughout S.E. Asia; Indo-C. Pacific; to 16 cm.

14 SANDPAPER SQUIRRELFISH
Sargocentron lepros (Allen & Cross, 1983)
Inhabits coral reef caves and ledges in 15-45 m depth; distinguished by stocky body shape, similar to **9** above, but has sharp spine on lower edge of cheek; offshore reefs of W.A. and scattered localities in E. Indian Ocean; to 7 cm. ★★

15 BLACKSPOT SQUIRRELFISH
Sargocentron melanospilos (Bleeker, 1858)
Inhabits coral reef caves and ledges; distinguished by yellow coloration and three black spots at bases of soft dorsal, anal, and caudal fins; often misidentified as *S. cornutum*; Great Barrier Reef, offshore reefs of W.A., and throughout S.E. Asia; Indo-C. Pacific; to 25 cm. ★★

16 SMALLMOUTH SQUIRRELFISH
Sargocentron microstoma (Günther, 1859)
Inhabits coral reef caves and ledges; distinguished by slender body, very long third anal spine, and alternating white and red stripes; Great Barrier Reef, offshore reefs of W.A., and throughout S.E. Asia; Indo-C. Pacific; to 19 cm. ★★

NIGHT SHIFT

By day the coral reef is literally a beehive of activity. Fishes of every description swarm over the reef. Keen observers with a mask and snorkel can effortlessly watch the fascinating drama of undersea life as the occupants of the reef engage in their daily activities. But what happens to the fishes at night? Most retire to the safety of a cave or crevice at dusk - but as darkness descends squirrelfishes (Plates 14-15), cardinalfishes (Plates 31-36) and other members of the night shift become active.

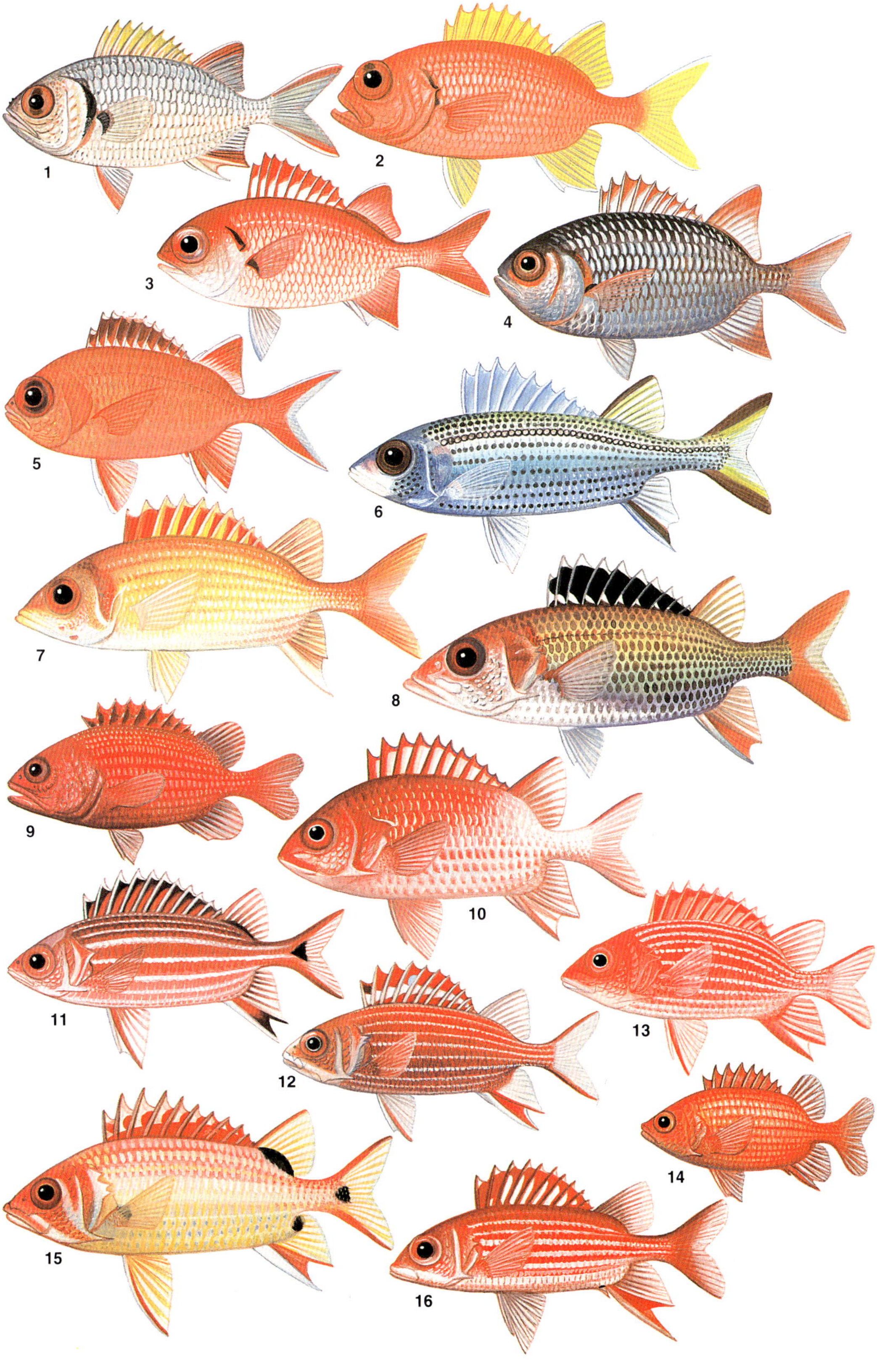
1
2
3
4
5
6
7
8
9
10
11
12
13
14
15
16

PLATE 16: KNIGHT FISHES, DORIES, FLUTEMOUTHS, ETC.

1 KNIGHT FISH
Cleidopus gloriamaris De Vis, 1882
Inhabits coastal reefs, usually in caves or under ledges; similar to **2** below, but has very conspicuous light-producing organ on each side of lower jaw (appearing as orange spot in daylight or a blue-green one at night), and scales are more strongly outlined by dark colouration; also known as Pineapple fish; Australia only, mainly in subtropical and temperate waters of east and west coasts; to 28 cm. (MONOCENTRIDAE)

2 JAPANESE PINEAPPLEFISH
Monocentrus japonicus (Houttuyn, 1782)
Inhabits deeper offshore reefs and trawling grounds; similar to **1** above, but light organs not as conspicuous, narrower dark margins around scales, and wider gap between eye and mouth; N.W. Australia, New S. Wales, and scattered localities in Indo-Malay region; Indo-W. Pacific; to 20 cm. (MONOCENTRIDAE)

3 LITTLE DORY
Cyttopsis cypho (Fowler, 1934)
Inhabits deeper trawling grounds of the continental shelf; general shape similar to **4** and **5** below, but a much smaller fish lacking filamentous dorsal fin rays; found throughout the region; mainly W. Pacific; to 18 cm. (ZEIDAE)

4 MIRROR DORY
Zenopsis nebulosus (Temminck & Schlegel, 1845)
Inhabits deeper trawling grounds of the continental shdf; similar to **5** below, but lacks scales (versus small scales present) and forehead profile distinctly concave (versus convex); found throughout the region; C. and W. Pacific; to 58 cm. (ZEIDAE) ★★★★

5 JOHN DORY
Zeus faber Linnaeus, 1758
Inhabits deeper trawling grounds of the continental shelf, although sometimes found close to the coast;similar to **4** above, but has small scales versus no scales and forehead profile is concave versus convex; Australia, mainly in temperate seas; tropical and temperate E. Atlantic and Indo-W. Pacific; to 75 cm. (ZEIDAE) ★★★★

6 PINK BOARFISH
Antigonia rhomboidea McCulloch, 1915
Inhabits deeper trawling grounds of the continental shelf; distinguished by diamond-shaped body; red or pink colouration with yellowish fins; Australia only - shelf areas off eastern and western coasts in tropical and temperate seas; to 15 cm. (CAPROIDAE)

7 HIGH-FINNED VEILFIN
Velifer hypselopterus Bleeker, 1879
Inhabits deeper trawling grounds of the continental shelf, although the young may appear in shallow coastal waters; distinguished by filamentous dorsal and anal fins and diffuse dark bars on side; shelf areas off eastern and western coasts of Australia in tropical and temperate seas, also Arafura Sea; Indo-W. Pacific; to 40 cm. (VELIFERIDAE)

8 SMOOTH FLUTEMOUTH
Fistularia commersonii Rüppell, 1838
Inhabits coastal waters in the vicinity of reefs; distinguished by long snout, trailing filament on tail, and greenish-brown colour of back; found throughout the region; Indo-E. Pacific; to 163 cm. (AULOSTOMIDAE)

9 ROUGH FLUTEMOUTH
Fistularia petimba Lacepède, 1803
Inhabits coastal waters, also found well offshore; similar to **8** above, but has row of bony plates along middle of back (absent in **8**) and is reddish or brownish-orange in colour (versus greenish-brown); found throughout the region; Atlantic and Indo-C. Pacific; to 185 cm. (FISTULARIIDAE)

10 PAINTED FLUTEMOUTH
Aulostomus chinensis (Linnaeus, 1766)
Inhabits coral reefs; roughly similar shape to **8** and **9** above, but has shorter snout, has small scales (versus no scales), row of feeble dorsal spines on back, different shaped fins, and lacks tail filament; a yellow variety is frequently seen; found throughout the region; Indo-E. Pacific; to 50 cm. (FISTULARIIDAE) Cairns 10/98

11 GHOST FLATHEAD
Hoplichthys regani Jordan & Richardson, 1908
Inhabits deep offshore trawling grounds; similar to flatheads (Plate 21) in body shape (i.e. greatly flattened), but lacks scales and has filamentous dorsal-fin rays; N.W. Australia and S. Indonesia; Indo-W. Pacific; to 20 cm. (HOPLICHTHYIDAE)

12 ORIENTAL SEAROBIN
Dactyloptaenia orientalis (Cuvier, 1829)
Inhabits coastal waters, usually on sand bottoms near coral reefs; distinguished by huge wing-like pectoral fins; found throughout the region; Indo-C. Pacific; to 38 cm. (DACTYLOPTERIDAE)

13 SLENDER SEAMOTH
Pegasus volitans Linnaeus, 1758
Inhabits sand or silt bottoms of bays and estuaries; distinguished by flattened head and tapered body encased in plate-like armour similar to seahorses, and fan-like pectoral fins; often identified as *Parapegasus natans*; found throughout the region; IndoW. Pacific; to 16 cm. (PEGASIDAE)

14 SHORT SEAMOTH
Eurypegasus draconis (Linnaeus, 1766)
Inhabits sand or silt bottoms, frequently in bays or estuaries; similar to **13** above, but wider body (when viewed from above), shorter snout and tail, and body is more "sculptured"; found throughout the region; Indo-C. Pacific; to 10 cm. (PEGASIDAE)

PLATE 17

1 GROOVED RAZORFISH
Centriscus scutatus Linnaeus, 1758
Inhabits coastal waters; distinguished by long snout and thin (highly compressed) body composed of bony plates; Great Barrier Reef and throughout S.E. Asia; Indo-W. Pacific; to 15 cm. (CENTRISCIDAE)

2 HARLEQUIN GHOST PIPEFISH
Solenostomus paradoxus (Pallas, 1770)
Inhabits inshore reefs and weed beds, sometimes in floating seaweed; distinguished by skin flaps on head and body, stripes on body and spotted fins; N.W. Australia, Great Barrier Reef, and throughout S.E. Asia; Indo-W. Pacific; to 12 cm. (SOLENOSTOMIDAE)

3 GHOST PIPEFISH
Solenostomus cyanopterus Bleeker, 1855
Inhabits inshore reef areas; distinguished by yellowish to green or brown colour; general shape similar to **2** above, but has fewer skin flaps and shorter tail base; N.W. Australia, Great Barrier Reef, and throughout S.E. Asia; Indo-W. Pacific; to 16 cm. (SOLENOSTOMIDAE)

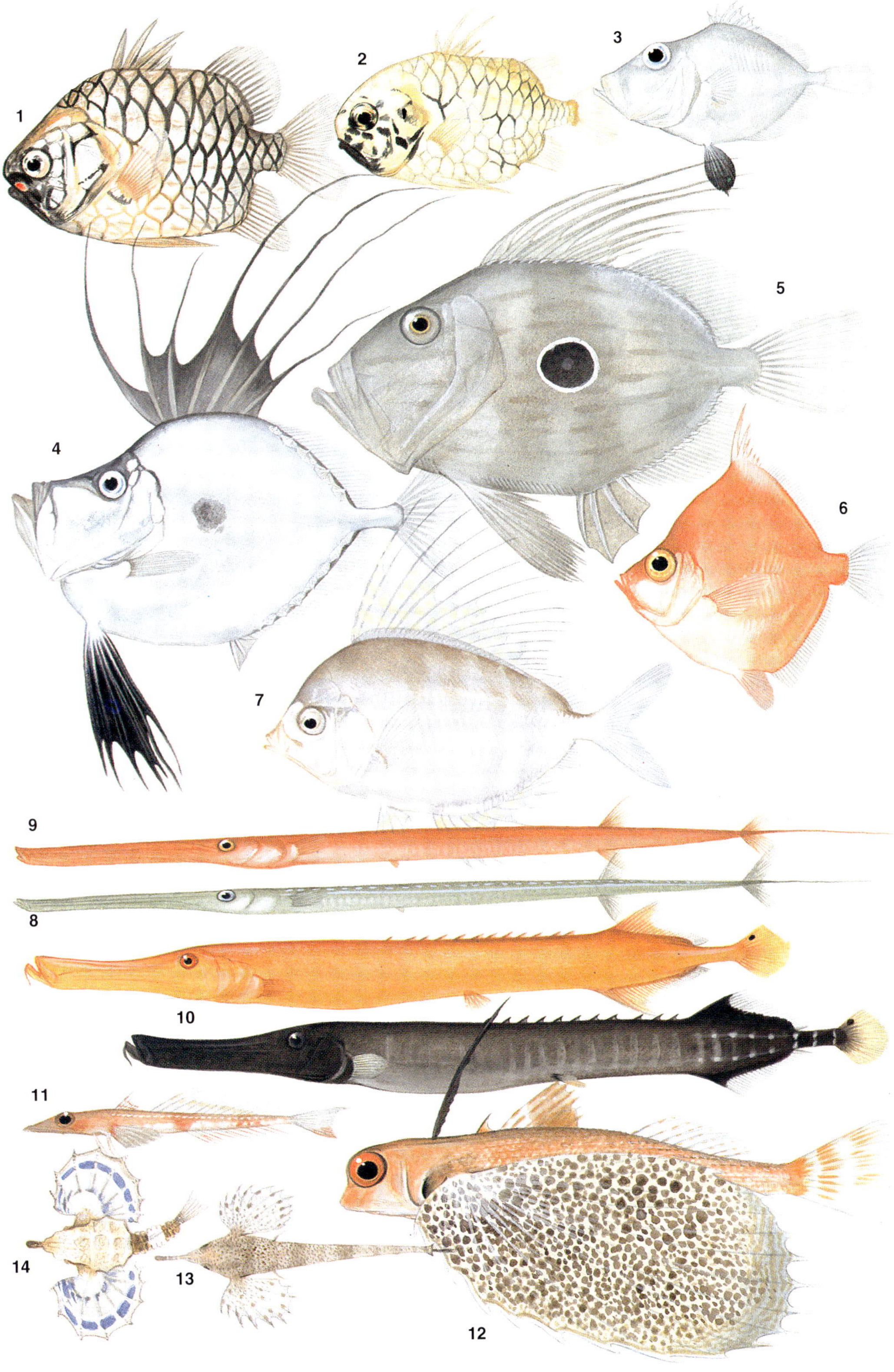
1
2
3
4
5
6
7
8
9
10
11
12
13
14

PLATE 17: SEAHORSES AND PIPEFISHES

1 GROOVED RAZORFISH
Centriscus scutatus Linnaeus. Text on page 70.

2 HARLEQUIN GHOST PIPEFISH
Solenostomus paradoxus (Pallas). Text on page 70.

3 GHOST PIPEFISH
Solenostomus cyanopterus Bleeker. Text on page 70.

4 SPOTTED SEAHORSE
Hippocampus kuda Bleeker, 1852
Inhabits sheltered bays and estuaries; variable in colour, either yellow, brown or blackish; found throughout the region; Indo-C. Pacific; to 30 cm. (SYNGNATHIDAE)

5 WESTERN AUSTRALIAN SEAHORSE
Hippocampus angustatus Günther, 1870
Inhabits sheltered bays; variable in colour, also has narrow lines across snout; N. Australia, except ranging south on W. Coast to Augusta; to 22 cm (illustration not shown at proper scale). (SYNGNATHIDAE)

6 SPINY SEAHORSE
Hippocampus hystrix Kaup, 1856
Inhabits coastal waters; distinguished by pronounced spiny ridges on head and body; colour variable; found throughout the region; Indo-W. Pacific; to 15 cm. (SYNGNATHIDAE)

7 LARSON'S PIPEHORSE
Acentronurus larsonae Dawson, 1984
Inhabits coral reefs or amongst sargassum weed; distinguished by seahorse shape and bulbous forehead; known only from Monte Bello Islands, W. Australia; to 3.5 cm. (SYNGNATHIDAE)

8 EEL PIPEFISH
Bulbonaricus brauni (Dawson & Allen, 1978)
Inhabits coral reefs amongst organ-pipe coral; distinguished by eel-like shape; Ningaloo Reef, W. Australia and Indo-Malay region; to 6 cm. (SYNGNATHIDAE)

9 MUIRON PIPEFISH
Choeroichthys latispinosus Dawson, 1978
Inhabits coral reefs; distinguished by relatively broad, elongate snout with upturned mouth, and short tail; known thus far only from South Muiron Island, W. Australia; to 3 cm. (SYNGNATHIDAE)

10 SHORT-BODIED PIPEFISH
Choeroichthys brachysoma (Bleeker, 1855)
Inhabits reefs and seagrass beds; distinguished by broad midsection tapering at head and tail; found throughout the region; Indo-C. Pacific; to 7 cm. (SYNGNATHIDAE)

11 BANDED PIPEFISH
Doryrhamphus dactyliophorus (Bleeker, 1853)
Inhabits coral reef crevices; distinguished by prominent light and dark bands found throughout the region; Indo-W. Pacific; to 18 cm. (SYNGNATHIDAE)

12 JANSS'S PIPEFISH
Doryrhamphus janssi (Herald & Randall, 1972)
Inhabits coral reef crevices; distinguished by red central section of body grading to blue on rear part and fan-shaped dark tail with pale centre and pale outer margin; N.W. Australia, Great Barrier Reef, and throughout S.E. Asia; mainly W. Pacific; to 13 cm. (SYNGNATHIDAE)

13 LADDER PIPEFISH
Festucalex scalaris (Günther, 1870)
Inhabits trawling grounds, amongst weeds; distinguished by short snout, and variegated pattern of light and dark spots, blotches and bars; central W. coast of W. Australia only; to 18 cm. (SYNGNATHIDAE)

14 TIGER PIPEFISH
Filicampus tigris (Castelnau, 1879)
Inhabits sand-weed areas; distinguished by diagonal dark stripes on head, diffuse dark bars, and abruptly white belly; Australia only - west and S.E. coast, also Spencer's Gulf, S. Australia; to 35 cm. (SYNGNATHIDAE)

15 TASSELLED PIPEFISH
Halicampus brocki (Herald, 1953)
Inhabits coral and rocky reefs; distinguished by skin flaps and branched tassles; N.W. Australia, Great Barrier Reef, and throughout S.E. Asia; mainly W. Pacific; to 11 cm. (SYNGNATHIDAE)

16 SHORT-NOSED PIPEFISH
Halicampus spinirostris (Dawson & Allen, 1981)
Inhabits coral reefs; distinguished by short snout, no skin flaps, and broad dark bars with narrow pale bars between them; W. Australia (Ningaloo Reef), Sri Lanka and Samoa; to 11 cm. (SYNGNATHIDAE)

17 RIBBONED PIPEFISH
Haliichthys taeniophorus Gray, 1859
Inhabits trawling grounds; distinguished by large size, elongate snout, bony knobs above eye, and prominent spines or knobs on body ridges; N. Australia and New Guinea; to 30 cm. (SYNGNATHIDAE)

18 MANGROVE PIPEFISH
Hippichthys penicillus (Cantor, 1850)
Inhabits mangrove estuaries; distinguished by relatively long snout, and small pale spots on front part of body; found throughout the region; N. Indian Ocean and W. Pacific; to 18 cm. (SYNGNATHIDAE)

19 PALLID PIPEFISH
Solegnathus hardwickii (Gray, 1830)
Inhabits trawling grounds; distinguished by large size, pale colouration and dark marks along edge of back; found throughout the region; mainly W. Pacific; to 50 cm. (SYNGNATHIDAE)

20 WHITE-SADDLED PIPEFISH
Micrognathus micronotopterus (Fowler, 1938)
Inhabits inshore reefs and tide pools, distinguished by short snout, small, usually unbranched skin flaps on head and body, and 10-12 pale saddles on back; N.W. Australia to Gulf of Carpentaria, also scattered Indo-Malay localites; to 7 cm. (SYNGNATHIDAE)

21 DOUBLE-ENDED PIPEFISH
Syngnathoides biaculeatus (Bloch, 1785)
Inhabits coastal waters, amongst weeds; distinguished by large size with deep, laterally compressed snout and prehensile tail; found throughout the region; Indo-W. Pacific; to 29 cm. (SYNGNATHIDAE)

22 SHORT-TAILED PIPE FISH
Trachyrhamphus bicoarctatus (Bleeker, 1857)
Inhabits sand, rubble or weed bottoms; distinguished by long thin tapering body, and tiny tail; found throughout the region; Indo-W. Pacific; to 40 cm. (SYNGNATHIDAE)

23 SLENDER PIPEFISH
Trachyrhamphus longirostris Kaup, 1856
Inhabits trawling grounds; similar to **22** above, but has thicker snout and fewer rings or body segments (41-53 behind anus versus 55-63); found throughout the region; Indo-W. Pacific; to 32 cm. (SYNGNATHIDAE)

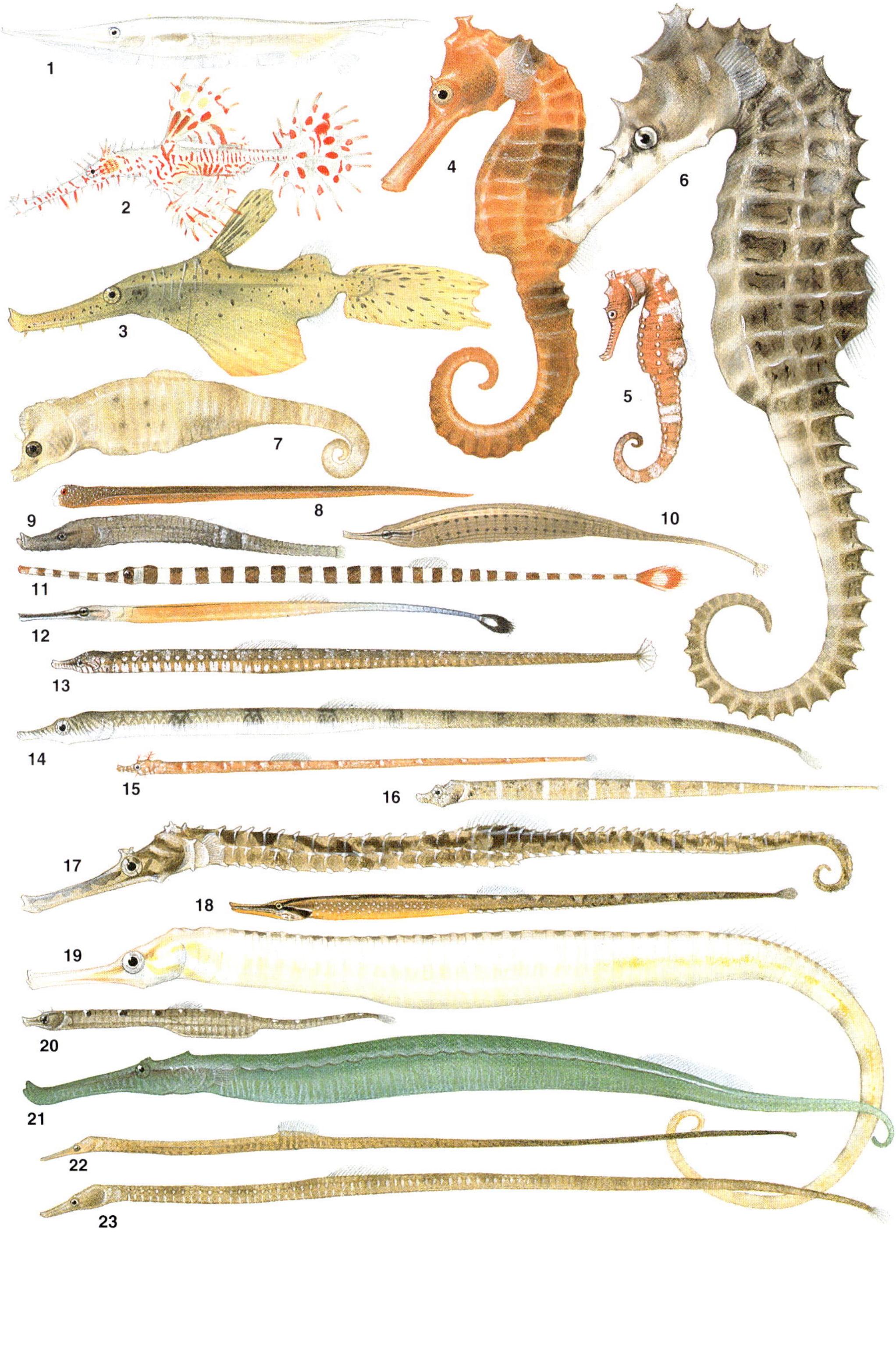
1
2
3
4
5
6
7
8
9
10
11
12
13
14
15
16
17
18
19
20
21
22
23

PLATE 18: PIPEFISHES AND SCORPIONFISHES

1 RAZORFISH
Aeoliscus strigatus (Günther, 1860)
Inhabits coral reefs forming schools that orient themselves vertically with head downward among branching corals or urchins; similar to *Centriscus strigatus* (Pl. 17-1), but tail has different shape; Great Barrier Reef and throughout S.E. Asia; Indo-W. Pacific; to 14 cm. (CENTRISCIDAE)

2 MULTI-BANDED PIPEFISH
Doryrhamphus multiannulatus (Regan, 1903)
Inhabits coral reef crevices; similar to *D. dactyliophorus* (Pl. 17-11), but has more numerous dark bands; N.W. Australia; Indian Ocean and Red Sea; to 18 cm. (SYNGNATHIDAE)

3 ORANGE-BANDED PIPEFISH
Doryrhamphus pessuliferus (Fowler, 1938)
Inhabits coral reef crevices; similar to *D. dactyliophorus* (Pl. 17-11), but has orange instead of white background colour; Indonesia (N. Sulawesi) and Philippines; to 15 cm. (SYNGNATHIDAE)

4 BLUE-STRIPED PIPEFISH
Doryrhamphus excisus excisus Kaup, 1856
Inhabits coral reef crevices; distinguished by reddish fan-shaped tail; found throughout the region; Indo-E. Pacific; to 7 cm. (SYNGNATHIDAE)

5 RUBBLE PIPEFISH
Corythoichthys haematopterus (Bleeker, 1851)
Inhabits coral reefs, usually on rubble bottoms; similar to **6** below, but dark markings are usually less distinct on rear part of body; N.W. Australia and throughout S.E. Asia; Indo-W. Pacific; to 20 cm. (SYNGNATHIDAE)

6 RINGED PIPEFISH
Corythoichthys intestinalis (Ramsay, 1881)
Inhabits coral reefs, usually on rubble bottoms; similar to **5** above, but dark markings are usually more distinct on rear part of body; Great Barrier Reef, N.W. Australia and throughout S.E. Asia; mainly W. Pacific; to 16 cm. (SYNGNATHIDAE)

7 SCHULTZ'S PIPEFISH
Corythoichthys schultzi Herald, 1953
Inhabits coral reefs, usually on rubble bottoms; similar to **5** and **6** above, but has much longer snout; Great Barrier Reef, offshore reefs of N.W. Australia and throughout S.E. Asia; Indo-W. Pacific; to 15 cm. (SYNGNATHIDAE)

8 BROWN-BANDED PIPEFISH
Corythoichthys amplex Dawson & Randall, 1975
Inhabits coral reefs; distinguished by broad brown bars with narrower white bars between; Great Barrier Reef and throughout S.E. Asia; Indo-W. Pacific; to 9 cm. (SYNGNATHIDAE)

9 GLITTERING PIPEFISH
Halicampus nitidus (Günther, 1873)
Inhabits coral reefs; distinguished by short snout and white background colour with numerous narrow brown bars; Great Barrier Reef, offshore reefs of N.W. Australia and throughout S.E. Asia; mainly W. Pacific; to 7.5 cm. (SYNGNATHIDAE)

10 WHITE PIPEFISH
Siokunichthys nigrolineatus Dawson, 1983
Inhabits coral reefs; lives among polyps of mushroom corals; distinguished by overall white colour; Indonesia, Philippines, and New Guinea; to 8 cm. (SYNGNATHIDAE)

11 CLEARFIN LIONFISH
Pterois radiata Cuvier, 1829
Inhabits coral reef caves and ledges; distinguished by broad brown bars on body separated by narrow white lines, horizontal band on tail base, and enlarged filamentous pectoral fins; offshore reefs of N.W. Australia and throughout S.E. Asia; Indo-C. Pacific; to 20 cm. (SCORPAENIDAE)

12 DOUBLESPOT LIONFISH
Dendrochirus biocellatus (Fowler, 1938)
Inhabits coral reef caves and ledges; distinguished by elongate "whiskers" and pair of pale-rimmed dark spots on rear part of dorsal fin; offshore reefs of N.W. Australia and throughout S.E. Asia; Indo-C. Pacific; to 13 cm. (SCORPAENIDAE)

13 WEEDY SCORPIONFISH
Rhinopias aphanes Eschmeyer, 1973
Inhabits coral reefs; distinguished by bold maze-like pattern and filamentous tentacles on head and body; Great Barrier Reef and New Guinea; W. Pacific; to 24 cm. (SCORPAENIDAE)

14 MCADAM'S SCORPIONFISH
Parascorpaena mcadamsi (Fowler, 1938)
Inhabits coral reef crevices; distinguished by 12 dorsal spines, second spine above upper jaw curves outward and hooks forward, and black spot at rear of spiny part of dorsal fin; offshore reefs of N.W. Australia and throughout S.E. Asia; Indo-W. Pacific; to 6 cm. (SCORPAENIDAE)

15 SHORT-FINNED SCORPIONFISH
Scorpaenodes parvipinnis (Garrett, 1864)
Inhabits coral reef crevices; distinguished by 13 dorsal spines and relatively low spiny part of dorsal fin; Great Barrier Reef, offshore reefs of N.W. Australia, and throughout S.E. Asia; Indo-C. Pacific; to 13 cm. (SCORPAENIDAE)

16 HAIRY SCORPIONFISH
Scorpaenodes hirsutus (Smith, 1957)
Inhabits coral reef crevices, usually on outer slopes; distinguished by 13 dorsal spines, dark blotch at front of dorsal fin, and another on basal half of pectoral fin; Great Barrier Reef and throughout S.E. Asia; Indo-C. Pacific; to 5 cm. (SCORPAENIDAE)

17 SMALLSCALE SCORPIONFISH
Scorpaenopsis oxycephala (Bleeker, 1849)
Inhabits coral reefs; rests in the open and is most commonly observed scorpionfish; similar to *S. venosa* (Pl. 20-10), but usually has 20 pectoral rays instead of 17-18; Great Barrier Reef, offshore reefs of N.W. Australia, and throughout S.E. Asia; Indo-W. Pacific; to 30 cm. (SCORPAENIDAE)

18 YELLOW-SPOTTED SCORPIONFISH
Sebastapistes cyanostigma (Bleeker, 1856)
Inhabits coral heads (often *Pocillopora*); distinguished by large yellow blotches and tiny pale spots on side; Great Barrier Reef, offshore reefs of N.W. Australia, and throughout S.E. Asia; Indo-C. Pacific; to 7 cm. (SCORPAENIDAE)

19 REEF STONEFISH
Synanceja verrucosa Bloch & Schneider, 1801)
Inhabits coral reefs, among rocks or under slabs of dead coral; wound from venomous dorsal and anal-fin spines may cause serious injury or death; similar to *S. horrida* (Pl. 19-8), but found more offshore in clear water; found throughout the region; Indo-W. Pacific; to 35 cm. (SCORPAENIDAE)

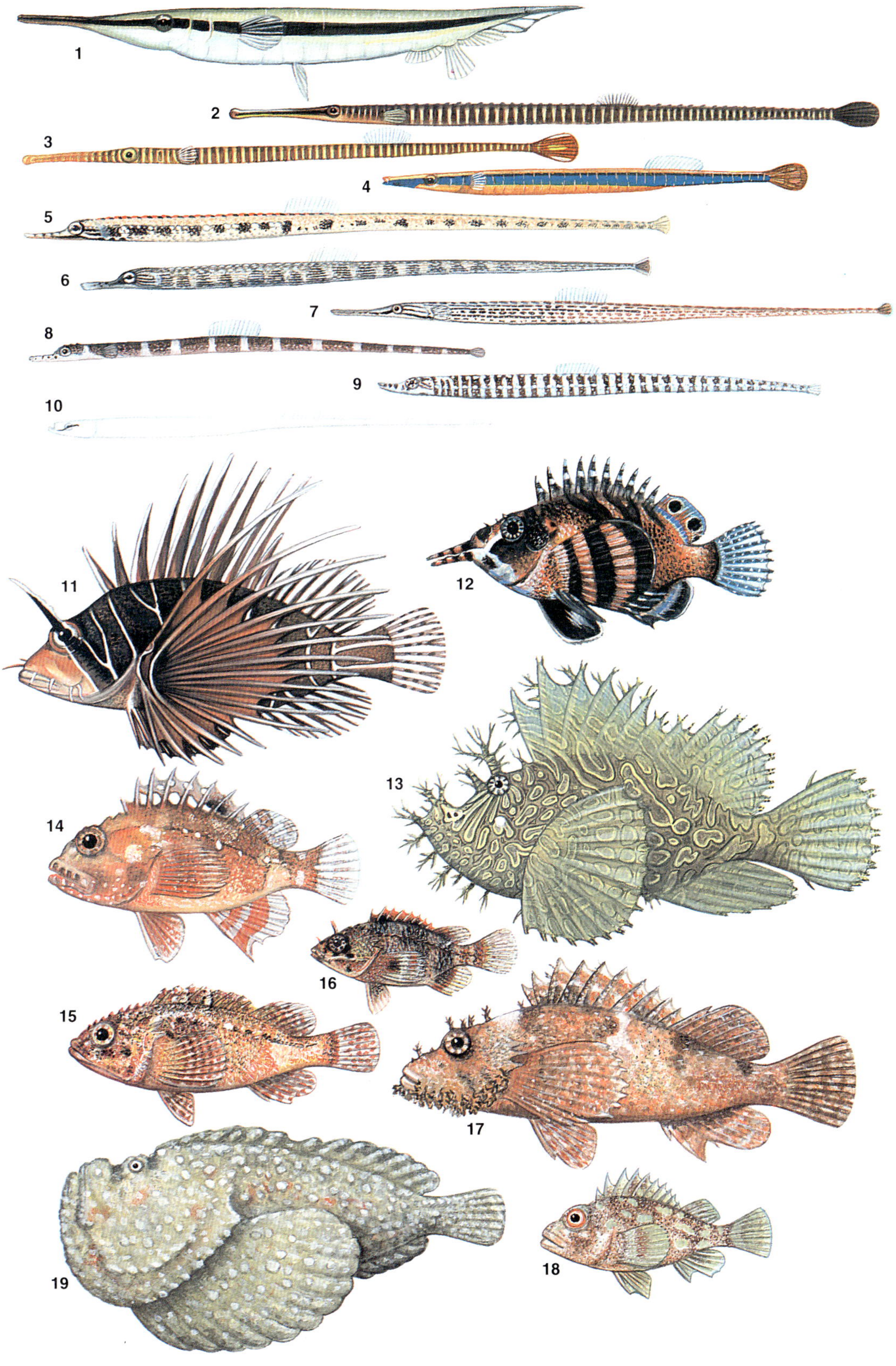
1
2
3
4
5
6
7
8
9
10
11
12
13
14
15
16
17
18
19

PLATE 19: SCORPIONFISHES (FAMILY SCORPAENIDAE)

1 DWARF LIONFISH
Dendrochirus brachypterus (Cuvier, 1829)
Inhabits coral reefs, distinguished by large pectoral fins without elongate, free filamantous rays, similar to **2** below, but has more prominent *curved* bands on pectoral fins and bars on side are less well defined; also known as Short-finned scorpionfish; found throughout the region; Indo-C. Pacific; to 15 cm.

2 ZEBRA LION FISH
Dendrochirus zebra (Cuvier, 1829)
Inhabits coral reefs; distinguished by large pectoral fins without elongate, free filamentous rays, similar to **1** above, but bands on pectoral fins poorly defined and bars on side more distinct, also known as Butterfly scorpionfish; found throughout the region; Indo-C Pacific; to 18 cm.

3 RAGGED-FINNED FIREFISH
Pterois antennata (Bloch, 1787)
Inhabits coral reefs, usually in caves and crevices distinguished by white filamentous pectoral rays, row of large dark spots at base of this fin and relatively few dark bars on head; also known as Spotfin lionfish; found throughout the region; Indo-C. Pacific; to 30 cm.

4 DEEPWATER FIREFISH
Pterois mombasae (Smith, 1957)
Inhabits offshore reefs, usually below 40 m depth; distinguished by relatively short filamentous tips on pectoral fin which is densely spotted and maze of dark and light bands on tail base; N.W. Australia and Indonesia; Indo-W. Pacific; to 16 cm.

5 SPOTLESS FIREFISH
Pterois russelli Bennett, 1831
Inhabits offshore reefs, usually below 20-30 m depth; distinguished by filamentous pectoral rays and lack of spots on dorsal, anal, and tail fins; found throughout the region; Indo-W. Pacific; to 30 cm.

6 RED FIREFISH
Pterois volitans (Linnaeus, 1758)
Inhabits coral and rocky reefs, usually in caves or crevices, distinguished by broad filamentous pectoral rays, similar to **5** above, but has spots on dorsal, anal, and tail fins; also known as Butterfly cod and Volitans lionfish; found throughout the region; Indo-C. Pacific; to 38 cm.

7 ROUNDFACE FIREFISH
Brachypterois serrulatus (Richardson, 1846)
Inhabits deeper trawling grounds, distinguished by dusky fan-like pectoral fins without free filamentous rays, lack of distinct bars on side, and faint spotting on dorsal, anal, and tail fins;N.W. Australia and Timor-Arafura seas; E. Indian Ocean and W. Pacific; to 10 cm.

8 ESTUARINE STONEFISH
Synanceja horrida (Linnaeus, 1766)
Inhabits estuaries and inshore areas on sand or mud bottoms, amongst rocks, or sometimes under dead coral slabs; distinguished by stone-like appearance and warty projections on body; a similar species, *S. verrucosa* (not shown), occurs on coral reefs, fin spines extremely venomous; found throughout the region; E. Indian Ocean and W. Pacific; to 47 cm.

9 MONKEYFISH
Erosa erosa (Langsdorf, 1829)
Inhabits trawling grounds; similar to **10** below, but has slight hump in front of dorsal fin, white spots on outer part of pectoral fin and narrow cross-bars on tail; also known as Pitted scorpionfish; found throughout the region; mainly W. Pacific; to 15 cm.

10 DAMPIE R STONEFISH
Dampierosa daruma Whitley, 1932
Inhabits coastal waters in the vicinity of reefs, sometimes under wharves; similar to **9** above, but has rounded head profile without hump in front of dorsal fin, broad white band (instead of spots) on outer pectoral fin and a single dark bar across middle part of tail; N.W. Australia only; to 13 cm.

11 COCKATOO WASPFISH
Ablabys taenianotus (Cuvier, 1829)
Inhabits coastal reefs; distinguished by thin, laterally compressed body, vertical profile of snout, and elevated rays at front of dorsal fin; found throughout the region; Indo-W. Pacific; to 10 cm.

12 BLACKSPOT WASPFISH
Cottapistus praepositus (Ogilby, 1903)
Inhabits trawling grounds; similar to **11** above, but has larger eye, lacks elevated rays at front of dorsal fin, and has prominent dark blotch above pectoral fin; N. Australia and Indonesia; E. Indian Ocean and W. Pacific; to 13 cm.

13 PLUMB-STRIPED STINGFISH
Minous versicolor Ogilby, 1910
Inhabits trawling grounds; distinguished by irregular stripes and blotches on upper side, and wavy cross bands on dorsal and tail fins, also has free lower-most pectoral ray; N. Australia only; to 11 cm.

14 SPOT FIN WASPFISH
Paracentropogon vespa Ogilby, 1910
Inhabits trawling grounds; similar to **11** and **12** above, but has large dark blotch on base of front part of dorsal fin; N. Australia only; to 9 cm.

15 DEMON STINGER
Inimicus didactylus (Pallas, 1769)
Inhabits rubble bottoms, frequently in the vicinity of coral reefs; distinguished by large upturned mouth, free pair of rays at lowermost part of pectoral fin, and prominent dorsal spines, similar to **16** below, but has yellow tail with dark submarginal bar and pale band across pectoral fin; N.W. Australia and throughout S.E. Asia; mainly Indo-Australian Archipelago; to 18 cm.

16 SPOTTED STINGER
Inimicus sinensis (Valenciennes, 1833)
Inhabits rubble bottoms; similar to **15** above, but tail mainly dusky or spotted (not yellow with dark cross bar) and lacks pale band on pectoral fin; N.W. Australia, Gulf of Carpentaria and throughout S.E. Asia; E. Indian Ocean and W. Pacific; to 25 cm.

WARNING! The fishes shown on Plates 19 and 20 possess venomous fin spines and handling of live or freshly dead specimens should be avoided. The Estuarine Stonefish is amongst the most venomous of all fishes and is capable of causing death. Symptoms of scorpionfish stings range from a bee-sting type sensation to violent pain and may lead to unconsciousness or extended coma. Immersing the wound in very hot water is an effective first aid treatment and a physician should be consulted immediately.

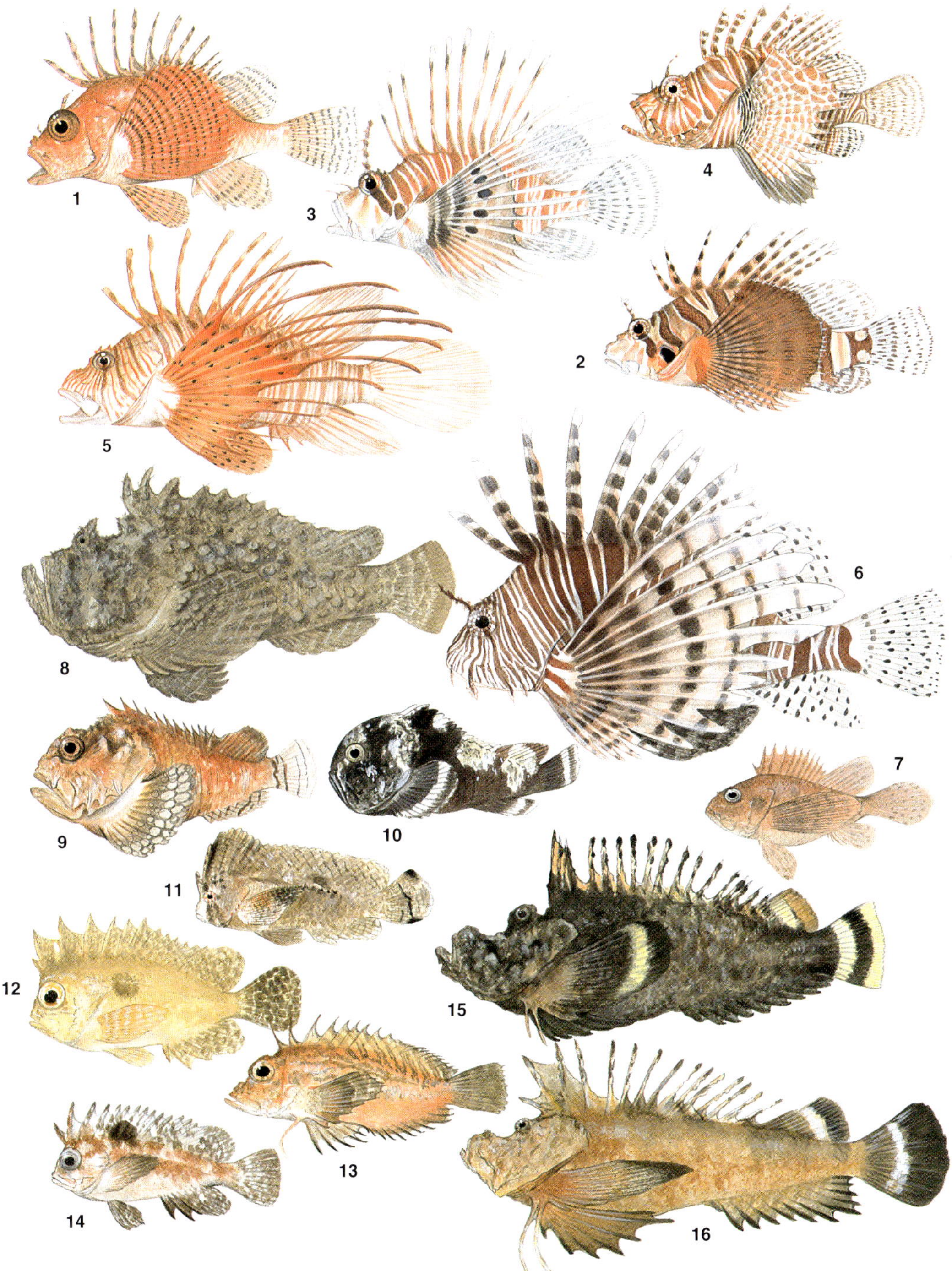
1
3
4
5
2
8
6
9
10
7
11
12
15
13
14
16

PLATE 20: SCORPIONFISHES (FAMILY SCORPAENIDAE)

1 SHORT-FINNED WASPFISH
Apistops coloundra (De Vis, 1886)
Inhabits trawling grounds; distinguished by 5 chin barbels, diffuse dark stripes on side and spot on dorsal fin, similar to **2** below, but pectoral fins shorter (do not reach the rear part of anal fin); N. Australia only; to 12 cm.

2 LONG-FINNED WASPFISH
Hypodytes carinatus (Bloch & Schneider, 1801)
Inhabits trawling grounds, similar to **1** above, but with 3 chin barbels, no stripes on side and pectoral fins usually longer, reaching to rear part of anal fin; found throughout the region; Indo-W. Pacific; to 18 cm.

3 COD VELVETFISH
Peristrominous dolosus Whitley, 1952
Inhabits trawling grounds; distinguished by elongate brown body without distinct markings and pointed snout; N. Australia only; to 9 cm. (APLOACTINIDAE)

4 MARBLED STINGFISH
Cottapistus cottoides (Cuvier, 1829)
Inhabits deep offshore reefs and trawling grounds; distinguished by forward position (over eye) of dorsal fin origin, hump on snout, and very small scales; N.W. Australia, Gulf of Carpentaria, and scattered localities in Indo-Malay region; mainly W. Pacific, to 12 cm.

5 NORTHERN SCORPIONFISH
Parascorpaena picta (Cuvier, 1829)
Inhabits crevices of coral and rocky reefs; distinguished by well camouflaged appearance with skin flaps and tentacles on head and body, 12 dorsal spines, and spine above upper jaw that curves forward; found throughout the region; Indo-W. Pacific; to 16 cm

6 GUAM SCORPIONFISH
Scorpaenodes guamensis (Quoy & Gaimard, 1824)
Inhabits coral reef crevices; distinguished by dark spot on upper part of gill cover and 13 dorsal spines; found throughout the region; Indo-C. Pacific; to 12 cm.

7 ORNATE SCORPIONFISH
Scorpaenodes varipinnis Smith, 1957
Inhabits coral reef crevices; distinguished by red colour with white blotches on head and along middle of sides, a curved dark band across pectoral fin and 13 dorsal spines; found throughout the region; Indo-W. Pacific; to 7 cm.

8 LITTLE SCORPIONCOD
Scorpaenodes sp.
Inhabits coral reef crevices; distinguished by red or brown spots on fins, pale bar across tail base, lack of dark spot on gill cover, and 13 dorsal spines; possibly a colour variation of **9** below; W. Australia; to 8 cm.

9 PYGMY SCORPIONFISH
Scorpaenodes scaber (Ramsay & Ogilby, 1886)
Inhabits coastal and estuaries reefs; similar to **6** above, but has dark spot (darker than shown) on lower edge of gill cover instead of upper part; found throughout the region; Indo-W. Pacific; to 8 cm.

10 RAGGY SCORPION FISH
Scorpaenopsis venosa (Cuvier, 1829)
Inhabits coral reef crevices; distinguished by skin flaps and tentacles on head and body; relatively tall dorsal fin, and 12 dorsal spines, juveniles usually more ornate as shown; N.W. Australia and throughout S.E. Asia; Indo-C. Pacific; to 18 cm.

11 FALSE STONEFISH
Scorpaenopsis diabolus (Cuvier, 1829)
Inhabits coral reef crevices and rubble bottoms; distinguished by well camouflaged appearance, humped back, and bright yellow-orange patch on inner surface (not shown) of pectoral fins; often confused with the true stonefish (*Synanceja*), which is far more venomous; found throughout the region; Indo-C. Pacific; to 18 cm.

12 WHITE-BELLIED ROUGEFISH
Richardsonichthys leucogaster (Richardson, 1848)
Inhabits trawling grounds; distinguished by laterally compressed body, forward position (above rear part of eye) of dorsal fin origin, relatively large eye, and complete lack of scales; found throughout the region; Indo-W. Pacific; to 8 cm.

13 LEAF SCORPIONFISH
Taenianotus triacanthus Lacepède, 1802
Inhabits offshore coral reefs; distinguished by thin leaf-like body and tall dorsal fin, several colour varieties encountered including ones that are predominantly reddish, yellow, or black; offshore reefs of W. Australia, Great Barrier Reef, and throughout S.E. Asia; Indo-C. Pacific; to 10 cm.

14 DEEPSEA SCORPIONFISH
Setarches guentheri Johnston, 1862
Inhabits offshore trawling grounds, distinguished by overall red colour without distinct marks, relatively pointed snout, very stout spines on edge of cheek, and 12 dorsal spines; N.W. Australia; worldwide in tropical seas; to 23 cm.

PLATE 21

1 SANDPAPER VELVETFISH
Adventor elongatus (Whitley, 1952)
Inhabits trawling grounds; distinguished by elongate body shape and bony knobs on head, similar to **5** below, but dorsal fin begins further back on head; N. Australia only; to 11 cm. (APLOACTINIDAE)

2 DUSKY VELVETFISH
Aploactis aspersa (Temminck & Schlegel, 1844)
Inhabits trawling grounds; has longer anal fin base than other velvet fishes on this page; colour sometimes brown; found throughout the region, but rare; E. Indian Ocean and W. Pacific; to 9 cm. (APLOACTINIDAE)

SCORPIONFISH COLOURS

The coloration of many of the scorpionfishes shown on this plate is extremely variable, depending on size, depth, and habitat; it is not unusual for the same species to exhibit very different patterns at a particular locality. For example the Northern Scorpionfish (5) is often mottled brown when found among rocks in shallow weedy areas, and red (as shown) if seen in caves on deeper sections of the reef.

SCORPIONFISHES AND ALLIED FAMILIES

The families featured on Plates 19-21 are members of the order Scorpaeniformes. They are distinguished by a bony ridge on the cheek and the head is frequently spiny or tassled. All are bottom living fishes that occur in a variety of depths and habitats. They often exhibit variegated colour patterns that blend well with their surroundings.

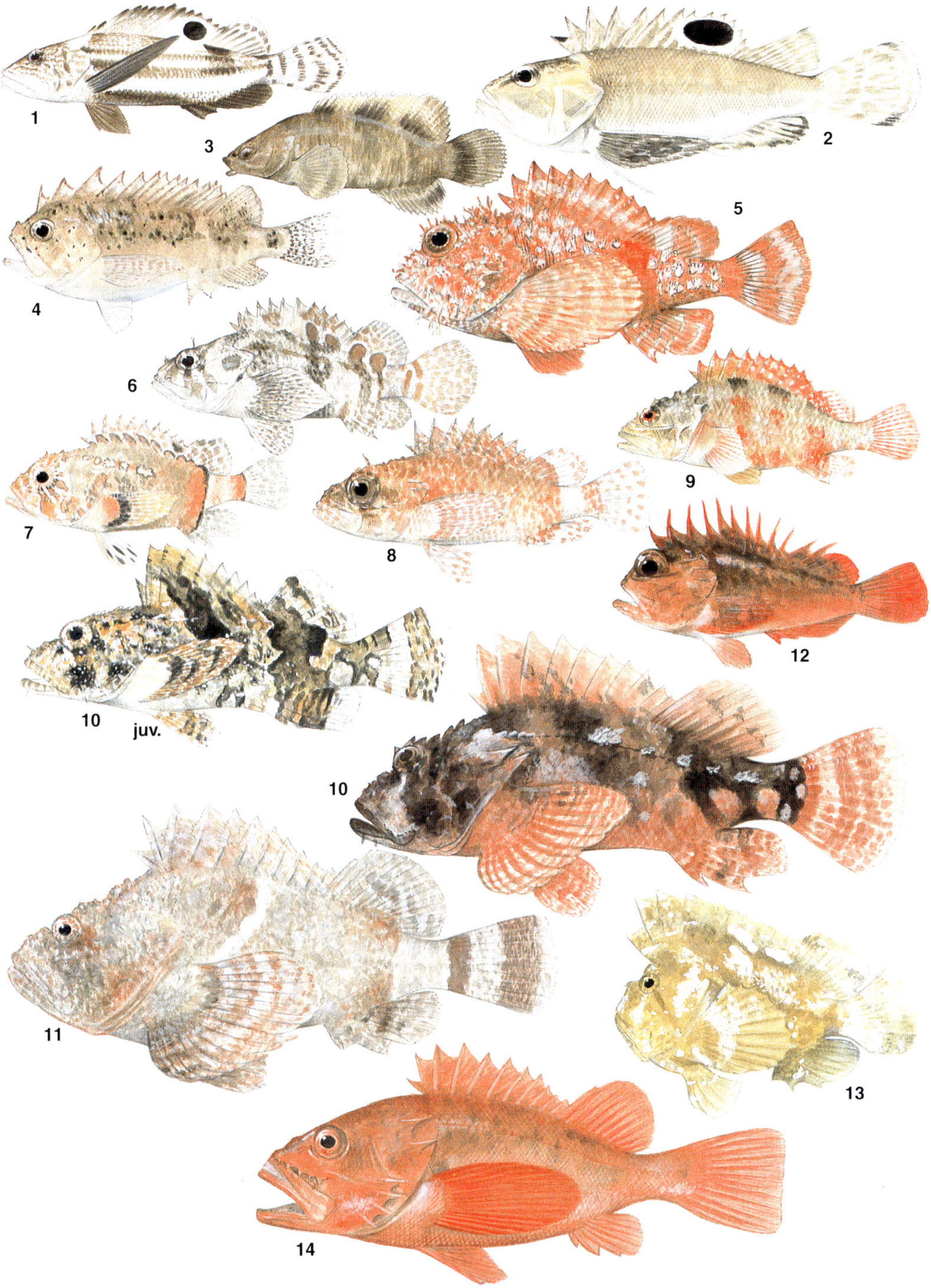
1
2
3
4
5
6
7
8
9
10
juv.
10
11
12
13
14

PLATE 21: VELVETFISHES, GURNARDS, AND FLATHEADS

1 SANDPAPER VELVETFISH
Adventor elongatus (Whitley, 1952). Text on previous page.

2 DUSKY VELVETFISH
Aploactis aspersa (Temminck & Schlegel, 1844). Text on previous page.

3 THIN VELVETFISH
Coccotropus sp.
Inhabits trawling grounds and sandy areas near reefs; distinguished by laterally compressed body, steep forehead, and tall anterior part of dorsal fin that begins above eye; similar to some scorpionfishes (Plates 19-20) but has bony knobs (versus spines) on head and body covered with prickles, possibly N.W. Australia only; to 5 cm. (APLOACTINIDAE)

4 THREEFIN VELVETFISH
Neoaploactis tridorsalis Eschmeyer & Allen, 1978
Inhabits sand or rubble bottoms near reefs, distinguished by 3 separate dorsal fins; N. Australia only; to 5 cm. (APLOACTINIDAE)

5 BEARDED VELVETFISH
Paraploactis intonsa Poss & Eschmeyer, 1978
Inhabits trawling grounds, similar to **1** and **2** above, but deeper bodied, steeper forehead, and dorsal fin begins above eye; *P. pulvinus* (not shown) is similar, but lacks prickles on ventral surface of lower jaw; known only from Shark Bay, W. Australia only; to 14 cm. (APLOACTINIDAE)

6 DARK-FINNED VELVETFISH
Kanekonia aniara (Thompson, 1967)
Inhabits continental shelf in about 200 m depth; distinguished by pale colouration except for blackish fins; N.W. Australia and E. Queensland; E. Indian Ocean and W. Pacific; to 10 cm. (APLOACTINIDAE)

7 LONG-FINNED GURNARD
Lepidotrigla argus Ogilby, 1910
Inhabits trawling grounds; distinguished by pair of short forward projecting spines on snout and enlarged fan-like pectoral fins with blue and yellow markings and blue-edged black spot; N. Australia only; to 18 cm. (TRIGLIDAE)

8 BLACK-FINNED GURNARD
Pterygotrigla leptacanthus (Günther, 1880)
Inhabits trawling grounds; similar to **7** above, but lacks scales, has longer forward projecting spines on snout, and blackish pectoral fins; N.W. Australia and Arafura Sea; to 15 cm. (TRIGLIDAE)

9 HALF-SPOTTED GURNARD
Pterygotrigla hemisticta (Temminck & Schlegel, 1844)
Inhabits trawling grounds, similar to **8** above, but has black spot on first dorsal fin and scattered brown spots on back; N.W. Australia and Indonesia; E. Indian Ocean and W. Pacific; to 25 cm. (TRIGLIDAE)

10 SLENDER ARMOURED-GURNARD
Peristedion liorhynchus (Günther, 1872)
Inhabits trawling grounds, distinguished by dark margin on dorsal fins and banded pectoral fins; shape from dorsal view similar to **11** below; found throughout the region; E. Indian Ocean and W. Pacific; to 25 cm. (TRIGLIDAE)

11 SPOTTED ARMOURED-GURNARD
Satyrichthys rieffeli (Kaup, 1859)
Inhabits trawling grounds; distinguished by black spotting on head, body and dorsal fin; shape from side view similar to **10** above; N.W. Australia and Indonesia; E. Indian Ocean and W. Pacific; to 20 cm. (TRIGLIDAE)

12 FRINGE-EYED FLATHEAD
Papilloculiceps nematophthalmus (Günther, 1860)
Inhabits sand bottoms; distinguished by 6-9 skin tentacles above eye (versus 0-1 in most flatheads), 7-8 dusky bands across nape and back extending on to sides, and strongly variegated pattern on fins; found throughout the region; Indo-Australian Archipelago; to 58 cm. (PLATYCEPHALIDAE)

13 DWARF FLATHEAD
Elates ransonnetti (Steindachner, 1877)
Inhabits sand bottoms, distinguished by 6 dorsal spines (versus 7-9 spines for other flatheads), filamentous upper lobe of tail, and semi-transparent appearance; found throughout the region; Indo-Australian Archipelago; to 19 cm. (PLATYCEPHALIDAE)

14 HARRIS'S FLATHEAD
Inegocia harrisii (McCulloch, 1914)
Inhabits sand bottoms, distinguished by overall orange-brown colour with fine brown spots on back and white below, irregular dark bars on pectoral fins, and elongate dark streaks on tail; N. Australia only; to 20 cm. (PLATYCEPHALIDAE)

15 SPINY FLATHEAD
Onigocia spinosa (Temminck & Schlegel, 1844)
Inhabits sand bottoms; distinguished by numerous small spines on head which is very broad (when viewed from above), outer half of spiny dorsal fin dark brown; irregular brown bars across back and sides, and largely blackish pelvic fins with yellowish tips; N.W. Australia and Indonesia; E. Indian Ocean and W. Pacific; to 9 cm. (PLATYCEPHALIDAE)

16 NORTHERN SAND FLATHEAD
Platycephalus arenarius Ramsay & Ogilby, 1886
Inhabits sand bottoms; distinguished by black stripes on tail; N. Australia and Indonesia; to 45 cm. (PLATYCEPHALIDAE) ★★★

17 BAR-TAILED FLATHEAD
Platycephalus endrachtensis Quoy & Gaimard, 1825
Inhabits sand bottoms; distinguished by black stripes on tail, similar to **16** above, but has fewer black stripes and a yellow blotch on upper part of tail; N. Australia and New Guinea; to 26 cm. (PLATYCEPHALIDAE) ★★★

18 RUSTY FLATHEAD
Inegocia japonica (Tilesius, 1812)
Inhabits sand bottoms; a reddish-brown flathead similar to **14** above in general appearance, but has definite dark spots on tail (versus elongate streaks); sometimes referred to as *S. isacathus*; found throughout the region; Indo-W. Pacific; to 20 cm. (PLATYCEPHALIDAE) ★★★

19 HEART-HEADED FLATHEAD
Sorsogona tuberculata (Cuvier, 1829)
Inhabits sand bottoms; distinguished by prominent black area on outer part of pectoral fins and strongly barred pelvic fins; found throughout the region; Indo-W. Padfic; to 50 cm. (PLATYCEPHALIDAE) ★★★

20 OLIVE-TAILED FLATHEAD
Rogadius asper (Cuvier, 1829)
Inhabits sand bottoms, distinguished from other flatheads by forward directed spine on lower edge of cheek, similar in colour to **19** above, but has broad dusky margin on spiny dorsal fin and lacks faint spotting on tail; found throughout the region; Indo-W. Pacific; to 17 cm. (PLATYCEPHALIDAE) ★★★

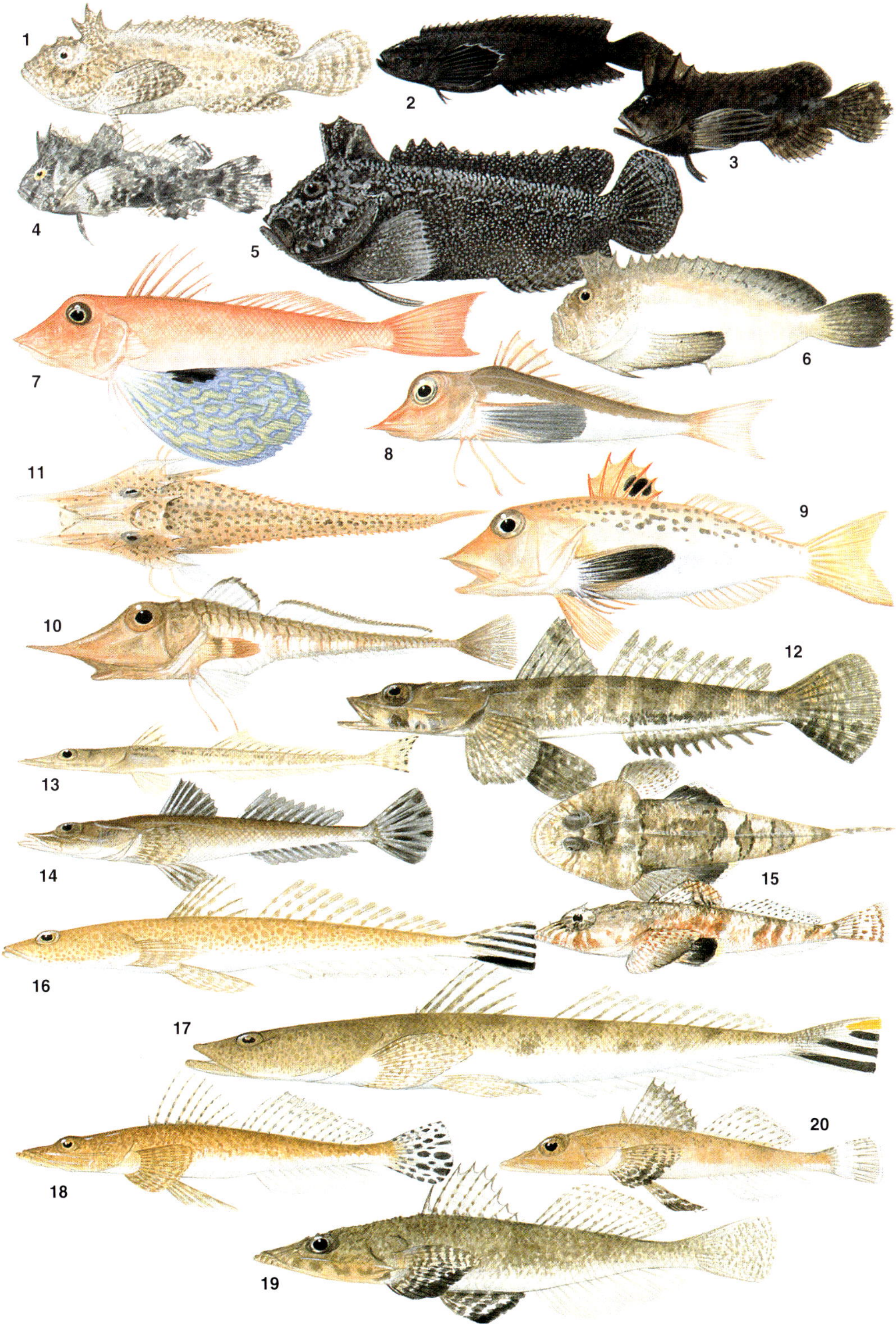
1
2
3
4
5
6
7
8
9
10
11
12
13
14
15
16
17
18
19
20

PLATE 22: GROPERS (FAMILY SERRANIDAE)

1 WHITE-LINED ROCKCOD
Anyperodon leucogrammicus (Valenciennes, 1828)
Inhabits coral reefs; distinguished by elongate shape and pattern of pale longitudinal stripes and numerous dark spots; N.W. Australia, Great Barrier Reef, and throughout S.E. Asia; Indo-C. Pacific; to 50 cm. ★★★

2 FALSE SCORPIONFISH
Centrogenys vaigiensis (Quoy & Gaimard, 1824)
Inhabits sand and weed flats, frequently around rocky outcrops; resembles members of scorpionfish family (Plates 19-20), but lacks numerous head spines and is *not* venomous (fin spines); found throughout the region; Indo-Australian Archipelago; to 15 cm.

3 BROWN-BANDED ROCKCOD
Cephalopholis boenack Bloch, 1790
Inhabits dead reefs in protected inshore waters; distinguished by overall brown colour with faint dark bars on sides and large blackish spot at rear of gill cover; formerly known as *C. pachycentron*; found throughout the region; Indo-W. Pacific; to 22 cm. ★★

4 PEACOCK ROCKCOD
Cephalopholis argus Bloch & Schneider, 1801
Inhabits caves and crevices of coral reefs; distinguished by numerous dark-edged blue spots on head, body and fins, also by whitish area in front of pectoral fin and 5-6 pale bars on posterior part of body frequently present; N.W. Australia, Great Barrier Reef, and throughout S.E. Asia; Indo-C. Pacific; to 50 cm. ★★★

5 CORAL COD
Cephalopholis miniata (Forsskål, 1775)
Inhabits caves and crevices of coral reefs; distinguished by numerous blue spots on orange-red to red-brown background; also known as Coral trout; found throughout the region; Indo-W. Pacific; to 41 cm; 1.6 kg. ★★★

6 BLUE-LINED ROCKCOD
Cephalopholis formosa (Shaw & Nodder, 1812)
Inhabits inshore coral reefs; distinguished by narrow dark blue stripes on head, body, and fins; N.W. Australia, Great Barrier Reef, and throughout S.E. Asia; Indo-W. Pacific; to 33 cm. ★★★

7 RED-SPOTTED ROCKCOD
Cephalopholis leopardus (Lacepède, 1801)
Inhabits caves and crevices of coral reefs; distinguished by oblique dark streaks on upper and lower lobe of tail, dark patch on upper edge of tail base, and dark spot at rear of gill cover; N.W. Australia, Great Barrier Reef, and throughout S.E. Asia; Indo-C. Pacific; to 20 cm. ★★

8 TOMATO ROCKCOD
Cephalopholis sonnerati (Valenciennes, 1828)
Inhabits coral and rock reefs, adults often in deep water (30-100 m); distinguished by orange-red to reddish brown colour, frequently with scattered whitish blotches; a brown variety is shown on Plate 23 (**12**); juveniles are pale pinkish (see opposite Plate); N.W. Australia, Great Barrier Reef, and throughout S.E. Asia; Indo-C. Pacific; to 58 cm. ★★★

9 FLAG-TAILED ROCKCOD
Cephalopholis urodeta (Schneider, 1801)
Inhabits coral reef crevices, distinguished by pale oblique streaks on lobes of tail; found throughout the region; Indo-C. Pacific; to 23 cm. ★★

10 BARRAMUNDI COD Cairns 10/98
Cromileptes altivelis (Valenciennes, 1828)
Inhabits caves and crevices of coral reefs; distinguished by small head, laterally compressed body and polkadot pattern; N.W. Australia, Great Barrier Reef, and throughout S.E. Asia; Indo-W. Pacific, to 70 cm and 4.8 kg. ★★★★

11 COOPER'S FAIRY BASSLET
Pseudanthias cooperi (Regan, 1902)
Inhabits offshore reefs to at least 160 m depth; male (shown here) has red spot on sides and elongate filaments on pelvic, anal, and tail fins; female lacks these features and has a red spot on the tip of each lobe of the tail; N.W. Australia, Great Barrier Reef, and throughout S.E. Asia; Indo-W. Pacific; to 13 cm.

12 PEARL-SPOTTED FAIRY BASSLET
Selenanthias analis Tanaka, 1918
Inhabits offshore trawling grounds; distinguished by pearly spots on sides and black spot on anal fin; N. Australia and Japan to Taiwan; to 16 cm.

13 CITRON PERCHLET
Plectranthias megalophthalmus Fourmanoir & Randall, 1979
Inhabits deep offshore trawling grounds to 360 m depth; distinguished by yellow colouration; New Caledonia and N.W. Australia; to 8 cm.

14 SPOTTED PERCHLET
Plectranthias wheeleri Randall, 1980
Inhabits deep offshore trawling ground to 230 m; distinguished by irregular blotches on sides; found throughout the region; Australia and Indonesia; to 10 cm.

15 JAPANESE PERCHLET
Plectranthias japonicus (Steindachner, 1884)
Inhabits deep trawling grounds to at least 200 m; distinguished by red-orange colouration, sometimes with darker blotches on sides; N.W. Australia and Japan; to 15 cm.

INTRODUCTION TO ROCKCODS

Rockcods and their relatives in the groper family (Serranidae) are a dominant element of the fish community on all coral reefs, particularly in the Indo-Pacific region. The larger species are important table fish at many localities and the small schooling members, for instance the numerous Anthias (or Fairy Basslets) no doubt play an important role in the reef's food chain. Members of the genus *Cephalopholis* are small to medium-sized gropers (also know as Cods or Rockcods in Australia) that occur in a variety of coral reef habitats. The Coral Cod (5), with its bright red coat studded with blue spots, is among the most colourful species. It grows to a reported length of 41 cm and is most often encountered in clear water of outer reefs to depths of at least 150 m. Like most of the larger groupers it feeds mainly on small fishes, supplemented by crustaceans. Feeding occurs mainly during the early morning and midafternoon. The favourite food appears to be small, schooling Anthias, which are skilfully captured with a quick head on rush.

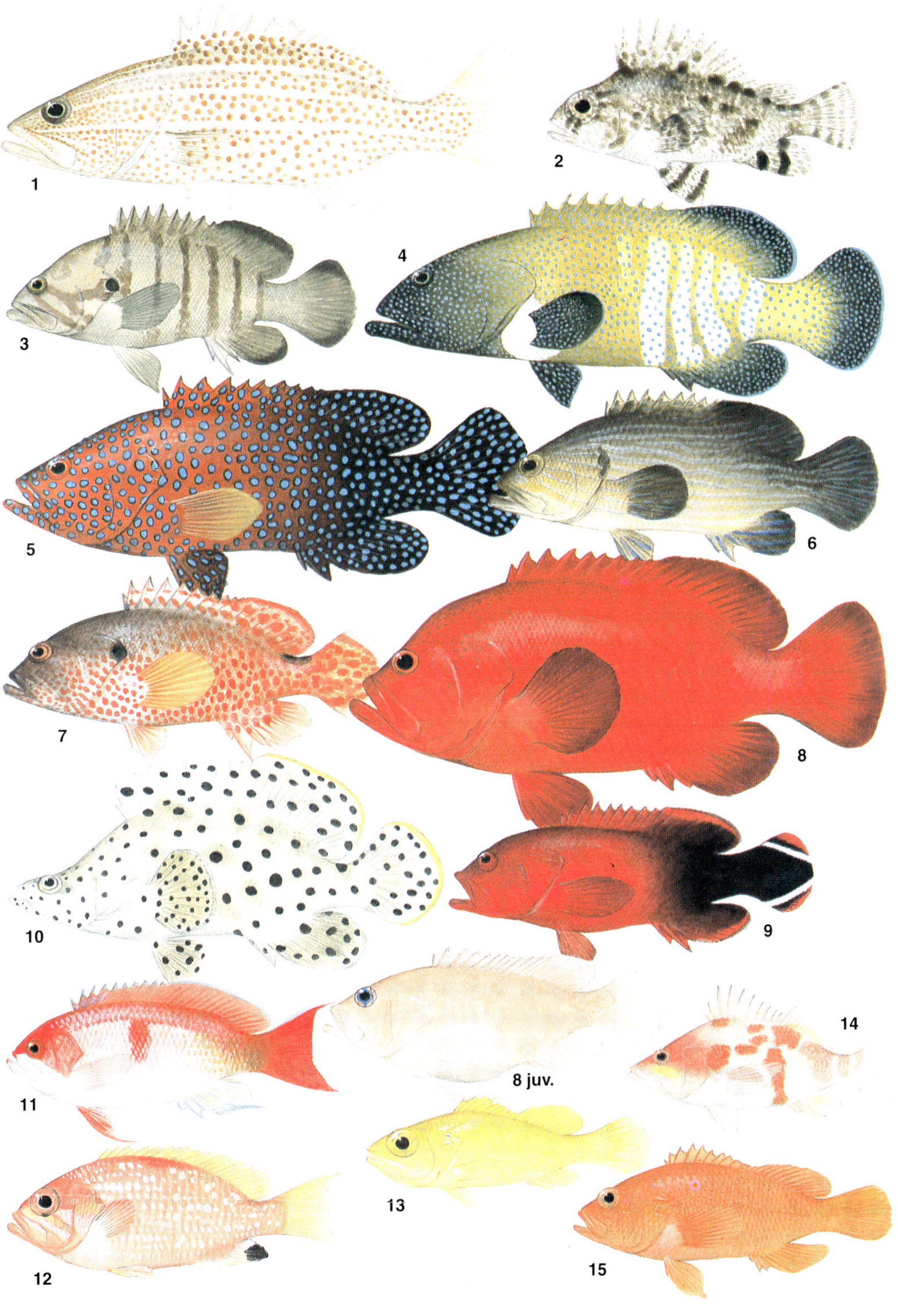
1
2
3
4
5
6
7
8
10
9
11
8 juv.
14
13
12
15

PLATE 23: GROPERS (FAMILY SERRANIDAE)

1 RED-FLUSHED ROCKCOD
Aethaloperca rogaa (Forsskål, 1775) Cairns 10/98
Inhabits coral reefs in the vicinity of caves; distinguished by elevated shape of body, dark colouration and white edge on tail; N.W. Australia, Great Barrier Reef, and throughout S.E. Asia; Indo-W. Pacific; to 60 cm. ★★★

2 BLUNT-HEADED ROCKCOD
Epinephelus amblycephalus (Bleeker, 1857)
Inhabits deeper offshore reefs; distinguished by 5 dark bars on body; N.W. Australia and throughout S.E. Asia; mainly Indo-Australian Archipelago; to 45 cm. ★★★

3 YELLOW-SPOTTED ROCKCOD
Epinephelus areolatus (Forsskål, 1775)
Inhabits inshore reefs, usually around small coral heads in sandy areas or among sea grass; distinguished by dense pattern of large round spots and truncate (not rounded) tail; N.W. Australia and throughout S.E. Asia; Indo-W. Pacific; to 35 cm. Cairns 10/98 ★★★

4 OCELLATED ROCKCOD
Epinephelus caeruleopunctatus (Bloch, 1790)
Inhabits coral reefs, near caves and crevices; similar to **2** on Plate 24, but spots more uniformly round and smaller; N.W. Australia, Great Barrier Reef, and throughout S.E. Asia; Indo-W. Pacific; to 60 cm. ★★★

5 CORAL ROCKCOD
Epinephelus corallicola (Valenciennes, 1828)
Inhabits shallow, silty reefs and estuaries; distinguished by round black spots on grey background; found throughout the region; W.Pacific; to 31cm. ★★★

6 BLACK-TIPPED COD
Epinephelus fasciatus (Forsskål, 1775)
Inhabits coral reefs and rocky bottoms to 100 m depth; distinguished by reddish bars and narrow black border on front part of dorsal fin; found throughout the region; Indo-C. Pacific; to 40 cm. ★★★

7 SPOTFIN ROCKCOD
Epinephelus latifasciatus (Temminck & Schlegel, 1842)
Inhabits sand and rock bottoms on the continental shelf between 20-200 m depth; distinguished by thin dark lines (or sometimes rows of faint spots) on sides, and spots on dorsal fin and tail; N.W. Australia; Indo-W. Pacific; to 70 cm. ★★★

8 FLOWERY COD
Epinephelus fuscoguttatus (Forsskål, 1775)
Inhabits coral reefs and rocky bottoms; similar to **11** below, but irregular brown blotches on sides generally more diffuse and spot at upper base of tail smaller; best means of separation is higher pectoral fin ray count (18-20, usually 19); N.W. Australia, Great Barrier Reef, and throughout S.E. Asia; Indo-W. Pacific; to 90 cm. ★★★

9 THREE-LINED ROCKCOD
Epinephelus heniochus Fowler, 1904
Inhabits offshore trawling grounds to at least 80 m depth; distinguished by overall pinkish-red colour and narrow stripes (often very faint) on head; N.W. Australia to Gulf of Carpentaria, also scattered localites in S.E. Asia; W. Pacific; to 30 cm. ★★★

10 HONEYCOMB COD
Epinephelus merra Bloch, 1793 Cairns 10/98
Inhabits protected inshore coral reefs; distinguished by dense network of large spots on body and fins; similar to **13** below, but spotting denser and lacks diagonal bands on breast; N.W. Australia, Great Barrier Reef, and throughout S.E. Asia; Indo-C. Pacific; to 28 cm. ★★★

11 SMALL-TOOTHED COD
Epinephelus polyphekadion (Bleeker, 1849)
Inhabits coral reefs, often around large bommies; similar to **8** above, but has more distinct oblique bands on back and head, and fewer pectoral rays (16-17, usually 17); *E. microdon* is a synonym; N.W. Australia, Great Barrier Reef, and throughout S.E. Asia; Indo-C. Pacific, to 61 cm. ★★★

12 TOMATO ROCKCOD
Cephalopholis sonnerati (Valenciennes, 1828)
Inhabits coral reefs in the vicinity of caves and crevices; the red variety and juvenile are shown on Plate 22 (**8**); brown variety shown here is sometimes seen at Ningaloo Reef, W. Australia in 12-20 m depth; N.W. Australia, Great Barrier Reef, and throughout S.E. Asia; Indo-C. Pacific; to 58 cm. ★★★

13 LONG-FINNED ROCKCOD
Epinephelus quoyanus (Valenciennes, 1830)
Inhabits silty inshore reefs; similar to **10** above, but spotting is less dense and has oblique dark bands just below and slightly in front of pectoral fin base; found throughout the region; mainly W. Pacific; to 35 cm. ★★★

14 CHINAMAN ROCKCOD
Epinephelus rivulatus (Valenciennes, 1830)
Inhabits inshore coral and rock reefs, usually around small coral heads or among weed; distinguished by oblique brown bars on sides and white blotches on head; W. and E. coasts of Australia and scattered locations in S.E. Asia; Indo-W. Pacific; to 35 cm. ★★★

15 RADIANT ROCKCOD
Epinephelus radiatus (Day, 1867)
Inhabits sand and rock bottoms, usually trawled between 80-160 m depth; distinguished by broad diagonal bands on head and body; N.W. Australia and outer edge of N. Great Barrier Reef; Indo-W. Pacific: to 70 cm. ★★★

SEX CHAMELEONS

Gropers are noted for their ability to change sex. Although relatively few species have been studied, gropers appear to be protogynous hemaphrodites. This is a fancy way of saying that an individual can function as an egg-producing female for one or more years, then changes sex and functions as a male. Adult fish are generally solitary in habit, but there are limited observations for a few species, which indicate they may aggregate at spawning times. Some species are known to migrate to specific spawning sites from distances up to several kilometres. Unfortunately, they are highly vulnerable to over-fishing at this time. Individual males may spawn several times during each reproductive period, which generally last one or two weeks. However, females only spawn once. The tiny (0.7-1.2 mm diameter) eggs float at the surface until hatching. The larvae have a characteristic kite-shaped body, imparted by the very elongate second dorsal spine and a pelvic spine, which is nearly as long. The larval stage lasts for about one or two months depending on the species involved. During this period the young fish, which have limited swimming powers, are dispersed by surface currents.

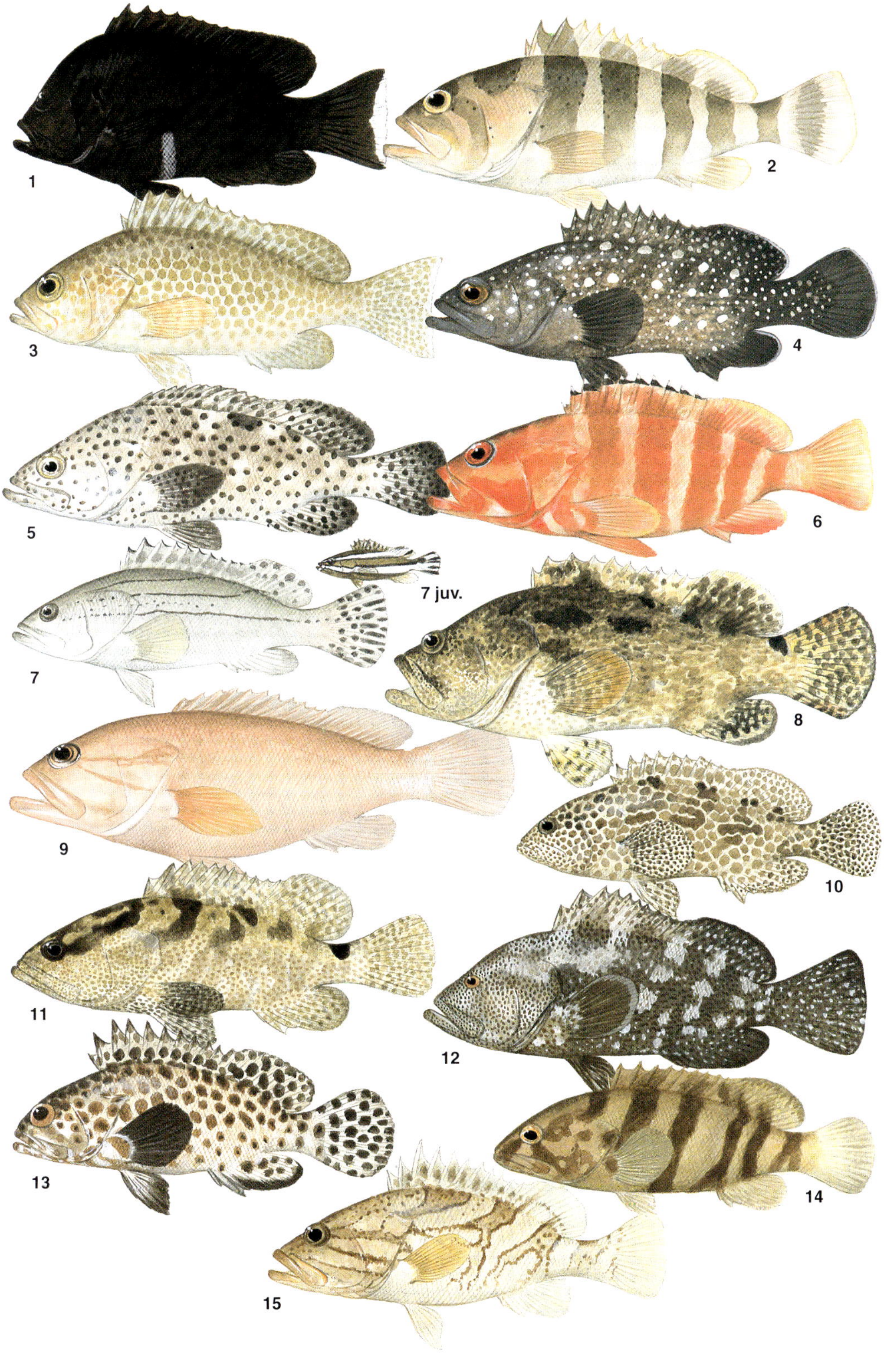
1
2
3
4
5
6
7 juv.
7
8
9
10
11
12
13
14
15

PLATE 24: GROPERS (FAMILY SERRANIDAE)

1 FROSTBACK COD
Epinephelus bilobatus Randall & Allen, 1987
Inhabits inshore coral reefs, usually where there is some sand; distinguished by white area on upper back and dark spots along base of dorsal fin; N.W. Australia only; to 40 cm. ★★★

2 WHITE-BLOTCHED ROCKCOD
Epinephelus multinotatus (Peters, 1876)
Inhabits inshore coral reefs and deeper offshore trawling grounds; distinguished by irregular white blotches; also known as Rankin's rockcod; coast of W. and N.W. Australia; mainly W. Indian Ocean; to 100cm and 9 kg. ★★★

3 SIX-BANDED ROCKCOD
Epinephelus sexfasciatus (Valenciennes, 1828)
Inhabits offshore trawling grounds to 70 m depth; distinguished by combination of bars on sides and spotted tail; found throughout the region; Indo-Australian Archipelago; to 26 cm. ★★★

4 ESTUARY COD
Epinephelus coioides (Hamilton, 1822)
Inhabits inshore coral reefs and estuaries; distinguished by oblique bands on sides overlaid with red-brown spotting; frequently misidentified as *E. malabaricus* or *E. tauvina*; also known as Greasy cod; found throughout the region; Indo-W. Pacific; to at least 95 cm. ★★★

5 POTATO COD
Epinephelus tukula Morgans, 1959
Inhabits coral reefs in the vicinity of caves and crevices; distinguished by large ovate spots; offshore reefs of N.W. Australia, Great Barrier Reef, and throughout S.E. Asia; Indo-W. Pacific, to 140 cm. ★★★

6 WOORE'S ROCKCOD
Triso dermopterus (Temminck & Schlegel, 1842)
Inhabits offshore trawling grounds; distinguished by 11 dorsal spines, rounded tail, 2 short canine teeth on each side at front of jaws, and ctenoid scales (i.e. rough to the touch); N.W. Australia, and C. Queensland; Australia and Japan to Taiwan; to 45 cm. ★★★

7 CORAL TROUT
Plectropomus leopardus (Lacepède, 1802)
Inhabits coral reefs; distinguished by numerous small round spots on head and body; also known as Leopard cod; N.W. Australia, Great Barrier Reef, and throughout most of S.E. Asia; mainly W. Pacific; to 75 cm; 15.5 kg. ★★★★

8 POLKADOT COD
Plectropomus areolatus (Rüppell, 1830)
Inhabits coral reefs; distinguished by numerous dark-edged round spots on head and body; N.W. Australia, Great Barrier Reef, and throughout S.E. Asia; Indo-W. Pacific; to 70 cm. ★★★

9 BAR-CHEEKED CORAL TROUT
Plectropomus maculatus (Bloch, 1790)
Inhabits coral reefs; similar to 7 above, but has fewer more widely-spaced spots and those on head are elongate in shape, also known as Coral cod; found throughout the region; Indo-Australian Archipelago; to 70 cm and 6 kg. ★★★

10 VERMICULAR COD
Plectropomus oligocanthus (Bleeker, 1854)
Inhabits offshore coral reefs; distinguished by bright pinkish-red colour, blue spots (some of which are elongated) on body and fins, and relatively tall dorsal (posterior half) and anal fins; offshore reefs of N.W. Australia, Great Barrier Reef, and throughout S.E. Asia; mainly W. Pacific; to 56 cm. ★★★

11 CORONATION TROUT
Variola louti (Forsskål, 1775)
Inhabits inshore coral reefs and deeper offshore reefs to 100 m; distinguished by bright colour pattern and distinct lunar-shaped tail; *V. albimarginata* (not shown) is a similar species occuring in the region, but has a narrow white margin on the tail instead of a relatively broad yellow one; N.W. Australia, Great Barrier Reef, and throughout S.E. Asia; Indo-C. Pacific; known to cause ciguatera poisoning in some areas; to 80 cm. ★★★

12 QUEENSLAND GROPER
Epinephelus lanceolatus (Bloch, 1790)
Inhabits coral reefs and rocky areas, often in the vicinity of caves; mainly distinguished by its huge size, is one of the largest bony fishes; placed in the genus *Promicrops* by some authors; N.W. Australia, Great Barrier Reef, and throughout S.E. Asia; Indo-W. Pacific; to 270 cm. ★★★

GIANTS OF THE REEF

Although most of the serranid fishes featured on Plates 22-26 have a maximum size well under one metre, three of the species featured on this plate are giants by comparison. The Queensland Grouper (12) holds the distinction of being the largest of all bony fishes occurring on coral reefs, which excludes sharks and rays. There are reports of particularly huge individuals weighing as much as 400 kg and measuring nearly three m. Although its common name suggests a limited distribution, it is actually found over a huge area of the tropical Indo-Pacific extending from East Africa to the islands of the central Pacific Ocean. These monsters are usually just curious towards divers. They either approach at close range for a brief moment, or may spend several minutes following them from a safe distance. Although the details are vague and none are fully documented, there are a few reports of fatal attacks on humans. One precaution that divers must take is to never try and hand-feed large individuals. At least one person learned this lesson the hard way on Australia's Great Barrier Reef. A young man offered a struggling fish he had just speared. The Giant Grouper's response was lightning quick - the fish was literally inhaled, along with the man's arm. Luckily he managed to jerk his arm out of the mouth, but in the process it was severely lacerated by the numerous rows of small, sharp teeth. Although not as formidable as the giant Queensland Groper, other large species that grow over one metre include the Estuary Cod (4), Potato Cod (5), and Malabar Groper (Plate 25-1).

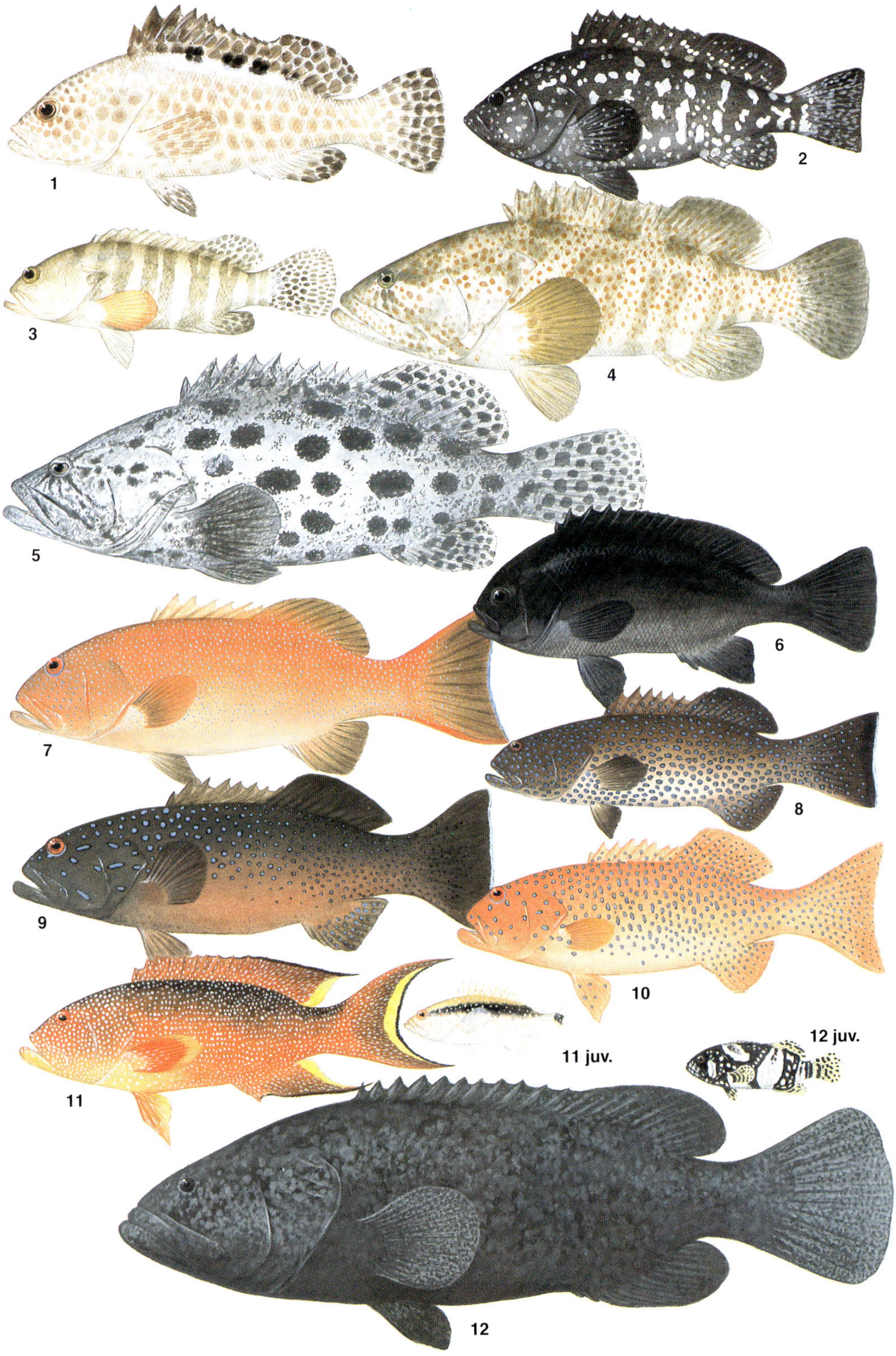
1
2
3
4
5
6
7
8
9
10
11
11 juv.
12 juv.
12

PLATE 25: GROPERS (FAMILY SERRANIDAE)

1 MALABAR GROPER
Epinephelus malabaricus (Bloch & Schneider, 1801)
Inhabits coral or rocky reefs, estuaries, mangrove swamps, tidepools, and sand-mud bottoms to depths of 150 m; similar to *E. coioides* (Pl. 24-4), but has dark brown or black spots with irregular white spots or blotches (orange or reddish brown spots and no white spots on *E. coioides*); found throughout the region; Indo-W. Pacific; to 115 cm and at least 25 kg. ★★★

2 BLACK-DOTTED GROPER
Epinephelus stictus Randall & Allen, 1987
Inhabits continental shelf on mud or sand bottoms at depths between 60-142 m; distinguished by light brown colouration and small dark spots on back; N.W. Australia, Java, and Vietnam to S. Japan; to 41 cm. ★★★

3 YELLOW-SPOTTED GROPER
Epinephelus timorensis Randall & Allen, 1987
Inhabits continental shelf on reefs or over sand-mud bottoms at depths between 73-210; distinguished by irregular bars on body and yellow spots on head; known only from W. Australia, Samoa, and Phoenix Islands; to 32 cm. ★★★

4 PLUMP GROPER
Epinephelus trophis Randall & Allen, 1987
Inhabits continental shelf on sand-mud bottoms to depths of at least 130 m; a plain charcoal-coloured fish without distinguishing marks; known on the basis of only 2 specimens collected from the base of an experimental drilling rig at Dillon Shoals in the Timor Sea; to at least 15 cm. ★★★

5 DUSKYTAIL GROPER
Epinephelis bleekeri (Vaillant, 1877)
Inhabits rocky banks on the continental shelf at depths between 30-104 m; distinguished by dusky tail and reddish, orange, or yellow spots covering head and body; N.W. Australia, N. Territory, and throughout S.E. Asia; Indo-W. Pacific; to 76 cm. ★★★

6 NETFIN GROPER
Epinephelus miliaris (Valenciennes, 1830)
Inhabits coral reefs, but juveniles found in seagrass beds, mangroves, or on mud bottoms; distinguished by pattern of fine spots on body and mesh-like spotting of fins; offshore reefs of N.W. Australia; Indo-W. Pacific, usually around oceanic islands; to 53 cm. ★★★

7 SPECKLED GROPER
Epinephelus magniscuttis Postel, Fourmanoir, and Guézé, 1963
Inhabits continental shelf and deep reefs around islands at depths between 128-300 m; N. Queensland, New Guinea, and Philippines; Indo-W. Pacific; 150 cm and 50 kg. ★★★

8 COMET GROPER
Epinephelus morrhua (Valenciennes, 1833)
Inhabits deep reefs of continental shelf and oceanic islands between depths of 80-370 m; distinguished by broad, curved bands on sides and diagonal bands on head; Queensland and a few scattered localities in Indo-Malay Archipelago; Indo-C. Pacific; to 80 cm. ★★★

9 SMALL-SCALED GROPER
Epinephelus polylepis Randall & Heemstra, 1991
Inhabits continental shelf on flat sand-rubble bottoms at depths between 30-100 m; distinguished by network of small close-set dark brown spots and several faint dark bars on side; N.W. Australia, Indian Ocean; to at least 61 cm. ★★★

10 SPECKLED-FIN ROCKCOD
Epinephelus ongus (Bloch, 1790)
Inhabits shallow coral reefs; distinguished by general brown coloration and pattern of undulating lines on sides and large white blotches; *E. summana* is a synonym; found throughout the region; Indo-W. Pacific; to 40 cm. ★★★

11 MAORI GROPER
Epinephelus undulostriatus (Peters, 1867)
Inhabits coral and rocky reefs, usually at depths between 25-80 m; distinguished by pattern of undulating and slanting fine bands on side; S. Queenlsand and New So. Wales; to 55 cm. ★★★

12 CHINESE FOOTBALLER
Plectropomus laevis (Lacepede, 1802)
Inhabits coral reefs, both in lagoons and on outer reefs; distinguished by white coloration, yellow fins, and prominent black bars or saddles; another commonly observed variety has numerous small dark-edged spots on the head and body with greyish bars and fins; Great Barrier Reef, offshore reefs of W.A. and throughout S.E. Asia; Indo-W. Pacific; to 100 cm. ★★★

RAINBOWS OF THE REEF
(See Plate 27)

Perhaps no other group of coral reef fishes exhibits such a myriad of dazzling colours as the members of the genus *Pseudanthias*, often commonly referred to as Anthias or Fairy Basslets. The vivid neon lustre of these fishes is greatly enhanced by their habit of forming large aggregations. Particularly notable in this respect is the Purple Anthias (*P. tuka*) which forms huge schools adjacent to outer reef dropoffs. Surprisingly, these graceful fishes are members of the family Serranidae which also contains the rockcods and coral trout. They constitute a separate subfamily, Anthiinae, which contains more than 100 species in tropical and subtropical seas. Most of these occur in the Indo-W. Pacific region. Perhaps the most interesting aspect of the biology of these fishes is their reproductive behaviour. They exhibit a harem-type social structure. Each male reigns supreme over a group of females which may include up to 10 or more fishes. Among the females there is a "peck-order" hierarchy of social dominance. If something should happen to the male fish, for example if it is eaten by a predator or experimentally removed, the highest ranking female changes sex and assumes control of the harem. The sex change is relatively rapid beginning within about 3-10 days and is completed within two weeks. The sex change is accompanied by a change in coloration. It is facilitated by the presence of both ovarian and testicular tissue within the gonads. During the initial or female phase of the sexual cycle only the ovarian tissue is functional, the assumption of social dominance triggers a hormonal reaction that stimulates the testicular tissue.

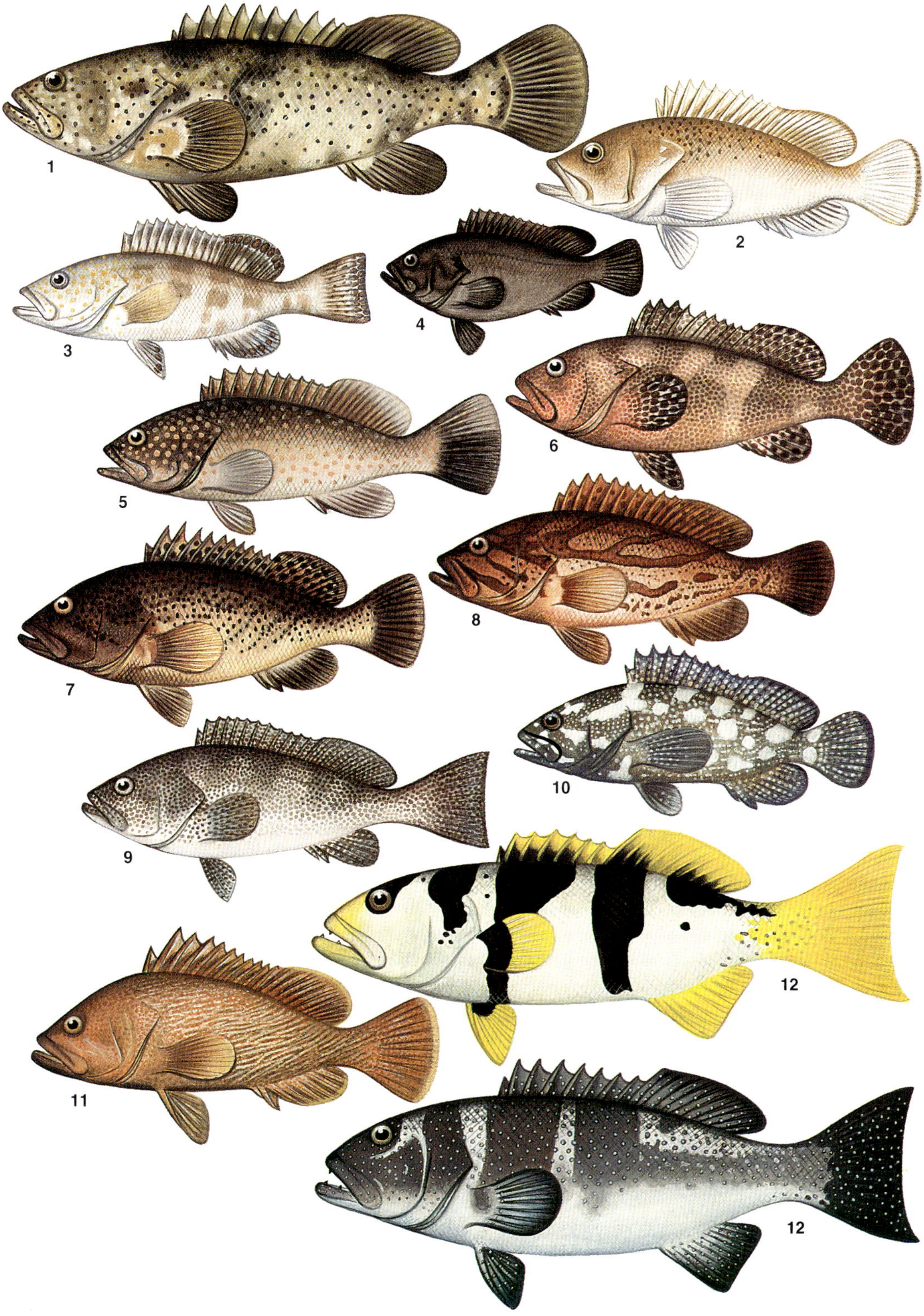
1
2
3
4
5
6
7
8
9
10
11
12
12

PLATE 26: GROPERS (FAMILY SERRANIDAE)

1 THINSPINE ROCKCOD
Gracila albomarginata (Fowler & Bean, 1930)
Inhabits outer slopes in 15-100 m; usually swims a short distance above the bottom; distinguished by large squarish white blotch on upper side and black spot at base of tail; juveniles plain with brilliant red markings on fins; Great Barrier Reef, offshore reefs of W. Australia, and throughout S.E. Asia; Indo-C. Pacific. to 40 cm. ★★★

2 DOTHEAD ROCKCOD
Cephalopholis microprion (Bleeker, 1852)
Inhabits inshore reefs, common on shallow silty reefs; distinguished by plain brown body and numerous small blue spots on head and breast; N. Great Barrier Reef and throughout S.E. Asia; W. Pacific to Andaman Sea; to 23 cm. ★★

3 GARISH ROCKCOD
Cephalopholis igarashiensis Katayama, 1957
Inhabits deeper reef areas, generally between 60-250 m; distinguished by bright pattern of yellow, red, and orange bars; S.E. Asia; W. Pacific; to 43 cm. ★★★

4 STRAWBERRY ROCKCOD
Cephalopholis spiloparaea (Valenciennes, 1828)
Inhabits steep outer reef slopes, usually in about 15-100 m depth; appears plain grey brown underwater, but is actually bright red orange with diagonal white markings near edge of tail; Great Barrier Reef, offshore reefs of W. Australia, and throughout S.E. Asia; Indo-C. Pacific; to 22 cm.

5 SIX-BLOTCH ROCKCOD
Cephalopholis sexmaculata (Rüppell, 1830)
Inhabits caves and ledges, usually seen on outer reef slopes below 10 m depth; distinguished by combination of bright red colouration, brilliant blue spots and lines, and series of dark bars on side that are darkest on upper back; Great Barrier Reef, offshore reefs of W. Australia, and throughout S.E. Asia; Indo-C. Pacific; to 47 cm. ★★★

6 BLUE-SPOTTED ROCKCOD
Cephalopholis cyanostigma (Valenciennes, 1828)
Inhabits coral reefs, often in lagoons or seagrass beds; distinguished by brown or orange-brown colouration with numerous blue spots on head, body, and fins, a white "halo" surrounding each spot on body; juveniles are plain grey brown with yellow fins; Great Barrier Reef, offshore reefs of W. Australia, and S.E. Asia; W. Pacific; to 35 cm. ★★★

7 TROUT COD
Epinephelus maculatus (Bloch, 1790)
Inhabits coral reefs, usually seen around coral bommies in lagoons; distinguished by overall dark grey colouration, pattern of dark spotting, and white saddles on forehead, middle of dorsal fin, and sometimes on upper tail base; Great Barrier Reef, offshore reefs of W. Australia, and throughout S.E. Asia; W. Pacific to Samoa and Marshall Islands; to 50 cm. ★★★

8 BLUE MAORI COD
Epinephelus cyanopodus (Richardson, 1846)
Inhabits sandy areas near coral reefs, usually in protected lagoons; distinguished by general pale colouration and numerous small dark spots on head, body, and fins; juveniles mainly yellow with blue-grey wash on head and front of body; Great Barrier Reef, offshore reefs of W. Australia, and throughout S.E. Asia; Indo-C. Pacific; to 100 cm. ★★★

9 HEXAGON ROCKCOD
Epinephelus hexagonatus (Bloch & Schneider, 1801)
Inhabits coral reefs, usually found in exposed outer reef areas in shallow water; distinguished by numerous brown spots separated by pale hexagonal "wire-netting" pattern; Great Barrier Reef, offshore reefs of W. Australia, and throughout S.E. Asia; Indo-C. Pacific; to 30 cm. ★★★

10 LARGE-SPOTTED ROCKCOD
Epinephelus macrospilos (Bleeker, 1855)
Inhabits coral reefs to depths of at least 44 m; distinguished by pattern of dark brown spots, which are very large and roughly hexagonal-shaped in juveniles and subadults, but become more numerous, smaller, and rounded in adults; Great Barrier Reef, offshore reefs of W. Australia, and throughout S.E. Asia; Indo-C. Pacific; to 43 cm.

11 REEF COD
Epinephelus tauvina (Forsskål, 1775)
Inhabits coral reefs in clear water; distinguished by pattern of brown spots on a whitish background; similar to **10**, but spots are larger and more widely spaced; Great Barrier Reef, northern W. Australia, and throughout S.E. Asia; Indo-C Pacific; to 70 cm.

PLATE 27

1 PRINCESS ANTHIAS
Pseudanthias smithvanizi (Randall & Lubbock, 1981)
Inhabits outer reef slopes, usually seen below depths of 25-30 m; occurs in aggregations; distinguished by small size, swollen upper lip (males), prolonged 3rd dorsal fin spine (males), numerous yellow spots on upper half of body, and strongly forked tail; Great Barrier Reef, offshore reefs of W. Australia, and throughout S.E. Asia; Indo-C. Pacific; to 9.5 cm.

2 LORI'S ANTHIAS
Pseudanthias lori (Lubbock & Randall, 1976)
Inhabits outer reef slopes, usually seen in aggregations at depths between 25-60 m; distinguished by bright red bars on back and broad red band on tail base; Great Barrier Reef, offshore reefs of W. Australia, and throughout S.E. Asia; C.-W. Pacific and E. Indian Ocean; to 12 cm.

3 THREADFIN ANTHIAS
Pseudanthias huchtii (Bleeker, 1857)
Inhabits coral reefs; a common inhabitant of lagoons, passes, and upper edge of outer reef slopes; similar to **4** below, but less brilliantly coloured, males have reddish rather than orange band across cheek and lack large dark spot on pectoral fin, females are greenish yellow rather than orange; Great Barrier Reef and throughout S.E. Asia; Philippines to N.E. Australia and Vanuatu; to 12 cm.

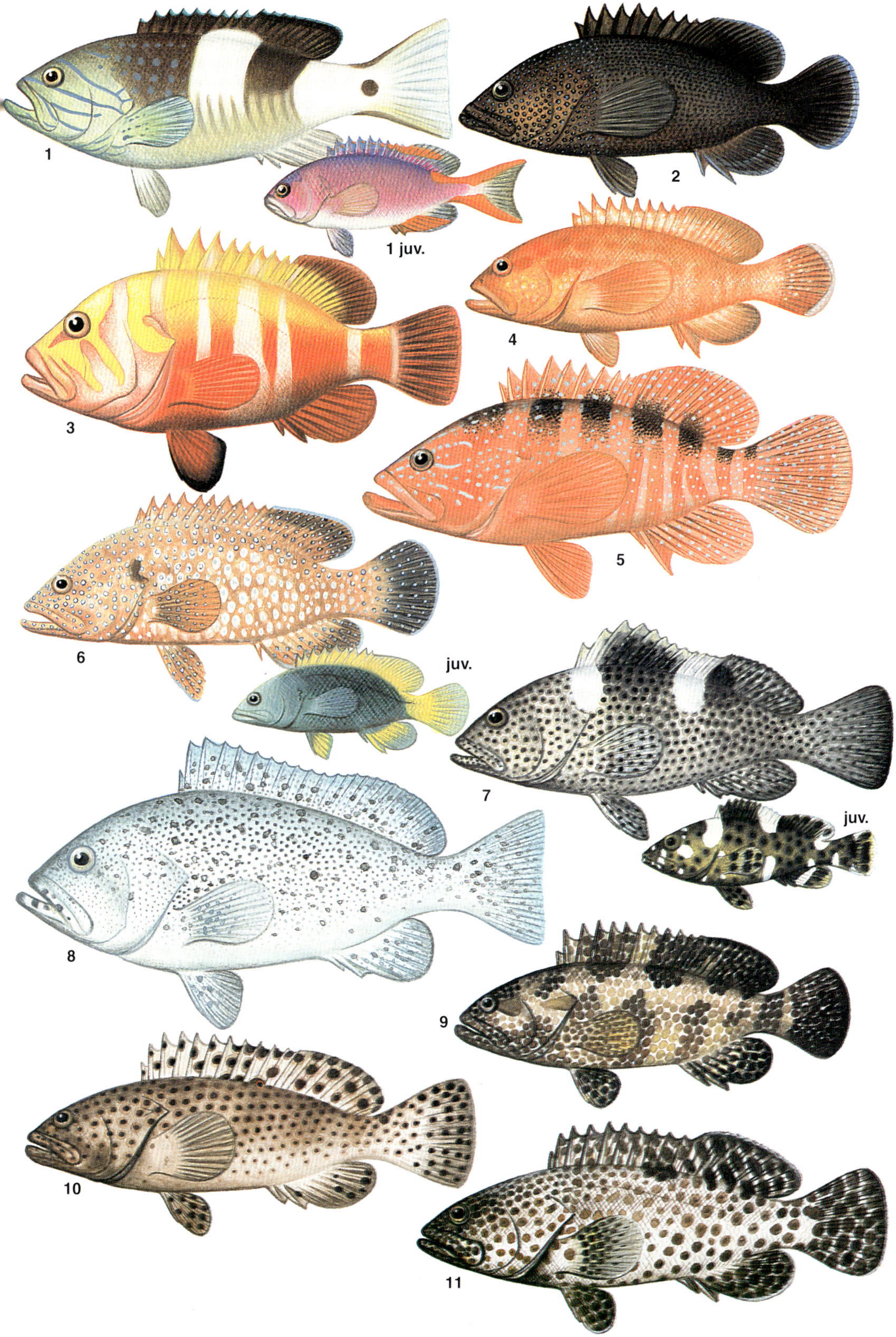
1
2
1 juv.
3
4
5
6
juv.
7
juv.
8
9
10
11

PLATE 27: ANTHIAS (FAMILY SERRANIDAE; SUBFAMILY ANTHIINAE)

1 PRINCESS ANTHIAS
Pseudanthias smithvanizi (Randall & Lubbock, 1981).
Text on page 90.

2 LORI'S ANTHIAS
Pseudanthias lori (Lubbock & Randall, 1976).
Text on page 90.

3 THREADFIN ANTHIAS
Pseudanthias huchtii (Bleeker, 1857).
Text on page 90.

4 SCALEFIN ANTHIAS
Pseudanthias squamipinnis (Peters, 1855)
Inhabits coral reefs, a common inhabitant of lagoons, passes, and upper edge of outer reef slopes; males distinguished by purple coloration, red spot on pectoral fins, and elongated 3rd dorsal spine, females bright yellow orange with diagonal bar across cheek; Great Barrier Reef, offshore reefs of W.A., and throughout S.E. Asia; Indo-W. Pacific; to 15 cm.

5 YELLOW-LINED ANTHIAS
Pseudanthias luzonensis (Katayama & Masuda, 1983)
Inhabits steep outer reef slopes, usually seen below 25-30 m depth; males distinguished by elongate 3rd dorsal spine, red spot on front half of dorsal fin and somewhat oblique rows of small orange spots on sides, females are pinkish, both males and females have diagonal orange stripe across cheek to pectoral fin base; Philippines and Indonesia to N.E. Australia and New Guinea; to 14.5 cm.

6 REDFIN ANTHIAS
Pseudanthias dispar (Herre, 1955)
Inhabits coral reefs, usually seen on the upper edge of outer reef slopes or passes in 1-15 m depth; forms aggregations which alternately feed on plankton or swarm into crevices in the reef; males distinguished by bright red dorsal fin, swollen protuberance on upper lip, and yellowish hue, females are delicate pinkish orange and always outnumber males; Great Barrier Reef, offshore reefs of W.A., and throughout S.E. Asia; W. and C. Pacific; to 9.5 cm

7 WAITE'S SPLITFIN
Luzonichthys waitei (Fowler, 1931)
Inhabits coral reefs, usually seen in aggregations on outer reef slopes in 10-50 m depth; several similar species in this genus occurring in the region; distinguished from similar coloured Anthias (genus *Pseudanthias*) by two instead of a single dorsal fin; Great Barrier Reef, offshore reefs of W.A., and throughout S.E. Asia; Indo-W. Pacific; to 7 cm.

8 RANDALL'S ANTHIAS
Pseudanthias randalli (Lubbock & Allen, 1978)
Inhabits steep outer reef dropoffs in 20-70 m depth, usually seen in small groups; distinguished by broad stripes of magenta and orange on body and dorsal fin, female is mainly yellow; Indo-Malay Archipelago to Marshall Islands and north to Yaeyama Islands; to 7 cm.

9 AMETHYST ANTHIAS
Pseudanthias pascalus (Jordan & Tanaka, 1927)
Inhabits coral reefs, usually seen in aggregations on outer reef slopes in 5-45 m depth; distinguished by bright purple coloration of both males and females; males have swollen protuberance on upper lip and elongated soft dorsal rays with red colour on outer part of fin; Great Barrier Reef; French Polynesia west to Queensland and in the N. Pacific from the Marshall Islands to S. Japan; to 17 cm.

10 PURPLE ANTHIAS
Pseudanthias tuka (Herre & Montalban, 1927)
Inhabits coral reefs, usually seen in spectacular large aggregations on the upper edge of outer reef slopes or in lagoon passes; distinguished by bright purple coloration; males have swollen protuberance on upper lip and elongated soft dorsal rays, females characterised by yellow stripe along back and yellow margins on tail; also known as Purple Queen; Great Barrier Reef, offshore reefs of W.A., and throughout S.E. Asia; Philippines to N.E. Australia; to 12 cm.

11 BICOLOR ANTHIAS
Pseudanthias bicolor (Randall, 1979)
Inhabits coral reefs, usually seen in deeper lagoon channels or on outer reef slopes below 20 m depth; forms aggregations over isolated coral formations; distinguished by elongate dorsal spines and "two-tone" colour of body which is readily apparent when viewed underwater; male and female is similar; Great Barrier Reef and throughout S.E. Asia; Indo-C. Pacific; to 13 cm.

12 SQUARESPOT ANTHIAS
Pseudanthias pleurotaenia (Bleeker, 1857)
Inhabits steep outer reef slopes, usually found below 20 m depth; males distinguished by large squarish violet area above pectoral fin and prolonged 3rd dorsal spine, females are bright yellow with a pair of violet bands running across cheek and continuing along ventral part of body; Great Barrier Reef and throughout S.E Asia, replaced by the similar *P. sheni* at offshore reefs of W. Australia; W. Pacific to Marshall Islands; to 20 cm.

13 LONGFIN ANTHIAS
Pseudanthias ventralis (Randall, 1979)
Inhabits outer slopes between 26-68 m depth, usually seen in small aggregations; males distinguished by irregular magenta and yellow streaks, females are more uniform in appearance with yellow back and fins; Great Barrier Reef; C. and W. Pacific, usually around islands or oceanic reefs; to 7 cm.

14 STOCKY ANTHIAS
Pseudanthias hypselosoma Bleeker, 1878
Inhabits coral reefs, usually seen on sheltered inshore reefs or in lagoon passes; typical habitat consists of sand bottom or rubble areas where it congregates around isolated coral outcrops or wreckage; male is distinguished by rounded profile of dorsal fin and elongate reddish blotch on anterior half of this fin, females are pinkish red with red tips on the tail; Great Barrier Reef and thoughout S.E. Asia; Samoa to Maldives; to 19 cm.

15 STRIPED ANTHIAS
Pseudanthias fasicatus (Kamohara, 1954)
Inhabits coral reefs, usually on outer slopes between 20-68 m depth; distinguished by yellow body with bright red stripe on middle of side; Great Barrier Reef northward to S. Japan; to 21 cm.

16 TWOSPOT ANTHIAS
Pseudanthias bimaculatus (Smith, 1961)
Inhabits coral reefs, usually seen on coastal reefs or in deep lagoons in about 20-60 m; often forms aggregations around isolated coral bommies; males distinguished by reddish-pink colouration, irregular red lines and bands on sides, and pair of red spots on dorsal fin (one at front and one at rear), females are duller; East Africa to the Maldive Islands and Indonesia (Sumatra, Java, and Bali); to 10 cm.

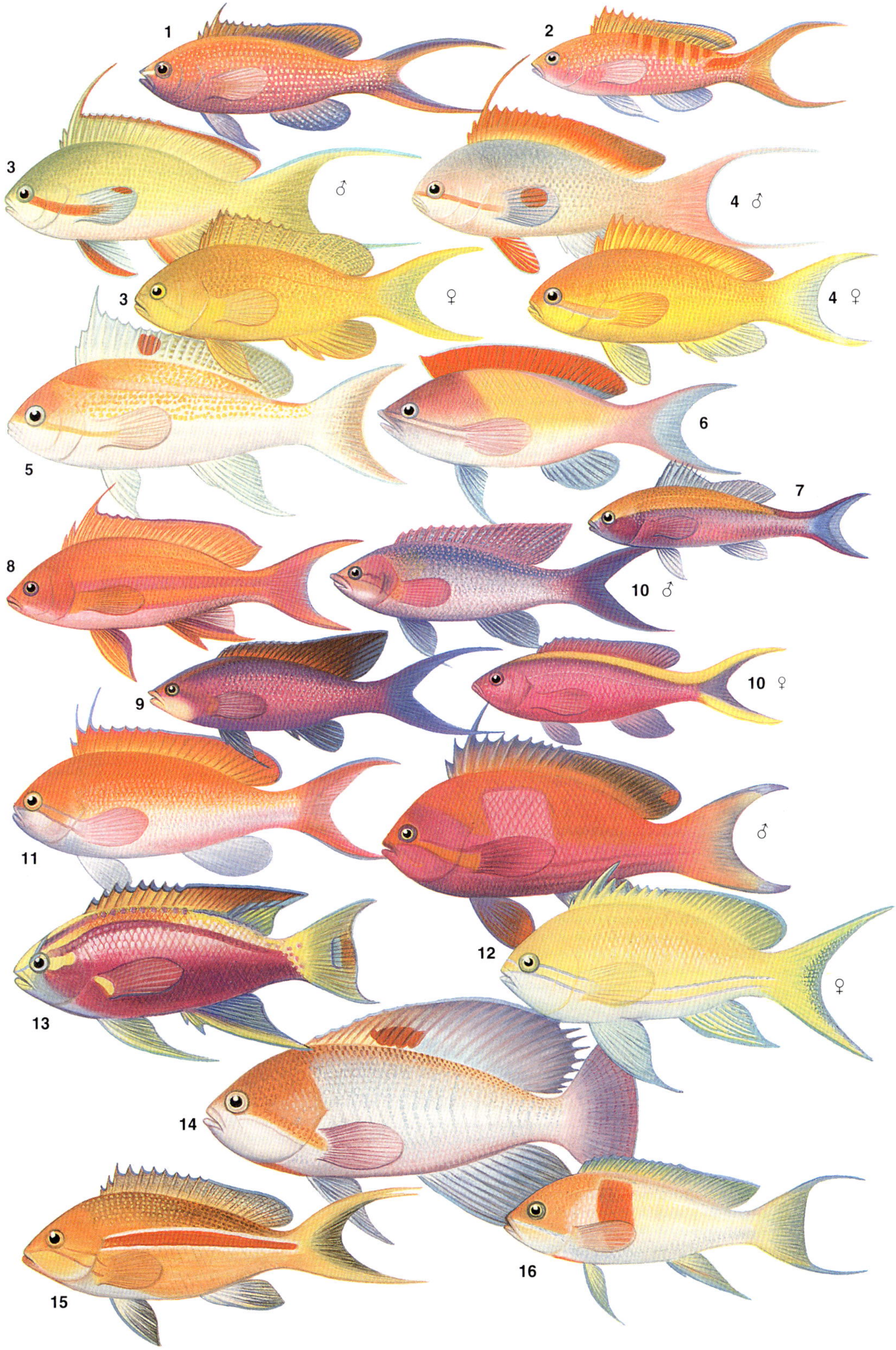
1
2
3
♂
4 ♂
3
♀
4 ♀
5
6
7
8
10 ♂
9
10 ♀
11
♂
12
♀
13
14
15
16

PLATE 28: LONGFINS, DOTTYBACKS AND RELATIVES

1 YELLOW EMPEROR
Diploprion bifasciatum Kuhl & Van Hasselt, 1928
Inhabits coral reefs; distinguished by bar through eye and broad dark bar across middle of body; also known as Two-banded soapfish; found throughout the region; N. Indian Ocean and W. Pacific; to 38 cm. (SERRANIDAE)

2 DEEPSEA FAIRY BASSLET
Pseudanthias rubrizonatus (Randall, 1983)
Inhabits offshore reefs, usually below 25 m depth; distinguished by dark saddle below dorsal spines and pale diagonal stripe across lower part of head; N.W. Australia, Great Barrier Reef and throughout S.E. Asia; Indo-Australian Archipelago; to 12 cm. (SERRANIDAE)

3 LITTLE FAIRY BASSLET
Sacura parva Heemstra & Randall, 1979
Inhabits offshore reefs, usually below 50 m depth; distinguished by elongate third dorsal spine and filament at front of soft dorsal fin; N.W. Australia only; to 12 cm. (SERRANIDAE)

4 RAINFORD'S PERCH
Rainfordia opercularis McCulloch, 1923
Inhabits coral reefs, usually in caves; distinguished by elongate body, flattened head and stripe pattern; N.W. Australia and Great Barrier Reef; N. Australia only; to 15 cm. (SERRANIDAE)

5 SIX-LINED PERCH
Grammistes sexlineatus (Thunberg, 1792)
Inhabits coral reefs, usually in caves or crevices; distinguished by narrow yellow stripes on side, has mucus that is toxic to other fishes; found throughout the region; Indo-C. Pacific; to 27 cm. (SERRANIDAE)

6 FALSE GRAMMA
Pseudogramma polyacantha (Bleeker, 1856)
Inhabits coral reefs, in caves and crevices; distinguished by small size, and dusky brown pattern with faint blotches; found throughout the region; Indo-C. Pacific; to 7 cm. (SERRANIDAE)

7 BANDED LONGFIN
Belonepterygium fasciolatum (Ogilby, 1889)
Inhabits coral reefs, in caves and crevices; distinguished by dark stripe through eye, narrow dark bars on side, and long thread-like pelvic fins; N.W. Australia and Great Barrier Reef; N. Australia only; to 5 cm. (ACANTHOCLINIDAE)

8 COMET
Calloplesiops altivelis (Steindachner, 1903)
Inhabits coral reefs, in caves and crevices; distinguished by white spots on head, body, and fins, and pale-edged black spot at rear of dorsal fin; N.W. Australia, Great Barrier Reef and throughout S.E. Asia; Indo-W. Pacific; to 16 cm. (PLESIOPIDAE)

9 RED-TIPPED LONGFIN
Plesiops verecundus Mooi, 1995
Inhabits coral reefs, in caves and crevices; distinguished by deeply incised dorsal fin profile and red margin on dorsal fin; N.W. Australia, Great Barrier Reef and throughout S.E. Asia; Indo-W. Pacific; to 8 cm. (PLESIOPIDAE)

10 LINED DOTTYBACK
Labracinus lineatus (Castelnau, 1875)
Inhabits coral reefs and rocky areas usually in crevices; distinguished by relatively large size and numerous narrow stripes on body and fins; also known as Lined cichlops; W. Australia only, between Jurien Bay and Broome; to 25 cm. (PSEUDOCHROMIDAE)

11 BROWN DOTTYBACK
Pseudochromis fuscus Müller & Troschel, 1849
Inhabits the vicinity of coral reefs; has two distinct colour phases, one entirely yellow and the other dusky brown or purplish; found throughout the region; E. Indian Ocean and W. Pacific; to 8 cm. (PSEUDOCHROMIDAE)

12 MARSHALL DOTTYBACK
Pseudochromis marshallensis Schultz, 1953
Inhabits coral reefs; distinguished by scales on sides having orange centres; N.W. Australia and throughout S.E. Asia; mainly W. Pacific; to 8 cm. (PSEUDOCHROMIDAE)

13 LONGFIN DOTTYBACK
Assiculus punctatus Richardson, 1846
Inhabits sandy areas with occasional coral or rocky outcrops; distinguished by dark bluish-brown colour and relatively tall dorsal and anal fins; also the dorsal fin begins farther forward than in similar species; N. Australia only, between Shark Bay, W. Australia and Gulf of Carpentaria; to 7.5 cm. (PSEUDOCHROMIDAE)

14 SPOTTED DOTTYBACK
Pseudochromis quinquedentatus McCulloch, 1926
Inhabits sand or rubble areas with occasional outcrops that serve as shelter; distinguished by small dark spots arranged in longitudinal rows on side; N. Australia only, between Murion Islands and Capricorn Group, Queensland; to 9 cm. (PSEUDOCHROMIDAE)

15 YELLOWHEAD DOTTYBACK
Pseudochromis cyanotaenia Bleeker, 1857
Inhabits coral reef crevices; male distinguished by yellowish area on head and breast with broad pale margin on tail and dorsal fin, females by yellow-edged red tail; often mis-identified as *P. tapeinosoma*; N.W. Australia, Great Barrier Reef and throughout S.E. Asia; E. Indian Ocean and W Pacific; to 6 cm. (PSEUDOCHROMIDAE)

16 YELLOWFIN DOTTYBACK
Pseudochromis wilsoni (Whitley, 1929)
Inhabits coral reef crevices and rubble areas; males distinguished by plain purplish colour and yellow-orange iris, females by yellowish dorsal fin and yellow edges on tail; N. Australia only, between Port Denison, W. Australia and Bargara, Queensland; to 8 cm. (PSEUDO-CHROMIDAE)

17 ROSE DEVILFISH
Pseudoplesiops rosae Schultz, 1943
Inhabits coral reef crevices; distinguished by small size, yellow-brown colour, large eye, and thread-like pelvic fins; N.W. Australia, Great Barrier Reef and throughout S.E. Asia; Indo-C Pacific; to 4 cm. (PSEUDOCHROMIDAE)

SLIMY SOAPFISHES

Most of the species illustrated on this plate live in caves and crevices. The Sixlined Perch (5) and its relatives (Pl. 29-2 and 3) are called soapfishes because of their slimy mucus coat. It is extremely bitter to the taste and probably offers some measure of protection from larger predatory species. Aquarium observations reveal it is an unpalatable mouthful when offered to a hungry Red Firefish (Pl. 19-6) - once tasted it is quickly spat out.

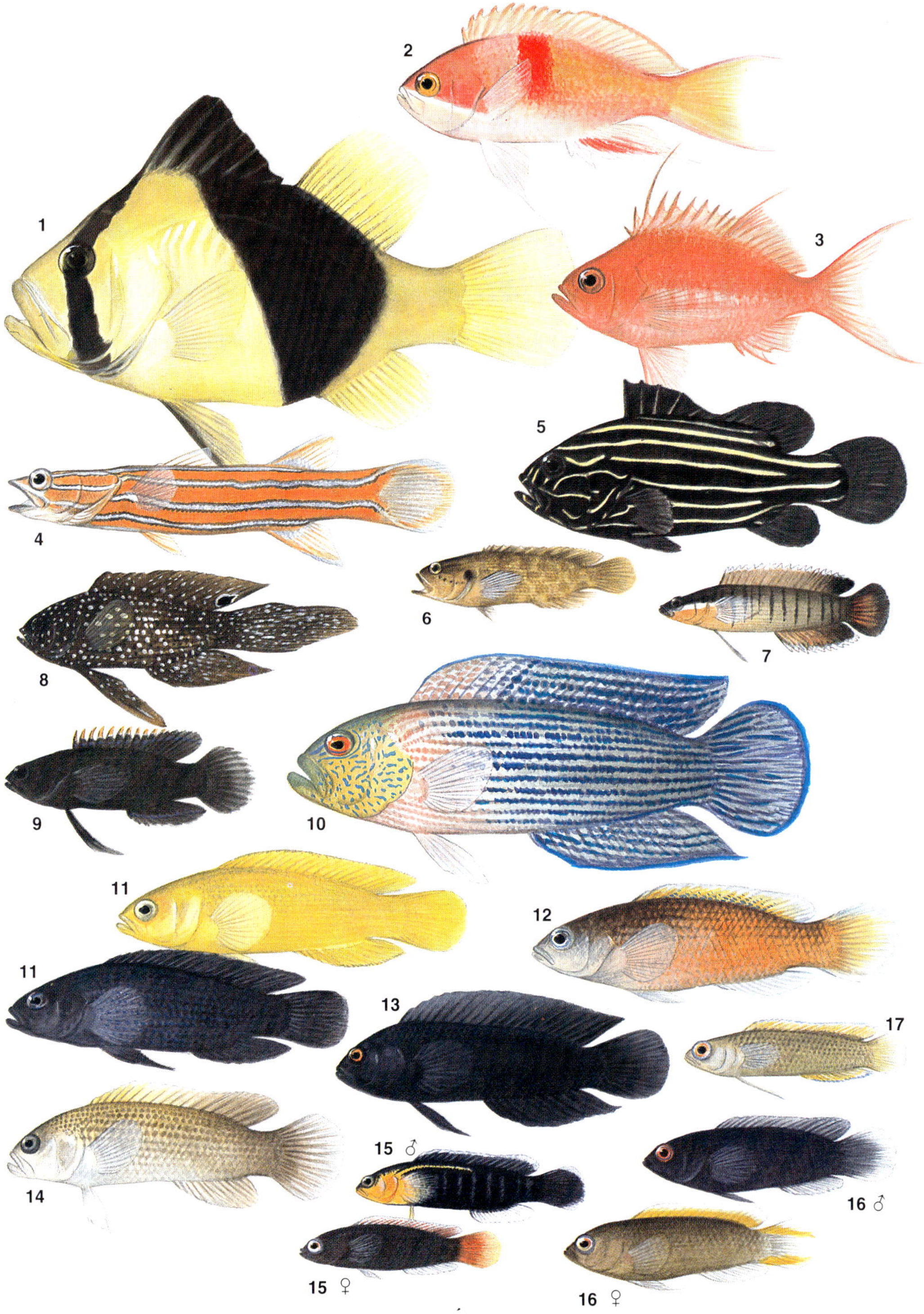
2
1
3
5
4
6
7
8
9
10
11
12
11
13
17
15 ♂
14
16 ♂
15 ♀
16 ♀

PLATE 29: GROPERS (SERRANIDAE) AND DOTTYBACKS (PSEUDOCHROMIDAE)

1 HAWK ANTHIAS
Serranocirrhitus latus Watanabe, 1949
Inhabits coral reefs, usually in caves or beneath ledges on outer reef slopes; distinguished by deep pink colour with yellow bands on head and yellow scale edges; N. Great Barrier Reef, Indonesia, and New Guinea; W. Pacific; to 13 cm. (SERRANIDAE)

2 ARROWHEAD SOAPFISH
Belonoperca chabanaudi Fowler & Bean, 1930
Inhabits outer reef slopes, usually seen in shadows of caves and ledges; distinguished by slender shape, dark spot on first dorsal fin, and bright yellow saddle on tail base; Great Barrier Reef and throughout S.E. Asia; Indo-C. Pacific; to 15 cm. (SERRANIDAE)

3 SPOTTED SOAPFISH
Pogonoperca punctata (Valenciennes, 1830)
Inhabits coral reefs, on outer slopes between 25-150 m; distinguished by skin flap on chin, numerous white spots, and black saddles on back; throughout S.E. Asia; Indo-W. Pacific; to 33 cm. (SERRANIDAE)

4 BLUE DEVILFISH
Assessor macneilli Whitley, 1935
Inhabits caves and ledges, often swims upside down; distinguished by slender shape, dark blue colour, and forked tail; Great Barrier Reef and New Caledonia; to 6 cm. (PLESIOPIDAE)

5 YELLOW DEVILFISH
Assessor flavissimus Allen & Kuiter, 1976
Inhabits caves and ledges, often swims upside down; distinguished by slender shape, yellow colour, and forked tail; N. Great Barrier Reef; to 5.5 cm. (PLESIOPIDAE)

6 FIRETAIL DOTTYBACK
Labracinus cyclophthalmus (Müller & Troschel, 1849)
Inhabits coral reefs in 1-15 m depth; colour varies from bright red to grey-green or with isolated red patch on side; N.W. Australia (Kimberley district and Ashmore Reef) and throughout S.E. Asia; W. Pacific; to 20 cm.

7 ROYAL DOTTYBACK
Pseudochromis paccagnellae Axelrod, 1973
Inhabits coral reef drop-offs, usually below 15 m depth; distinguished by brilliant bicolour pattern; Great Barrier Reef, offshore reefs of N.W. Australia, and Indonesia to New Guinea; W. Pacific; to 7 cm.

8 MAGENTA DOTTYBACK
Pseudochromis porphyreus Lubbock & Goldman, 1974
Inhabits coral reef drop-offs, usually below 10 m depth; distinguished by brilliant magenta colour; E. Indonesia and Philippines; W. Pacific; to 6 cm.

9 PURPLETOP DOTTYBACK
Pseudochromis diadema Lubbock & Randall, 1978
Inhabits coral reef drop-offs, usually below 10 m depth; distinguished by yellow colour with bright magenta area on back; Malaysia, Borneo, and Philippines; to 6 cm.

10 SPLENDID DOTTYBACK
Pseudochromis splendens Fowler, 1931
Inhabits coral reefs in 3-30 m depth, usually seen with sponges; distinguished by close-set rows of yellow-orange spots, white snout, and dark bar through eye; Indonesia (Lesser Sunda Islands and Banda Sea); to 10 cm.

11 STEENE'S DOTTYBACK
Pseudochromis steenei Gill & Randall, 1993
Inhabits sand-rubble slopes, around small rocky outcrops with crinoids; has narrow white bar behind eye; orange head of male is lacking in female, which is mainly dark with yellow tail; Indonesia (Bali to Flores); to 12 cm.

12 LONG-FINNED DOTTYBACK
Pseudochromis polynemus Fowler, 1931
Inhabits coral reef slope with abundant crevices and sponges; distinguished by light brown colour with yellow dot on each scale and long pelvic fins with orange spot at base; N. Indonesia and Philippines; to 12 cm.

13 HOWSON'S DOTTYBACK
Pseudochromis howsoni Allen, 1995
Inhabits sand-rubble bottoms around low rocky outcrops; found in pairs; similar to **11** above, but lacks narrow white bar behind eye and female is entirely dark brown; Timor Sea (Ashmore Reef); to 10 cm.

14 MOORE'S DOTTYBACK
Pseudochromis moorei Fowler, 1931
Inhabits sand-rubble bottoms around low rocky outcrops; found in pairs; female entirely dark brown and male orange with dark spot on each scale of upper half of body and dark "ear" spot; Philippines; to 10 cm.

15 IMPERIAL DOTTYBACK
Pseudochromis sp.
Inhabits steep outer reef slopes; distinguished by blue and yellow colour scheme and red spot on pelvic fins; (Malaysia and Borneo); to 8 cm.

16 DOUBLE-STRIPED DOTTYBACK
Pseudochromis bitaeniata (Fowler, 1931)
Inhabits coral reefs drop-offs in 5-30 m depth; distinguished by pair of broad dark brown stripes separated by white; Great Barrier Reef, offshore reefs of N.W. Australia, and throughout S.E. Asia; to 8 cm.

17 OBLIQUE-LINED DOTTYBACK
Cypho purpurescens (De Vis, 1884)
Inhabits coral reef crevices; distinguished by reddish colour with blue scale edges, also "eye-spot" on dorsal fin; Great Barrier Reef; S.W. Pacific; to 7.5 cm.

18 MULTI-COLOURED DOTTYBACK
Ogilbyina novaehollandiae (Steindachner, 1880)
Inhabits coral reefs; female is mainly reddish and male is reddish on head and front of body, but purplish on rear half; S. Great Barrier Reef; to 10 cm.

19 QUEENSLAND DOTTYBACK
Ogilbyina queenslandiae (Saville-Kent, 1893)
Inhabits coral reefs; similar to **18** above, but female with brown bars on back; Great Barrier Reef; to 15 cm.

20 SAILFIN DOTTYBACK
Ogilbyina velifera (Lubbock, 1980)
Inhabits coral reefs; distinguished by overall pale grey colour and lanceolate tail, adult has bluish forehead compared to yellowish of juvenile; Great Barrier Reef; to 12 cm.

21 BLACK-STRIPED DOTTYBACK
Pseudochromis perspicillatus Günther, 1862
Inhabits sand-rubble bottoms around low rocky outcrops; found in pairs; distinguished by white to yellow-brown colour with oblique black stripe; Indonesia and Philippines; to 12 cm.

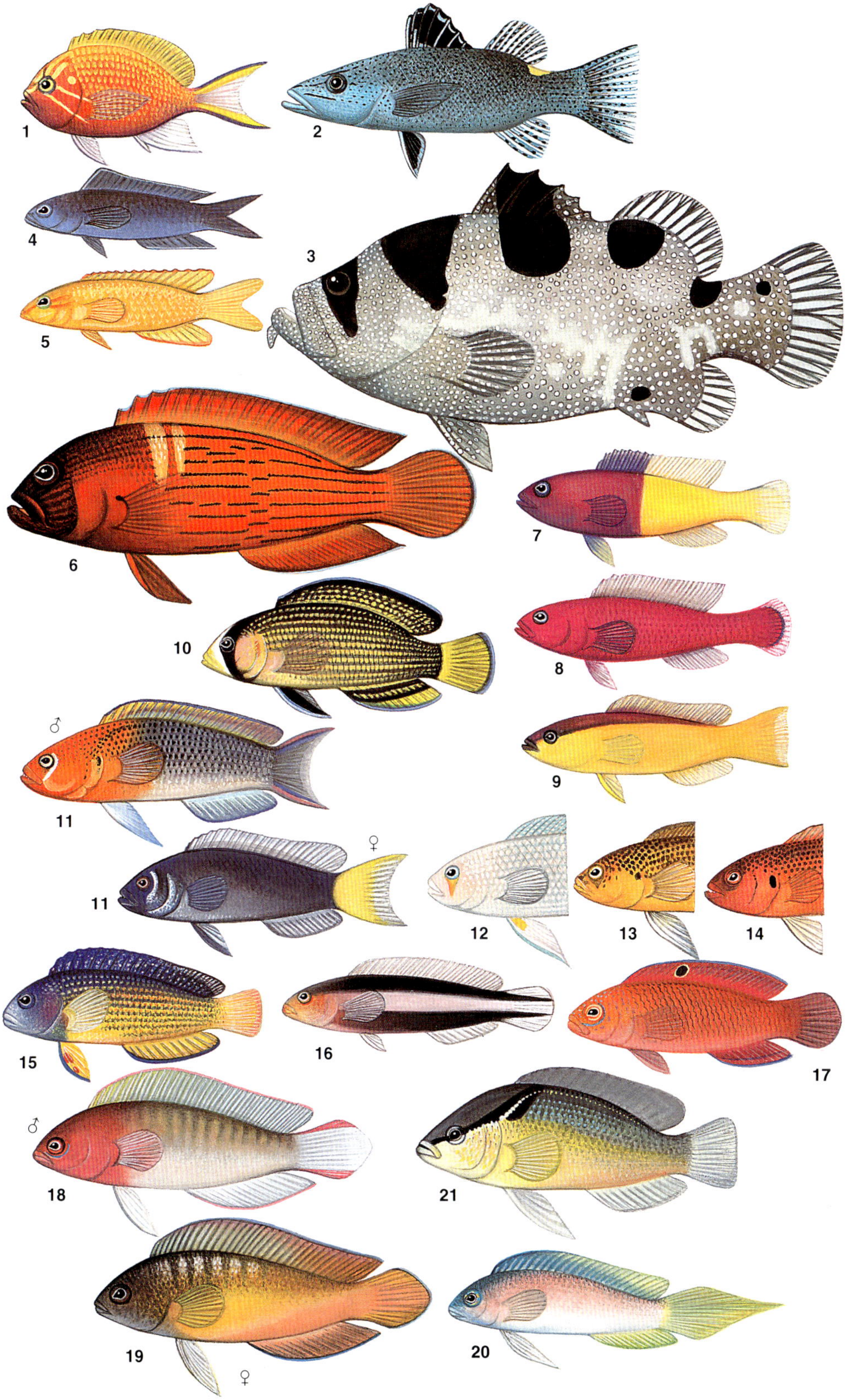

1
2
4
3
5
6
7
10
8
♂
9
11
♀
11
12
13
14
15
16
17
♂
18
21
19
♀
20

PLATE 30: PERCHLETS (CHANDIDAE), GRUNTERS (TERAPONTIDAE), AND WHITINGS (SILLAGINIDAE)

1 BARRAMUNDI
Lates calcarifer (Bloch, 1790)
Inhabits coastal waters, entering estuaries and fresh water; highly esteemed as a food fish and for its angling qualities, distinguished by silvery appearance, deeply notched dorsal fin, 'humped' back, and glassy eye; found at localities throughout the region which have extensive mangrove coasts, for example N. Australia and S. New Guinea; N. Indian Ocean and W. Pacific, to 150 cm and at least 20 kg. (CENTROPOMIDAE) ★★★★

2 SAND BASS
Psammoperca waigiensis (Cuvier, 1828)
Inhabits rocky or coral reefs, frequently in weedy areas; distinguished by barramundi-like appearance and glassy eyes; colour ranges from light silvery grey to very dark brown; found throughout the region; N. Indian Ocean and W. Pacific; to 47 cm. (CENTROPOMIDAE) ★★

3 SPIKY BASS
Hypopterus macropterus (Günther, 1859)
Inhabits sand-weed areas; has similar shape to **2** above, although a much smaller fish and deeper-bodied; colour is strongly mottled; W. Australia only, between Jurien Bay and Onslow; to 14 cm. (CENTROPOMIDAE)

4 SAILFIN PERCHLET
Ambassis interruptus Bleeker, 1852
Inhabits brackish bays and estuaries, also mangrove-lined tidal creeks; distinguished by semi-transparent appearance, tall dorsal fin, and white tips on pelvic fins and front of anal fin; found throughout the region; Andaman Islands and Indo-Australian Archipelago; to 10 cm.

5 SCALLOPED PERCHLET
Ambassis nalua (Hamilton, 1822)
Inhabits brackish bays and estuaries, similar to **4** above, but has slight hump on snout, dorsal fin is slightly shorter, and lacks white tip on pelvic and anal fins; found throughout the region; E. Indian Ocean and Indo-Australian Archipelago; to 12 cm.

6 TELKARA PERCHLET
Ambassis vachelli Richardson, 1846
Inhabits brackish bays, estuaries and tidal creeks similar to **4** and **5** above, but has more slender body and lower dorsal fin; frequently misidentified as *A. dussumieri*; found throughout the region; E. Indian Ocean and Indo-Australian Archipelago; to 7 cm.

7 DEEPSEA JEWFISH
Glaucosoma burgeri Richardson, 1845
Inhabits trawling grounds; distinguished by silvery appearance, large eye, and single dorsal fin that is elevated on rear portion, frequently with narrow stripes which may disappear in larger fish; closely related to the jewfish (*G. hebraicum*) of S.W. Australia; N.W. Australia and scattered localities in S.E. Asia; mainly W. Pacific; to 45 cm. (GLAUCOSOMIDAE) ★★★

8 THREADFIN PEARL-PERCH
Glaucosoma magnificum (Ogilby, 1915)
Inhabits trawling grounds, similar to **7** above, but has elongate dorsal fin filaments and brown bars on head; N. Australia and S. New Guinea only; to 32 cm. (GLAUCOSOMIDAE) ★★★

9 YELLOWTAIL TRUMPETER
Amniataba caudovittatus (Richardson, 1845)
Inhabits estuaries over sand-weed bottoms; distinguished by yellow fins and conspicuous stripes and spots on tail; also known as Yellowtail grunter and Yellow-tailed perch; W. and N. Australia and S. New Guinea; to 28 cm. ★★

10 TRUMPETER
Pelates quadrilineatus (Bloch, 1790)
Inhabits coastal waters, entering estuaries; distinguished by 6-8 straight dark stripes on side and none on tail; found throughout the region; Indo-W. Pacific; to 20cm. ★

11 CRESCENT PERCH
Terapon jarbua (Forsskål, 1775)
Inhabits coastal waters, entering estuaries and lower reaches of freshwater streams; distinguished by 3 or 4 curved dark stripes on side; found throughout the region; Indo-W. Pacific; to 32 cm. ★

12 THREE-LINED GRUNTE R
Terapon puta Cuvier, 1829
Inhabits coastal waters, entering brackish estuaries; distinguished by 3 or 4 straight dark stripes on side also known as Spiny-cheeked grunter; found throughout the region; N. Indian Ocean and Indo-Australian Archipelago; to 16 cm. ★

13 BANDED GRUNTER
Terapon theraps Cuvier, 1829
Inhabits coastal waters; similar to **12** above, but has wider stripes and scales much larger (46-56 in lateral line versus 70-85), also is deeper bodied; found throughout the region; Indo-W. Pacific; to 28 cm. ★

14 GOLDEN-LINED WHITING
Sillago analis Whitley, 1943
Inhabits sandy bottoms near shore; distinguished by golden-silver to golden-yellow stripe along middle of side, also known as Rough-scale whiting; found throughout the region; N. Australia and S. New Guinea; to 45 cm. ★★★

15 STOUT WHITING
Sillago robusta Stead, 1908
Inhabits sandy bottoms near shore; distinguished by yellow blotch on cheek and silvery stripe on middle of side; most of Australia except southern coast; to 30 cm. ★★★

16 TRUMPETER WHITING
Sillago maculata burrus Richardson, 1842
Inhabits sandy bottoms near shore, distinguished by irregular dark blotches on side; N W Australia, New Guinea, and Indonesia; to 30 cm. ★★★

17 NORTHERN WHITING
Sillago sihama (Forsskål, 1775)
Inhabits sandy bottoms near shore, a plain uniform coloured whiting without distinguishing marks; also known as Sand smelt; found throughout the region; Indo-W Pacific; to 31 cm. ★★★

18 WESTERN SCHOOL WHITING
Sillago vittata McKay, 1985
Inhabits sandy bottoms near shore; distinguished by dark diagonal lines on back; W Australia only, between Rottnest Island and Pt. Maud; to 30 cm. ★★★

BARRA FACTS
The Barramundi (1) spawns in estuaries and coastal shallows between September and March. Growth of the young is rapid and sexual maturity is reached in 3-4 years at a length of 55-70 cm.

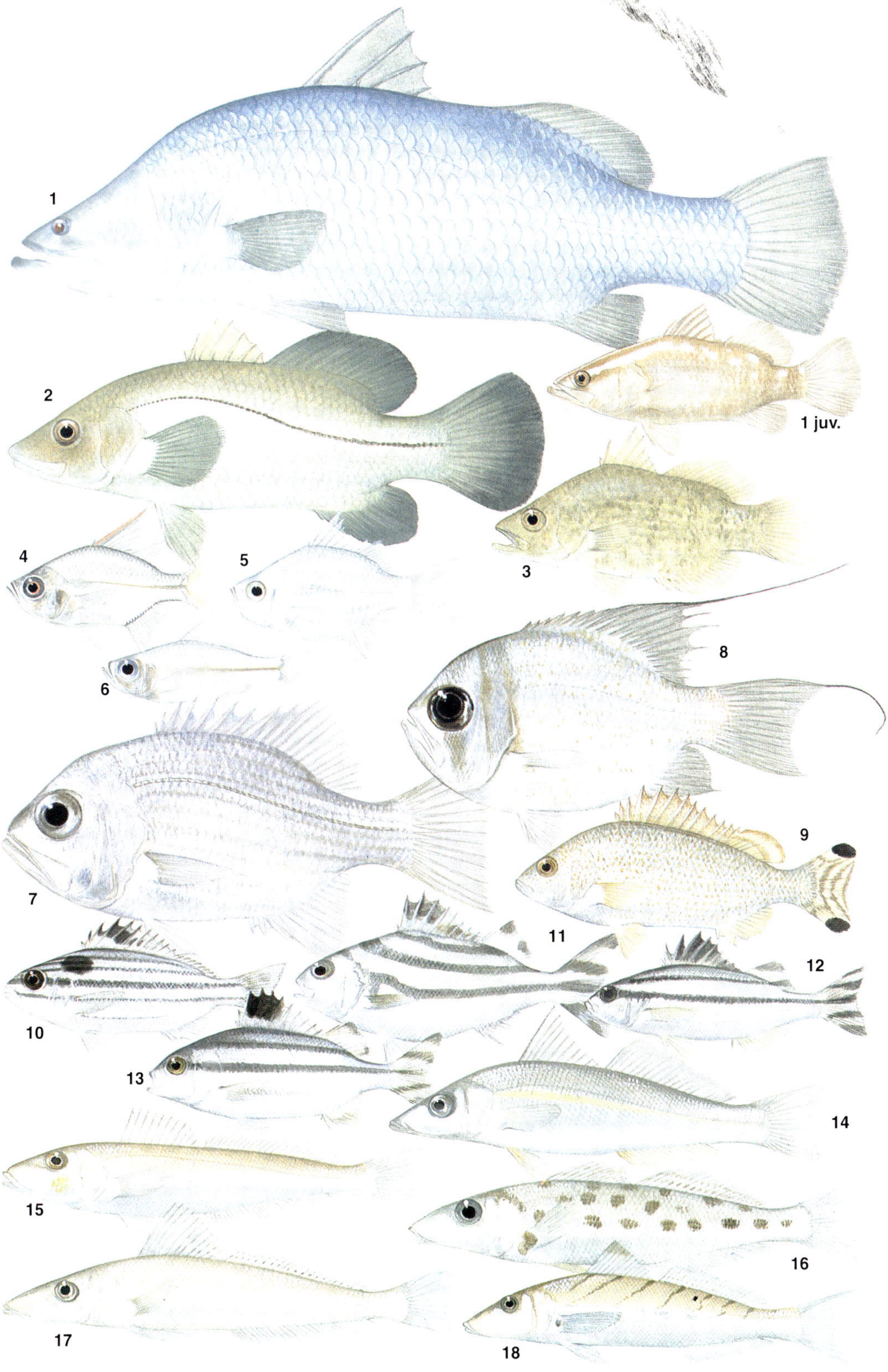
1
2
1 juv.
4
5
3
8
6
7
9
11
12
10
13
14
15
16
17
18

PLATE 31: BIGEYES AND CARDINALFISHES

1 WHITE-BARRED BIGEYE
Pristigenys niphonia (Cuvier, 1829)
Inhabits deeper offshore reefs and trawling grounds; distinguished by white bars on sides; found throughout the region; Indo-W. Pacific; to 26 cm. (PRIACANTHIDAE) ★★★

2 DUSKYFIN BIGEYE
Heteropriacanthus cruentatus (Lacepède, 1801)
Inhabits coral reefs, usually in caves except at night when seen in the open; similar to **4** below, but has lighter pelvic fin, spots (sometimes faint) on dorsal, anal and tail fins, and tail not crescentic; spot pattern shown is not always evident; also known as Glass bigeye; N.W. Australia, Great Barrier Reef, and throughout S.E. Asia; Indo-C. Pacific; to 30 cm. (PRIACANTHIDAE) ★★★

3 RED BIGEYE
Priacanthus macracanthus Cuvier, 1829
Inhabits inshore and offshore reefs; distinguished by relatively short dorsal and anal fins which are distinctly spotted; similar to **7** below, which lacks distinct spotting; also known as Large-spined bigeye; N.W. Australia, Great Barrier Reef, and throughout S.E. Asia; mainly W. edge of Pacific; to 35 cm. (PRIACANTHIDAE) ★★★

4 LUNAR-TAILED BIGEYE
Priacanthus hamrur (Forsskål, 1775)
Inhabits coral reefs and rocky bottoms; similar to **2** above, but has darker pelvic fins, crescentic tail, and lacks spotting on fins except a dark spot present at base of pelvic fins; N.W. Australia, Great Barrier Reef, and throughout S.E. Asia; Indo-C. Pacific: to 40 cm. (PRIACANTHIDAE) ★★★

5 THREADFIN BIGEYE
Priacanthus tayenus Richardson, 1846
Inhabits coral reefs and rocky bottoms to at least 200 m depth; distinguished by relatively tall dorsal and anal fins and filamentous tips on tail; N.W. Australia, Great Barrier Reef, and throughout S.E. Asia; N. Indian Ocean and W. edge of Pacific; to 35 cm. (PRIACANTHIDAE) ★★★

6 ROBUST BIGEYE
Priacanthus sagittarius Starnes, 1988
Inhabits rocky bottoms in 60-100 m depth; distinguished by tall dorsal and anal fins, large pelvic fins, and dark membrane between first and second dorsal spines; N.W. Australia, S.E. Asia, and S. New Guinea; to 40 cm. (PRIACANTHIDAE) ★★★

7 DEEPSEA BIGEYE
Priacanthus fitchi Starnes, 1988
Inhabits offshore waters between 150-400 m depth; similar to **3** above, but lacks distinct spotting on fins; N.W. Australia, W. Indonesia, and Philippines; E. Indian Ocean and W. Pacific; to 25 cm. (PRIACANTHIDAE) ★★★

8 LONG-FINNED BIGEYE
Cookeolus boops (Forster, 1801)
Inhabits trawl grounds and rocky bottoms; distinguished by very large pelvic fins and tall dorsal and anal fins; found throughout the region; Indo-C. Pacific; to 60 cm. (PRIACANTHIDAE) ★★★

9 WOLF CARDINALFISH
Cheilodipterus artus Smith, 1961
Inhabits coral reefs, in caves and crevices; distinguished by 8-10 stripes, yellow area with black spot at tail base (which can be "switched" off), and large canine teeth; N.W. Australia, Great Barrier Reef, and throughout S.E. Asia; Indo-C. Pacific; to 12 cm. (APOGONIDAE)

10 EIGHT-LINED CARDINALFISH
Cheilodipterus macrodon (Lacepède, 1801)
Inhabits coral reefs, in caves and crevices; distinguished by striped pattern, lack of yellow on tail base and large canine teeth; N.W. Australia, Great Barrier Reef, and throughout S.E. Asia; Indo-C. Pacific; to 22 cm. (APOGONIDAE)

11 FIVE-LINED CARDINALFISH
Cheilodipterus quinquelineatus Cuvier, 1828
Inhabits coral reefs, in caves and crevices; distinguished by 5 stripes on side, spot at tail base, and large canine teeth; found throughout the region; Indo-C. Pacific; to 12 cm. (APOGONIDAE)

12 SAILFIN CARDINALFISH
Pterapogon mirifica (Mees, 1966)
Inhabits coral reef caves and ledges; distinguished by blackish colour, white tail and enlarged fins; W. Australia only, between Ningaloo Reef and Kimberley coast; to 14 cm. (APOGONIDAE)

13 WEED CARDINALFISH
Foa brachygramma (Jenkins, 1903)
Inhabits sand-weed areas, usually under dead coral slabs; distinguished by mottled pattern, incomplete lateral line and presence of teeth on palate; N.W. Australia, Great Barrier Reef, and throughout S.E. Asia; Indo-C. Pacific; to 6 cm. (APOGONIDAE)

14 VARIEGATED CARDINALFISH
Fowleria variegata (Valenciennes, 1832)
Inhabits coral reef crevices; distinguished by spot on gill cover, and irregular spotting on body and fins; found throughout the region; Indo-C. Pacific; to 8 cm. (APOGONIDAE)

15 AURITA CARDINALFISH
Fowleria aurita (Valenciennes, 1831)
Inhabits coral reef crevices; distinguished by spot on gill cover and lack of markings on body and fins; found throughout the region; Indo-C. Pacific; to 8 cm. (APOGONIDAE)

16 STRIPED SIPHONFISH
Siphamia majimai Matsubara & Iwai, 1958
Inhabits coral reefs, usually found among the spines of sea urchins; distinguished by silver stripe on belly and alternating black and white stripes on side or may be entirely blackish; found throughout the region; mainly W. Pacific; to 5 cm. (APOGONIDAE)

17 PINK-BREASTED SIPHONFISH
Siphamia roseigaster (Ramsay & Ogilby, 1886)
Inhabits coastal reefs; distinguished by silver stripe on belly, plain pale colouration and rosy fins; N. Australia only; to 7 cm. (APOGONIDAE)

BIGEYES

Bigeyes are similar in appearance to the squirrelfishes and soldierfishes (Plates 14-15), but are distinguished by smaller scales, a larger, upturned mouth, and lack of spines on the head. They also exhibit similar behaviour, being essentially nocturnal and spending the daylight hours in caves. At night they emerge to feed on cephalopods, crustaceans, and fishes. The approximately 20 known species are mainly found in the Indo-Pacific region.

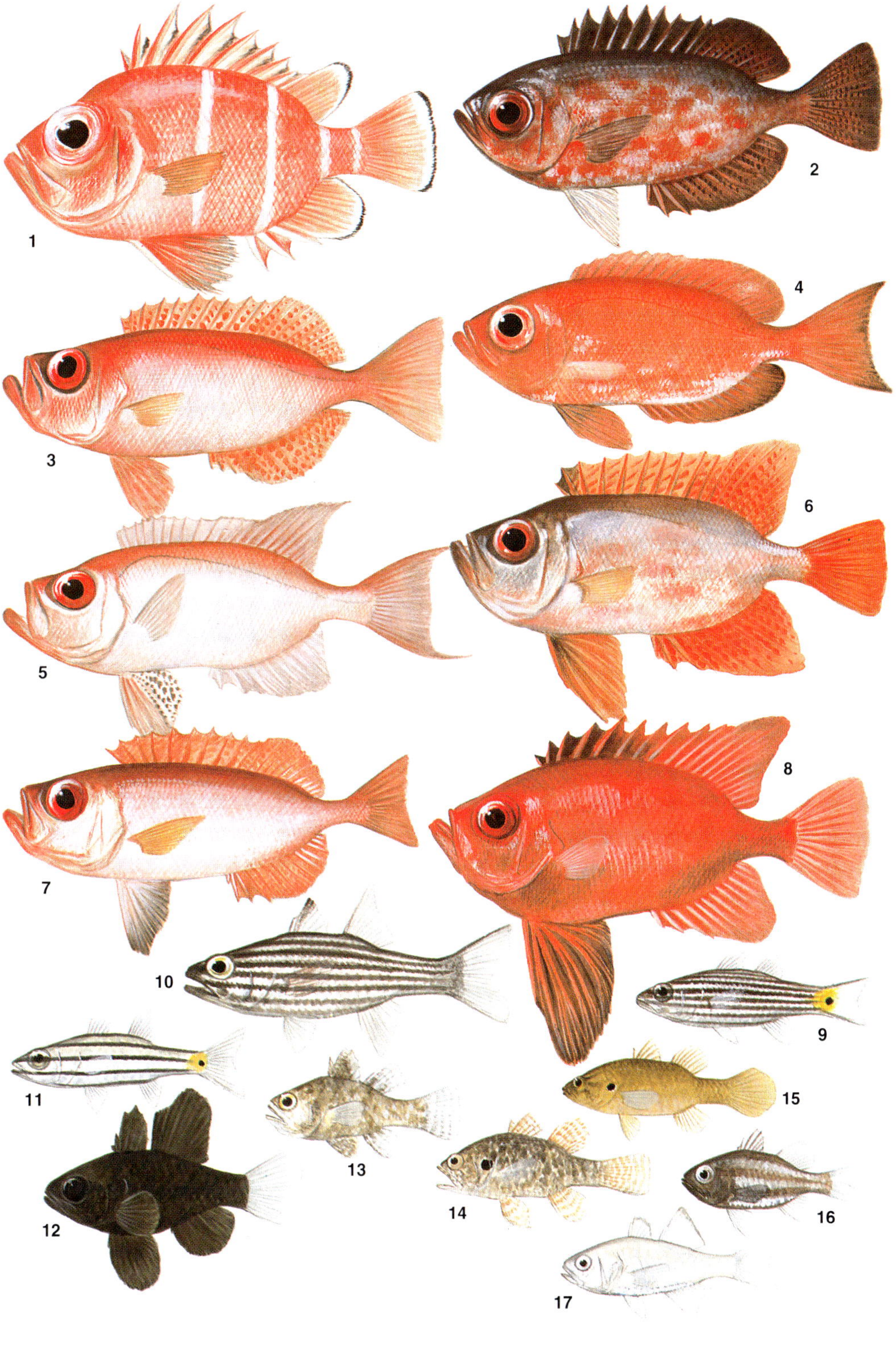
1
2
3
4
5
6
7
8
9
10
11
12
13
14
15
16
17

PLATE 32: CARDINALFISHES (FAMILY APOGONIDAE)

1 STRIPED CARDINALFISH
Apogon angustatus (Smith & Radcliffe, 1911)
Inhabits coral reef crevices, usually below 10-15 m depth; distinguished by relatively thin black stripes and distinct spot at base of tail; similar to **2** below, but found in deeper water; N.W. Australia, Great Barrier Reef, and throughout S.E. Asia; mainly Indo-C. Pacific; to 9 cm

2 REEF-FLAT CARDINALFISH
Apogon taeniophorus Regan, 1908
Inhabits shallow reef flats exposed to wave action, often found under boulders; similar to **1** above, but no black spot at tail base; found throughout the region; Indo-W. Pacific; to 10 cm

3 DOEDERLEIN'S CARDINALFISH
Apogon doederleini Jordan & Snyder, 1901
Inhabits coral reef crevices and southern rocky reefs; distinguished by thin red to black stripes on side and spot at tail base; N.W. Australia, Great Barrier Reef, and throughout S.E. Asia; mainly W. Pacific; to 9 cm.

4 CANDYSTRIPE CARDINALFISH
Apogon endekataenia Bleeker, 1852
Inhabits coral reef crevices; similar to **5** below, but stripes are thinner and lacks short (incomplete) stripe behind upper corner of eye; N.W. Australia and S.E. Asia; mainly W. Pacific; to 10 cm.

5 COOK'S CARDINALFISH
Apogon cooki Mackay, 1881
Inhabits shallow reefs; frequently in rocky crevices in weedy areas; distinguished from other striped species by incomplete stripe behind upper corner of eye; found throughout the region; Indo-W. Pacific; to 10 cm.

6 BLACK-TIPPED CARDINALFISH
Apogon semilineatus Temminck & Schlegel, 1843
Inhabits offshore trawling grounds; distinguished by black tip on dorsal fin and pair of narrow stripes along back; N.W. Australia and S.E. Asia; mainly W. Pacific; to 10 cm.

7 PALE-STRIPED CARDINALFISH
Apogon pallidofasciatus Allen, 1987
Inhabits shallow inshore reefs, in crevices often located in weedy areas; distinguished by dusky brown colour with faint stripes on side; N. Australia only; to 13 cm.

8 BROAD-BANDED CARDINALFISH
Apogon quadrifasciatus Cuvier, 1828
Inhabits flat sand bottoms with rocky outcrops, distinguished by dark mid-lateral stripe and 1-2 thinner stripes above; *A. kiensis* (Plate 36-15) is a similar species, but has 6 instead of 7 spines in the first dorsal fin; found throughout the region; Indo-W. Pacific; to 10 cm

9 EVERMANN'S CARDINALFISH
Apogon evermanni Jordan & Snyder, 1904
Inhabits caves, frequently on steep outer reef slopes; distinguished by red colouration, dark stripe behind eye, and small white spot behind second dorsal fin; N.W. Australia and throughout S.E. Asia; worldwide circumtropical; to 9 cm.

10 THREE-SADDLE CARDINALFISH
Apogon bandanensis Bleeker, 1854
Inhabits coral reefs, frequently among branching corals; similar to **11** below, but dark bar completely encircles tail base; found throughout the region; mainly W. and C. Pacific; to 10 cm.

11 SAMOAN CARDINALFISH
Apogon fuscus Quoy & Gaimard, 1825
Inhabits coral reefs, similar to **10** above, but bar at tail base not complete; *A. savayensis* is a synonym; found throughout the region; Indo-C. Pacific; to 10 cm.

12 SEVEN-BANDED CARDINALFISH
Apogon septemstriatus Günther, 1880
Inhabits flat sand bottoms and offshore trawling grounds; similar to **8** above, but has extra stripe on midline of forehead from snout to dorsal fin; found throughout the region; W. Pacific and E. Indian Ocean; to 9 cm.

13 SPINY-EYED CARDINALFISH
Apogon fraenatus Valenciennes, 1832
Inhabits coral reef crevices; distinguished by black stripe along middle of side and spot at tail base; N.W. Australia, Great Barrier Reef, and throughout S.E. Asia; Indo-C. Pacific; to 10 cm.

14 BLUE-STRIPED CARDINALFISH
Apogon cyanosoma Bleeker, 1853
Inhabits shallow reef crevices, distinguished by yellow-orange stripes; N.W. Australia, Great Barrier Reef, and throughout S.E. Asia; Indo-W. Pacific; to 8 cm.

15 IRIDESCENT CARDINALFISH
Apogon kallopterus Bleeker, 1856
Inhabits caves and coral reef crevices; distinguished by broad dark stripe along middle of side, dusky colour on back, and spot at base of tail; N.W. Australia, Great Barrier Reef, and throughout S.E. Asia; Indo-C. Pacific; to 15 cm.

16 YELLOW-LINED CARDINALFISH
Apogon chrysotaenia Bleeker, 1851
Inhabits coral reef crevices; distinguished by blue stripes on head, faint pale stripes on side, and spot on tail base; N.W. Australia and S.E. Asia; mainly Indo-Australian Archipelago; to 10 cm.

17 GOBBLEGUTS
Apogon rueppelli Günther, 1859
Inhabits inshore reefs and weedy areas, also in estuaries; distinguished by row of small black spots on upper side; W. and N. Australia and S. New Guinea only; to 12 cm.

18 TIMOR CARDINALFISH
Apogon timorensis Bleeker, 1854
Inhabits inshore reef flats with weed and sand, found under rocks; distinguished by overall dusky colour, sometimes with faint dark bars, and a narrow stripe below eye; found throughout the region; Indo-W. Pacific; to 8 cm.

CARDINALFISHES

Cardinalfishes (family Apogonidae) are shown on Plates 32-36. They are small, reef dwelling fishes inhabiting tropical and temperate seas. Some species also frequent estuaries and freshwater streams. The majority of the estimated 250 species are found on Indo-Pacific coral reefs. Most are nocturnally active and spend daylight hours in the dark recesses of caves and ledges They mainly feed on fishes and crustaceans. A few species, for example members of the genus *Rhabdamia*, form dense open-water shoals that literally envelope entire coral formations. Male cardinalfishes incubate a fertilised egg mass in their mouth for several days until hatching.

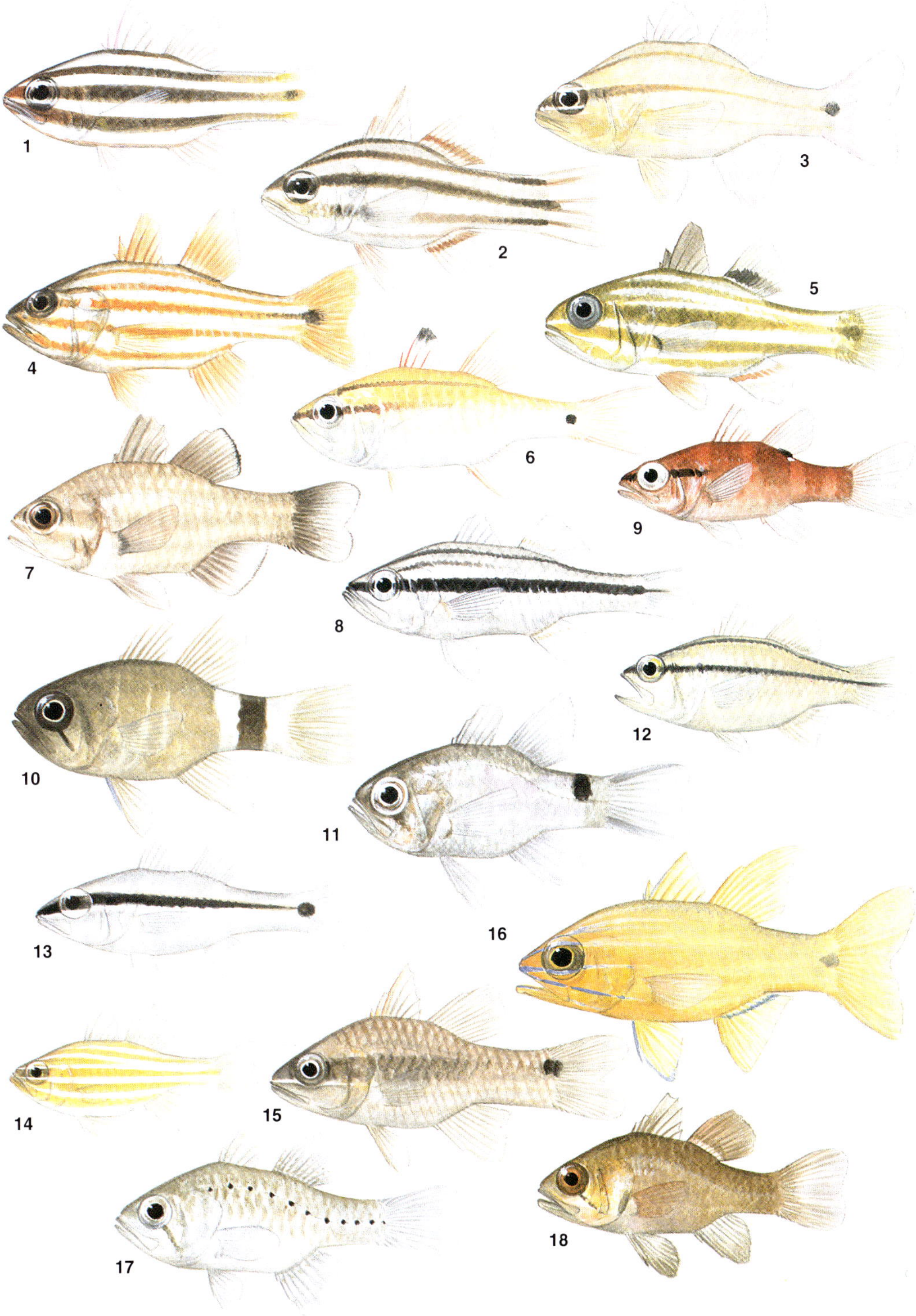
1
2
3
4
5
6
7
8
9
10
11
12
13
14
15
16
17
18

PLATE 33: CARDINALFISHES (FAMILY APOGONIDAE)

1 MOLUCCAN CARDINALFISH
Apogon moluccensis Valenciennes, 1832
Inhabits coral reef crevices; distinguished by broad, faint stripe on side and pearly spot just behind dorsal fin (not shown); N.W. Australia and throughout S.E. Asia; mainly Indo-Australian Archipelago; to 9 cm.

2 CAVITE CARDINALFISH
Apogon cavitiensis (Jordan & Seale, 1907)
Inhabits silty or sandy areas around rocky outcrops; distinguished by yellow-orange stripes on side and black spot on tail base; *A. virgulatus* is a synonym; found at scattered sites throughout the region; Indo-Australian Archipelago; to 8 cm.

3 OBLIQUE-BANDED CARDINALFISH
Apogon semiornatus Peters, 1876
Inhabits coral reef caves and crevices; distinguished by semi-transparent appearance and oblique dark bands; N.W. Australia, Great Barrier Reef, and throughout S.E. Asia; Indo-C. Pacific; to 7 cm.

4 TWO-EYED CARDINALFISH
Apogon nigripinnis Cuvier, 1828
Inhabits trawling grounds and inshore reefs; distinguished by dark colour and pale-edged black spot on middle of side; found throughout the region; Indo-W. Pacific; to 8 cm.

5 LITTLE RED CARDINALFISH
Apogon coccineus Rüppell, 1835
Inhabits coral reef crevices; distinguished by semi-transparent appearance and overall red colour; similar to **6** below, but tail base is shorter; found throughout the region; Indo-C. Pacific; to 5 cm.

6 BIG RED CARDINALFISH
Apogon unicolor Doderlein, 1884
Inhabits coral reef crevices; similar to **5** above, but grows much larger and has longer tail base; found throughout the region; Indo-C. Pacific; to 12 cm.

7 MANY-BANDED CARDINALFISH
Apogon brevicaudatus Weber, 1909
Inhabits trawling grounds and inshore reefs; distinguished by 8-9 narrow stripes on side and black spot at base of second dorsal fin; N. Australia and S. New Guinea only; to 9 cm.

8 FALSE THREE-SPOT CARDINALFISH
Apogon rhodopterus Bleeker, 1852
Inhabits coral reef crevices; very similar to *A. trimaculatus* (Pl. 35-12), but lacks dark spot on gill cover and dark bar or saddle on back below junction of first and second dorsal fins;offshore reefs of N.W. Australia and Indo-Malay Archipelago to New Guinea; mainly W. Pacific; to 15 cm.

9 FLAGFIN CARDINALFISH
Apogon ellioti Day, 1878
Inhabits trawling grounds; distinguished by black outer half of first dorsal fin and stripe across middle of anal and second dorsal fins; found throughout the region; Indo-W. Pacific; to 12 cm.

10 PEARLY-FINNED CARDINALFISH
Apogon poecilopterus Cuvier, 1828
Inhabits trawling grounds; similar to **11** below, but lacks dark spot at rear base of second dorsal fin; found throughout the region; mainly W. Pacific; to 12 cm.

11 OCELLATED CARDINALFISH
Apogon carinatus Cuvier, 1828
Inhabits trawling grounds; distinguished by dark spot at rear base of second dorsal fin; N.W. Australia and S.E. Asia; mainly W. Pacific, to 12 cm.

12 RING-TAILED CARDINALFISH
Apogon aureus (Lacepède, 1802)
Inhabits coral reef caves and crevices; distinguished by golden-orange colour and black ring around tail base; N.W. Australia, Great Barrier Reef, and throughout S.E. Asia; Indo-C. Pacific; to 15 cm.

13 SPINDLE-EGG CARDINALFISH
Apogon melanopus Weber, 1911
Inhabits trawling grounds and inshore reefs; distinguished by dark bar below dorsal fin and one or more rows of spots on rear half of body, these markings are often faint, this fish has a very unusual spindle-shaped egg; *A. fusovatus* is a synonym; N.W Australia and Arafura Sea only; to 12 cm.

14 CREAM-SPOTTED CARDINALFISH
Apogon albimaculosus Kailola, 1976
Inhabits trawling grounds; distinguished by cluster of large white spots on side and dark spotting on fins; N. Australia and S. New Guinea only; to 10 cm.

15 BLACK-NOSED CARDINALFISH
Rhabdamia cypselurus Weber, 1909
Inhabits coral reefs, forming large aggregations; distinguished by transparent appearance; similar to **16** below, but has black stripe on side of snout and dark margins on tail; N.W. Australia, Great Barrier Reef, and throughout S.E. Asia; Indo-C. Pacific; to 6 cm.

16 SLENDER CARDINALFISH
Rhabdamia gracilis (Bleeker, 1856)
Inhabits coral reefs, forming large aggregations; similar to **15** above, but lacks stripe on snout and dark margins on tail (sometimes has black spot at tips of lobes); N.W. Australia, Great Barrier Reef, and throughout S.E. Asia; Indo-C. Pacific; to 6 cm.

17 NARROW-LINED CARDINALFISH
Archamia fucata (Cantor, 1850)
Inhabits coral reefs and inshore rocky areas; similar to **18** below, but has spot on tail base and lacks dark blotch above upper edge of gill cover; found throughout the region; Indo-C. Pacific; to 9 cm.

18 BLACKSPOT CARDINALFISH
Archamia melasma Lachner & Taylor, 1960
Inhabits coral reefs and inshore rocky areas; similar to **17** above, but has dark blotch above upper edge of gill cover and lacks spot on tail base; N. Australia and S. New Guinea only; to 9 cm.

NEW DISCOVERIES

One of the exiting aspects of the region's fish fauna is that many new species have been discovered in recent years. More efficient methods of collecting including the widespread application of SCUBA techniques and increased ease of access to remote areas are contributing factors in these discoveries. Several of the cardinalfishes on Plates 32-36 were discovered and subsequently described for the first time during the past 10-12 years (refer to year of description after author names for each species).

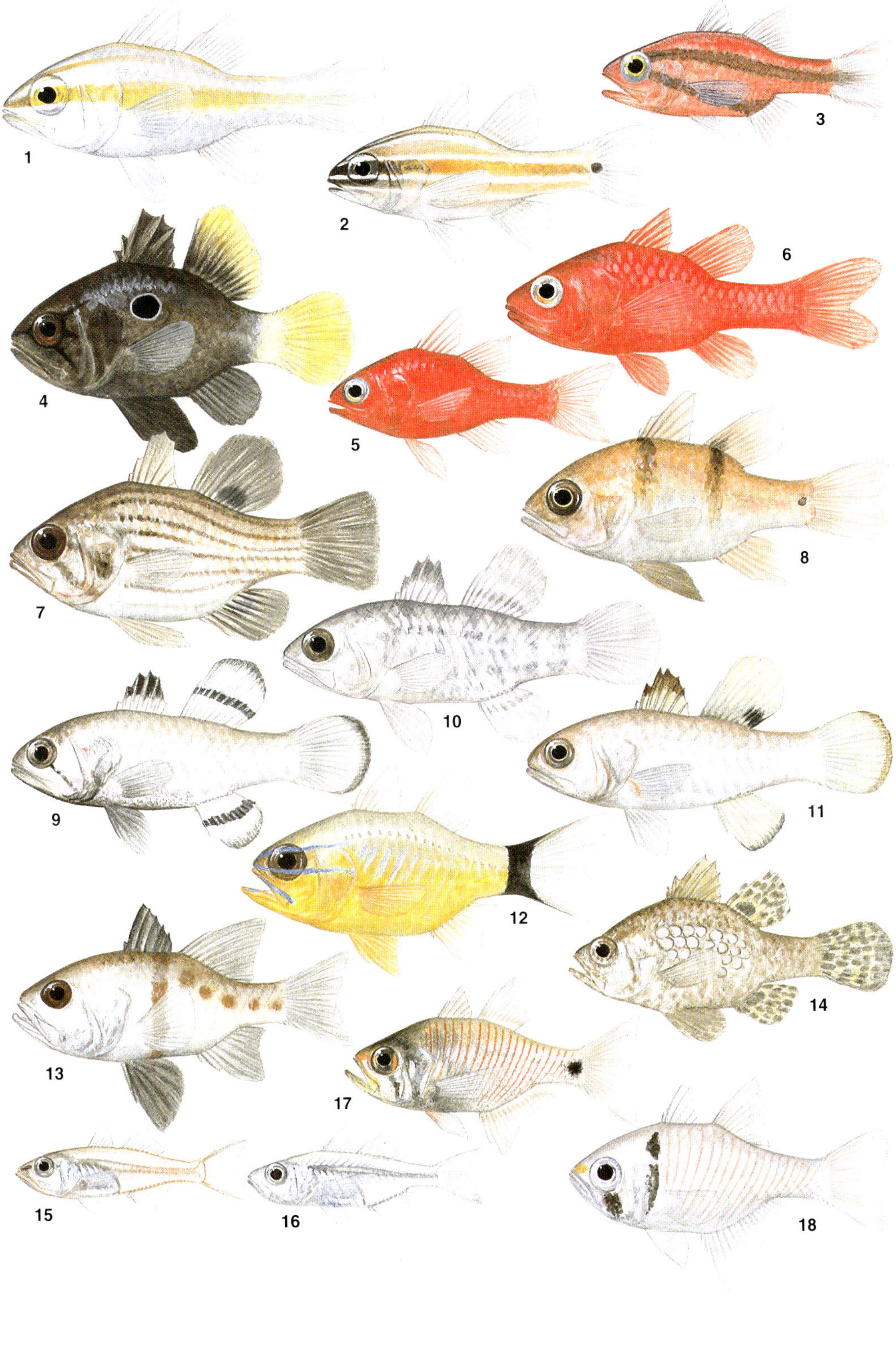
1
2
3
4
5
6
7
8
9
10
11
12
13
14
15
16
17
18

PLATE 34: CARDINALFISHES (FAMILY APOGONIDAE)

1 INTERMEDIATE CARDINALFISH
Cheilodipterus intermedius Gon, 1993
Inhabits coral reefs; similar in appearance to *C. artus* (Pl. 31-9), but differs in having 7-9 developed gill rakers instead of 10-17; Great Barrier Reef and New Guinea; W. Pacific; to 11 cm.

2 BLACK-STRIPED CARDINALFISH
Cheilodipterus nigrotaeniatus Smith & Radcliffe, 1912
Inhabits coral reefs; active during daylight, mimics the poison-fanged blenny *Meiacanthus grammistes*; Philippines and Sulu Sea; to 8 cm.

3 SINGAPORE CARDINALFISH
Cheilodipterus singapurensis Bleeker, 1859
Inhabits reef caves and crevices; nocturnal; distinguished by large, fang-like teeth, relatively large size, and fewer stripes than other large *Cheilodipterus* (stripes sometimes obscure and fish appears entirely dusky); offshore reefs of N.W. Australia and throughout S.E. Asia; W. Pacific; to 20 cm.

4 ALLEN'S CARDINALFISH
Cheilodipterus alleni Gon, 1993
Inhabits reef caves and crevices; nocturnal; distinguished by pattern of narrow dark stripes and black botch covering outer half of first dorsal fin; Indo-Malay Archipelago to New Guinea; to 12 cm.

5 MIMIC CARDINALFISH
Cheilodipterus parazonatus Gon, 1993
Inhabits coral reefs; active during daylight, mimics the poison-fanged blenny *Meiacanthus vittatus*; distinguished by broad, black stripe on side; Great Barrier Reef, New Guinea, and Solomon Islands; to 7 cm.

6 YELLOWBELLY CARDINALFISH
Cheilodipterus zonatus Smith & Radcliffe, 1912
Inhabits coral reefs; active during daylight, mimics the poison-fanged blenny *Meiacanthus geminatus*; distinguished by black stripe on side with broad white stripe above and yellow below; Malaysia, Philippines, and Solomon Islands; to 7 cm.

7 BANGGAI CARDINALFISH
Pterapogon kauderni (Koumans, 1933)
Inhabits seagrass areas near shore in less than 3 m depth; associated with *Diadema* sea urchins; male broods eggs and young in mouth; known only from Banggai Island, off central-eastern Sulawesi; to 6.5 cm.

8 PEARL-JAWED CARDINALFISH
Pseudamia amblyuropterus (Bleeker, 1856)
Inhabits estuaries and mangrove shores, but occasionally found in clear water on coral reefs; nocturnal; distinguished by pattern of fine, broken lines on side, blackish blotch on tail base, and pearly-white area on rear half of lower jaw; Indo-Malay Archipelago and W. Melanesia; W. Pacific; to 9 cm.

9 GELATINOUS CARDINALFISH
Pseudamia gelatinosa Smith, 1955
Inhabits reef caves and crevices usually in relatively clear water; nocturnal; similar appearance to **8** above, but rear half of lower jaw generally brownish instead of white, also developed gill rakers usually 8 instead of 11; Great Barrier Reef, N.W. Australia, and throughout S.E. Asia; Indo-C. Pacific; to 8 cm.

10 PADDLEFIN CARDINALFISH
Pseudamia zonata Randall, Lachner & Fraser, 1985
Inhabits reef caves and crevices usually in clear water of outer slopes below 10 m depth; nocturnal; distinguished by broad black bars and fan-shaped tail; Philippines and New Guinea; W. Pacific; to 9 cm.

11 ESTUARY CARDINALFISH
Pseudamia nigra Allen, 1992
Inhabits mangrove-lined shores and estuaries; nocturnal; distinguished by overall blackish appearance, but may have numerous fine, black longitudinal lines on side; N. Australia only; to 6 cm.

12 SPOTLESS CARDINALFISH
Fowleria vaiulae (Jordan & Seale, 1906)
Inhabits coral reef crevices, often found under dead coral slabs or rocks; distinguished from other *Fowleria* cardinalfishes by the lack of a pale-rimmed dark spot on the gill cover; Great Barrier Reef, offshore reefs of W. Australia, and throughout S.E. Asia; Indo-W. Pacific; to 5 cm. F. abocellata is a synonym.

13 BARRED CARDINALFISH
Fowleria marmorata (Alleyne & Macleay, 1877)
Inhabits coral reef crevices; distinguished by about 10 brown bars on side and pale-rimmed dark spot on gill cover; Great Barrier Reef, offshore reefs of W. Australia, and throughout S.E. Asia; Indo-C. Pacific; to 6 cm.

14 SPOTTED CARDINALFISH
Fowleria punctulata (Rüppell, 1838)
Inhabits coral reef crevices; distinguished by small black spots arranged in horizontal rows and pale-rimmed dark spot on gill cover; Great Barrier Reef, offshore reefs of W. Australia, and throughout S.E. Asia; Indo-C. Pacific; to 6 cm.

15 EIGHT-SPINED CARDINALFISH
Neamia octospina Smith & Radcliffe, 1912
Inhabits coral reef crevices; distinguished by overall pale colouration and trio of brown bands around rear half of eye; Great Barrier Reef, offshore reefs of W. Australia, and throughout S.E. Asia; Indo-W. Pacific; to 5 cm.

16 GLASSY CARDINALFISH
Rhabdamia spilota Allen & Kuiter, 1995
Inhabits coral reefs, usually on outer slopes between 20-55 m depth; forms aggregations around coral outcrops and gorgonian fans; distinguished by semi-transparent body with several darks spots on middle of side just behind head; Indonesia (Bali and Sulawesi); to 5 cm.

17 GIRDLED CARDINALFISH
Archamia zosterophora (Bleeker, 1856)
Inhabits coral reefs, often seen in aggregations among branching corals; distinguished by broad dark bar across middle of body; Great Barrier Reef, offshore reefs of W. Australia, and throughout S.E. Asia; mainly W. Pacific; to 8 cm.

18 DUSKY-TAILED CARDINALFISH
Archamia macroptera (Cuvier, 1828)
Inhabits coral reefs, often seen in aggregations among branching corals; similar to *A. fucata*, but has entire tail base dusky rather than a distinct spot; Indo-Malay and Melanesian archipelagos; W. Pacific; to 9 cm.

19 TWINSPOT CARDINALFISH
Archamia biguttata Lachner, 1951
Inhabits coral reef crevices, usually in aggregations; distinguished by large dark blotch behind head and small dark spot on tail base; Indo-Malay and Melanesian archipelagos; W. Pacific; to 9 cm.

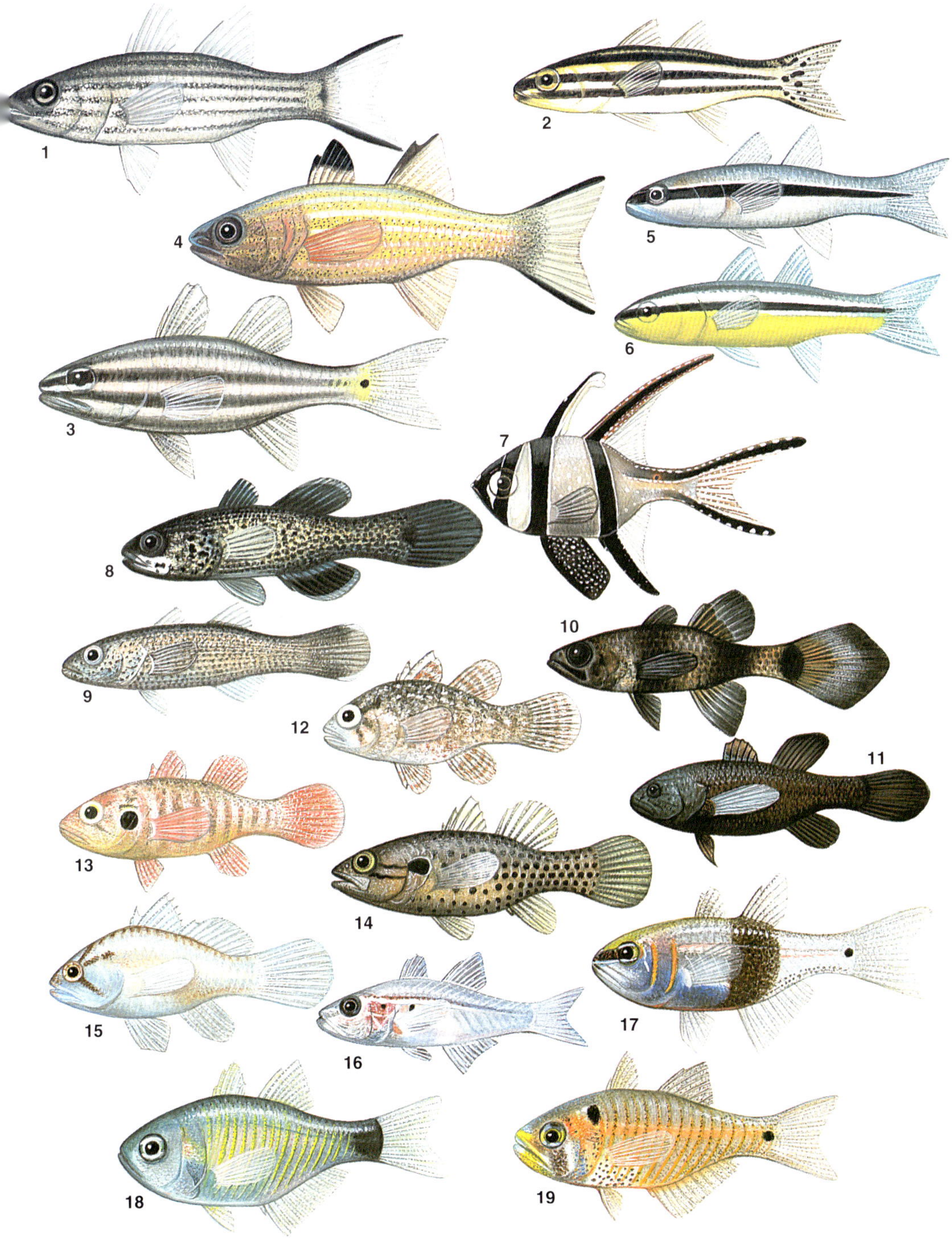
1
2
4
5
6
3
7
8
10
9
12
11
13
14
15
17
16
18
19

PLATE 35: CARDINALFISHES (FAMILY APOGONIDAE)

1 GOLDBELLY CARDINALFISH
Apogon apogonides (Bleeker, 1856)
Inhabits coral crevices or forms small schools around coral formations, usually below 15-20 m depth; distinguished by strong yellow hue on lower half of body, pair of blue lines through eye, and scattered blue spots on side; Great Barrier Reef and throughout S.E. Asia; Indo-W. Pacific; to 10 cm.

2 CORAL CARDINALFISH
Apogon properupta (Whitley, 1964)
Inhabits coral reefs, usually in silty coastal areas; often confused with *A. cyanosoma*, which has narrower yellow stripes; Great Barrier Reef, N.W. Australia, and throughout S.E. Asia; W. Pacific; to 6 cm.

3 SPLIT-BANDED CARDINALFISH
Apogon compressus (Smith & Radcliffe, 1911)
Inhabits branching *Acropora* corals, forming aggregations; distinguished by bold stripe pattern and neon-blue iris; young have black stripes and yellow tail base with isolated black spot; Great Barrier Reef, offshore reefs of N.W. Australia, and throughout S.E. Asia; W. Pacific; to 10 cm.

4 NARROWSTRIPE CARDINALFISH
Apogon exostigma (Jordan & Starks, 1906)
Inhabits coral crevices; similar to *A. frenatus*, but small black spot on tail base is above level of mid-lateral black stripe; Great Barrier Reef, offshore reefs of N.W. Australia, and throughout S.E. Asia; Indo-C.. Pacific; to 11 cm.

5 SINGLE-STRIPED CARDINALFISH
Apogon unitaeniata Allen, 1995
Inhabits shallow bays and trawling grounds between 2-18 m depth; distinguished by thin black, mid-lateral stripe; NW. Australia only, between Kimberley district and vicinity of Darwin; to 6 cm.

6 MANY-LINED CARDINALFISH
Apogon multilineatus (Bleeker, 1865)
Inhabits coral reef crevices; distinguished by numerous narrow strripes on side, also body and fins frequently with yellow hue; Indo-Malay Archipelago to New Guinea; W. Pacific; to 10 cm.

7 FRAGILE CARDINALFISH
Apogon fragilis Smith, 1961
Inhabits coral reefs, usually in aggregations among branching corals; distinguished by semi-transparent body and small black spot on tail base; Great Barrier Reef, offshore reefs of N.W. Australia, and throughout S.E. Asia; Indo-W. Pacific; to 6 cm.

8 THREADFIN CARDINALFISH
Apogon leptacanthus Bleeker, 1856
Inhabits coral reefs, usually in aggregations among branching corals; distinguished by blue streaks on head and elongate dorsal fin; Great Barrier Reef, offshore reefs of N.W. Australia, and throughout S.E. Asia; Indo-W. Pacific; to 6 cm.

9 BLACKSTRIPE CARDINALFISH
Apogon nigrofasciatus Lachner, 1953
Inhabits coral reef crevices; similar to **10** below, but stripes do not converge on tail; Great Barrier Reef, offshore reefs of N.W. Australia, and throughout S.E. Asia; Indo-C. Pacific; to 8 cm.

10 NINE-BANDED CARDINALFISH
Apogon novemfasciatus Cuvier, 1828
Inhabits coral reefs, usually in lagoon shallows; similar to **9** above, which is found in deeper water, but has dark stripes converging on tail base; Great Barrier Reef, offshore reefs of N.W. Australia, and throughout S.E. Asia; mainly W. and C. Pacific; to 9 cm.

11 SANGI CARDINALFISH
Apogon sangiensis Bleeker, 1857
Inhabits coral reef crevices, sometimes seen in open near branching corals; distinguished by dark band through eye and dark mark at front of first dorsal fin; Great Barrier Reef and throughout S.E. Asia; W. Pacific; to 8 cm.

12 THREE-SPOT CARDINALFISH
Apogon trimaculatus (Cuvier, 1828)
Inhabits coral reef crevices; very similar to *A. rhodopterus* (Pl. 33-8), but has dark spot on gill cover and dark bar or saddle on back below junction of first and second dorsal fins; N.W. Australia, Great Barrier Reef, and throughout S.E. Asia; mainly W. Pacific; to 15 cm.

13 FROSTFIN CARDINALFISH
Apogon hoeveni Bleeker, 1854
Inhabits coral reefs, usually associated with sponges, crinoids or *Diadema* sea urchins in weedy areas; distinguished by brilliant white posterior edge on first dorsal fin; Great Barrier Reef and throughout S.E. Asia; Indo-Australian Archipelago; to 5 cm.

14 BELLY-BARRED CARDINALFISH
Apogon ventrifasciatus Allen, Randall & Kuiter, 1995
Inhabits coral reef crevices in 3-30 m depth; usually seen in pairs or aggregations; distinguished by faint brown bars on lower sides, white lines through eye, and small white spot below base of last dorsal rays; Indo-Malay Archipelago to Solomon Islands; to 6 cm.

15 HOOKFIN CARDINALFISH
Apogon griffini (Seale, 1910)
Inhabits coral reefs, usually in sheltered areas near shore; distinguished by peach-coloured fins and distinct hooked shape of second dorsal fin, due to elongate anterior rays; *A. sabahensis* is a synonym; Borneo and Philippines; to 12 cm.

16 YELLOW CARDINALFISH
Apogon flavus Allen & Randall, 1993
Inhabits coral and rocky reefs in 2-25 m depth; often seen in large aggregations; similar to **17** below, but lacks blue stripes on head and dusky wavy bars on back; S. Great Barrier Reef and S.W. Coral Sea; to 12 cm.

17 CAPRICORN CARDINALFISH
Apogon capricornis Allen & Randall, 1993
Inhabits coral and rocky reefs in 2-15 m depth; similar to **16** above, but has blue stripes on each side of eye and narrow dusky bars on back; S. Great Barrier Reef and S. Coral Sea south to Sydney; to 8 cm.

18 LARVAL CARDINALFISH
Apogon neotes Allen, Randall & Kuiter, 1995
Inhabits coral reefs of lagoons and outer slopes in 15-25 m depth; forms aggregations among soft corals and gorgonian fans; distinguished by small size and larval-like appearance with semi-transparent body; Indo-Malay Archipelago to New Guinea and Palau; to 3 cm.

19 TINY CARDINALFISH
Apogon nanus Allen, Randall & Kuiter, 1995
Inhabits coral reefs of inshore silty areas; occurs in aggregations between 5-16 m depth; distinguished by small size and semi-transparent body with dusky orange mid-lateral stripe; Indo-Malay Archipelago to New Guinea; to 3.5 cm.

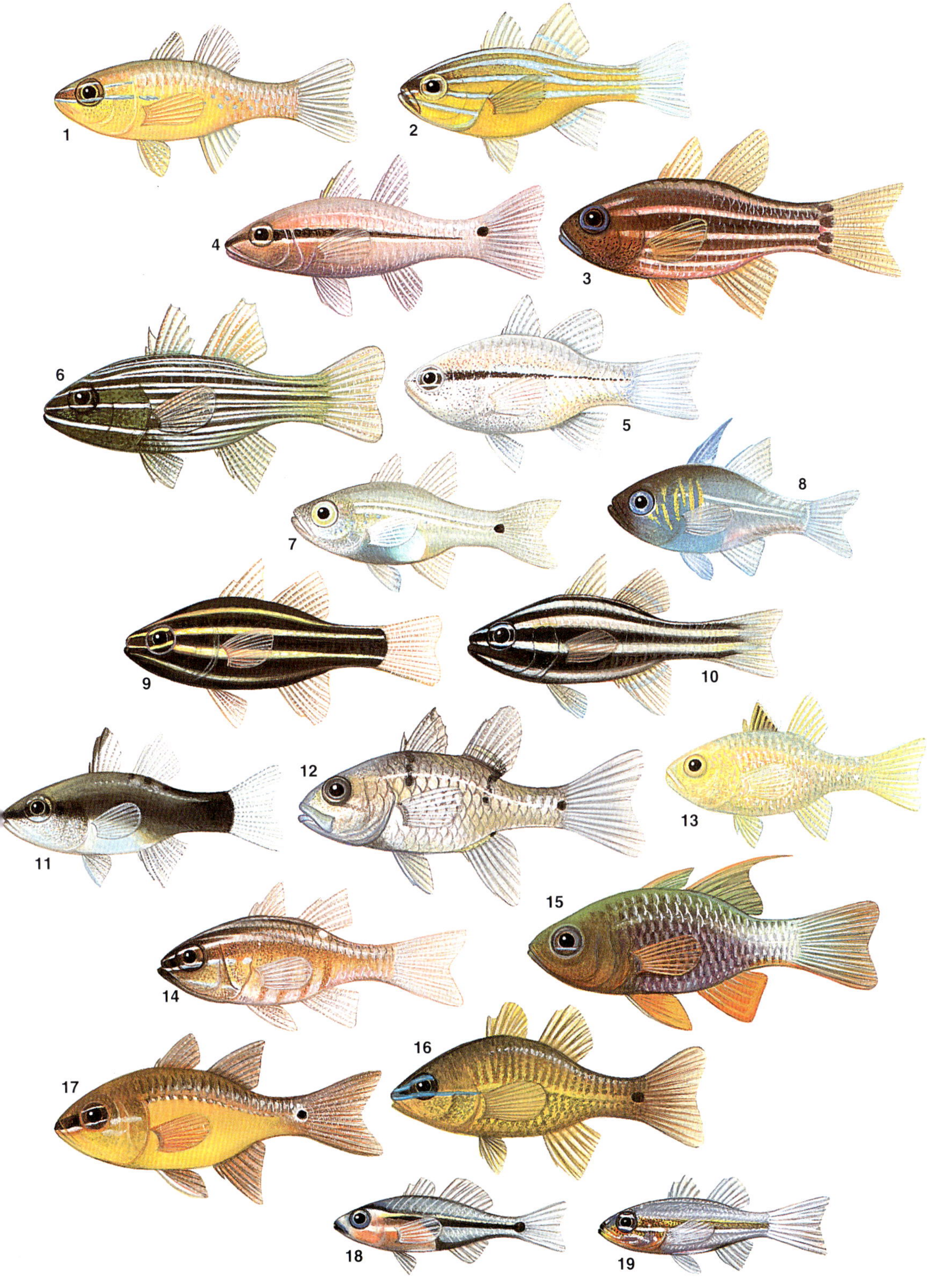
1
2
4
3
6
5
8
7
9
10
12
11
13
15
14
16
17
18
19

PLATE 36: CARDINALFISHES (FAMILY APOGONIDAE)

1 TAILSPOT CARDINALFISH
Apogon ocellicaudus Allen, Randall & Kuiter, 1995
Inhabits coral reefs under ledges at base of coral formations; distinguished by pale-edged dark spot at base of tail and absence of stripes on body; Indonesia (Flores) and Ashmore Reef, Timor Sea; to 6 cm.

2 FLORES CARDINALFISH
Apogon franssedai Allen, Randall & Kuiter, 1995
Inhabits steep outer slopes between 13-40 m depth; brownish stripes on side and large black spot on tail base; Indonesia, Philippines, and Palau; to 6.5 cm.

3 REDSPOT CARDINALFISH
Apogon parvulus Smith & Radcliffe, 1912
Inhabits coral reefs; distinguished by semi-transparent body, and bright red spot on tail base; Indo-Malay Archipelago; W. Pacific; to 4.5 cm.

4 DISPAR CARDINALFISH
Apogon dispar Fraser & Randall, 1976
Inhabits coral reefs, usually on steep drop-offs below 20 m; distinguished by semi-transparent body, red mid-lateral stripe, small reddish spot at middle of tail base with prominent white spot just above; Indo-Malay Archipelago to New Guinea and Solomon Islands; to 5.5 cm.

5 BROWN-SPOTTED CARDINALFISH
Apogon fuscomaculatus Allen & Morrison, 1996
Inhabits trawling grounds; distinguished by large dusky spots on pale background; N.W. Australia only; to 6 cm.

6 BLACK CARDINALFISH
Apogon melas Bleeker, 1848
Inhabits coral reef crevices; distinguished by overall dark colour and pale-edged black spot on basal half of second dorsal fin; Indo-Malay Archipelago to New Guinea; W. Pacific; to 13 cm.

7 SEALE'S CARDINALFISH
Apogon sealei (Fowler, 1918)
Inhabits coral reefs; see **8** below, also juveniles with trio of bold black stripes on side (2 stripes in *A. chrysopoma*); Timor Sea (Cartier-Ashmore Reefs) and Indo-Malay Archipelago; to 9 cm.

8 SPOTTED-GILL CARDINALFISH
Apogon chrysopomus Bleeker, 1854
Inhabits coral reefs, often in aggregations among branching corals; very similar to **7** above, but has spots on gill cover rather than vertical bands; Indo-Malay Archipelago to New Guinea and Solomon Islands; to 9 cm.

9 HARTZFELD'S CARDINALFISH
Apogon hartzfeldii Bleeker, 1852
Inhabits coral reefs, usually in sheltered bays and lagoons; often among spines of *Diadema* sea urchins; distinguished by purple-brown colour with silvery-white stripes on each side of eye and along back; juveniles have several similar stripes on sides; Indo-Malay Archipelago; to 12 cm.

10 GILBERT'S CARDINALFISH
Apogon gilberti Jordan & Seale, 1905
Inhabits coral reefs; similar to **Pl. 35-7** and **11** below, but entire tail base is dusky brown and dark brown "ear spot" often present; offshore reefs of N.W. Australia, Indonesia, Philippines, and E. Caroline Islands; to 5.5 cm.

11 PEARLY CARDINALFISH
Apogon perlitus Fraser & Lachner, 1985
Inhabits branching corals; similar to *A. fragilis* (Pl. 35-7), but has blackish strip along base of anal fin; offshore reefs of N.W. Australia, Indonesia, Philippines, and E. Caroline Islands; to 5.5 cm.

12 RED-STRIPED CARDINALFISH
Apogon margaritophorus Bleeker, 1854
Inhabits shallow coral reefs in lagoons and sheltered bays; distinguished by broad white mid-lateral stripe with narrower stripes of red and white, forming "window" pattern on lower side; Indo-Malay Archipelago to Solomon Islands; to 5.5 cm.

13 HUMP-BACKED CARDINALFISH
Apogon hyalosoma Bleeker, 1852
Inhabits mangrove shores, estuaries, and lower reaches of freshwater streams; distinguished by hump-backed appearance of adults and large dark spot at base of tail; N. Australia and throughout S.E. Asia; Indo-W. Pacific; to 20 cm.

14 MANGROVE CARDINALFISH
Apogon lateralis Valenciennes, 1832
Inhabits mangrove shores and weed beds; distinguished by thin black mid-lateral stripe, and small dark spot on tail base; N.W. Australia and Indo-Malay Archipelago to New Guinea; E. Indian Ocean and W. Pacific; to 9 cm.

15 RIFLE CARDINALFISH
Apogon kiensis Jordan & Snyder 1901
Inhabits sandy bottoms near coral reefs; similar to *A. quadrifasciatus* (Pl. 32-8), but has 6 instead of 7 spines in the first dorsal fin; Great Barrier Reef and throughout S.E. Asia; Indo-W. Pacific; to 9 cm.

16 SPOTNAPE CARDINALFISH
Apogon notatus (Houttuyn, 1782)
Inhabits coral reefs; distinguished by small pale-rimmed black spot on each side of forehead; Great Barrier Reef, Timor Sea (Ashmore Reef), and Indonesia (Komodo and nearby islands); W. Pacific; to 10 cm.

17 BAND-SPOT CARDINALFISH
Apogon selas Randall & Hayashi, 1990
Inhabits coral reefs, usually in sheltered lagoons between 15-35 m depth; distinguished by yellow upper iris, broad dusky stripe on side, and large rounded spot on tail base; Philippines, New Guinea, and Solomon Islands; W. Pacific; to 5.5 cm.

18 THERMAL CARDINALFISH
Apogon thermalis Valenciennes, 1829
Inhabits sandy areas with occasional coral outcrops in 15-40 m depth; distinguished by thin, dark mid-lateral stripe, dark spot on tail base, and narrow bars on lower half of side; throughout S.E. Asia; Indo-W. Pacific; to 7 cm.

19 PAJAMA CARDINALFISH
Sphaeramia nematoptera (Bleeker, 1856)
Inhabits rich coral areas in protected bays and lagoons; distinguished by broad dark bar across middle of side, large spots on rear half of body, and filament at beginning of second dorsal fin; Great Barrier Reef and Indo-Malay Archipelago to New Guinea and Solomon Islands; W. Pacific; to 8 cm.

20 ORBICULAR CARDINALFISH
Sphaeramia orbicularis (Cuvier, 1828)
Inhabits coastal reefs and mangrove shores, frequently seen around wharves; similar to **19** above, but pattern is duller; found throughout S.E. Asia to New Guinea and Solomon Islands; Indo-W. Pacific; to 11.5 cm.

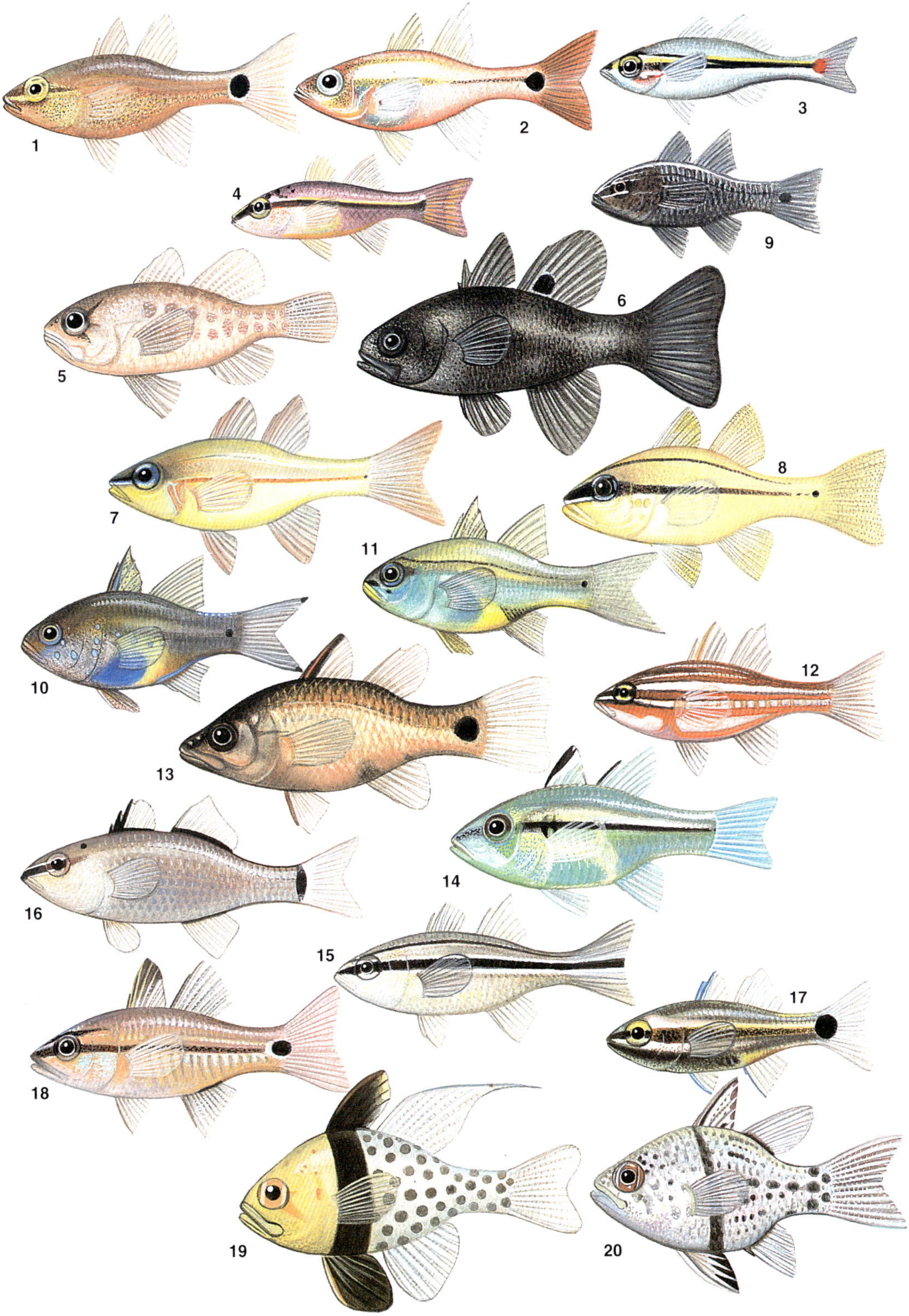

1
2
3
4
9
5
6
7
8
11
10
12
13
14
16
15
17
18
19
20

PLATE 37: TREVALLIES (FAMILY CARANGIDAE)

1 PENNANTFISH
Alectis ciliaris (Bloch, 1788)
Inhabits coastal reefs; juvenile has long filamentous fin rays, both juvenile and adult distinguished from **2** below by more rounded head profile and eye closer to mouth; found throughout the region; worldwide tropical seas; to 130 cm and 13 kg. ★★

2 DIAMOND TREVALLY
Alectis indicus (Rüppell, 1828)
Inhabits coastal reefs; similar to **1** above, but has more angular head profile and wider space between eye and mouth; both species differ from other trevallies by their scaleless skin; found throughout the region; Indo-W. Pacific; to 150 cm. ★★

3 FRINGE-FINNED TREVALLY
Pantolobus radiatus (Macleay, 1881)
Inhabits coastal waters, sometimes in estuaries or river mouths; distinguished by yellow tail with black upper tip; female is shown here, male (Plate 38-17) has long filamentous rays on the dorsal and anal fins; found throughout the region; Indo-Australian Archipelago; to 40 cm. ★★

4 SMALL MOUTH SCAD
Alepes sp.
Inhabits coastal waters; distinguished by clear fleshy eyelid covering rear half of eye and often has dark tips on lobes of tail; found throughout the region; Indo-Australian Archipelago; to 35 cm.

5 YELLOWTAIL SCAD
Atule mate (Cuvier, 1833)
Inhabits coastal waters, forming large schools; distinguished by clear fleshy eyelid covering most of eye except narrow slit in centre; found throughout the region; Indo-C. Pacific; to 30 cm. ★★

6 CLUB-NOSED TREVALLY
Carangoides chrysophrys (Cuvier, 1833)
Inhabits coastal waters; distinguished by gently sloping head profile except abruptly vertical at tip of snout and has scaleless area on breast extending on to pectoral fin base; found throughout the region; Indo-W. Pacific; to 44 cm. ★★

7 ONION TREVALLY
Carangoides caeruleopinnatus (Rüppell, 1830)
Inhabits coastal waters; distinguished by relatively deep body, short dorsal and anal fin lobes, small black blotch on upper margin of gill cover, and small yellow spots on body; found throughout the region; Indo-C. Pacific; to 40 cm. ★★

8 BLUE TREVALLY
Carangoides ferdau (Forsskål, 1775)
Inhabits coastal waters and offshore reefs; distinguished by separate scaleless areas on breast and base of pectoral fin, a bluntly rounded snout, and frequently has 5-6 dusky bars on sides; found throughout the region; Indo-C. Pacific; to 70 cm. ★★

9 GOLD-SPOTTED TREVALLY
Carangoides fulvoguttatus (Forsskål, 1775)
Inhabits coastal waters; distinguished by relatively elongate shape and many gold or brassy spots on side (mainly on back), similar to **15** below, but has eye higher above mouth, more tapered snout, and more yellow spots; also known as Turrum and Yellow-spotted trevally; found throughout the region; Indo-W. Pacific; to 130 cm and 12 kg. ★★★

10 WHITEFIN TREVALLY
Carangoides equula (Temminck & Schlegel, 1844)
Inhabits coastal waters; distinguished by blackish or dusky submarginal band on second dorsal fin and sometimes on anal fin, by very short lobes at front of dorsal and anal fins, and a fully scaled breast; found throughout the region; Indo-W. Pacific; to 37 cm. ★★

11 EPAULET TREVALLY
Carangoides humerosus (McCulloch, 1915)
Inhabits coastal waters; distinguished by large eye (about equal to distance from eye to snout tip); blackish first dorsal fin, and moderately long dorsal and anal lobes; found throughout the region; Indo-Australian Archipelago; to 25 cm. ★★

12 BUMP-NOSED TREVALLY
Carangoides hedlandensis (Whitley, 1934)
Inhabits coastal waters; distinguished by long filamentous extensions on dorsal and anal fin rays, this feature also present in adult males of **3** above, but that species much more slender and has black tip on upper lobe of tail, also known as Port Hedland trevally; found throughout the region; Indo-C. Pacific; to 32 cm. ★★

13 WHITE-TONGUED TREVALLY
Carangoides talamparoides Bleeker, 1852
Inhabits coastal waters; distinguished by a white or pale grey tongue, relatively deep body, steep head profile, and extensive scaleless area encompassing breast, pectoral fin base, and small area above pectoral fin; found throughout the region; N. Indian Ocean and Indo-Australian Archipelago; to 32 cm. ★★

14 THICKLIP TREVALLY
Carangoides orthogrammus (Jordan & Gilbert, 1882)
Inhabits coastal areas; distinguished by several ovate yellow spots on middle of side and well separated scaleless areas on breast and base of pectoral fin; also known as False bluefin trevally; found throughout the region; Indo-C. Pacific; to 70 cm. ★★

15 BLUDGER TREVALLY
Carangoides gymnostethus (Cuvier, 1833)
Inhabits coastal areas in the vicinity of coral or rocky reefs; distinguished by relatively elongate body and a few brown or golden spots often present on side; similar to **9** above, but has eye closer to level of mouth, fewer yellow spots, and a steeper snout profile; found throughout the region; Indo-W. Pacific; to 90 cm and at least 11 kg. ★★

16 MALABAR TREVALLY
Carangoides malabaricus (Bloch & Schneider, 1801)
Inhabits coastal waters; similar to **13** above, but tongue grey-brown to brown instead of whitish; found throughout the region; Indo-W. Pacific; to 28 cm. ★★

17 JAPANESE TREVALLY
Carangoides uii Wakiya, 1924
Inhabits coastal waters; distinguished by relatively deep body and thread-like filament at front of second dorsal (and often anal) fin; found throughout the region; Indo-W. Pacific to 25 cm. ★★

PLATE 38

1 GIANT TREVALLY
Caranx ignobilis (Forsskål, 1775)
Inhabits coastal and offshore waters in the vicinity of reefs; the largest of the trevallies, distinguished by steep forehead profile and silvery to dusky colouration; also known as Lowly trevally; found throughout the region; Indo-C. Pacific to 170 cm and at least 35 kg. ★★

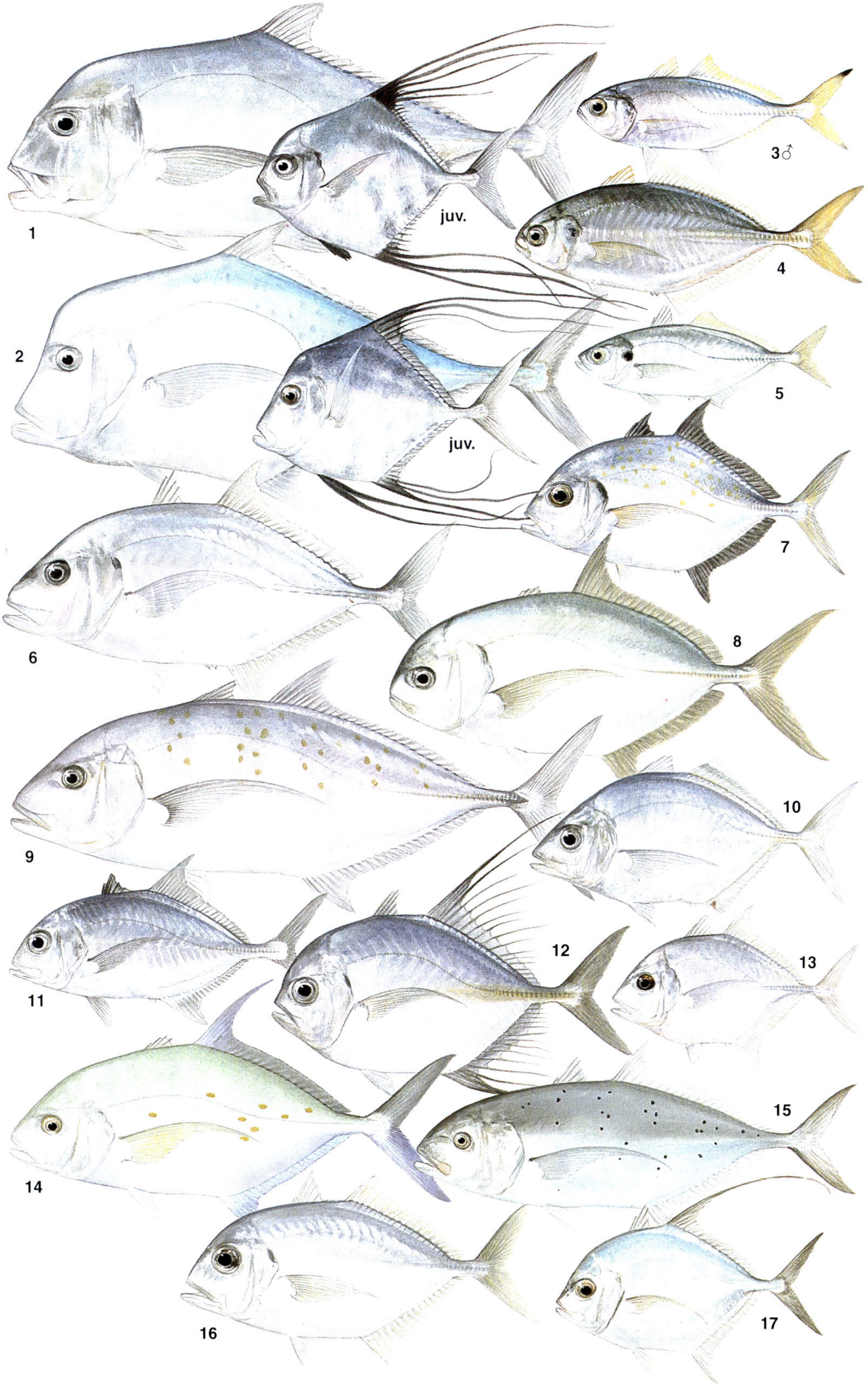
1
juv.
3♂
4
2
juv.
5
7
6
8
9
10
11
12
13
14
15
16
17

PLATE 38: TREVALLIES (FAMILY CARANGIDAE)

1 GIANT TREVALLEY
Caranx ignobilis (Forsskål, 1775). Text on previous page.

2 BLACK TREVALLY
Caranx lugubris Poey, 1861
Inhabits mainly offshore waters in the vicinity of coral reefs; distinguished by dark colouration ranging from brown to nearly black; found throughout the region; worldwide tropical seas; to 80 cm. ★★

3 BLUEFIN TREVALLY
Caranx melampygus Cuvier, 1833
Inhabits coastal and offshore waters in the vicinity of reefs; distinguished by blue fins and dark speckling on upper half of body; found throughout the region; Indo-E. Pacific; to 100 cm. ★★★

4 BRASSY TREVALLY
Caranx papuensis Alleyne & Mackay, 1877
Inhabits coastal and offshore waters in the vicinity of reefs; similar to **1** above, but forehead not as steep and has white margin on lower lobe of tail, also usually with scattered dark spots on upper side; also known as Papuan trevally; found throughout the region; E. Indian Ocean and W. Pacific, to 75 cm. ★★

5 TILLE TREVALLY
Caranx tille Valenciennes, 1833
Inhabits coastal and offshore waters in the vicinity of reefs; distinguished by relatively slender shape, well developed gelatinous membrane covering much of eye, and blackish spot on upper corner of gill cover; found throughout the region; Indo-W. Pacific; to 70 cm. ★★

6 BANDED SCAD
Caranx para Cuvier, 1833
Inhabits coastal waters; distinguished by small size, exaggerated profile of belly (i.e. ventral profile more convex than dorsal profile), and black spot on upper corner of gill cover; found throughout the region; Indo-W. Pacific; to 18 cm. ★★

7 BLUE-SPOTTED TREVALLY
Caranx bucculentus Alleyne & Macleay, 1877
Inhabits coastal waters; distinguished by steep forehead profile, large eye, dark spot at upper pectoral fin base, and blue spots on upper side; juveniles with 6 dark bars which develop into 3 horizontal rows of square blotches; N. Australia and New Guinea only; to 66 cm. ★★

8 BIGEYE TREVALLY
Caranx sexfasciatus Quoy & Gaimard, 1825
Inhabits coastal and offshore waters in the vicinity of reefs; distinguished by relatively large eye with well developed gelatinous membrane, and white tip on dorsal fin lobe; juveniles have 5-6 dark bars on body and sometimes occur in fresh water and estuaries; found throughout the region; Indo-E. Pacific; to 78 cm and at least 4 kg. ★★

9 ROUGHEAR SCAD
Decapterus tabl Berry, 1968
Inhabits coastal waters, occurring in schools; a slender, silvery fish distinguished from other mackerel scad by a red tail and 4-10 scales in straight part of lateral line immediately preceding expanded bony scutes on rear part of body; found throughout the region; Atlantic and Indo-C. Pacific, to 50 cm. ★★

10 REDTAIL SCAD
Decapterus kurroides Bleeker, 1855
Inhabits coastal waters, occurring in schools; similar to **9** above, including red tail, but lacks a straight section in the lateral line (i.e. between curved anterior part and expanded bony scules on rear portion); found throughout the region; Indo-W. Pacific; to 50 cm. ★★

11 MACKEREL SCAD
Decapterus macarellus (Cuvier, 1833)
Inhabits coastal waters, occurring in schools; distinguished by yellow-green tail and 18-32 scales in straight part of lateral line in front of bony scutes; found throughout the region; worldwide tropical seas; to 32 cm and 1.65 kg. ★★

12 RUSSELL'S MACKEREL SCAD
Decapterus russelli (Rüppell, 1828)
Inhabits coastal waters, occurring in schools; distinguished by clear to dusky tail and 0-4 scales in straight part of lateral line in front of bony scutes; also known as Indian scad; found throughout the region; Indo-W. Pacific; to 38 cm and 1.8 kg. ★★★

13 LONG-BODIED SCAD
Decapterus macrosoma Bleeker, 1851
Inhabits coastal waters, occurring in schools; distinguished by clear to dusky tail and 14-29 scales in straight part of lateral line in front of bony scutes; found throughout the region; Indo-E. Pacific; to 32 cm. ★★

14 GOLDEN TREVALLY
Gnathanodon speciosus (Forsskål, 1775)
Inhabits coastal and offshore waters, usually near reefs, sometimes occurring in schools; distinguished by large fleshy lips, lack of discernible teeth (unlike other trevallies), and golden belly; juvenile (not shown) is bright yellow with dark bars and dark tips on tail; found throughout the region; Indo-E. Pacific; to 111 cm and 15 kg. ★★★

15 FINNY SCAD
Megalaspis cordyla (Linnaeus, 1758)
Inhabits coastal waters; distinguished by series of separate finlets behind dorsal and anal fins and greatly expanded bony scutes along middle of side; found throughout the region; Indo-C. Pacific; to 80 cm. ★★

16 PILOT FISH
Naucrates ductor (Linnaeus, 1758)
Inhabits oceanic waters, usually in company with sharks, rays, turtles, or large fishes; juveniles may occur in floating weed or with jellyfishes; distinguished by prominent dark bars; found throughout the region; worldwide tropical seas; to 70 cm. ★★

17 FRINGE-FINNED TREVALLY
Pantolobus radiatus (Macleay, 1881)
Inhabits coastal waters; distinguished by orange-yellow dorsal and tail fins, and black tip on upper lobe of tail; male (shown here) has filamentous dorsal and anal rays; female shown on Plate 37-3; found throughout the region; Indo-Australian Archipelago; to 40 cm. ★★

18 BLACK POMFRET
Parastromateus niger (Bloch, 1795)
Inhabits offshore waters of continental shelf, frequently in schools; distinguished by equal-shaped dorsal and anal fin with triangular anterior lobe; colour ranges from silvery-grey to bluish-brown; found throughout the region; Indo-W. Pacific; to 55 cm and 2.55 kg.

19 SILVER TREVALLY
Pseudocaranx dentex. Text on page 116.

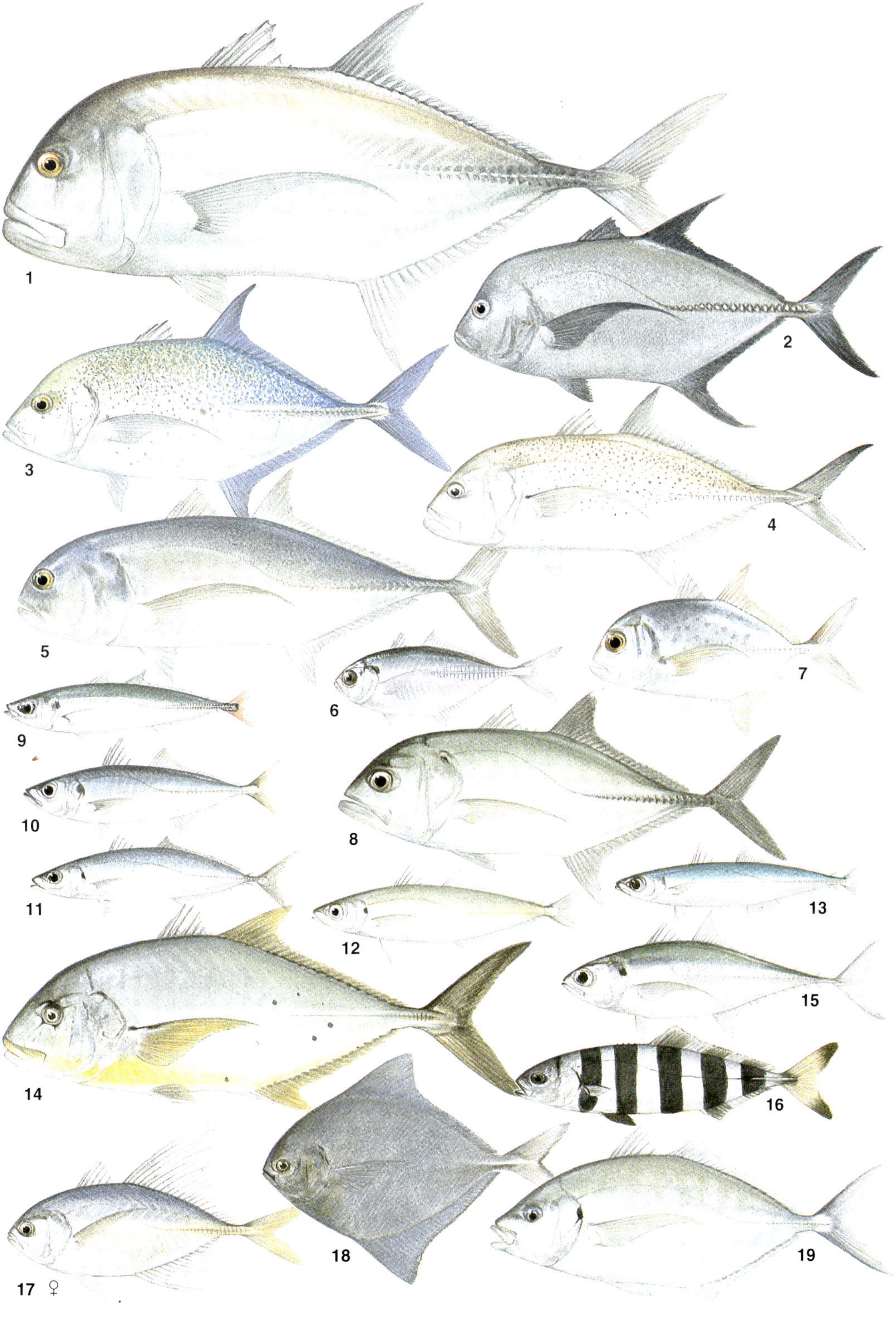
1
2
3
4
5
6
7
8
9
10
11
12
13
14
15
16
17 ♀
18
19

PLATE 39: DOLPHINFISHES AND TREVALLIES

1 COMMON DOLPHINFISH
Coryphaena hippurus (Linnaeus, 1758)
Inhabits mainly oceanic waters well offshore; distinguished by elongate, compressed body, long-based dorsal and anal fins, and steep forehead profile; female (not shown) is less colourful and lacks a distinct hump on the forehead; also known as Mahimahi and Dorado; found throughout the region; worldwide tropical and subtropical seas; to 200 cm and 22.4 kg. (CORYPHAENIDAE) ★★★

2 RAINBOW RUNNER
Elegatis bipunnulata (Quoy & Gaimard, 1825)
Inhabits coastal waters and offshore reefs, usually in schools; distinguished by slender shape, yellow and blue stripes, and isolated finlets on tail base; found throughout the region; worldwide tropical and subtropical seas; to 120 cm and at least 5 kg. (CARANGIDAE) ★★★

3 NEEDLESKIN QUEENFISH
Scomberoides tol (Cuvier, 1832)
Inhabits coastal waters, often in small schools; distinguished by 5-8 round spots, the first 4-5 making contact with lateral line; also known as Slender leatherskin; found throughout the region; Indo-W. Pacific; to 60 cm. (CARANGIDAE) ★★

4 BARRED QUEENFISH
Scomberoides tala (Cuvier, 1832)
Inhabits coastal waters; distinguished by 4-8 vertically elongate blotches most of which make contact with lateral line; also known as Deep leatherskin; found throughout the region; Indian Ocean and W. edge of Pacific; to 75 cm. (CARANGIDAE) ★★

5 TALANG QUEENFISH
Scomberoides commersonnianus Lacepède, 1801
Inhabits coastal waters, usually near reefs, but occasionally in estuaries; distinguished by 5-8 blotches that are mainly above lateral line (first two may contact lateral line); also known as Leatherskin; found throughout the region; Indo-W. Pacific; to 120 cm and at least 11.4 kg. (CARANGIDAE) ★★

6 DOUBLE-SPOTTED QUEENFISH
Scomberoides lysan (Forsskål, 1775)
Inhabits coastal waters, often in schools; distinguished by double row of 6-8 dark spots on side; found throughout the region; Indo-C. Pacific; to 70 cm. (CARANGIDAE) ★★

7 OXEYE SCAD
Selar boops (Cuvier, 1833)
Inhabits coastal waters in large schools; distinguished by large eye covered over by clear fleshy eyelid except for central portion, and by yellow or bronze hue on back; found throughout the region; W. Pacific and E. Indian Ocean; to 25 cm. (CARANGIDAE) ★★

8 PURSE-EYED SCAD
Selar crumenthalmops (Bloch, 1793)
Inhabits coastal waters in large schools; similar to **7** above, but lacks bronze or yellow hue, instead usually has broad yellowish mid-lateral stripe; found throughout the region; worldwide tropical seas; to 30 cm. (CARANGIDAE) ★★

9 SMOOTH-TAILED TREVALLY
Selaroides leptolepis (Kuhl & van Hasselt, 1833)
Inhabits coastal waters forming large schools over soft sand or mud bottoms; similar to **7** and **8** above, but has smaller plate-like scales (scutes) along rear part of lateral line, lacks teeth in upper jaw, and has clear fleshy eyelid covering rear half of eye only; found throughout the region; N. Indian Ocean and W. edge of Pacific; to 20 cm. (CARANGIDAE) ★★

10 BLACK-SPOTTED DART
Trachinotus bailloni (Lacepède, 1801)
Inhabits coastal waters, frequently in surge off sandy beaches; distinguished by strongly forked tail and 1-5 small black spots along middle of sides; found throughout the region; Indo-C. Pacific; to 54 cm. (CARANGIDAE) ★★

11 SNUB-NOSED DART
Trachinotus blochii (Lacepède, 1801)
Inhabits coastal waters, frequently in surge off sandy beaches; distinguished by strongly forked tail broadly rounded snout profile, and lack of spots on side; found throughout the region; Indo-C. Pacific; to 65 cm and at least 9 kg. (CARANGIDAE) ★★

12 COMMON DART
Trachinotus botla (Shaw, 1803)
Inhabits coastal waters, frequently in surge off sandy beaches; distinguished by strongly forked tail and 1-5 large spots along middle of side; found throughout the region; Indo-W. Pacific; to 61 cm and at least 3.5 kg. (CARANGIDAE) ★★

13 BLACK-BANDED KINGFISH
Seriolina nigrofasciata (Rüppell, 1829)
Inhabits oceanic waters usually well offshore; distinguished by forward slanting dark bars or large blotches arranged in ventral rows; found throughout the region; Indo-W. Pacific; to 70 cm. (CARANGIDAE) ★★

14 ALMACO JACK
Seriola rivoliana Valenciennes, 1833
Inhabits oceanic waters usually well offshore, occasionally visiting coastal areas; distinguished by diagonal dark band above eye and has deeper body than other *Seriola*; found throughout the region; worldwide tropical and temperate seas; to 70 cm. (CARANGIDAE) ★★

15 YELLOWTAIL KINGFISH
Seriola lalandi Valenciennes, 1833
Inhabits coastal and offshore waters, sometimes in schools; similar to **16** below, but has narrower upper jaw and yellow tail; found throughout the region; world-wide temperate and tropical seas; to 180 cm and at least 47 kg. (CARANGIDAE) ★★

16 AMBERJACK
Seriola dumerili (Risso, 1810)
Inhabits mainly offshore waters in the vicinity of reefs, sometimes adjacent to dropoffs; similar to **15** above, but lacks yellow tail and rear part of jaw broadly expanded; found throughout the region; Atlantic and Indo-C. Pacific; to 188 cm and at least 39.5 kg. (CARANGIDAE) ★★

PLATE 38

19 SILVER TREVALLY
Pseudocaranx dentex (Bloch & Schneider, 1801)
Inhabits coastal waters, including estuaries, usually in schools; silvery white to pale bluish with or without diffuse dark bars on side, back greenish or bronzy, a dark spot on upper corner of gill cover and sometimes a yellow stripe on middle of sides; mainly southern half of Australia; largely anti-tropical distribution in Atlantic and Indo-C. Pacific; to 94 cm and 10 kg. (CARANGIDAE) ★★

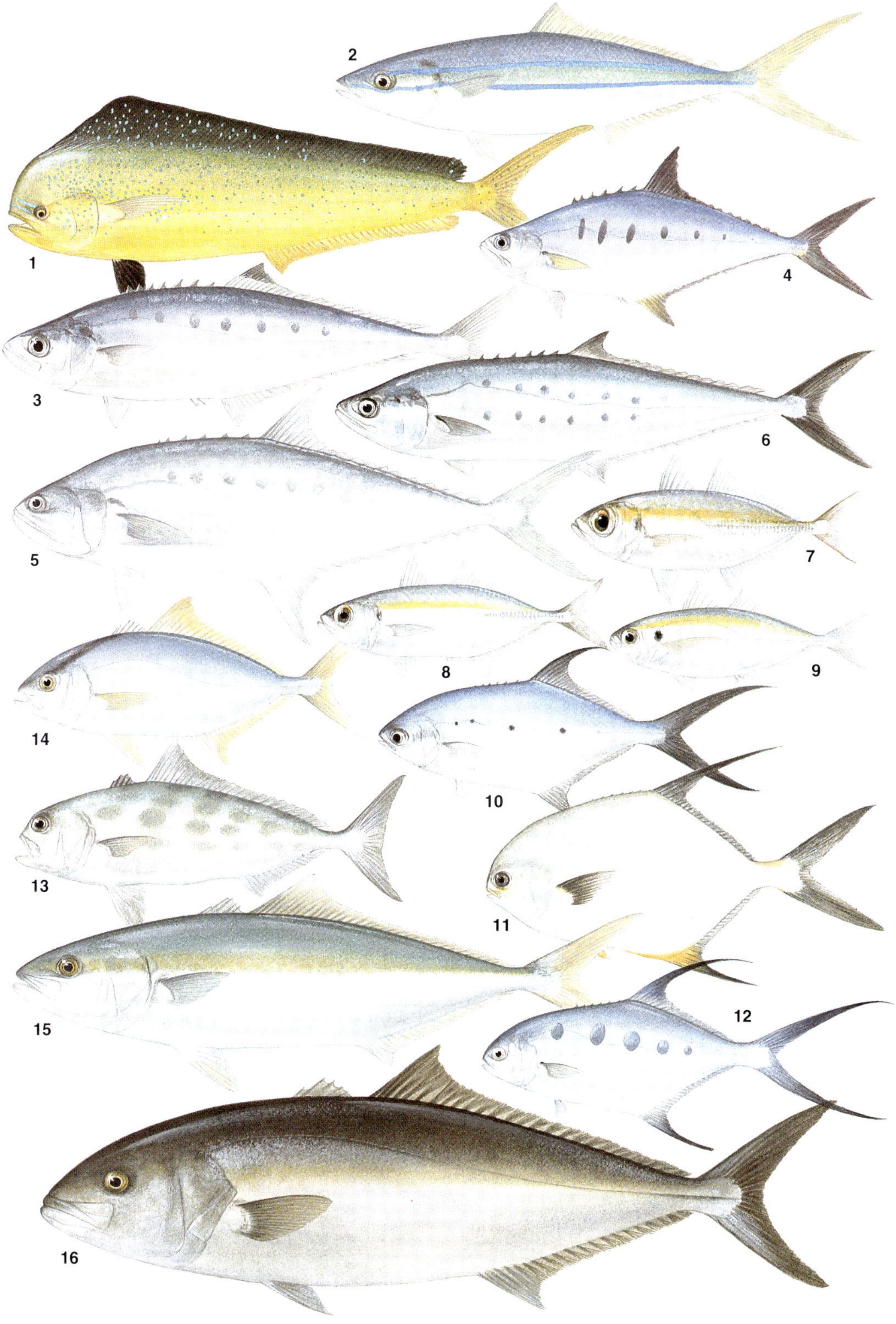

1
2
3
4
5
6
7
8
9
10
11
12
13
14
15
16

PLATE 40: TREVALLIES, PONYFISHES, AND MISCELLANEOUS FAMILIES

1 BLACK-CRESTED TREVALLY
Ulua aurochs (Ogilby, 1915)
Inhabits coastal waters; distinguished by extremely long gill rakers that project into mouth along side of tongue and prominent chin which protrudes well ahead of upper jaw in older specimens (not shown), similar to **2** below, but has longer dorsal fin filament, diffuse dark bars on side and deeper body; N. Australia and New Guinea; to 50 cm. (CARANGIDAE) ★★

2 CALE CALE TREVALLY
Ulua mentalis (Cuvier, 1833)
Inhabits coastal waters; similar to **1** above, but has shorter dorsal fin filament, lacks broad diffuse bars on side (may have narrower chevron-shaped ones), and body shape is slightly more slender; found throughout the region; Indo-W. Pacific; to 90 cm. (CARANGIDAE) ★★

3 BASSETT-HULLS TREVALLY
Uraspis uraspis (Gunther, 1860)
Inhabits coastal waters; distinguished by white tongue and floor of mouth vividly contrasted against rest of mouth which is blue-black, also general colouration of body and fins dusky to black shading to grey on lower part of body; found throughout the region; Indo-W. Pacific; to 30 cm. (CARANGIDAE) ★★

4 FLAGTAIL BLANQUILLO
Malacanthus brevirostris Guichenot, 1848
Inhabits sand and rubble areas in the vicinity of coral reefs, often occurs in pairs; distinguished by elongate shape with long-based dorsal and anal fins, and pair of dark stripes on tail; N.W. Australia, Great Barrier Reef, and throughout S.E. Asia; Indo-C. Pacific; to 30 cm. (MALACANTHIDAE) (Also see Plate 46)

5 TAILOR
Pomatomus saltator (Linnaeus, 1766)
Inhabits estuaries and inshore waters; related to trevallies and kingfishes (Pl. 39-16), but has larger, more obvious scales and very sharp teeth; colour silvery with green or bluish tinge; southern half of Australia, barely penetrating tropical latitudes; temperate and subtropical Atlantic and Indo-W. Pacific; to 120 cm and at least 12.1 kg. (POMATOMIDAE) ★★★

6 COBIA
Rachycentron canadus (Linnaeus, 1766)
Inhabits coastal waters; distinguished by white stripe on side (may fade after death) and very small dorsal spines; also known as Black kingfish and Sergeant fish; found throughout the region; Atlantic and Indo-C. Pacific; to 202 cm and 61.5 kg. (RACHYCENTRIDAE) ★★★

7 SLENDER SUCKERFISH
Echeneis naucrates Linnaeus, 1758
Inhabits inshore reefs and offshore oceanic waters; distinguished by striped pattern and flattened head with sucking-disk structure on top; disc is used for attaching itself to larger fishes such as sharks, rays and mackerels; found throughout the region; worldwide temperate and tropical seas; to 100 cm. (ECENEIDAE) ★★★

8 REMORA
Remora remora (Linnaeus, 1758)
Inhabits coastal and oceanic waters; similar in shape and with same sucking-disk apparatus as **7** above, but lacks stripes and is overall brownish-black to grey in colour; found throughout the region; worldwide temperate and tropical seas; to 45 cm. (ECENEIDAE)

9 MOONFISH
Mene maculata (Bloch & Schneider, 1801)
Inhabits deeper coastal waters, sometimes enters estuaries; distinguished by oval shape, protrusible jaws, very low dorsal and anal fins, and silvery colour; found throughout the region; Indo-W. Pacific; to 24 cm. (MENIDAE) ★★

10 TOOTHPONY
Gazza minuta (Bloch, 1797)
Inhabits coastal waters to 40 m depth; distinguished from other ponyfishes on this page by canine-like teeth in jaws; found throughout the region; Indo-W. Pacific; to 14 cm. (LEIOGNATHIDAE)

11 ORANGEFIN PONYFISH
Leiognathus bindus (Valenciennes, 1835)
Inhabits coastal waters to 40 m depth, occurs in schools; distinguished by protrusible jaws and black-edged orange tip of spiny dorsal fin; found throughout the region; Indo-W. Pacific; to 11 cm. (LEIOGNATHIDAE)

12 COMMON PONYFISH
Leiognathus equulus (Forsskål, 1874)
Inhabits coastal waters, sometimes enters estuaries, occurs in schools; distinguished by protrusible jaws and thin vertical lines on back; *L. splendens* (not shown) is similar, but has a black tip on the spiny dorsal fin; found throughout the region; Indo-W. Pacific; to 24 cm. (LEIOGNATHIDAE)

13 SLENDER PONYFISH
Leiognathus elongatus (Günther, 1874)
Inhabits coastal waters to 40 m depth, occurs in schools; distinguished by protrusible jaws and very elongate shape; found throughout the region; Indo-W. Pacific; to 12 cm. (LEIOGNATHIDAE)

14 WHIPFIN PONYFISH
Leiognathus leuciscus (Günther, 1860)
Inhabits coastal waters, occurs in schools; distinguished by protrusible jaws, elongate filament on spiny dorsal fin, and dark vermiculations on back; *L. moretoniensis* (not shown) has similar shape and markings, but has small scales on the cheek (versus no scales); *L. fasciatus* (not shown) is also similar, but has dark vertical lines on back and is deeper-bodied; found throughout the region; Indo-W. Pacific; to 12 cm. (LEIOGNATHIDAE)

15 SMITHURST'S PONYFISH
Leiognathus smithursti (Ramsay & Ogilby, 1886)
Inhabits coastal waters to 40 m depth, occurs in schools; distinguished by protrusible jaws and elongate filaments at front of both dorsal and anal fins; found throughout the region; Indo-W. Pacific; to 16 cm. (LEIOGNATHIDAE)

16 PUGNOSE PONYFISH
Secutor ruconius (Hamilton, 1822)
Inhabits coastal waters, sometimes entering estuaries and rivers, occurs in schools; distinguished by jaw that when protruded point upwards (those of other ponyfishes on this page point either slightly downwards or straight ahead), differs from *S. insidiator* (not shown) by having scales present on cheek; found throughout the region; Indo-W. Pacific; to 8 cm. (LEIOGNATHIDAE)

17 TRIPLETAIL
Lobotes surinamensis (Bloch, 1790)
Inhabits mangrove estuaries and lower reaches of freshwater streams; distinguished by steep, sloping humped forehead and large rounded posterior lobes of dorsal and anal fins that are about equal to the tail in size; found throughout the region; worldwide in tropical and subtropical seas; to 100 cm and at least 7 kg. (LOBOTIDAE) ★★★★

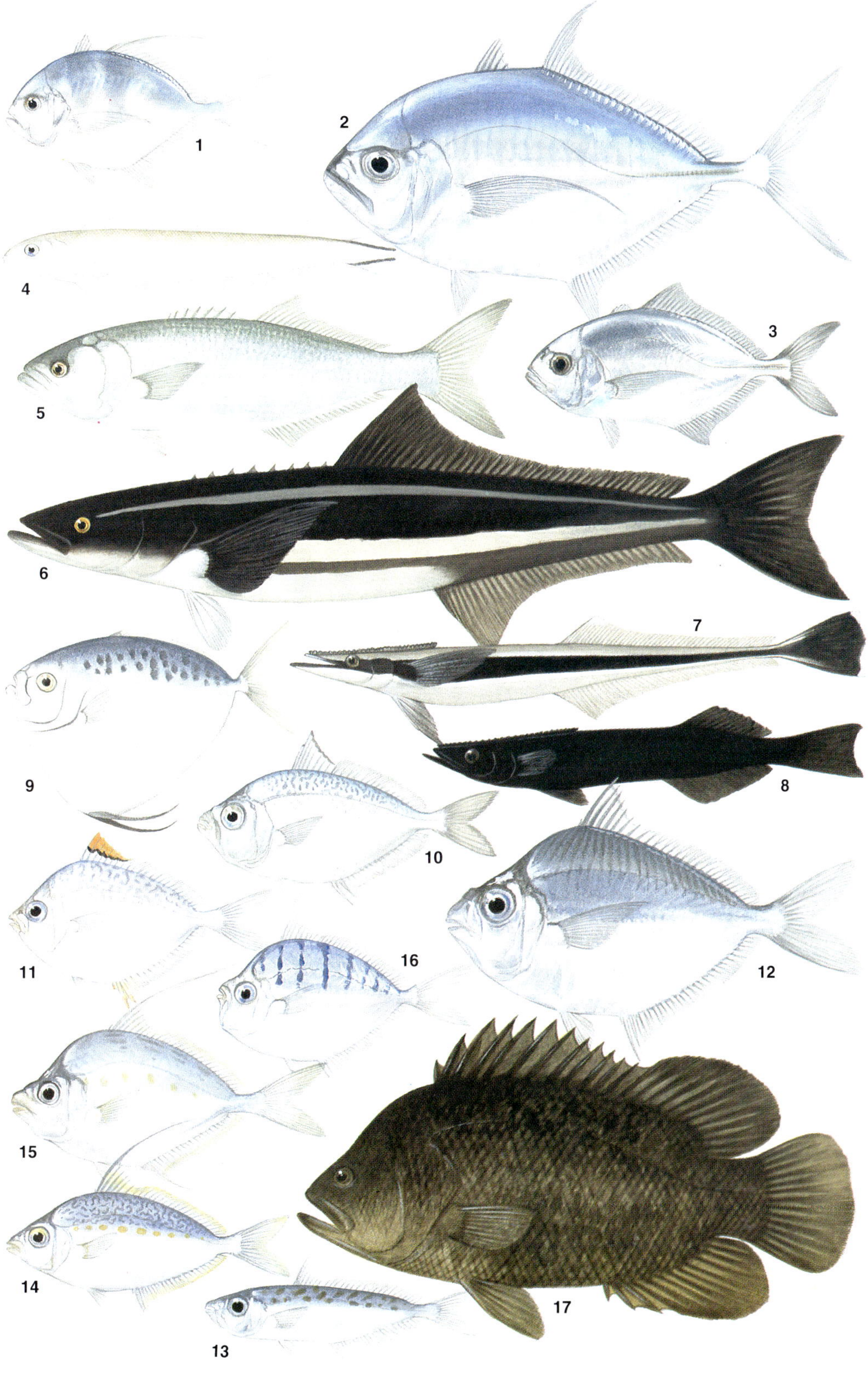
1
2
4
5
3
6
7
9
8
10
11
16
12
15
14
17
13

PLATE 41: SEAPERCHES OR SNAPPERS (FAMILY LUTJANIDAE)

1 SMALL-TOOTHED JOBFISH
Aphareus furca (Lacepède, 1802)
Inhabits reefs and rocky bottoms to at least 100 m depth; distinguished by elongate body shape and deeply forked tail; similar to *A. rutilans* (Pl. 43-1), but has dark margin on edge of cheek and gill cover and body lacks yellow hue; N.W. Australia, Great Barrier Reef, and throughout S.E. Asia; Indo-C. Pacific: to 80 and at least 4 kg. ★★★

2 GREEN JOBFISH
Aprion virescens Valenciennes, 1830
Inhabits reef areas to at least 100 m depth; distinguished by dark green to blue-grey colour and dark patches along base of dorsal fin, snout blunter and tail less forked than **1** above; found throughout the region; Indo-C. Pacific: to 100 cm and at least 12 kg. ★★★

3 RUBY SNAPPER
Etelis carbunculus Cuvier, 1828
Inhabits rocky offshore reefs usually between 90-300 m depth; distinguished from **4** below by brighter red colouration and less forked tail; N.W. Australia, N. Queensland, and throughout S.E. Asia; Indo-C. Pacific; to 80 cm. ★★★

4 PALE SNAPPER
Etelis radiosus Anderson, 1981
Inhabits rocky offshore reefs, usually between 90-200 m depth; distinguished from 3 above by pale (pinkish) colour and more forked tail; N.W. Australia, N. Queensland, and throughout S.E. Asia; W. Pacific and Indian Ocean; to 60 cm. ★★★

5 TANG'S SNAPPER
Lipocheilus carnolabrum (Chan, 1970)
Inhabits rocky bottoms of continental shelves between 90-300 m depth; distinguished by thick fleshy protrusion at the front of the upper lip; N.W. Australia, N. Queensland, and throughout S.E. Asia; Indo-W. Pacific; to 60 cm. ★★★

6 ROSY SNAPPER
Pristipomoides filamentosus (Valenciennes, 1830)
Inhabits rocky bottoms, mainly between 90-360 m depth; distinguished by rosy colour and elongate tips at rear of dorsal and anal fins; N.W. Australia, N. Queensland, and throughout S.E. Asia; Indo-W. Pacific; to 80 cm and at least 6 kg. ★★★★

7 SHARPTOOTH JOB FISH
Pristipomoides typus (Bleeker, 1852)
Inhabits rocky bottoms between about 40-100 m depth; similar to **8** below, but lacks orange stripes below eye; N.W. Australia, N. Queensland, and throughout S.E. Asia; mainly western Pacific; to 70 cm. ★★★

8 GOLD-BANDED JOBFISH
Pristipomoides multidens (Day, 1870)
Inhabits rocky bottoms between 40-200 m depth; similar to **7** above, but has pair of orange stripes below eye; N.W. Australia, N. Queensland, and throughout S.E. Asia; Indo-W. Pacific; to 90 cm and at least 6 kg. ★★★

9 CHINAMAN FISH
Symphorus nematophorus (Bleeker, 1860)
Inhabits inshore coral reefs and deeper offshore areas to at least 50 m; recognised by distinctive shape and olive grey-brown to reddish (when freshly caught) colour of adults, young have elongate filaments on the rear part of the dorsal fin; considered dangerous to eat due to its susceptibility to ciguatera; found throughout the region; mainly western Pacific; to 80 cm and at least 18 kg.

10 MANGROVE JACK
Lutjanus argentimaculatus (Forsskål, 1775)
Inhabits estuaries and inshore and offshore reefs to 100 m depth; young and sub-adults found in mangrove estuaries; distinguished from **12** below by taller dorsal fin, lack of stripes on side, and absence of black on fins, young have pronounced bars; found throughout the region; Indo-W. Pacific; to 120 cm and at least 12 kg. ★★★

11 INDONESIAN SEAPERCH
Lutjanus bitaeniatus (Valenciennes, 1830)
Inhabits deeper offshore reefs between about 40-70 m; distinguished by reddish colouration and absence of stripes, although juvenile has a black stripe along middle of side; N.W. Australia and Indonesia; to 30 cm. ★★★

12 RED BASS Cairns 10/98
Lutjanus bohar (Forsskål, 1775)
Inhabits coral reefs to at least 70 m depth; distinguished by reddish hue, faint stripes on lower side, and blackish colour on dorsal and anal fins (also tip of pelvic fins and upper edge of pectoral fins), juveniles and subadults with white spot below rear part of dorsal fin; reported to be good eating, but best avoided due to the possibility of ciguatera poisoning; found throughout the region; Indo-C. Pacific; to 75 cm and at least 11 kg.

FISH POISONING

Although relatively few cases of a form of fish poisoning known as Ciguatera have been reported from the region, those of us, particularly anglers, who are regular consumers of fresh seafood should be aware of its dangers. This topic is included here because two of the snappers shown on this plate, **9** and **12**, have been frequently implicated in various regions of the Indo-Pacific. The incidence of poisonous fishes is alarmingly high in some areas, particularly around some Polynesian islands. The exact nature of the causative agent has long puzzled scientists. The same species of fish can be poisonous at one locality, but safe to eat from a nearby reef. Apparently the poisonous properties are caused by a toxic dinoflagellate (*Gambierdiscus toxicus*), that lives on dead coral or among benthic algae and is first consumed by herbivorous fishes who are eventually eaten by larger predatory fishes. The toxin is apparently accumulative and the largest fishes are potentially the most dangerous. Young and small adults of a particular species may be eaten with impunity while large adults may be toxic.

The symptoms from eating ciguatoxic fish appear from one to 10 hours later and range from mild dizziness, diarrhoea, and a numb sensation of the lips, hands and fingers to extreme nausea, coma and total respiratory failure causing death. The degree of poisoning depends on the amount of fish consumed and the concentration of toxin it contains.

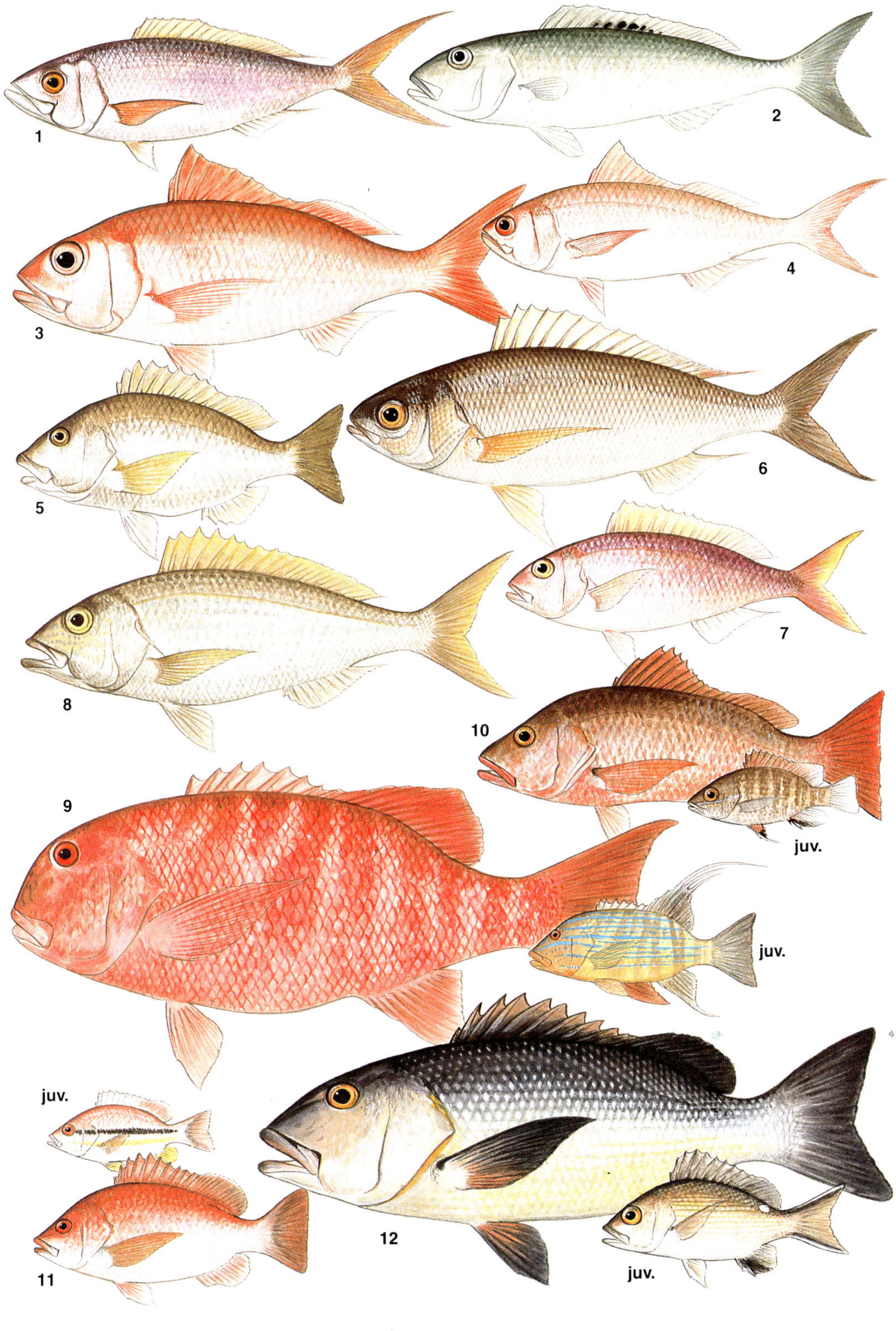
1
2
3
4
5
6
7
8
10
9
juv.
juv.
juv.
12
11
juv.

PLATE 42: SEAPERCHES OR SNAPPERS (FAMILY LUTJANIDAE)

1 CRIMSON SEAPERCH
Lutjanus erythropterus Bloch, 1790
Inhabits trawling grounds and reefs to depths of at least 100 m; similar to **4** below, but head and mouth much smaller; also known as Saddle-tailed seaperch; found throughout the region; Indo-W. Pacific; to 100 cm. ★★★

2 STRIPEY SEAPERCH
Lutjanus carponotatus (Richardson, 1842)
Inhabits coral reefs in sheltered lagoons and outer reef areas; distinguished by striped pattern; found throughout the region; mainly Indo-Australian Archipelago; to 40 cm. ★★★

3 CHECKERED SEAPERCH
Lutjanus decussatus (Cuvier, 1828)
Inhabits coral reefs to at least 30 m depth; distinguished by chequered pattern on the upper sides and black spot at the base of the tail; N.W. Australia, Great Barrier Reef, and throughout S.E. Asia; mainly Indo-Australian Archipelago; to 30 cm.

4 SADDLE-TAILED SEAPERCH
Lutjanus malabaricus (Bloch & Schneider, 1801)
Inhabits coastal and offshore reefs, also flat-bottom trawling grounds; similar to **1** above, but head and mouth much larger; also known as Scarlet seaperch; found throughout the region; Indo-W. Pacific; to 100 cm and at least 7 kg. ★★★

5 YELLOW-LINED SEAPERCH
Lutjanus rufolineatus (Valenciennes, 1830)
Inhabits offshore coral reefs to depths of 50 m; distinguished by series of faint yellow stripes on the sides and a deep notch in the rear margin of the cheek, often has dark spot on back; previously confused with *L. bouton* (Pl. 43-5); N.W. Australia, Great Barrier Reef, and throughout S.E. Asia; mainly W. Pacific; to 30 cm. ★★★

6 BLUE-STRIPED SEAPERCH
Lutjanus kasmira (Forsskål, 1775)
Inhabits inshore coral reefs; distinguished from **7** below by its white belly and lack of a fifth blue stripe; found throughout the region; Indo-C. Pacific; to 35 cm. ★★★

7 FIVE-LINED SEAPERCH
Lutjanus quinquelineatus (Bloch, 1790)
Inhabits sheltered lagoon reefs and outer reef areas; similar to **6** above, but has extra blue stripe on body; also known as Blue-banded seaperch; found throughout the region; Indo-W. Pacific; to 38 cm. ★★★

8 BLACK-SPOT SEAPERCH
Lutjanus fulviflamma (Forsskål, 1775)
Inhabits coral reefs to at least 35 depth; distinguished by yellow stripes on sides and elongate black spot on back, snout blunter than **12** below; found throughout the region; Indo-C. Pacific; to 35 cm. ★★★

9 MAORI SEAPERCH
Lutjanus rivulatus (Cuvier, 1828)
Inhabits inshore coral reefs and also deeper offshore waters; distinguished by "blubbery" lips and wavy lines on head; found throughout the region; Indo-C. Pacific; to 65 cm. ★★★

10 RED EMPEROR
Lutjanus sebae (Cuvier, 1828)
Inhabits the vicinity of coral reefs, often over adjacent sand and rubble flats; distinguished by shape and overall red-pink colour, juveniles and subadults have distinctive pattern of dark bars; juveniles sometimes found among spines of sea urchins; found throughout the region; Indo-W. Pacific; to 100 cm and at least 16 kg. ★★★

11 STRIPED SEAPERCH
Lutjanus vitta (Quoy & Gaimard, 1824)
Inhabits the vicinity of coral reefs, also flat bottoms with coral outcrops, sponge and sea whips; distinguished by brown or blackish stripe along middle of sides; found throughout the region; Indo-W. Pacific; to 40 cm. ★★★

12 MOSES PERCH
Lutjanus russelli (Bleeker, 1849)
Inhabits inshore rock or coral reefs, also deeper offshore reefs to at least 80 m depth; distinguished by reddish colouration and black spot (sometimes faint) on back, has more pointed snout than **8** above; found throughout the region; Indo-W. Pacific; to 45 cm. ★★★

13 BIGEYE SEAPERCH
Lutjanus lutjanus Bloch, 1790
Inhabits offshore coral reefs and trawling grounds to at least 90 m depth; distinguished from **11** by lighter midlateral stripe and much narrower space between eye and upper jaw; formerly known as *L. lineolatus*; found throughout the region; Indo-W. Pacific; to 30 cm. ★★★

14 DARK-TAILED SEAPERCH
Lutjanus lemniscatus (Valenciennes, 1828)
Inhabits inshore coral reefs and deeper offshore reefs; distinguished by blackish tail; found throughout the region; mainly Indo-Australian Archipelago; to 65 cm. ★★★

TROPICAL SNAPPERS

The members of the family Lutjanidae are generally known worldwide as snappers. However, this name sometimes causes confusion in Australia, because it is also used for at least two other common fishes belonging to two different families. The common Snapper of Australia's southern half belongs to the family Sparidae and is perhaps the best known of the two. The other fish is a member of the emperor family Lethrinidae (Plate 47-48). Although it is officially known as the Spangled emperor (Pl. 47-8), it is often referred to as North West snapper by anglers in Western Australia.

Snappers or seaperches featured on Plates 41-44 are primarily inhabitants of tropical reefs. Most species are distributed in the Indo-Pacific region, although they also occur in the Atlantic Ocean. *Lutjanus* is by far the largest genus, containing 66 species, many of which are brightly coloured. They are active predators that feed mainly on fishes, but crabs, shrimps, gastropods, cephalopods, and planktonic organisms are also eaten. The larger, deep bodied snappers (*Lutjanus* for example) usually have well developed canine teeth, adapted for seizing and holding larger prey items. More slender snappers such as *Pristipomoides* and *Etelis* have weaker dentition and consume a significant amount of plankton. The larger species of *Lutjanus*, particularly the "red snappers" are favourite angling fishes and also commercially important.

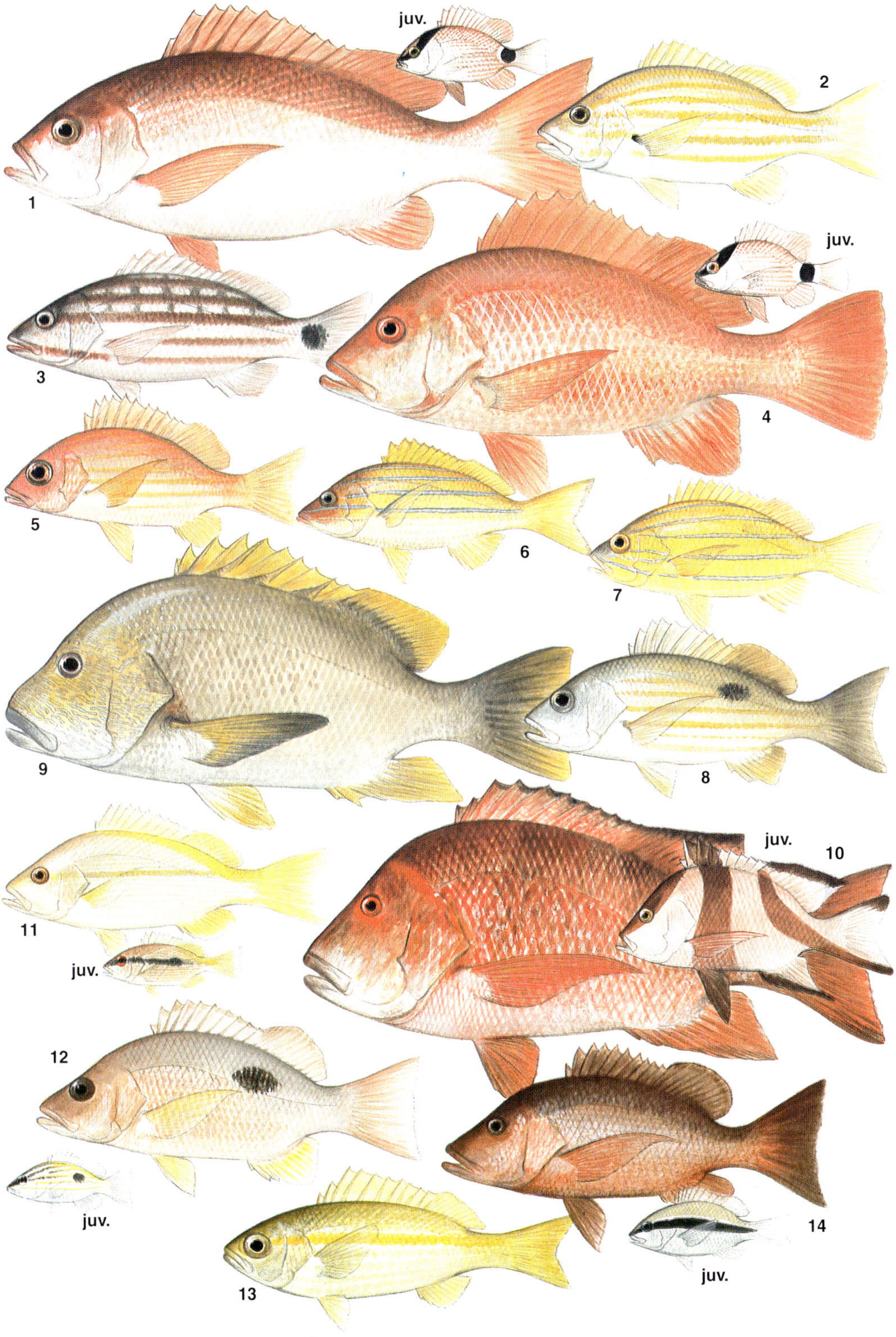
juv.
2
1
juv.
3
4
5
6
7
9
8
juv.
10
11
juv.
12
juv.
14
13
juv.

PLATE 43: SEAPERCHES OR SNAPPERS (FAMILY LUTJANIDAE)

1 RUSTY JOBFISH
Aphareus rutilans Cuvier, 1830
Inhabits reefs and rocky bottoms to at least 100 m depth; similar to *A. furca* (Plate 41-1), but lacks distinct black outline on edge of cheek and gill cover, also has strong yellow hue on body; N.W. Australia, Great Barrier Reef, and throughout S.E. Asia; to 80 cm. ★★★

2 HUSSAR
Lutjanus adetii (Castelnau, 1873)
Inhabits coral reefs, sometimes forming large daytime aggregations; distinguished by yellow iris, faint oblique lines on back, and yellow mid-lateral stripe; E. Australia, Coral Sea, and New Caledonia; to 50 cm.

3 BENGAL SEAPERCH
Lutjanus bengalensis (Bloch, 1790)
Inhabits coral reefs between about 10-25 m depth; similar to *L. kasmira* (Plate 42-6), but lacks faint horizontal lines on belly; W. Indonesia; mainly N. Indian Ocean; to 30 cm.

4 TWO-SPOT BANDED SEAPERCH
Lutjanus biguttatus (Valenciennes, 1930)
Inhabits coral reefs between 5-25 m depth, usually seen in rich coral areas of protected lagoons; distinguished by slender shape, midlateral white stripe, and pair of white spots below dorsal fin; offshore reefs of N. Great Barrier Reef and throughout S.E. Asia; E. Indian and W. Pacific oceans; to 20 cm. ★★

5 MOLUCCAN SEAPERCH
Lutjanus boutton (Lacepede, 1803)
Inhabits coral reefs to at least 20 m depth; found alone or in groups; distinguished by overall pink or reddish hue with yellow lower parts, also has deep notch on rear margin of cheek; E. Indonesia, New Guinea, and New Britain; mainly W. Pacific; to 30 cm. ★★★

6 EHRENBERG'S SEAPERCH
Lutjanus ehrenbergi Peters, 1869
Inhabits coral reefs and inshore areas, sometimes forms aggregations under wharves; similar to *L. fulviflamma* (Plate 42-8), but has much narrower space between eye and upper jaw, and scale rows on the back are parallel to the lateral line instead of rising obliquely; N. Great Barrier Reef and throughout S.E. Asia; Indo-W. Pacific; to 35 cm. ★★★

7 YELLOW-MARGINED SEAPERCH
Lutjanus fulvus (Schneider, 1801)
Inhabits coral reefs; distinguished by dark tail and soft dorsal fin both of which have a narrow white margin, and yellow anal, pelvic, and pectoral fins; offshore reefs of N.W. Australia, Great Barrier Reef, and throughout S.E. Asia; Indo-C. Pacific; to 40 cm. ★★★

8 PADDLETAIL
Lutjanus gibbus (Forsskål, 1775)
Inhabits coral reefs; distinguished by forked caudal fin with rounded lobes, a deep notch in rear margin of cheek, and obliquely oriented scale rows both above and below the lateral line; frequently implicated in ciguatera fish poisoning and therefore not recommended for eating; offshore reefs of N.W. Australia, Great Barrier Reef, and throughout S.E. Asia; Indo-C. Pacific; to 50 cm.

9 FINGERMARK SEAPERCH
Lutjanus johnii (Bloch, 1792)
Inhabits coastal areas, especially where mangroves are prevalent in estuaries and tidal rivers; distinguished by metallic pale yellow to silvery colour with dusky scale edges; smaller fish usually have a round, dark smudge on back below rear half of dorsal fin; found throughout the region; Indo-W. Pacific; to 70 cm. ★★★

10 LUNARTAIL SEAPERCH
Lutjanus lunulatus (Park, 1797)
Inhabits coral reefs between about 10-30 m depth; distinguished by black lunar-shaped marking on tail; S.E. Asia, New Guinea, Solomon Islands, and Vanuatu; N. Indian Ocean and Indo-Malay and Melanesian archipelagos; to 35 cm. ★★★

11 INDIAN SEAPERCH
Lutjanus madras (Valenciennes, 1831)
Inhabits coral and rocky reefs between 5-90 m depth; similar to *L. lutjanus* (Pl. 42-13), but space between eye and upper jaw is much wider; Indo-Malay Archipelago; Indo-W. Pacific; to 30 cm. ★★★

12 ONESPOT SEAPERCH
Lutjanus monostigma (Cuvier, 1828)
Inhabits coral reefs; distinguished by overall light grey to yellowish coloration, pinkish head and upper back, no longitudinal stripes - black spot on upper side may be faint or absent in adults; sale of this species is forbidden in Tahiti because of frequent ciguatera poisoning, but usually safe to eat in our region; N.W. Australia, Great Barrier Reef, and throughout S.E. Asia; Indo-C. Pacific; to 50 cm. ★★★

13 BLACK-BANDED SEAPERCH
Lutjanus semicinctus Quoy & Gaimard, 1824
Inhabits coral reefs, usually between 10-30 m depth; distinguished by series of dark bars on side and large dark spot at tail base; N. Great Barrier Reef, Philippines, E. Indonesia, and Melanesian archipelago to Tahiti; to 35 cm. ★★★

14 TIMOR SEAPERCH
Lutjanus timorensis (Quoy & Gaimard, 1824)
Inhabits coral reefs, usually below 15 m depth; distinguished by pearl-coloured spot just behind dorsal fin and black axil ("armpit") of pectoral fin; Indo-Malay Archipelago to Samoa; to 50 cm. ★★★

15 BLACK AND WHITE SNAPPER
Macolor niger Forsskål, 1775
Inhabits coral reefs, most commonly encountered on outer reef slopes to at least 90 m depth; feeds on plankton; distinguished from **16** below by juvenile and adult colour pattern differences shown here; N.W. Australia, Great Barrier Reef, and throughout S.E. Asia; Indo-C. Pacific; to 60 cm. ★★★

16 MIDNIGHT SNAPPER
Macolor macularis Fowler, 1931
Inhabits coral reefs, usually on outer reef slopes; small juveniles of this species and **15** above often seen in the vicinity of feather stars (crinoids); feeds on plankton; similar to **15**, but has more spots on back in juveniles and subadults, yellow eye, and network of bluish lines and spots on head; N.W. Australia, Great Barrier Reef, and throughout S.E. Asia; W. Pacific; to 60 cm. ★★★

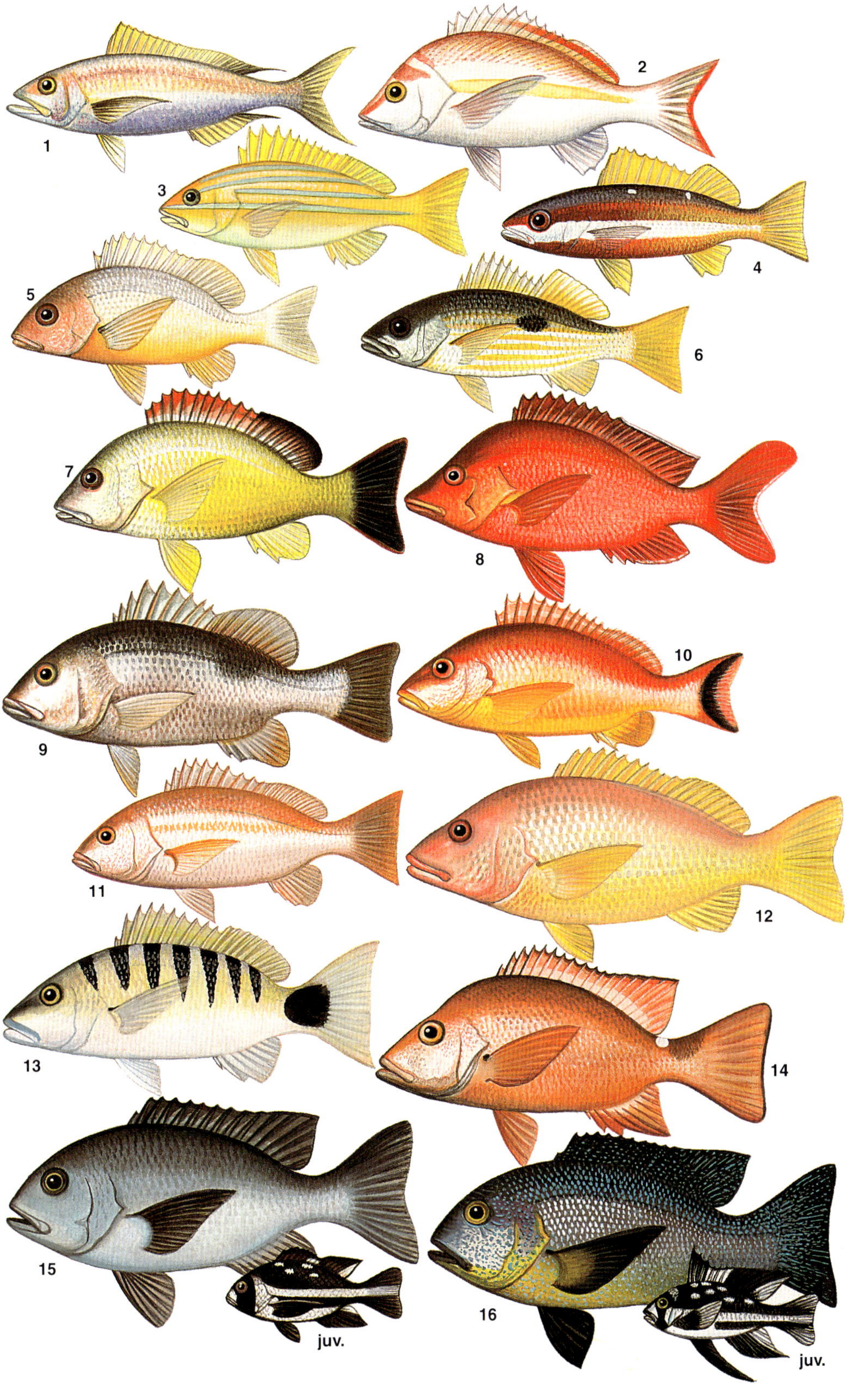
1
2
3
4
5
6
7
8
9
10
11
12
13
14
15
juv.
16
juv.

PLATE 44: SEAPERCHES AND SWEETLIPS

1 PINJALO
Pinjalo pinjalo (Bleeker, 1850)
Inhabits reefs and rocky bottoms to depths of 60 m; similar to *P. lewisi* (not shown), but lacks pearly bar on caudal peduncle and has yellowish hue on pelvic and anal fins; S.E. Asia and New Guinea; Indo-W. Pacific; to 50 cm. (LUTJANIDAE) ★★★

2 SAILFIN SNAPPER
Symphorichthys spilurus (Günther, 1874)
Inhabits sand bottoms in the vicinity of coral reefs; distinguished by unusual shape, elevated dorsal fin, and black saddle on upper tail base; juvenile has white-edged black stripe on middle of side; offshore reefs of W. Australia, Great Barrier Reef, and throughout S.E. Asia; mainly W. Pacific; to 60 cm. (LUTJANIDAE) ★★★

3 BLUE SNAPPER
Paracaesio sordidus Abe & Shinohara, 1962
Inhabits rocky bottoms and steep outer reef sopes between 30-200 m; distinguished by blue colouration and reddish (appears brown underwater) lower caudal fin lobe; S.E. Asia to Melanesian Archipelago; E. Indian Ocean and W. Pacific; to 40 cm. (LUTJANIDAE) ★★★

4 YELLOWTAIL BLUE SNAPPER
Paracaesio xanthurus (Bleeker, 1869)
Inhabits rocky bottoms and outer reef slopes between about 20-150 m; distinguished by brilliant yellow back and tail; similar to some species of fusiliers (Pls. 45-46), but has 10-11 soft dorsal rays and 8-9 soft anal rays (versus 14-16 and 11-12 respectively for fusiliers); N.W. Australia, Great Barrier Reef, and throughout S.E. Asia; Indo-W. Pacific; to 40 cm. (LUTJANIDAE) ★★★

5 GOLDFLAG JOBFISH
Pristipomoides auricilla (Jordan, Evermann, & Tanaka, 1927)
Inhabits rocky bottoms in about 90-360 m depth; distinguished by narrow yellow bars on body and yellow upper lobe on tail; S.E. Asia and Melanesian Archipelago; mainly W. Pacific; to 45 cm. (LUTJANIDAE) ★★★

6 GOLDENEYE JOBFISH
Pristipomoides flavipinnis Shinohara, 1963
Inhabits rocky bottoms in about 90-360 m depth; distinguished by yellowish dorsal fin, yellow edge on tail, and yellow iris; N. Queensland and Indo-Malay Archipelago to Samoa; mainly W. Pacific; to 60 cm. (LUTJANIDAE) ★★★

7 OBLIQUE-BANDED SNAPPER
Pristipomoides zonatus (Valenciennes, 1830)
Inhabits rocky bottoms in about 70-300 m depth; distinguished by broad yellow bands on red background colour; N. Queensland and Indo-Malay Archipelago to Tahiti; Indo-W. Pacific; to 50 cm. (LUTJANIDAE) ★★★

8 STRIPED SWEETLIPS
Plectorhinchus lessoni (Cuvier, 1830)
Inhabits coral reefs; distinguished by combination of dark stripes on upper side and spots on fins, also has prominent dark spot on upper pectoral fin base; *P. diagrammus* is a synonym; Great Barrier Reef and throughout S.E. Asia to Melanesian Archipelago; W. Pacific; to 40 cm. (HAEMULIDAE) ★★

9 YELLOWDOT SWEETLIPS
Diagramma pictum (Thunberg, 1792)
Inhabits sandy areas in the vicinity of coral reefs; similar to *D. labiosum* (Pl. 45-6), but adults have yellow spots and juvenile pattern also differs as shown; Indo-Malay Archipelago; mainly W. Pacific; to 90 cm. (HAEMULIDAE) ★★

10 ORIENTAL SWEETLIPS
Plectorhinchus orientalis (Bloch, 1793)
Inhabits coral reefs; similar to **13** below, but black stripes are horizontal rather than oblique; throughout S.E. Asia to New Guinea; Indo-W. Pacific; to 50 cm. (HAEMULIDAE) ★★

11 GIANT SWEETLIPS Cairns 10/98
Plectorhinchus obscurum (Günther, 1871)
Inhabits coral reefs; usually recognised by its large size, blubbery lips, and black markings on fins; *P. harrawayi* is a synonym; Great Barrier Reef and throughout S.E. Asia to Melanesian Archipelago; Indo-W. Pacific; to at least 100 cm. (HAEMULIDAE) ★★

12 DOTTED SWEETLIPS
Plectorhinchus picus (Cuvier, 1830)
Inhabits coral reefs; distinguished by pattern of small dark spots on head, body, and fins, also has dark upper lip; juvenile has distinctive black and white pattern; Great Barrier Reef and throughout S.E. Asia to Melanesian Archipelago; Indo-W. Pacific; to 50 cm. (HAEMULIDAE) ★★

13 DIAGONAL-BANDED SWEETLIPS
Plectorhinchus lineatus (Linnaeus, 1758)
Inhabits coral reefs; distinguished by combination of oblique black bands, black-spotted yellow fins, and yellow lips; *P. goldmanni* is a synonym; Great Barrier Reef and throughout S.E. Asia to Melanesian Archipelago; W. Pacific; to 50 cm. (HAEMULIDAE) ★★

DEEPWATER SNAPPERS

Some of the best eating fishes live well below the depths penetrated by most recreational anglers, but their existence is no secret to villagers in South-east Asia, who have utilised this food source for centuries. Volcanic island shores drop quickly into deep water throughout the Malay-Indonesian-Melanesian Archipelago. In many cases it requires only a few hundred metres of paddling in a small boat to reach this habitat. But getting there is only half the battle. Fishers utilise extraordinary lengths of monofilament handline, weighted with rocks. Depths between 50 and 100 metres are routinely fished, but a in some cases much deeper water is penetrated. However, as the depth increases, so does the risk of sharks. If the exposure time after hooking a fish is lengthy, say five or 10 minutes, the catch will certainly be devoured by sharks on the way up.

Snappers (also called jobfishes) in the genera *Paracaesio* and *Pristipomoides* are prime targets of handline anglers. Their mild-tasting flesh has a delicate consistency and is highly prized. At Hawaii and other Pacific islands the Rosy Snapper (Plate 41-6) is captured commercially in 100-300 metres depth with handlines and bottom long-lines that are hauled with motorised winches. This fish, localy known as Opakapaka, retails for more than US $25 per kilogram in Hawaii. In Australia, stocks of this fish and several other deep-dwelling snappers have only recently been discovered on the edge of the continental shelf.

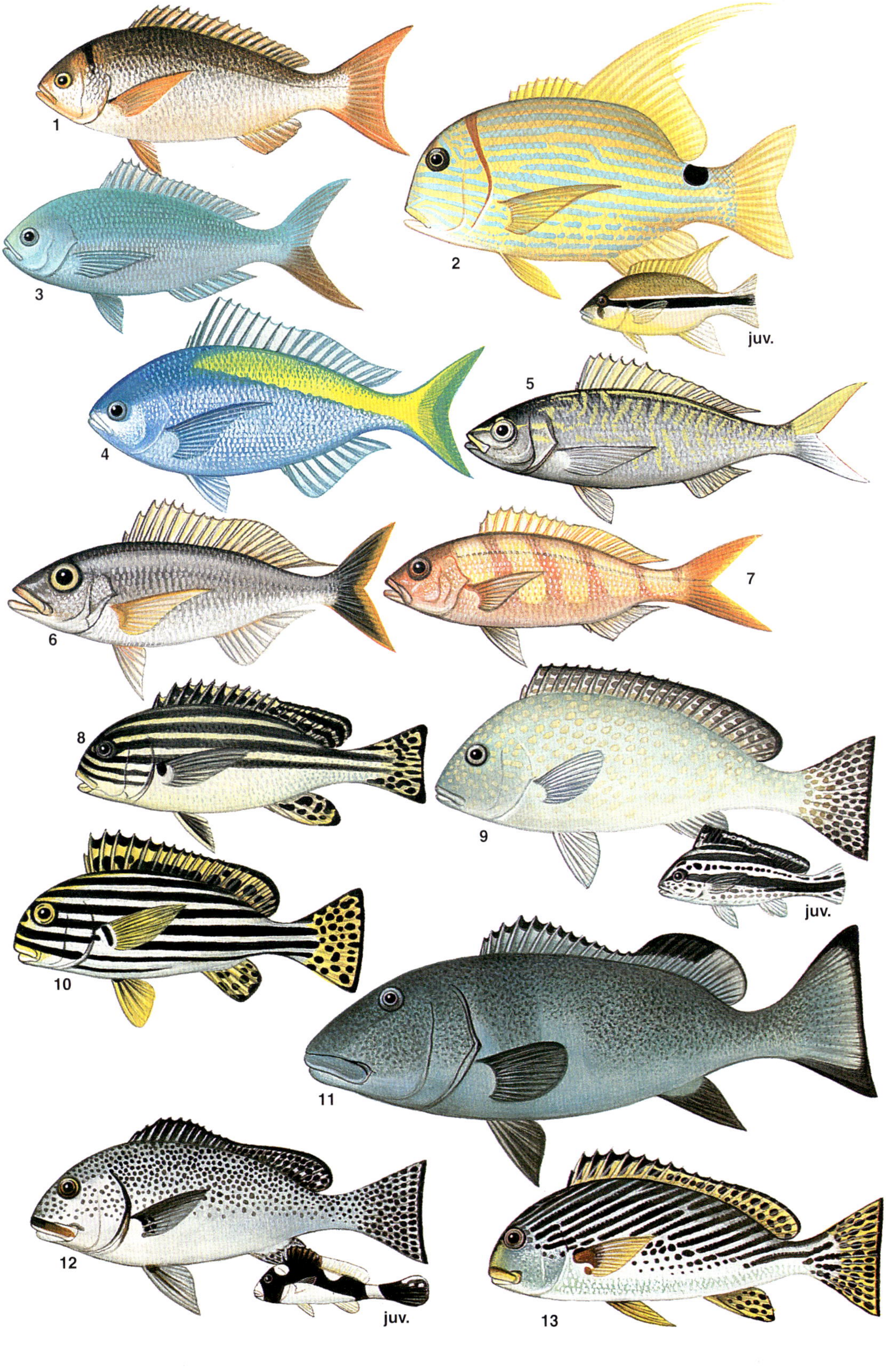
1
2
3
juv.
5
4
6
7
8
9
juv.
10
11
12
juv.
13

PLATE 45: SWEETLIPS, FUSILERS, AND BANJOFISHES

1 MINSTREL SWEETLIPS
Plectorhinchus schotaf (Forsskål, 1775)
Inhabits caves and crevices of coral reefs; similar to **5** below, but distinguished by 11 or 12 dorsal spines (versus 13 or 14); N.W. Australia, Great Barrier Reef, and throughout S.E. Asia; Indo-W Pacific; to 60 cm. (HAEMULIDAE) ★★

2 MANY-SPOTTED SWEETLIPS
Plectorhinchus chaetodontoides Lacepède, 1800
Inhabits caves and crevices of coral reefs; distinguished by spotted pattern; juvenile has peculiar undulating motion; N.W. Australia, Great Barrier Reef, and throughout S.E. Asia; Indo-W. Pacific; to 60 cm and at least 7kg. (HAEMULIDAE) ★★

3 MANY-LINED SWEETLIPS
Plectorhinchus multivittatum (Macleay, 1878)
Inhabits coral reefs; distinguished by white spots arranged in horizontal rows and yellow fins; N.W. Australia and throughout S.E. Asia; Indo-W. Pacific; to 50 cm. (HAEMULIDAE) ★★

4 CELEBES SWEETLIPS
Plectorhinchus chrysotaenia (Bleeker, 1855)
Inhabits coral reefs; distinguished by yellow stripes on head and sides; N.W. Australia, Great Barrier Reef, and throughout S.E. Asia; W. Pacific; to 40 cm.
P. celebicus is a synonym. (HAEMULIDAE) ★★

5 BROWN SWEETLIPS
Plectorhinchus gibbosus (Lacepède, 1802)
Inhabits coral reefs; similar to **1** above, but distinguished by 13 or 14 dorsal spines (versus 11 or 12); formerly known as *P. nigrus*; found throughout the region; Indo-W. Pacific: to 60 cm and at least 6 kg. (HAEMULIDAE) ★★

6 PAINTED SWEETLIPS
Diagramma labiosum Macleay, 1883
Inhabits coral reefs; sometimes seen in schools, adults distinguished by silvery body colour, smaller specimens have spots or stripes, but differs from **2** above in having 9 or 10 dorsal spines (versus 12) and lacks notch in dorsal fin profile; found throughout the region; Indo-W. Pacific; to 90 cm and at least 6 kg. (HAEMULIDAE) ★★

7 RIBBON SWEETLIPS
Plectorhinchus polytaenia (Bleeker, 1852)
Inhabits coral reefs; distinguished by vivid pattern of dark-edged orange and white stripes; N.W. Australia and throughout S.E. Asia; Indo-Australian Archipelago; to 40 cm. (HAEMULIDAE) ★★

8 SPOTTED JAVELINFISH
Pomadasys kaakan (Cuvier, 1830)
Inhabits estuaries and coastal waters usually over sand or mud bottoms; distinguished by series of spots arranged in vertical rows on upper side and spots on dorsal fin; formerly known as *P. hasta*; found throughout the region; Indo-W. Pacific; to 38 cm and 6 kg. (HAEMULIDAE) ★★★

9 BLOTCHED JAVELINFISH
Pomadasys maculatum (Bloch, 1797)
Inhabits estuaries and coastal waters usually over sand or mud bottoms, distinguished by large dark blotches mainly on upper sides; found throughout the region; Indo-W. Pacific; to 45 cm. (HAEMULIDAE) ★★★

10 LINED JAVELINFISH
Hapalogenys kishinouyei Smith & Pope, 1907
Inhabits coastal waters, usually over mud or sand bottoms; distinguished by brown stripes on upper part of back; N.W. Australia, Indonesia, and Philippines; W. Pacific; to 30 cm. (HAEMULIDAE) ★★★

11 RED-BELLIED FUSILER
Caesio cuning (Bloch, 1791)
Inhabits coral reefs, forming midwater schools; distinguished by yellow tail, similar to **12** below, but lacks black tips of tail; found throughout the region; Indo-W. Pacific; to 43 cm. (CAESIONIDAE) ★★★

12 BLUE FUSILER
Caesio lunaris Cuvier, 1830
Inhabits coral reefs, forming midwater schools; similar to **11** above, but has blue tail with black tips; N.W. Australia, Great Barrier Reef, and throughout S.E. Asia; Indo-W. Pacific; to 38 cm. (CAESIONIDAE) ★★★

13 BLACK-TIPPED FUSILER
Pterocaesio digramma (Bleeker, 1865)
Inhabits coral reefs, forming midwater schools; distinguished from **11** and **12** above by more slender shape and pair of yellow stripes on upper side; found throughout the region; to 28 cm. (CAESIONIDAE) ★★

14 BANJOFISH
Banjos banjos (Richardson, 1846)
Inhabits coastal waters usually over sand bottoms; distinguished by stout dorsal and anal spines, angular forehead and black margin on soft dorsal and caudal fins; found throughout the region; Indo-W. Pacific; to 30 cm. (BANJOSIDAE)

SWEETLIPS, ETC.

Most of the fishes on Plate 45 belong to the family Haemulidae, commonly known as Grunts. Most members of this family found in reef habitats in the region belong to the genus *Plectorhinchus* and are referred to as Sweetlips. Perhaps the most notable aspect of these fishes, besides their use as food, are the dramatic changes which many of the species undergo. Juveniles often exhibit striking patterns that gradually change with advancing age. Two examples (2 and 6) of these transformations are shown on Plate 45. The young of *Plectorhinchus chaetodontoides* is thought to mimic unpalatable soft-bodied invertebrates such as nudibranchs or turbellarians, and thus enjoys some measure of freedom from predation. Young grunts feed on zooplankton; whereas adults eat a variety of benthic invertebrates. Maximum size is about 1 m TL, but many species are under 40-50 cm. Worldwide the family contains about 175 species and is represented in all tropical seas.

Fusilers in the family Caesionidae are closely related to Seaperches or Snappers (Plates 41-44). The family contains 20 species which are distributed across the Indo-Pacific region. They frequently form schools containing up to several hundred individuals that feed in midwater on plankton above coral reefs.

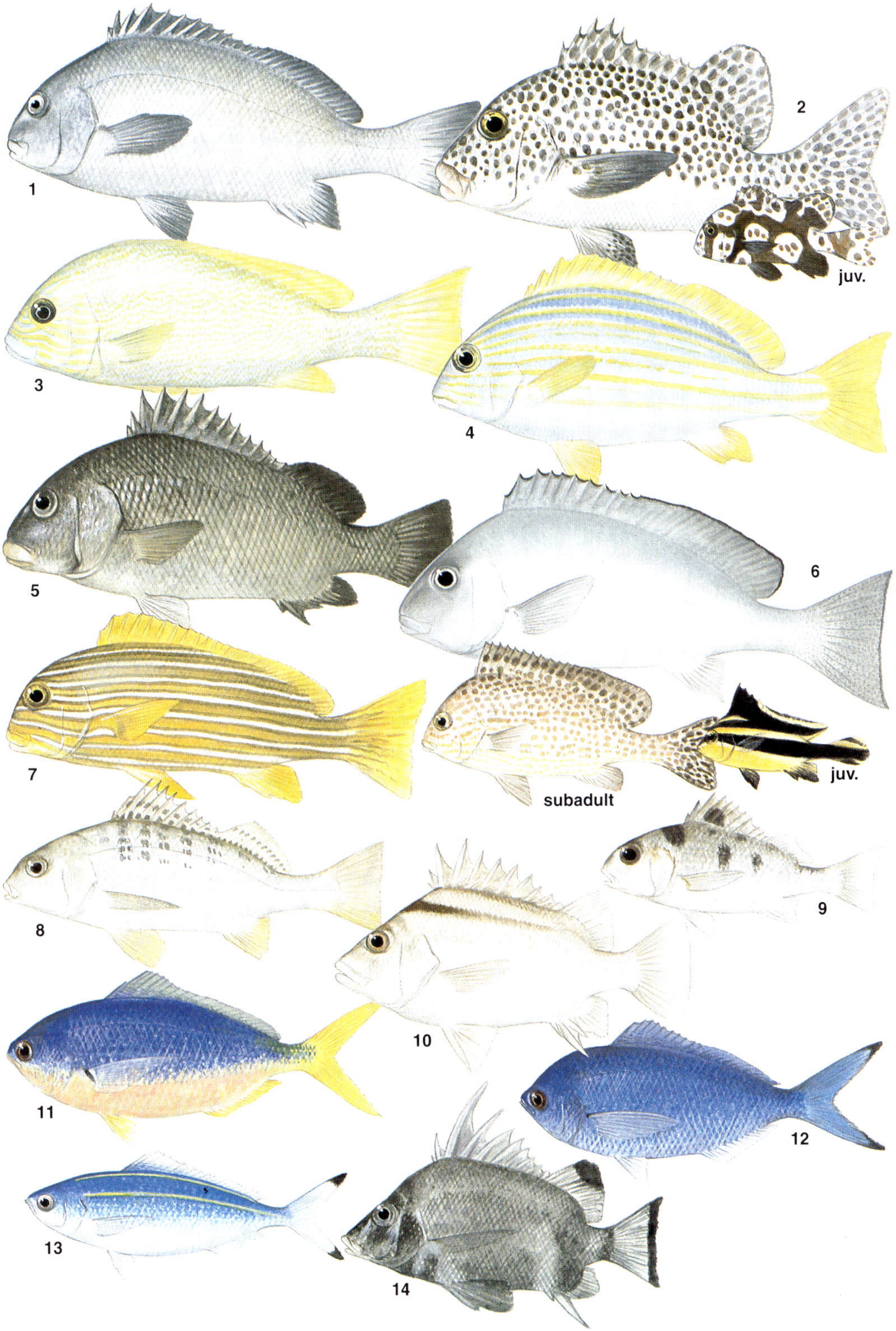

1
2
juv.
3
4
5
6
7
subadult
juv.
8
9
10
11
12
13
14

PLATE 46: FUSILIERS AND TILEFISHES

1 YELLOW AND BLUEBACK FUSILIER
Caesio teres, Seale, 1906
Inhabits coral reefs, forming mid-water schools; distinguished by yellow area encompassing dorsal fin, upper back, and tail; offshore reefs of N.W. Australia; Great Barrier Reef, and throughout S.E. Asia; Indo-C. Pacific; to 40 cm. (CAESIONIDAE) ★★

2 YELLOWBACK FUSILIER
Caesio xanthonota Bleeker, 1853
Inhabits coral reefs, forming mid-water schools; similar to 1 above, but yellow area on back more extensive, extending forward onto head; S. Indonesia; mainly Indian Ocean; to 40 cm. (CAESIONIDAE) ★★

3 BLUE AND GOLD FUSILIER
Caesio caerulaurea Lacepède, 1801
Inhabits coral reefs, forming mid-water schools; distinguished by yellow stripe above lateral line and dark streak along edge of upper and lower lobe of tail; N.W. Australia, Great Barrier Reef, and throughout S.E. Asia; Indo-W. Pacific; to 35 cm. (CAESIONIDAE) ★★

4 THREE-STRIPED FUSILIER
Pterocaesio trilineata Carpenter, 1987
Inhabits coral reefs, forming mid-water schools; distinguished by trio of alternating dark and light stripes on upper back; offshore reefs of N.W. Australia, Great Barrier Reef, Philippines, and E. Indonesia; W. Pacific; to 20 cm. (CAESIONIDAE) ★★

5 MARR'S FUSILIER
Pterocaesio marri Schultz, 1953
Inhabits coral reefs, forming mid-water schools; distinguished by pair of thin yellow stripes on side, the lower covering lateral line for most of its length; Great Barrier Reef and throughout S.E. Asia; Indo-C. Pacific; to 25 cm. (CAESIONIDAE) ★★

6 DARK-BANDED FUSILIER
Pterocaesio tile (Cuvier, 1830)
Inhabits outer reef slopes, forming mid-water schools; distinguished by dark stripe on upper side and neon blue band on middle of sides, also has dark streak on each lobe of tail; offshore reefs of N.W. Australia, Great Barrier Reef, and throughout S.E. Asia; Indo-C. Pacific; to 30 cm. (CAESIONIDAE) *Cairns 10/98* ★★

7 WIDEBAND FUSILIER
Pterocaesio lativittata Carpenter, 1987
Inhabits coral reefs, forming mid-water schools; distinguished by broad yellow stripe on middle of side and dark tips on tail; Indo-Malay Archipelago to New Guinea; C. and E. Indian Ocean and W. Pacific; to 20 cm. (CAESIONIDAE) ★★

8 RANDALL'S FUSILIER
Pterocaesio randalli Carpenter, 1987
Inhabits coral reefs, forming mid-water schools; distinguished by elongate blotch of golden yellow on anterior part of body; Indo-Malay Archipelago; to 25 cm. (CAESIONIDAE) ★★

9 ONE-STRIPED FUSILIER
Pterocaesio tessellata Carpenter, 1987
Inhabits coral reefs, forming mid-water schools; distinguished by narrow yellow stripe along lateral line, dusky scale edges giving appearance of wavy stripes on side, and dark tips on tail; Indo-Malay and Melanesian archipelagos; E. Indian Ocean and W. Pacific; to 25 cm. (CAESIONIDAE) ★★

10 GOLDBAND FUSILIER
Pterocaesio chrysozona (Cuvier, 1830)
Inhabits coral reefs, forming mid-water schools; similar to 7 above, but has narrower yellow stripe that is mainly below lateral line (except on tail base); found throughout the region; Indo-W. Pacific; to 21 cm. (CAESIONIDAE) ★★

11 BANANA FUSILIER
Pterocaesio pisang (Bleeker, 1853)
Inhabits coral reefs, forming mid-water schools; distinguished by lack of yellow stripe on side and dark (often red) tips on tail, overall colouration is sometimes reddish; Indo-Malay and Melanesian archipelagos; Indo-W. Pacific; to 21 cm. (CAESIONIDAE) ★★

12 MOTTLED FUSILIER
Dipterygonatus balteatus (Valenciennes, 1830)
Inhabits coral reefs, forming mid-water schools; distinguished by small size, very slender shape, and narrow tan-coloured stripe just above lateral line with pair of light, wavy lines immediately above, also lack of dark tips on tail; found throughout the region; Indo-W. Pacific; to 14 cm. (CAESIONIDAE) ★★

13 SLENDER FUSILIER
Gymnocaesio gymnopterus (Bleeker, 1856)
Inhabits coral reefs, forming mid-water schools; similar in appearance to some species of *Pterocaesio*, but lacks scales on dorsal and anal fins; Indo-Malay and Melanesian archipelagos; Indo-W. Pacific; to 18 cm. (CAESIONIDAE) ★

14 BLUE BLANQUILLO
Malacanthus latovittatus (Lacepède, 1801)
Inhabits sand-rubble bottoms, usually near edge of coral reefs; distinguished by elongate shape, bluish colouration, and broad black stripe which extends onto tail; offshore reefs of W. Australia, Great Barrier Reef, and throughout S.E. Asia; Indo-C. Pacific; to 35 cm. (MALACANTHIDAE)

15 YELLOW-BLOTCHED TILEFISH
Hopololatilus fourmanoiri Smith, 1963
Inhabits silt-sand bottoms in 40-60 m depth; like other members of the genus it is frequently found in pairs that share a burrow; distinguished by light grey body with scattered yellow patches on head and body, a large dark "ear" spot, and longitudinal nearly triangular black spot on middle of tail; *H. luteus* from Indonesia (Flores) is similar, but is entirely yellow and lacks the dark marking on tail; South China Sea to Solomon Islands; to 14 cm. (MALACANTHIDAE)

16 GREY TILEFISH
Hoplolatilus cuniculus Randall & Dooley, 1974
Inhabits sand-rubble bottoms, usually between 25-100 m depth; retreats to sandy burrow if disturbed; colour entirely grey; Great Barrier Reef and throughout S.E. Asia; Indo-W. Pacific; to 15 cm. (MALACANTHIDAE)

17 BLUE TILEEFISH
Hoplolatilus starcki Randall & Dooley, 1974
Inhabits steep outer reef slopes below about 20 m depth, often found in pairs; retreats to sandy burrow if disturbed; distinguished by slender shape, blue head, and forked yellow tail; young fish entirely blue; offshore reefs of W. Australia, Great Barrier Reef, and throughout S.E. Asia; W. and C. Pacific; to 15 cm. (MALACANTHIDAE)

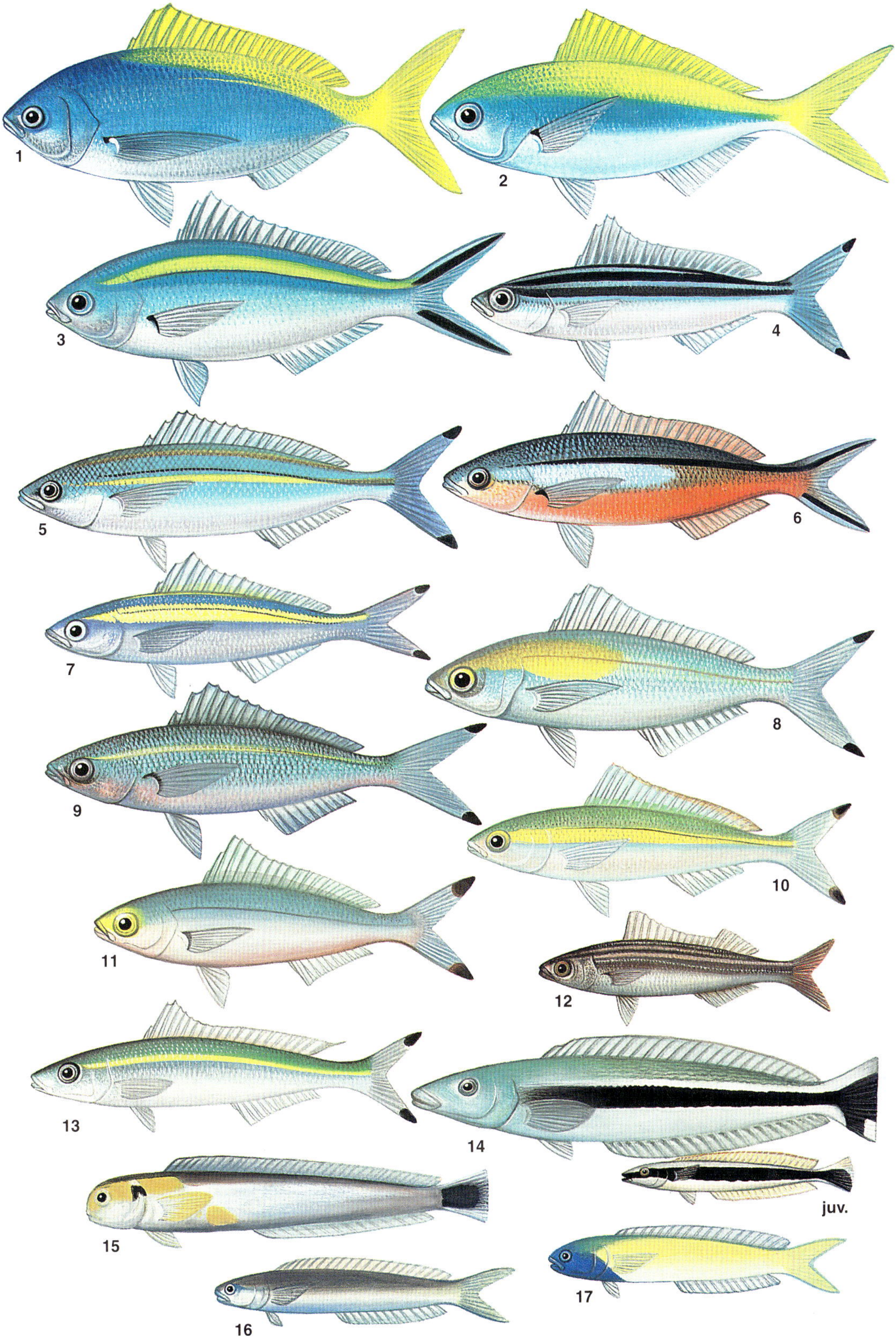
1
2
3
4
5
6
7
8
9
10
11
12
13
14
juv.
15
16
17

PLATE 47: EMPERORS AND BREAMS

1 HUMPNOSE BIG-EYE BREAM
Monotaxis grandoculis (Forsskål, 1775)
Inhabits coral reefs, frequently over adjacent sand or rubble areas; distinguished by large eye and bluntly rounded snout, juvenile with dark bars that are reduced to saddles on back of adults; found throughout the region; Indo-C. Pacific; to 60 cm. (LETHRINIDAE) ★★★

2 SWALLOWTAIL SEABREAM
Gymnocranius elongatus Senta, 1973
Inhabits sand bottoms and trawling grounds; distinguished by 5 narrow bars on side and strongly forked tail; found throughout the region; E. Indian Ocean and W. Pacific; to 18 cm. (LETHRINIDAE) ★★

3 NAKED-HEADED SEABREAM
Gymnocranius griseus (Schlegel, 1844)
Inhabits sand bottoms and trawling grounds; similar to **2**, but bars generally wider and more diffuse and tail not as forked; N.W. Australia and throughout S.E. Asia; mainly W. Pacific; to 35 cm. (LETHRINIDAE) ★★

4 ROBINSON'S SEABREAM
Gymnocranius grandoculus (Valenciennes, 1830)
Inhabits sand bottoms and trawling grounds; distinguished by wavy blue lines on cheek and snout; also known as Blue-lined seabream; found throughout the region; Indo-W. Pacific; to 80 cm and at least 5 kg. (LETHRINIDAE) ★★★

5 LONG-NOSED EMPEROR
Lethrinus olivaceus Valenciennes, 1830
Inhabits coral reefs; distinguished by long painted snout and overall greyish colour; found throughout the region; Indo-C. Pacific; to 100 cm and 10 kg. (LETHRINIDAE) ★★★

6 SWEETLIP EMPEROR
Lethrinus miniatus (Schneider, 1801)
Inhabits coral reefs; distinguished by orange area around eyes, bright red dorsal fin, and red patch at base of pectoral fin; sometimes a series of dark bars on side; formerly known as *L. chrysostomus*; N. Australia, Coral Sea, New Caledonia, and Ryukyu Is. to Taiwan; to 90 cm and at least 7 kg. (LETHRINIDAE) ★★★★

7 GRASS EMPEROR
Lethrinus laticaudis Alleyne & Macleay, 1877
Inhabits coral reefs and adjacent sand-rubble areas; similar to **8** below, but has blue spots (versus blue bars) on cheek, whitish oblique bar behind eye, and brownish band from eye to mouth; often assumes barred pattern when under stress; S. Indonesia, N. Australia, New Guinea, and Solomon Islands; to 80 cm. (LETHRINIDAE) ★★★★

8 SPANGLED EMPEROR
Lethrinus nebulosus (Forsskål, 1775)
Inhabits coral reefs, usually over adjacent sandy areas; distinguished by blue spots on scales, similar to **7** above, but has blue bars (versus spots) on cheek and lacks other marks on head as described above; also known as North-West snapper; found throughout the region; Indo-W. Pacific; to 86 cm; 6.577 kg. (LETHRINIDAE) ★★★★

9 BLUE-LINED EMPEROR
Lethrinus sp.
Inhabits coastal reefs, usually over adjacent sand or rubble areas; similar to **7** and **8** above, but has dark streak on each scale instead of blue spots, and has short blue lines radiating from eye of which several cross forehead and connect with opposite eye; an undescribed species; N. Australia; to 60 cm. (LETHRINIDAE) ★★★★

10 THREADFIN EMPEROR
Lethrinus genivittatus Valenciennes, 1830
Inhabits sand-weed areas, sometimes in estuaries; distinguished by dark blotch above pectoral fin and elongate filament near front of dorsal fin, also known as Longspine emperor and Lancer; found throughout the region; E. Indian Ocean and W. Pacific; to 25 cm. (LETHRINIDAE) ★★

11 PURPLE-HEADED EMPEROR
Lethrinus lentjan (Lacepède, 1802)
Inhabits sandy areas next to coral reefs; distinguished by pale spots on scales, red margin on gill cover, and red spot at base of pectoral fin; head of fresh caught specimens usually turns to purple colour; also known as Pink-eared emperor; found throughout the region; Indo-W. Pacific; to 50 cm. (LETHRINIDAE) ★★★

12 SPOTCHEEK EMPEROR
Lethrinus rubrioperculatus Sato, 1978
Inhabits coral reefs and trawling grounds; distinguished by elongate shape and red or brown blotch on upper margin of gill cover; colour varies from uniform greyish to blotchy pattern as shown (most emperors can quickly assume a similar pattern); found throughout the region; Indo-C. Pacific; to 50 cm. (LETHRINIDAE) ★★★

13 VARIEGATED EMPEROR
Lethrinus variegatus Valenciennes, 1830
Inhabits sand-weed areas near coral reefs; distinguished by elongate shape and blotchy pattern with faint cross bands on tail, similar to **12** above, but lacks blotch on upper gill cover; found throughout the region; Indo-W. Pacific; to 20 cm. (LETHRINIDAE) ★★

14 NORTHWEST BLACK BREAM
Acanthopagrus palmaris (Whitley, 1935)
Inhabits coastal reefs, sometimes in estuaries; similar to **15** below, but is generally darker and lacks yellow on fins; N.W. Australia only; to 40 cm. (SPARIDAE) ★★★

15 WESTERN YELLOWFIN BREAM
Acanthopagrus latus (Houttuyn, 1782)
Inhabits coastal reefs, frequently in schools; similar to **14** above, but lighter in colour and with yellow fins; found throughout the region; N.W. Austrlaia and S.E. Asia; N. Indian Ocean and W. Pacific; to 45 cm. (SPARIDAE) ★★★

16 LONG-SPINED SNAPPER
Argyrops spinifer (Forsskål, 1775)
Inhabits coastal waters and deeper trawling grounds; distinguished by filamentous dorsal spines and angular forehead profile; found throughout the region; Indo-W. Pacific; to 65 cm and at least 2 kg. (SPARIDAE) ★★★

17 DEEPSEA SNAPPER
Dentex tumifrons (Temminck & Schlegel, 1842)
Inhabits trawling grounds; similar to **16** above, but lacks filamentous dorsal spines and forehead is more rounded; N.W. Australia and S. Indonesia; mainly W. Pacific; to 35 cm. (SPARIDAE) ★★★

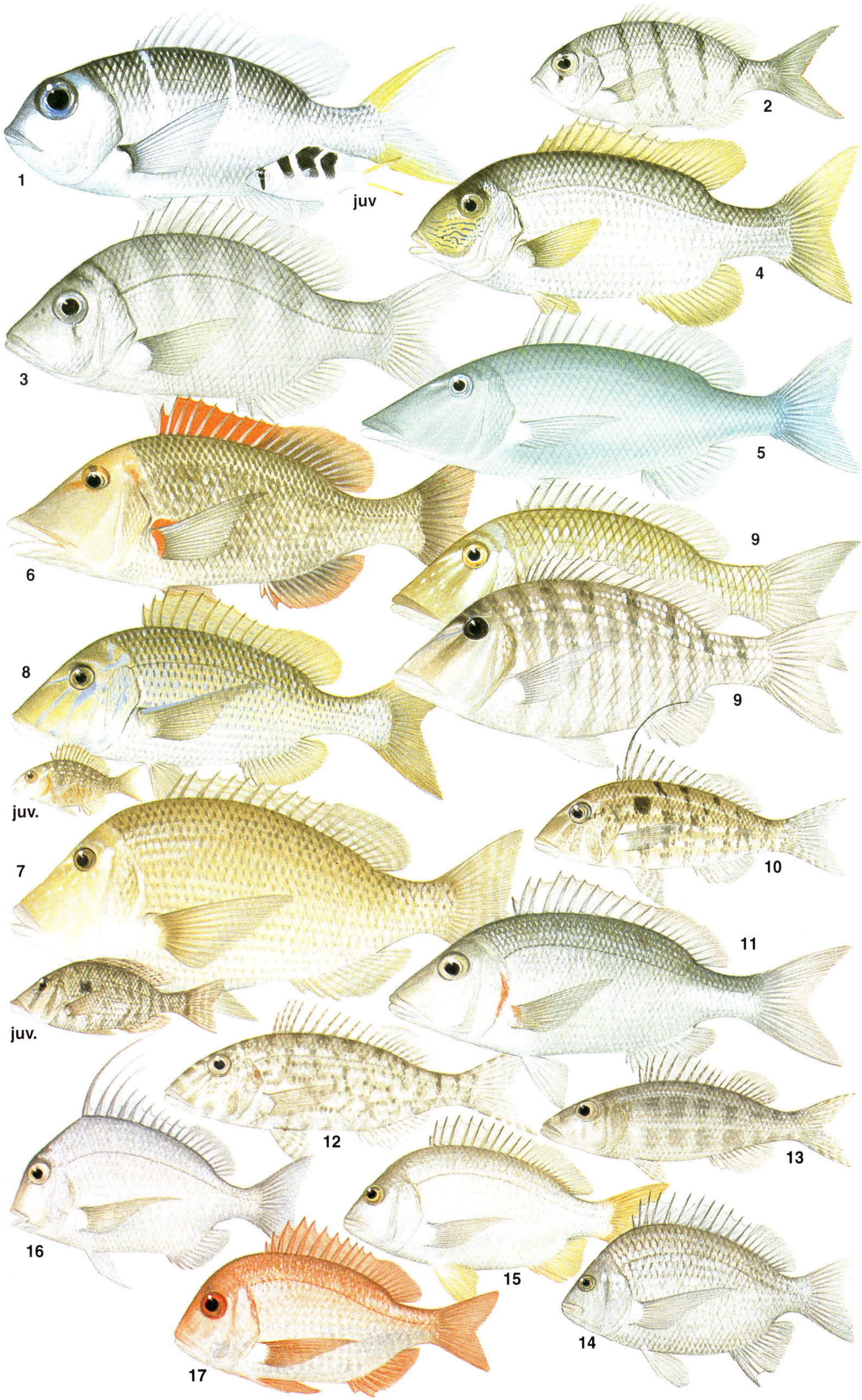
1
juv
2
4
3
5
6
9
8
9
juv.
10
7
11
juv.
12
13
16
15
14
17

PLATE 48: EMPERORS (FAMILY LETHRINIDAE)

1 STRIPED LARGE-EYE BREAM
Gnathodentex aurolineatus (Lacepède, 1802)
Inhabits coral reefs, sometimes seen in groups; distinguished by orange-brown stripes on side and yellow patch below rear part of dorsal fin; Great Barrier Reef, offshore reefs of W. Australia, and throughout S.E. Asia; Indo-C. Pacific; to 30 cm.

2 COLLARED LARGE-EYE BREAM
Gymnocranius audleyi Ogilby, 1916
Inhabit sand-rubble fringe of coral reefs; distinguished by pale-edged dark mark above and behind eye; S. Great Barrier Reef; S. Queensland only; to 40 cm.

3 JAPANESE LARGE-EYE BREAM
Gymnocranius euanus Günther, 1979
Inhabit sand-rubble fringe of coral reefs; distinguished by dark flecks on otherwise pale (often silvery) body; Great Barrier Reef, New Guinea, Philippines, and S. China Sea; W. Pacific; to 45 cm.

4 BLACKNAPE LARGE-EYE BREAM
Gymnocranius sp.
Inhabit sand-rubble fringe of coral reefs; distinguished by dusky back, dark mark immediately above eye, and faint dark bar below eye; Great Barrier Reef, Coral Sea, New Caledonia, and Japan; to 45 cm.

5 BLUE-SPOTTED LARGE-EYE BREAM
Gymnocranius microdon (Bleeker, 1851)
Inhabit sand-rubble fringe of coral reefs; distinguished by blue spots on head; Indo-Malay Archipelago; W. Pacific; to 45 cm.

6 YELLOWSNOUT LARGE-EYE BREAM
Gymnocranius frenatus Bleeker, 1873
Inhabits sand and mud bottoms between 20-80 m depth; distinguished by yellow snout and diagonal blue bands on head; Indo-Malay Archipelago and S. China Sea; to 35 cm.

7 ORANGE-STRIPED EMPEROR
Lethrinus obsoletus (Forsskål, 1775)
Inhabits coral reefs; distinguished by yellow stripe on lower side, can also assume mottled pattern; Great Barrier Reef, offshore reefs of W. Australia, and throughout S.E. Asia; Indo-W. Pacific; to 40 cm.

8 SMALLTOOTH EMPEROR
Lethrinus microdon Valenciennes, 1830
Inhabits coral reefs; similar to *L. olivaceus* (Pl. 47-5), but snout slightly shorter and 4.5 scales between lateral line and base of middle dorsal spine instead of 5.5 scales; N.W. Australia and Indo-Malay Archipelago to New Guinea; Indo-W. Pacific; to 50 cm.

9 ORNATE EMPEROR
Lethrinus ornatus Valenciennes, 1830
Inhabits coral reefs; distinguished by yellow stripes on side and red margin on cheek and gill cover; Great Barrier Reef and Indo-Malay Archipelago to New Guinea; E. Indian Ocean and W. Pacific; to 40 cm.

10 YELLOWLIP EMPEROR
Lethrinus xanthocheilus Valenciennes, 1830
Inhabits coral reefs; distinguished by relatively large size, elongate shape, and yellowish lips; N. Great Barrier Reef and throughout S.E. Asia; Indo-C. Pacific; to 60 cm.

11 YELLOW-SPOTTED EMPEROR
Lethrinus erythracanthus Valenciennes, 1830
Inhabits coral reefs, usually seen on outer slopes; distinguished by overall dusky grey colour (often with scattered white spots), yellowish fins, and large fleshy lips; juvenile yellowish with narrow blue lines or horizontal rows of blue spots; N. Great Barrier Reef and throughout S.E. Asia; Indo-W. Pacific; to 60 cm.

12 LONGFIN EMPEROR
Lethrinus erythropterus Valenciennes, 1830
Inhabits coral reefs; distinguished by dark band from eye to snout and light bars usually on rear half of body; Indo-Malay Archipelago to New Guinea; Indo-W. Pacific; to 45 cm.

13 AMBON EMPEROR
Lethrinus amboinensis Bleeker, 1854
Inhabits sand-rubble fringe of coral reefs; distinguished by elongate shape and yelllow pectoral fins; N.W. Australia, and throughout S.E. Asia; W. and C. Pacific; to 60 cm.

14 REDSNOUT EMPEROR
Lethrinus reticulatus Valenciennes, 1830
Inhabits sand-mud bottoms, often caught by trawlers; a nondescript emperor with dark head and tail, and yellowish pectoral fins; Indonesia and Philippines; Indo-W. Pacific; to 40 cm.

15 THUMBPRINT EMPEROR
Lethrinus harak (Forsskål, 1775)
Inhabits sand-rubble bottoms, weedy areas, and mangroves near coral reefs; distinguished by dark blotch on middle of side; N. Queensland and throughout S.E. Asia; Indo-W. Pacific; to 50 cm.

16 BLACK-BLOTCH EMPEROR
Lethrinus semicinctus Valenciennes, 1830
Inhabits sand-rubble bottoms near coral reefs; distinguished by irregular dark bars on side, including enlarged dark blotch below rear part of dorsal fin; found throughout the region; E. Indian Ocean and W. Pacific; to 35 cm.

EMPERORS

The members of the family Lethrinidae, commonly known as emperors and seabreams, are mainly reef fishes that prefer sand and rubble bottoms. Most species occur in relatively shallow water (5-30 m), but a few range to the edge of continental shelves to depths of 100 metres or more. Their diet consists of small fishes and a variety of invertebrates including polychaete worms, crabs, shrimps, squid, octopus, and starfish. Some species have well developed molar-like teeth adapted for crushing hard-shelled prey such as molluscs, crustaceans and sea urchins.

Mature adults are initially female, but change to the male sex later in life. There is little information about spawning other than a few observations by fishermen. Most species are believed to spawn after dark, following a local migration to particular areas near a reef, either in sheltered lagoons or on the edge of exposed oceanic reefs. Spawning apparently occurs in large aggregations, which mill about near the surface, or near the bottom of reef slopes. Reproductive activity generally peaks around the time of the new moon. The tiny (0.6-0.8) fertilised eggs rise to the surface and float with the currents until hatching about 20 to 40 hours later. The larvae measure 1.3-1.7 mm in length and lead a pelagic existence for several weeks. Growth rings on scales and ear bones (otoliths) have been used to determine the age of emperors. Most species appear to live at least 15 years and some may reach twice this age.

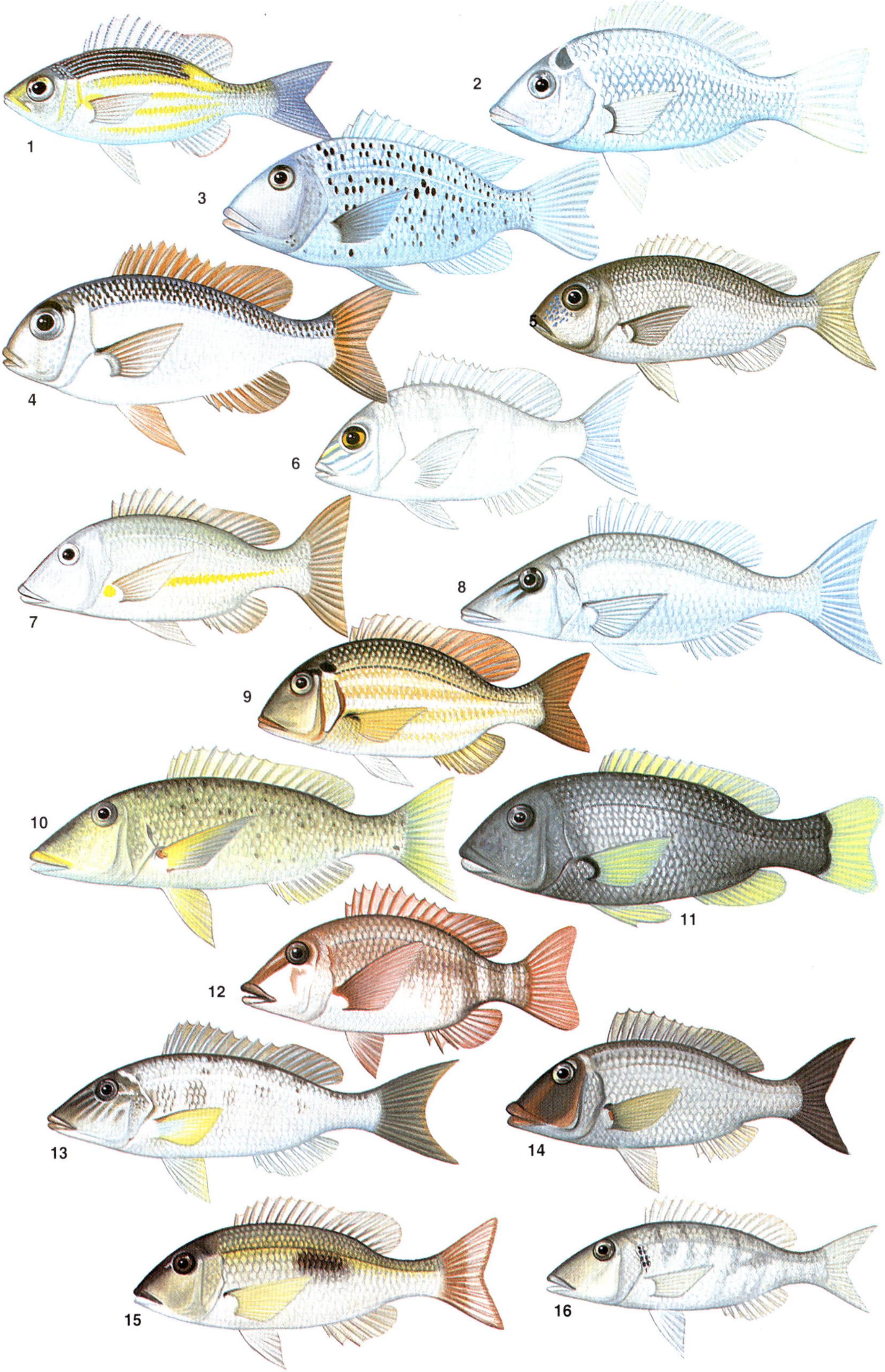
1
2
3
4
5
6
7
8
9
10
11
12
13
14
15
16

PLATE 49: THREADFIN-BREAMS (FAMILY NEMIPTERIDAE)

1 BALI THREADFIN-BREAM
Nemipterus balinensis (Bleeker, 1859)
Inhabits trawling grounds; distinguished by single yellow band on side and slender shape; Indo-Malay Archipelago to 24 cm. ★★

2 YELLOWBELLY THREADFIN-BREAM
Nemipterus bathybius Snyder, 1911
Inhabits trawling grounds; distinguished by 2 yellow bands on side and pair of yellow stripes along ventral profile of head and body; N.W. Australia, and throughout S.E. Asia; W. Pacific and E. Indian Ocean; to 20 cm. ★★

3 FIVE-LINED THREADFIN-BREAM
Nemipterus celebicus (Bleeker, 1854)
Inhabits trawling grounds; distinguished by 4-5 silvery yellow stripes on side, equal sized tail lobes and reddish tip on upper lobe of tail; N. Australia, S. Indonesia, and S. New Guinea; to 22 cm. ★★

4 ROSY THREADFIN-BREAM
Nemipterus furcosus (Valenciennes, 1830)
Inhabits trawling grounds; distinguished by series of faint dark blotches on upper back and uniform fins without stripes or spots; found throughout the region; W. Pacific and E. Indian Ocean; to 35 cm. ★★

5 ORNATE THREADFIN-BREAM
Nemipterus hexodon (Quoy & Gaimard, 1824)
Inhabits trawling grounds; distinguished by broad yellow stripe on upper side (with narrower stripes below), oblique yellow stripe on dorsal fin and, yellow tip on upper lobe of tail; found throughout the region; W. Pacific and E. Indian Ocean; to 30 cm. ★★

6 TWIN-LINED THREADFIN-BREAM
Nemipterus isacanthus (Bleeker, 1873)
Inhabits trawling grounds; distinguished by 2 broad yellow stripes along side, bright yellow band along ventral profile, and yellow patch below eye; N.W. Australia and throughout S.E. Asia; Indo-Australian Archipelago; to 24 cm. ★★

7 JAPANESE THREADFIN-BREAM
Nemipterus japonicus (Bloch, 1791)
Inhabits trawling grounds; distinguished by broad yellow band along base of dorsal fin, bright pink-red "shoulder" spot, yellow stripe along ventral profile, and elongate yellow filament on upper lobe of tail, colour generally more pink (less blue than shown); N.W. Australia and throughout S.E. Asia; W. Pacific and E. Indian Ocean; to 32 cm. ★★

8 SLENDER YELLOW-TIPPED THREADFIN-BREAM
Nemipterus mesoprion (Bleeker, 1853)
Inhabits trawling grounds, similar to **6** above, but lacks yellow patch below eye; S. Indonesia and Gulf of Thailand; to 20 cm. ★★

9 YELLOW-CHEEKED THREADFIN-BREAM
Nemipterus zysron (Bleeker, 1856)
Inhabits trawling grounds; distinguished by broad yellow stripe below eye and elongate yellow filament on upper lobe of tail; N.W. Australia and throughout S.E. Asia; Indo-W. Pacific; to 28cm. ★★

10 DOUBLEWHIP THREADFIN-BREAM
Nemipterus nematophorus (Bleeker, 1853)
Inhabits trawling grounds; distinguished by elongate filaments at front of dorsal fin and on upper lobe of tail (these features not shown, but usually present); found throughout most of S.E. Asia; E. Indian Ocean and Indo-Malay Archipelago; to 25 cm. ★★★

11 NOTCHED THREADFIN-BREAM
Nemipterus peronii (Valenciennes, 1830)
Inhabits trawling grounds; distinguished by deep notches along margin of dorsal fin between spines; found throughout the region; Indo-W. Pacific; to 30 cm. ★★

12 PINKFIN THREADFIN-BREAM
Nemipterus sp.
Inhabits trawling grounds; distinguished by 3 yellow bands between eye and upper lip, 6-7 yellow stripes on side, a yellow margin on dorsal fin and single yellow stripe on anal fin; Indo-Malay Archipelago, to 30 cm. ★★

13 YELLOW-LIPPED THREADFIN-BREAM
Nemipterus virgatus (Houttuyn, 1782)
Inhabits trawling grounds; distinguished by bright yellow lips (not shown), thin yellow stripes on sides; pair of thin stripes on dorsal and anal fins; N.W. Australia and Philippines; W. Pacific and E. Indian Ocean; to 40 cm. ★★

14 YELLOW-TIPPED THREAD FIN-BREAM
Nemipterus nematopus (Bleeker, 1851)
Inhabits trawling grounds; distinguished by pair of yellow stripes on side, lowermost wider and curved slightly upwards above base of pectoral fin, also has vivid yellow spot on tip of upper tail lobe and frequently with pair of yellow stripes along ventral surface of head and body; N Australia, Indonesia, and Philippines; W. Pacific and E. Indian Ocean; to 40 cm. ★★

THREADFIN AND MONOCLE BREAMS

The Sea Breams of the family Nemipteridae (Plates 48-49) are confined to the Indo-west Pacific region, primarily in tropical and subtropical latitudes. The family contains about 45 species, several of which are still undescribed. The largest genera are *Nemipterus* and *Scolopsis*, each having about 15-18 species. Both are commonly offered in South-east Asian fish markets, although they are relatively small in size (usually less than 40 cm TL). *Scolopsis* and *Pentapodus*, are closely associated with coral or rubble reefs, being particularly abundant in sandy areas between reefs. *Nemipterus* and *Parascolopsis* generally occur in deeper water, frequently between 30 and 100 m depth. They form a significant portion of trawl catches on continental shelves. Breams forage for worms, crustaceans, molluscs, and other invertebrates; small fishes are also sometimes eaten.

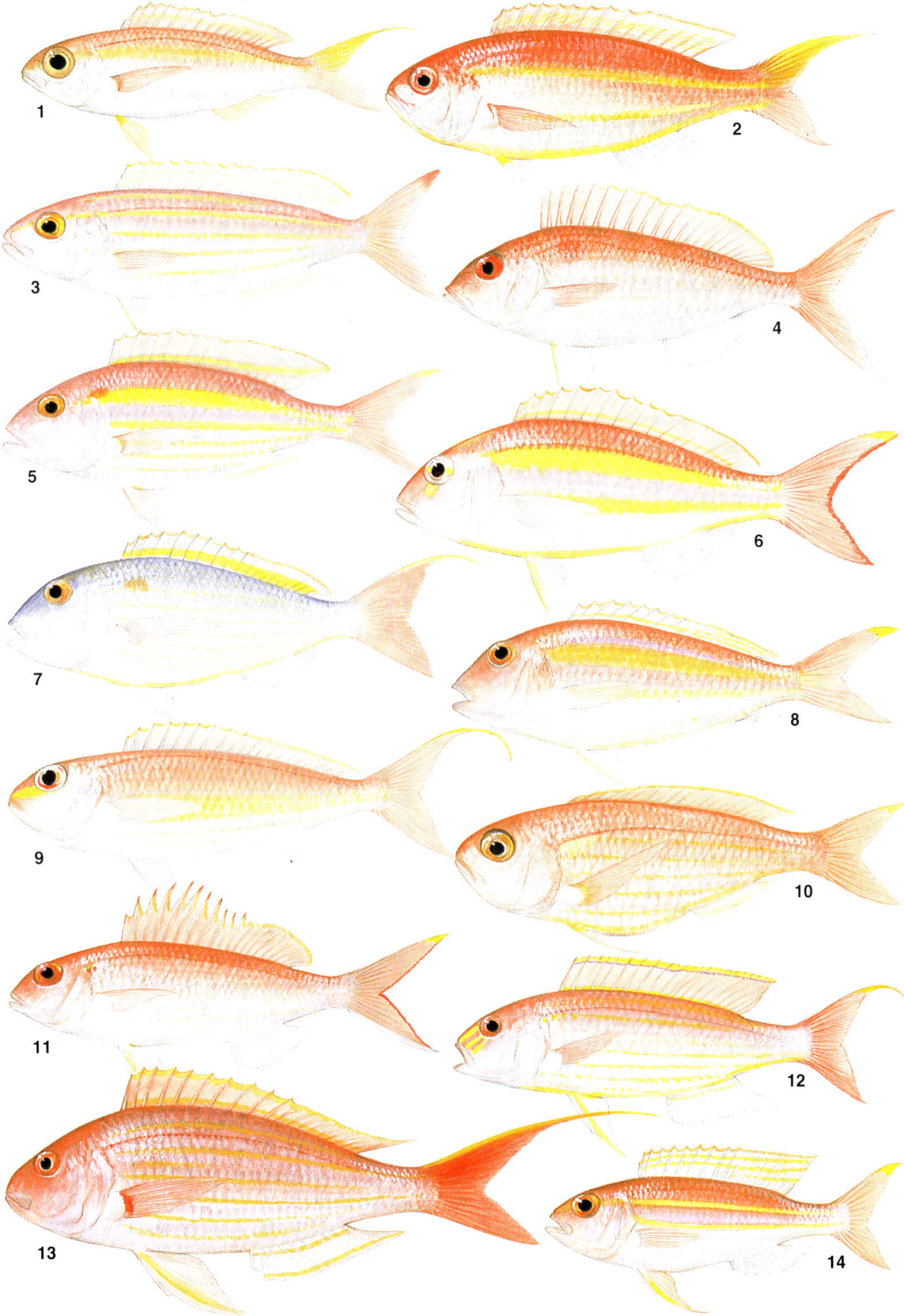
1
2
3
4
5
6
7
8
9
10
11
12
13
14

PLATE 50: MONOCLE BREAMS AND SILVER BIDDIES

1 ROSY DWARF MONOCLE BREAM
Parascolopsis eriomma Jordan & Richardson, 1909
Inhabits offshore trawling grounds; distinguished by rosy colour on back with diffuse yellow longitudinal band along middle of side; N.W. Australia, and throughout S.E. Asia; Indo-W. Pacific; to 30 cm. (NEMIPTERIDAE) ★★

2 SADDLED DWARF MONOCLE BREAM
Parascolopsis tanyactis Russell, 1986
Inhabits offshore trawling grounds in 45-80 m depth; distinguished by 2 dark bars on back, becoming diffuse on lower side and additional bar across tail base, also 4th and 5th soft dorsal rays elongated (not shown); *P. rufomaculatus* (not shown) is similar but soft dorsal rays not elongated and has broad golden stripe along middle of side; N.W. Australia, Indonesia, and Philippines; Indo-Australian Archipelago and W. Pacific; to 20 cm. (NEMIPTERIDAE) ★★

3 PURPLE THREADFIN-BREAM
Pentapodus emeryii (Richardson, 1843)
Inhabits coastal reefs; distinguished by bright yellow stripe from eye to tail, adults more bluish than shown and have long filaments at the tip of each caudal lobe; N.W. Australia, Indonesia, and Philippines; to 25 cm. (NEMIPTERIDAE) ★★

4 JAPANESE BUTTERFISH
Pentapodus nagasakiensis (Tanaka, 1915)
Inhabits sand rubble areas adjacent to deeper offshore reefs; distinguished by orange stripe from eye to tail base with pearly stripe just below it and often a narrower pale stripe along base of dorsal fin; N.W. Australia, Indonesia, and Philippines; mainly W. Pacific; to 20 cm. (NEMIPTERIDAE) ★★

5 FALSE WHIPTAIL
Pentapodus porosus (Valenciennes, 1830)
Inhabits sand and rubble bottoms frequently near coastal reefs; distinguished by broad longitudinal brown band on middle of sides and <-shaped blue mark and small black spot at tail base; N.W. Australia and S. New Guinea only; to 25 cm. (NEMIPTERIDAE) ★★

6 WESTERN BUTTERFISH
Pentapodus vitta Quoy & Gaimard, 1824
Inhabits sand-weed areas or rocky bottoms close to shore, often in schools; distinguished by dark stripe bordered by blue from snout to tail base; W. Australia only, between Geographe Bay and Dampier Archipelago; to 31 cm. (NEMIPTERIDAE) ★★

7 MONOCLE BREAM
Scolopsis monogramma (Kuhl & van Hasselt, 1830)
Inhabits sandy areas in the vicinity of coral reefs; distinguished by 3 blue stripes on snout (often with orange between them), generally pale body except may have series of slanting dotted lines on middle of side which sometimes form a solid broad stripe when viewed underwater, and blue edged tail with prolonged filament on upper lobe; often misidentified as *S. temporalis*, a similar species from New Guinea and other Melanesian islands; also known as Barred-face spinecheek; found throughout the region; mainly W. Pacific; to 30 cm. (NEMIPTERIDAE) ★★

8 BRIDLED MONOCLE BREAM
Scolopsis bilineatus (Bloch, 1793)
Inhabits coral reefs; distinguished by curved black-edged white stripe from below eye to midbase of dorsal fin; also known as Bridled spinecheek and Double-lined monocle bream; found throughout the region; E. Indian Ocean and W. Pacific; to 23 cm. (NEMIPTERIDAE) ★★

9 WHITECHEEK MONOCLE BREAM
Scolopsis vosmeri (Bloch, 1792)
Inhabits coastal reefs; distinguished by relatively deep body and white bar behind eye; N.W. Australia and throughout S.E. Asia; Indo-W. Pacific; to 25 cm. (NEMIPTERIDAE) ★★

10 REDSPOT MONOCLE BREAM
Scolopsis taeniopterus (Kuhl & van Hasselt, 1830)
Inhabits sandy areas in the vicinity of coral reefs; similar to 7 above, but has small red spot in axil ("armpit") of pectoral fin, lacks distinct blue stripes on snout, but has blue bands between eyes, and lacks dark spots or stripes on side; found throughout the region; mainly Indo-Australian Archipelago; to 23 cm. (NEMIPTERIDAE) ★★

11 PEARL-STREAKED MONOCLE BREAM
Scolopsis xenochrous Günther, 1872
Inhabits sand or rubble areas near coral reefs; distinguished by diagonal pearl-blue streak bordered with black spots above pectoral fin and series of slanted rows of dots on middle of side followed by broad white streak; found throughout the region; E. Indian Ocean and Indo-Australian Archipelago; to 18 cm. (NEMIPTERIDAE) ★★

12 CORAL MONOCLE BREAM
Scaevius milii (Bory de Saint-Vincent, 1823)
Inhabits coastal waters, frequently on sand or rubble bottoms; distinguished by 2 blue stripes on upper side ending on top of tail base, and wider orange stripe below along middle of side with small pale-edged black spot on upper tail base; also known as Jurgen; Australia only from Abrolhos Islands, W. Australia to Gulf of Carpentaria; to 25 cm. (NEMIPTERIDAE) ★★

13 LONG-FINNED SILVER BIDDY
Pentaprion longimanus (Cantor, 1850)
Inhabits coastal waters over sand bottoms; similar to 14 - 16 below, but has much larger anal fin base (anal fin contains 12-14 soft rays versus 7-8 in others); found throughout the region; Indo-W. Pacific; to 15 cm. (GERREIDAE) ★

14 WHIPFIN SILVER-BIDDY
Gerres filamentosus Cuvier, 1829
Inhabits coastal waters including estuaries, usually on sand or mud bottoms; distinguished by long filament at front of dorsal fin and series of spots in vertical rows on side; numerous species of similar silver-biddies found in the region, only three of the most common shown here; this species found throughout the region; IndoW. Pacific; to 25 cm. (GERREIDAE) ★

15 COMMON SILVER BIDDY
Gerres oyena (Forsskål, 1775))
Inhabits coastal waters including estuaries, usually on sand or mud bottoms; distinguished by relatively elongate body and lack of markings; also known as Silverbelly; found throughout the region; Indo-W. Pacific; to 25 cm. (GERREIDAE) ★

16 ROACH
Gerres subfasciatus Cuvier, 1830
Inhabits coastal waters entering bays and estuaries, usually on sand or mud bottoms; distinguished by narrow dark bars on side; also known as Banded silver-biddy; N. Australia, ranging down east and west coasts into subtropical and temperate seas; 22 cm. (GERREIDAE) ★

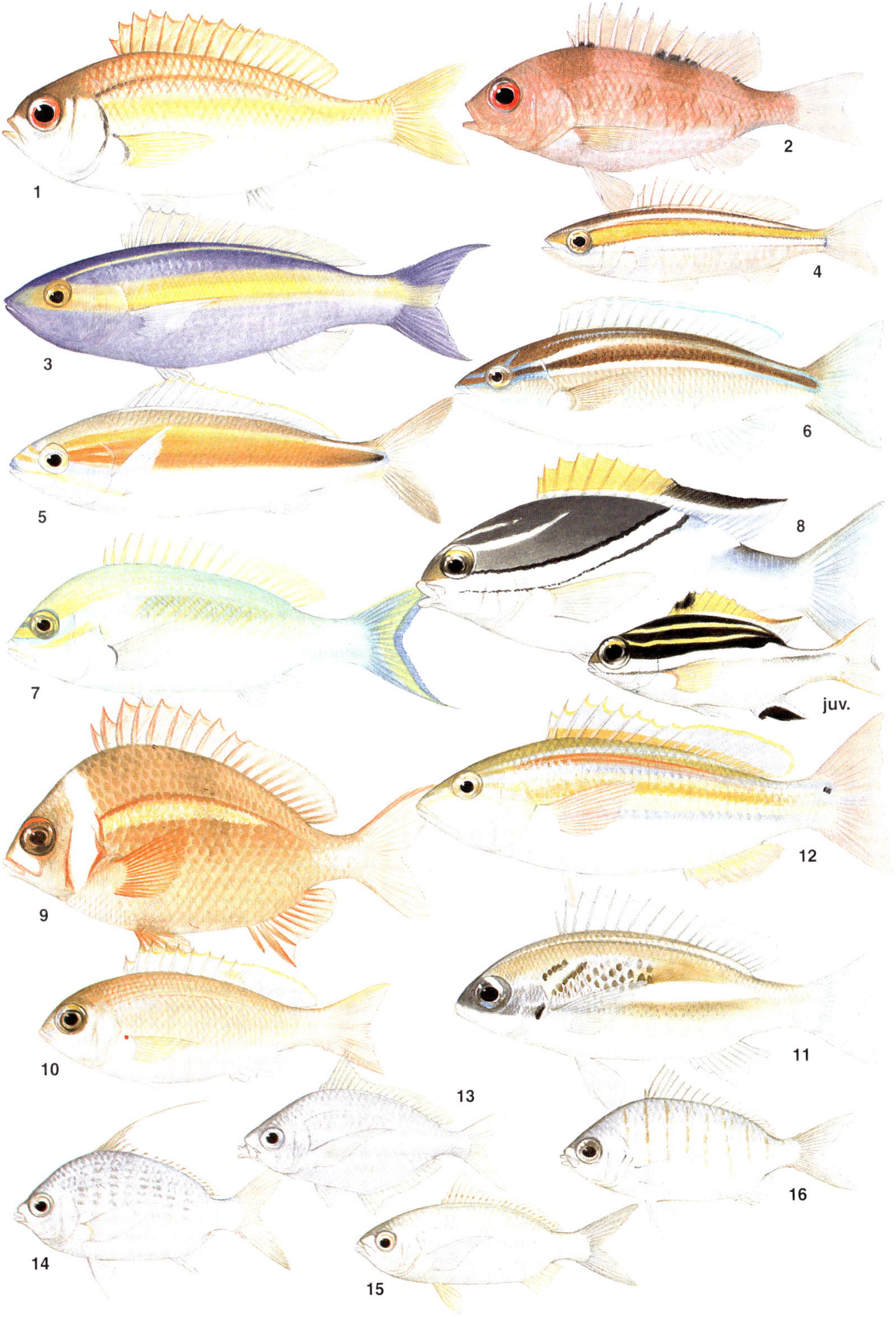
1
2
4
3
6
5
8
7
juv.
12
9
11
10
13
16
14
15

PLATE 51: MONOCLE BREAM AND GOATFISHES

1 WHITE-SHOULDERED WHIPTAIL
Pentapodus bifasciatus (Bleeker, 1848)
Inhabits coral reefs; distinguished by pair of pale stripes and narrow white bar between the two pale stripes, just behind head; W. Indonesia and Philippines; to 18 cm. (NEMIPTERIDAE) ★★

2 SMALL-TOOTHED WHIPTAIL
Pentapodus caninus (Cuvier, 1830)
Inhabit sand-rubble fringe of coral reefs; distinguished by single pale stripe, similar to **5** below, but stripe is angled slightly instead of horizontal; throughout S.E. Asia; W. Pacific; to 20 cm. (NEMIPTERIDAE) ★★

3 PARADISE WHIPTAIL
Pentapodus paradiseus (Günther, 1859)
Inhabit sand-rubble fringe of coral reefs; distinguished by diagonal pale stripe on side, small blue-edged dark spot at middle of tail base, and long filament on upper lobe of tail; Great Barrier Reef, Gulf of Carpentaria, and E. New Guinea to Solomon Islands; to 25 cm. (NEMIPTERIDAE) ★★

4 BUTTERFLY WHIPTAIL
Pentapodus setosus (Valenciennes, 1830)
Inhabits sand-rubble fringe of coral reefs; nearly identical in appearance to **3** above, but distributions do not overlap; Indo-Malay Archipelago; to 20 cm. (NEMIPTERIDAE) ★★

5 BLUE BUTTERFISH
Pentapodus sp.
Inhabits sand-rubble fringe of coral reefs; distinguished by blue colour on upper half of body with pair of narrow yellow stripes; juvenile is brilliant blue with pair of well-contrasted yellow stripes; Great Barrier Reef, New Guinea, E. Indonesia, and Philippines; W. Pacific; to 18 cm. (NEMIPTERIDAE) ★★

6 THREE-STRIPED WHIPTAIL
Pentapodus trivittatus (Bloch, 1791)
Inhabits coral reefs, frequently seen in lagoons; distinguished by pattern of light and dark stripes, narrow pale crossbars on back, and diagonal dark band below eye; throughout S.E. Asia; W. Pacific; to 22 cm. (NEMIPTERIDAE) ★★

7 PETER'S MONOCLE BREAM
Scolopsis affinis Peters, 1877
Inhabits sand-rubble fringe of coral reefs; similar to **8** below, but head scales reach forward to level of rear nostril (do not reach nostril in **8**); Great Barrier Reef, offshore reefs of N.W. Australia, and throughout S.E. Asia; to 22 cm. (NEMIPTERIDAE) ★★

8 YELLOWSTRIPE MONOCLE BREAM
Scolopsis auratus (Park, 1797)
Inhabits sand-rubble fringe of coral reefs; distinguished by pale yellow stripe on side, similar to some species of *Pentapodus* shown on this page, but is deeper-bodied; Indonesia (Sumatra, Java, Bali, Lombok, Sumbawa); C. and E. Indian Ocean; to 22 cm. (NEMIPTERIDAE)★★

9 WHITESTREAK MONOCLE BREAM
Scolopsis ciliatus (Lacepède, 1802)
Inhabits lagoon and inshore reefs, usually in sand or silt-bottom areas; distinguished by pearly streak just below dorsal fin and yellow spots on scales of side; throughout S.E. Asia; to 16 cm. (NEMIPTERIDAE) ★★

10 LINED MONOCLE BREAM
Scolopsis lineatus Quoy & Gaimard, 1824
Inhabits sandy areas near coral reefs; distinguished by bold black stripes and square blotches on upper half of body; *S. cancellatus* is a synonym; Great Barrier Reef, offshore reefs of N.W. Austraila, and throughout S.E. Asia; mainly W. Pacific; to 20 cm. (NEMIPTERIDAE) ★★

11 PEARLY MONOCLE BREAM
Scolopsis margaritifer (Cuvier, 1830)
Inhabits inshore and lagoon reefs; distinguished by pearly colour on sides and belly, dusky back, and yellowish pectoral fins; juveniles have a single black stripe on middle of side; Great Barrier Reef and throughout S.E. Asia; W. Pacific; to 20 cm. (NEMIPTERIDAE) ★★

12 RAINBOW MONOCLE BREAM
Scolopsis temporalis (Cuvier, 1830)
Inhabits sand-rubble areas near coral reefs; similar to *S. monogramma* (Pl. 50-7), but has blue-rimmed pale spot above and behind eye; E. Indonesia, New Guinea, and Solomon Islands; to 35 cm. (NEMIPTERIDAE) ★★

13 THREE-LINED MONOCLE BREAM
Scolopsis trilineatus Kner, 1868
Inhabits sandy or rubble fringe of coral reefs; distinguished by pale stripe from lower corner of eye that joins diagonal pale band running between eye and base of posterior part of dorsal fin; Great Barrier Reef, offshore reefs of N.W. Australia, and throughout S.E. Asia; W. Pacific; to 18 cm. (NEMIPTERIDAE) ★★

14 YELLOWSTRIPE GOATFISH
Mulloidichthys flavolineatus (Lacepède, 1801)
Inhabits sandy areas adjacent to coral reefs, often seen in schools; similar to **15** below, but lacks yellow fins and has dark spot above pectoral fin; Great Barrier Reef, offshore reefs of N.W. Australia, and throughout S.E. Asia; Indo-C. Pacific; to 30 cm. (NEMIPTERIDAE) ★★★

15 YELLOWFIN GOATFISH
Mulloidichthys vanicolensis (Valencennes, 1831)
Inhabits sandy areas adjacent to coral reefs, often seen in schools; distinguished by bright yellow fins and yellow stripe on middle of side; Great Barrier Reef, offshore reefs of N.W. Australia, and throughout S.E. Asia; Indo-C. Pacific; to 33 cm. (MULLIDAE) ★★★

16 DASH-DOT GOATFISH
Parupeneus barberinus (Lacepède, 1801)
Inhabits sand-rubble bottoms near coral reefs; distinguished by black stripe from snout through eye then continuing on upper side, and large black spot on tail base; Great Barrier Reef, offshore reefs of N.W. Australia, and throughout S.E. Asia; Indo-W. Pacific; to 30 cm. (MULLIDAE) ★★★

17 STRIPE-SPOT GOATFISH
Parupeneus macronema (Lacepède, 1801)
Inhabits coral reefs; similar to *P. pleurostigma* (Pl.52-9), but has broad black stripe on head and front of body, and black spot on middle of tail base; Indo-Malay Archipelago; Indo-W. Pacific; to 35 cm. (MULLIDAE) ★★★

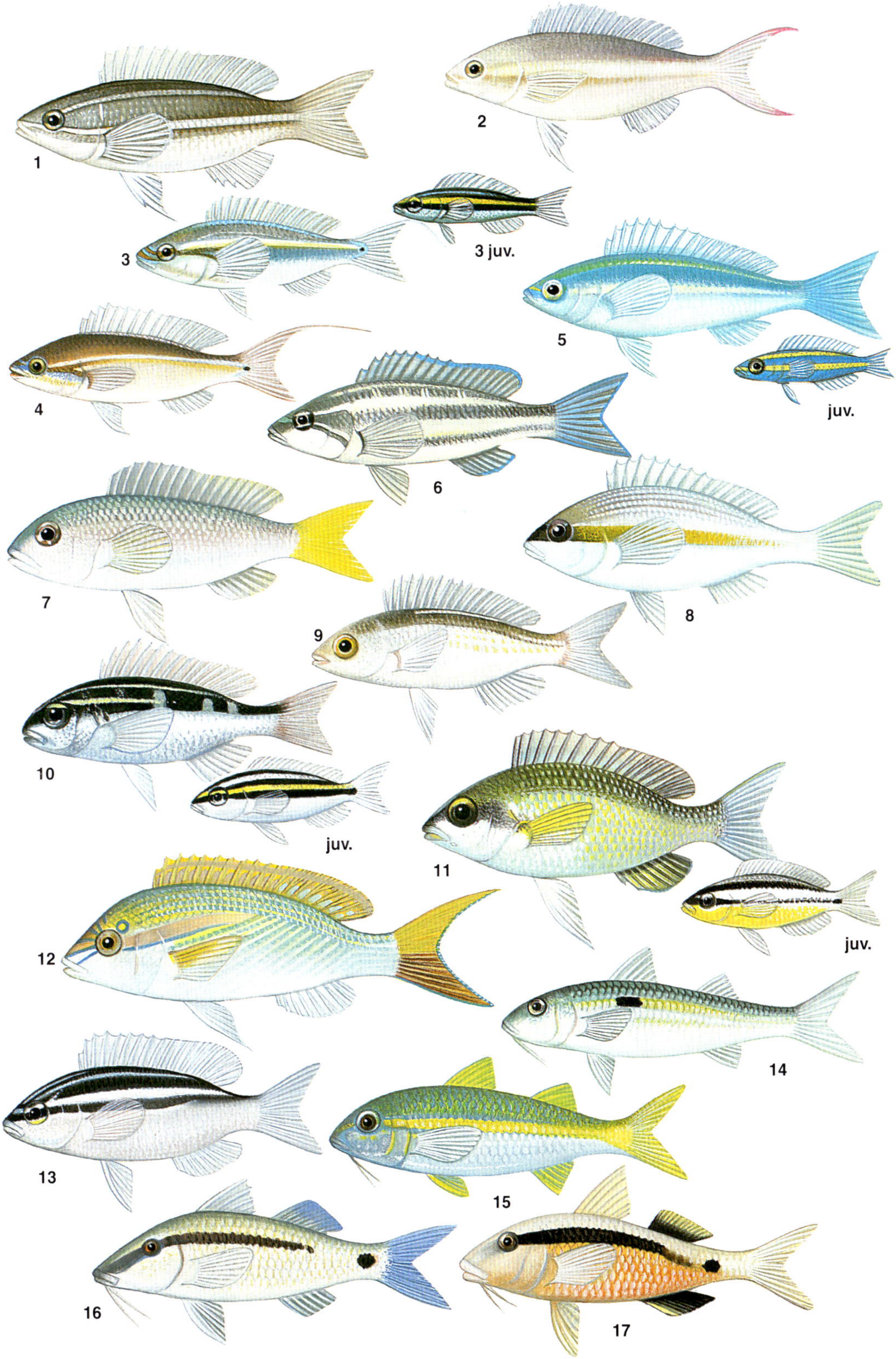
1
2
3
3 juv.
4
5
juv.
6
7
8
9
10
juv.
11
juv.
12
13
14
15
16
17

PLATE 52: GOATFISHES (FAMILY MULLIDAE)

1 GOLD-SADDLED GOATFISH
Parupeneus cyclostomus (Lacepède, 1801)
Inhabits coral reefs; has 2 colour varieties: one is entirely yellow, the other is purplish-pink or bluish with gold saddle on top of tail base; found throughout the region; Indo-C. Pacific; to 50cm. ★★★

2 SWARTHY-HEADED GOATFISH
Parupeneus barberinoides (Bleeker, 1801)
Inhabits weed-sand areas near coral reefs; distinguished by dark anterior half and pale rear portion with black spot in front of tail base; found throughout the region; mainly W. Pacific; to 25 cm. ★★★

3 DOUBLEBAR GOATFISH
Parupeneus bifasciatus (Lacepède, 1801)
Inhabits coral reefs; distinguished by 2 dark saddles on upper half of body; found throughout the region; Indo-C. Pacific; to 35 cm. ★★★

4 YELLOW STRIPED GOATFISH
Parupeneus chrysopleuron (Schlegel, 1843)
Inhabits sandy bottoms near rocky areas and coral reefs; distinguished by reddish colour and yellow stripe on upper side, **12** below is similar but has patches of teeth on roof of mouth; N.W. Australia and scattered localities in Indo-Malay region; W. Pacific and E. Indian Ocean; to 33 cm. ★★★

5 BLACKSPOT GOAT FISH
Parupeneus spilurus (Bleeker, 1854)
Inhabits sand-weed areas adjacent to coral reefs and southern rocky reefs; distinguished by alternating light and dark stripes on head and body, and black spot on upper half of tail base; *P. signatus* is a synonym; N. Australia, New Guinea, and N. New Zealand; to 47 cm. ★★★

6 INDIAN GOATFISH
Parupeneus indicus (Shaw, 1903)
Inhabits sand-weed areas in the vicinity of coral reefs; distinguished by elongate yellow blotch on middle of upper sides and black spot on tail base; found throughout the region; Indo-C. Pacific; to 40 cm. ★★★

7 SPOTTED GOLDEN GOATFISH
Parupeneus heptacanthus (Lacepède, 1801)
Inhabits sand-weed areas in the vicinity of coral or rocky reefs; distinguished by reddish-pink colour and yellow stripes on side, often has small red to brown spot on upper side above pectoral fin; found throughout the region; Indo-W. Pacific; to 30 cm. ★★★

8 BANDED GOATFISH
Parupeneus multifasciatus (Quoy & Gaimard, 1825,)
Inhabits weed-sand areas in the vicinity of coral reefs; distinguished by broad dark bars on side and large spot or bar on tail base; found throughout the region; Indo-C. Pacific; to 30 cm. ★★★

9 SIDESPOT GOATFISH
Parupeneus pleurostigma (Bennett, 1830)
Inhabits sand and rubble areas in the vicinity of coral reefs, distinguished by black spot on middle of upper sides followed by pearly white patch; found throughout the region; Indo-C. Pacific; to 33 cm. ★★★

10 GOLDBAND GOATFISH
Upeneus moluccensis (Bleeker, 1855)
Inhabits sandy or weed covered areas; distinguished by yellow stripe from eye to tail and narrow stripes on both dorsal fins and upper lobe of tail; N.W. Australia and throughout S.E. Asia; Indo-W. Pacific; to 20 cm. ★★★

11 SUNRISE GOATFISH
Upeneus sulphureus Cuvier, 1829
Inhabits sandy or weed covered areas; distinguished by black tip on first dorsal fin and 2-3 gold stripes on side, similar to **13** below, but lacks stripes on tail; N.W. Australia and throughout S.E. Asia; Indo-W. Pacific; to 23 cm. ★★★

12 OCHRE-BANDED GOATFISH
Upeneus sundaicus (Bleeker, 1855)
Inhabits sandy or weed covered areas; distinguished by yellow stripe on upper sides, similar to **4** above, but has patches of teeth on roof of mouth, also rear margin of lower lobe of tail usually dark; N.W. Australia and throughout S.E. Asia; N. Indian Ocean and Indo-Australian Archipelago; to 22 cm. ★★★

13 STRIPED GOAT FISH
Upeneus vittatus (Forsskål, 1775)
Inhabits sandy or weed covered areas; distinguished by 4 orange-yellow stripes on side, black tip on first dorsal fin, and stripes on tail; N.W. Australia and throughout S.E. Asia; Indo-C. Pacific; to 28 cm. ★★★

14 BAR-TAILED GOATFISH
Upeneus tragula Richardson, 1846
Inhabits sandy or weed covered areas; similar to **15** below, but stripe on side dark (reddish brown to blackish), sides mottled with spots, and 8 spines (versus 7) in first dorsal fin; found throughout the region; Indo-W. Pacific; to 30 cm. ★★★

15 ASYMMETRICAL GOATFISH
Upeneus asymmetricus Lachner, 1954
Inhabits sandy or weed covered areas; similar to **14** above, but stripe on side yellow, fewer small spots on sides, and 7 spines (versus 8) in first dorsal fin; N.W. Australia and throughout S.E. Asia; Indo-W. Pacific; to 30 cm. ★★★

GOATFISHES

The goatfishes (family Mullidae) are found mainly in tropical and subtropical seas, usually in the vicinity of reefs. Worldwide about 50-60 species are known. They are characterised by a relatively elongate body, two widely separated dorsal fins, and the presence of a pair of long chin barbels that are used for detecting food. The barbels are also used by males to attract females during courtship. When they are not being used the barbels are tucked tightly under the chin. Goatfishes feed on small fishes and a variety of mainly sand and weed dwelling organisms including worms, shrimps, crabs, molluscs, and echinoderms. They are frequently seen in large schools and most species are considered to be good eating. The maximum size is about 60 cm, but most species are much smaller. The majority of goatfishes are distributed in the Indo-west Pacific region, but there are also a few representatives in the Atlantic and eastern Pacific Oceans.

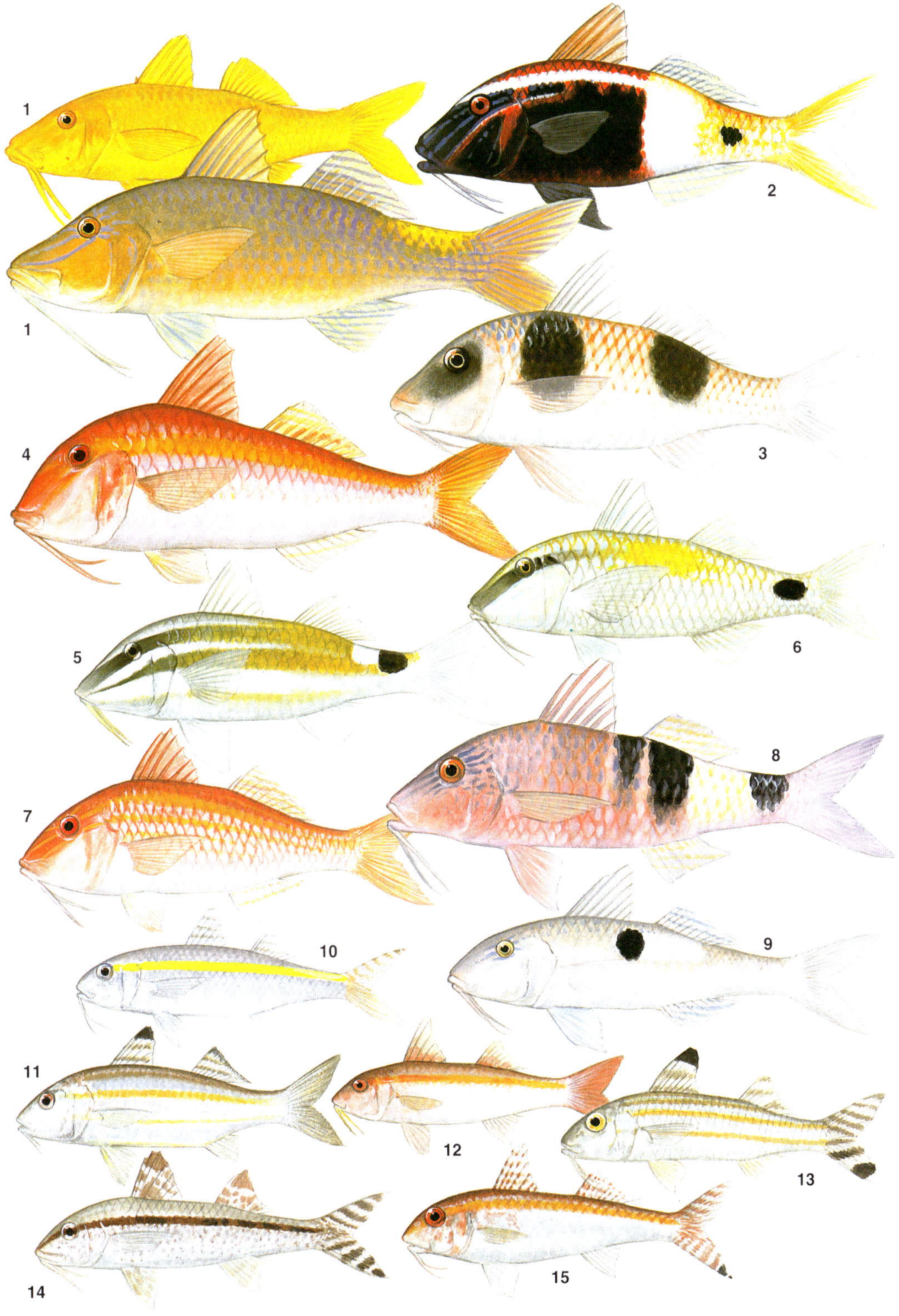
1
2
1
3
4
6
5
8
7
10
9
11
12
13
14
15

PLATE 53: CROAKERS, BULLSEYES AND DRUMMERS

1 BLACK JEW
Protonibea diacanthus (Lacepède, 1802)
Inhabits coastal waters, sometimes entering estuaries; distinguished by large size and overall grey to blackish colour, young fish have black spots on the back, dorsal fin, and tail; also known as Spotted croaker or Spotted jewfish; found throughout the region; Indo-W. Pacific; to 150 cm and at least 16 kg. (SCIAENIDAE) ★★★

2 ORANGE CROAKER
Atrobucca brevis Sasaki & Kailola, 1988
Inhabits coastal waters in 60-112 m depth; distinguished by faint orange tinge on lower third of body when fresh, otherwise without distinctive marks; N. Australia and S. New Guinea only; to 35 cm. (SCIAENIDAE) ★★

3 COITOR CROAKER
Johnius coitor (Hamilton, 1822)
Inhabits coastal waters; distinguished by bulging snout with small mouth below, similar to **4** below, but has dark blotch on gill cover, more soft dorsal rays (30-32 versus 23-26) and more gill rakers on the lower limb of the first gill arch (11 or 12 versus 6-10); found throughout the region; Indo-W. Pacific; to 16 cm. (SCIAENIDAE) ★★

4 GREEN-RACKED CROAKER
Johnius amblycephalus (Bleeker, 1855)
Inhabits coastal waters; distinguished by bulging snout with small mouth below, similar to **3** above, but distinguished by differences mentioned under that species, also has taller first dorsal fin; found throughout the region; W. Pacific and E. Indian Ocean; to 20 cm. (SCIAENIDAE) ★★

5 LITTLE JEWFLSH
Johnius vogleri (Bleeker, 1853)
Inhabits coastal waters including estuaries; distinguished by black outer portion of first dorsal fin and diffuse dark stripe often present on basal part of second dorsal fin; found throughout the region; E. Indian Ocean and Indo-Australian Archipelago, to 22cm. (SCIAENIDAE) ★★

6 DIAMOND FISH
Monodactylus argenteus (Linnaeus, 1758)
Inhabits estuaries and lower reaches of freshwater streams, also in harbours around wharves and jetties; distinguished by diamond shape and silvery colour; also known as Silver batfish; found throughout the region; Indo-W. Pacific; to 27 cm. (MONODACTYLIDAE) ★

7 BEACH SALMON
Leptobrama mulleri Steindachner, 1879
Inhabits coastal waters, frequently in schools off sandy beaches; distinguished by steel-blue colour on back and a single black-tipped dorsal fin that is set well back on the body; N. Australia and S. New Guinea only; to 43 cm. (LEPTOBRAMIDAE) ★★

8 SLENDER BULLSEYE
Parapriacanthus ransonneti Steindachner, 1870
Inhabits coral reefs, usually in large aggregations in caves; distinguished by semi-transparent appearance, yellowish head, silvery belly and a single dorsal fin; *P. unwini* is a synonym; found throughout the region; Indo-W. Pacific; to 7.5 cm. (PEMPHERIDAE) ★

9 BRONZE BULLSEYE
Pempheris analis Waite, 1910
Inhabits coral or rocky reefs, usually in caves similar to **11** below, but has smaller scales and blackish tips on dorsal and anal fins, and sometimes on lobes of tail; W. Australia, S. Great Barrier Reef, Lord Howe Island, and Kermadec Islands; to 17cm. (PEMPHERIDAE)

10 OUALAN BULLSEYE
Pempheris oualensis Cuvier, 1831
Inhabits coral or rocky reefs, usually in caves; distinguished by relatively large scales and series of dark stripes on side; found throughout the region; Indo-C. Pacific; to 20 cm. (PEMPHERIDAE)

11 STRIPED BULLSEYE
Pempheris schwenkii Bleeker, 1855
Inhabits coral or rocky reefs, usually in caves, similar to **9** above, but has larger scales, a yellowish tail, blackish anal fin base, and lacks dark fin tips; found throughout the region; Indo-W. Pacific; to 15 cm. (PEMPHERIDAE)

12 SOUTHERN DRUMMER
Kyphosus bigibbus Lacepède, 1802
Inhabits coral reefs, usually in large schools; similar to **14** below, but lacks well pronounced narrow stripes on side and has 12-13 soft dorsal rays (versus 14 or 15); sometimes misidentified as *K. gibsoni*; found throughout the region; Indo-C. Pacific; to 58 cm. (KYPHOSIDAE) ★

13 WESTERN BUFFALO BREAM
Kyphosus cornelii (Whitley, 1944)
Inhabits coral and rocky reefs, occurring in large schools; distinguished by dark streak on each tail lobe with white outer margin; W. Australia only, between Cape Leeuwin and Coral Bay; to 60 cm. (KYPHOSIDAE) ★

14 LOW-FINNED DRUMMER
Kyphosus vaigiensis (Quoy & Gaimard, 1825)
Inhabits coral reefs, frequently in schools; similar to **12** above, but has pronounced narrow stripes on side and l4-15 soft dorsal rays (versus 12-13); found throughout the region; Indo-C. Pacific; to 50 cm. (KYPHOSIDAE) ★

15 STRIPEY
Mirocanthus strigatus (Cuvier, 1831)
Inhabits rocky areas and coral reefs, sometimes forms schools around jetties; distinguished by bold pattern of slanting stripes; E. and W. coasts of Australia, Taiwan, S. Japan, and Hawaii; to 16 cm. (MICROCANTHIDAE)

CROAKERS, DRUMMERS, ETC.

Croakers of the family Sciaenidae live in a variety of marine and estuarine habitats. They can produce a "drumming" sound with the aid of their swim bladder. The maximum size is about 2 m, but most are under 50 cm. Some species are important food fishes.

The Diamond Fish (family Monodactylidae) appears to be equally at home in salt or fresh water. It is sometimes found in large aggregations. Young specimens make excellent aquarium pets.

Bulleyes of the Indo-Pacific family Pempheridae are cave and crevice dwellers that often occur in large aggregations. They feed mainly on crustaceans, but also consume worms and cephalopods.

Drummers of the family Kyphosidae are common inhabitants of reefs and weed beds in tropical and temperate seas. They feed mainly on plants and most are less than 50 cm TL. Although eaten in some localities the flesh is usually not very highly regarded. Some species of *Kyphosus* are reported to producc a mild hallucinogenic effect when consumed.

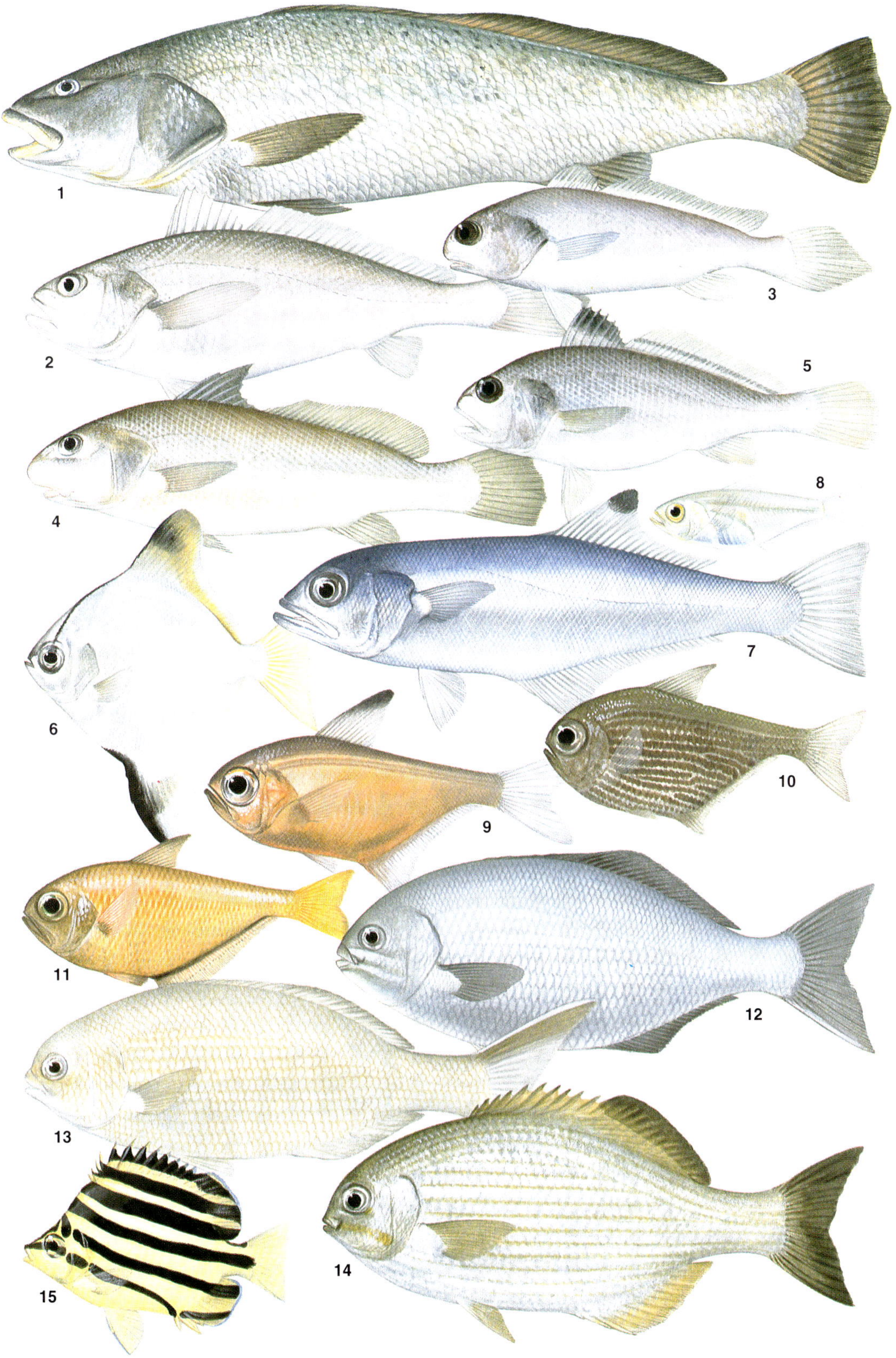
1
2
3
4
5
6
7
8
9
10
11
12
13
14
15

PLATE 54: ARCHERFISHES AND BATFISHES

1 SPOTTED ARCHERFISH
Toxotes chatareus (Hamilton, 1822)
Inhabits mangrove estuaries and freshwater streams; has ability to knock insects from overhanging vegetation by squirting jets of water from mouth; distinguished from **2** below, by spotted pattern; found throughout the region; Indo-Australian Archipelago and Andaman Sea; to 30 cm. (TOXOTIDAE) ★★★

2 BANDED ARCHERFISH
Toxotes jaculatrix (Pallas, 1767)
Inhabits brackish mangrove estuaries; distinguished from **1** above by 4 dorsal spines (versus 5) and barred pattern; found throughout the region; Indo-Australian Archipelago and Andaman Sea; to 20 cm. (TOXOTIDAE) ★★★

3 SICKLEFISH
Drepane punctata (Linnaeus, 1758)
Inhabits coastal reefs; distinguished by its triangular shape and vertical rows of small spots; found throughout the region; Indo-W. Pacific; to 50 cm. (DREPANIDAE) ★★★

4 HUMP-HEADED BATFISH
Platax batavianus Cuvier, 1831
Inhabits coastal reefs; distinguished by humped forehead and relatively elongate body of adult; found throughout the region; small juveniles (not shown) have spectacular zebra-like markings; Indo-Australian Archipelago; to 50 cm. (EPHIPPIDAE) ★★

5 LONG-FINNED BATFISH
Platax pinnatus (Linnaeus, 1758)
Inhabits coral reefs; adult distinguished by slightly protruding snout; found throughout the region; mainly Indo-Australian Archipelago; to 50 cm. (EPHIPPIDAE) ★★

6 TEIRA BATFISH
Platax teira (Forsskål, 1775)
Inhabits coral reefs; distinguished by dark blotch below pectoral fin and second blotch above front of anal fin, juvenile with very long dorsal, anal and pelvic fins; found throughout the region; Indo-W. Pacific; to 60 cm. (EPHIPPIDAE) ★★

7 ORBICULAR BATFISH
Platax orbicularis (Forsskål, 1775)
Inhabits coral reefs and inshore areas; similar to **6** above, but spot below pectoral fin generally more intense and second spot absent, juvenile effectively mimics dead leaves; found throughout the region; Indo-C. Pacific; to 50 cm. (EPHIPPIDAE) ★★

8 SHORT-FINNED BATFISH
Zabidius novemaculeatus (McCulloch, 1916)
Inhabits coastal reefs; similar to **4** above, but has 9 dorsal spines (versus 7) and lacks pronounced hump on forehead of adults; N. Australia and S. New Guinea only; to 45 cm. (EPHIPPIDAE) ★★

9 THREADFIN SCAT
Rhinoprenes pentanemus Munro, 1964
Inhabits coastal seas; distinguished by bulbous snout, laterally compressed body and long dorsal and anal fin filaments; N.W. Australia and S. New Guinea; to 15 cm. (RHINOPRENIDAE) ★★

10 SPOTTED SCAT
Scatophagus argus (Linnaeus) Text on page 148

11 STRIPED BUTTERFISH
Selenotoca multifasciata (Richardson) Text on page 148

AMAZING ARCHERFISHES

Archerfishes of the family Toxotidae exhibit one of nature's most remarkable feeding adaptations. They are renowned for their ability to knock down insects with a squirt of water at a distance of two metres or more. Archers produce their aqueous bullets by suddenly pulling in their gill flaps, forcing water through a tube formed by a groove on the roof of the mouth and the tongue. This clever mechanism is similar to that employed by a toy water pistol. The fish's accuracy is even more amazing, considering that it must compensate for the angle of refraction of light. In other words, due to the bending of light its target is not where it actually appears to be.

Archerfishes are common residents of South-east Asia and northern Australian. The family contains six species (all in the genus *Toxotes*). It is distributed in fresh and brackish waters from India eastward to Australia and the Melanesian Archipelago. The Spotted Archerfish (**1**) is one of the most common representatives. It inhabits rivers and creeks, usually near the coast, but may penetrate more than 200 kilometres upstream in larger rivers. It usually occurs in schools which patrol the shoreline for insects and other prey items. A specimen from the Fly River in Papua New Guinea measured 50 centimetres, but Australian fish are usually about half this size. Juveniles make excellent aquarium pets.

BEAUTIFUL BATFISHES

The graceful batfishes are a favourite of underwater photographers. Five species occur in the Indo-Pacific area, all belonging to the genus *Platax*. The handsome juveniles make excellent aquarium pets. They become tame very quickly and will accept feedings by hand. The young have highly exaggerated dorsal, anal, and pelvic fins. For example, an 8 cm long Teira Batfish may measure over 25 cm from dorsal fin tip to anal fin tip. As the fish grows older the fins become proportionally smaller, until the shape is approximately round. Young fish occur singly or in small groups and never stray far from shelter. They are usually seen hovering under coral heads or in the shadows of boat moorings and wharves. At least two species employ clever disguises in their youthful stage to avoid predation. The Round-faced Batfish (*Platax orbicularis*) lies on its side and drifts back and forth with the waves, appearing very much like a waterlogged leaf. The tiny young of the Pinnate Batfish (*Platax pinnatus*) swims on its side with a strange undulating motion - apparently mimicking a toxic polyclad flatworm.

Recent research indicates that batfishes are related to the surgeonfishes (Plates 93-95)

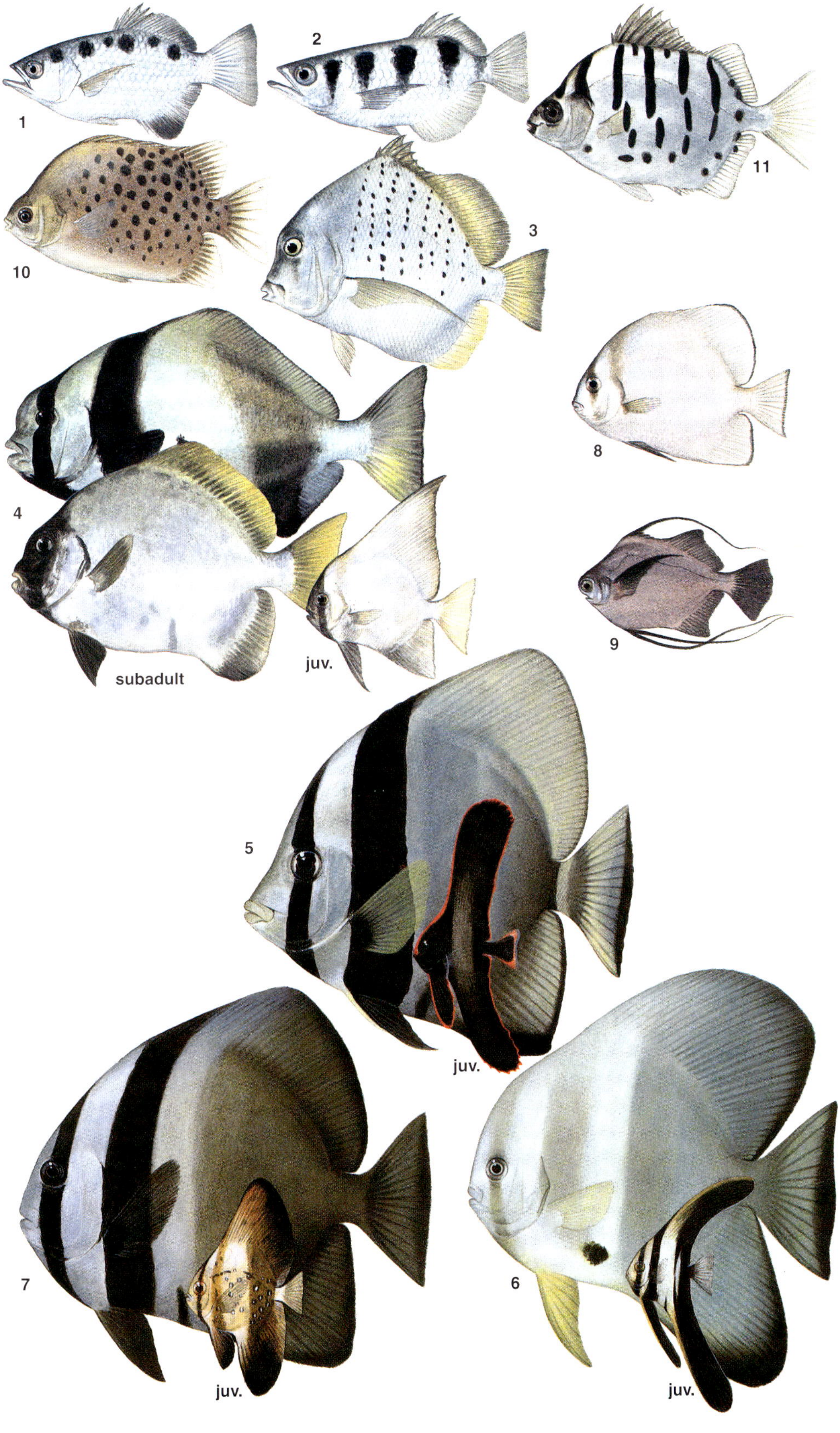

1
2
11
10
3
8
4
9
subadult
juv.
5
juv.
7
6
juv.
juv.

PLATE 55: SCATS AND BUTTERFLYFISHES

1 SPOTTED SCAT
Scatophagus argus (Linnaeus, 1766)
Inhabits brackish mangrove estuaries and freshwater streams; similar to **2** below, but has numerous round spots on side; found throughout the region; Indo-W. Pacific; to 33 cm. (SCATOPHAGIDAE) ★★

2 STRIPED BUTTERFISH
Selenotoca multifasciata (Richardson, 1846)
Inhabits brackish mangrove estuaries and freshwater streams; similar to **1** above, but has more silvery body colour and combination of bars and spots, also known as Spotband Scat; found throughout the region; Indo-Australian Archipelago; to 28 cm. (SCATOPHAGIDAE) ★★

3 PHILIPPINE BUTTERFLYFISH
Chaetodon adiergastos Seale, 1910
Inhabits coral reefs; distinguished by broad eye bar and narrow diagonal stripes on side; N.W. Australia and throughout Indo-Malay region; Indo-Australian Archipelago; to 15 cm. (CHAETODONTIDAE)

4 WESTERN BUTTERFLYFISH
Chaetodon assarius Waite, 1905
Inhabits rocky reefs and sand-weed flats; distinguished by several rows of small dots on sides; W. Australia only, between Recherche Archipelago and Monte Bello Islands; to 16 cm. (CHAETODONTIDAE)

5 THREADFIN BUTTERFLYFISH
Chaetodon auriga Forsskål, 1775
Inhabits coral reefs and sand-weed flats; distinguished by chevron markings and yellow posterior; N.W. Australia, Great Barrier Reef, and throughout S.E. Asia; Indo-C. Pacific; to 23 cm.

6 GOLDEN-STRIPED BUTTERFLYFISH
Chaetodon aureofasciatus Macleay, 1878
Inhabits coral reefs, distinguished by overall yellow orange colour and pair of orange stripes on head and anterior part of body; N. Australia and New Guinea; to 14 cm. (CHAETODONTIDAE)

7 SPECKLED BUTTERFLYFISH
Chaetodon citrinellus Cuvier, 1831
Inhabits shallow coral reefs, where live coral is sparse; distinguished by small dots on yellow background; N.W. Australia, Great Barrier Reef, and throughout S.E. Asia; Indo-C. Pacific; to 13 cm. (CHAETODONTIDAE)

8 SADDLED BUTTERFLYFISH
Chaetodon ephippium Cuvier, 1831
Inhabits coral reefs; distinguished by large black saddle; N.W. Australia, Great Barrier Reef, and throughout S.E. Asia; Indo-C. Pacific; to 23 cm. (CHAETODONTIDAE)

9 LINED BUTTERFLYFISH
Chaetodon lineolatus Cuvier, 1831
Inhabits coral reefs; distinguished by narrow black stripes on sides and broad black area along dorsal fin base; N.W. Australia, Great Barrier Reef, and throughout S.E. Asia; Indo-C. Pacific; to 3 cm. (CHAETODONTIDAE)

10 MEYER'S BUTTERFLYFISH
Chaetodon meyeri Bloch & Schneider, 1801
Inhabits coral reefs; distinguished by black diagonal stripes; N.W. Australia, Great Barrier Reef, and throughout S.E. Asia; Indo-W. Pacific; to 20 cm. (CHAETODONTIDAE)

11 KLEIN'S BUTTERFLYFISH
Chaetodon kleinii Bloch, 1790
Inhabits coral reefs; distinguished by combination of bars on front of body and golden brown area on rear half; N.W. Australia, Great Barrier Reef, and throughout S.E. Asia; Indo-C. Pacific; to 13 cm. (CHAETODONTIDAE)

12 BLUESPOT BUTTERFLYFISH
Chaetodon plebeius Cuvier, 1831
Inhabits coral reefs; distinguished by ovate blue spot on yellow background; N.W. Australia, Great Barrier Reef, and throughout S.E. Asia; Indo-W. Pacific; to 13 cm. (CHAETODONTIDAE)

13 RACOON BUTTERFLYFISH
Chaetodon lunula Lacepède, 1803
Inhabits coral reefs, distinguished by black "mask" and diagonal black bar behind head with golden colour on lower sides; N.W. Australia, Great Barrier Reef, and throughout S.E. Asia; Indo-C. Pacific; to 21 cm.

14 ORNATE BUTTERFLYFISH
Chaetodon ornatissimus Cuvier, 1831
Inhabits coral reefs; distinguished by brown-orange diagonal stripes; N.W. Australia, Great Barrier Reef, and throughout S.E. Asia; mainly W. and C. Pacific; to 18 cm. (CHAETODONTIDAE)

COLOURFUL BUTTERFLYFISHES

Butterflyfishehes of the family Chaetodontidae (Plates 55-58) are renowned for their striking colour patterns, delicate shape and graceful swimming movements The family contains about 120 species which occur mainly in tropical seas around coral reefs. Most of the species dwell in depths of less than 20 m, but some are restricted to deeper sections of the reef, to at least 200 m. Butterflyfishes are active during daylight hours and seek shelter close to the reef's surface during the night. They often assume a drab nocturnal colour pattern. Most species are restricted to a relatively small area of the reef, perhaps an isolated patch reef or part of a more extensive reef system. They travel extensively throughout their home range foraging for food. Many species feed on live coral polyps, others consume a mixed diet consisting of small invertebrates and algae. A few species, for example *Heniochus diphreutes,* feed in midwater on zooplankton. Young butterflyfishes are highly prized as aquarium pets. Most species grow to a maximum length under 30 cm.

Butterflyfishes are one of the most conspicuous inhabitants of tropical reefs and have attracted the attention of behavioural scientists. Some species are generally found solitarily or occasionally in small groups. Others form pairs and depending on the frequency of their pairing tendencies are termed weakly or strongly paired. The weak category includes such species 4, 7, 13 and 14. Examples of strong pairing species are 8 on Plate 55 and 2, 5, and 6 on Plate 56. There is strong evidence that fishes in the latter category form lifetime relationships.

Scats (1-2) are related to surgeonfishes. They generally occur around wharves in harbours, mangrove estuaries and the lower reaches of streams. Juveniles are popular aquarium fishes.

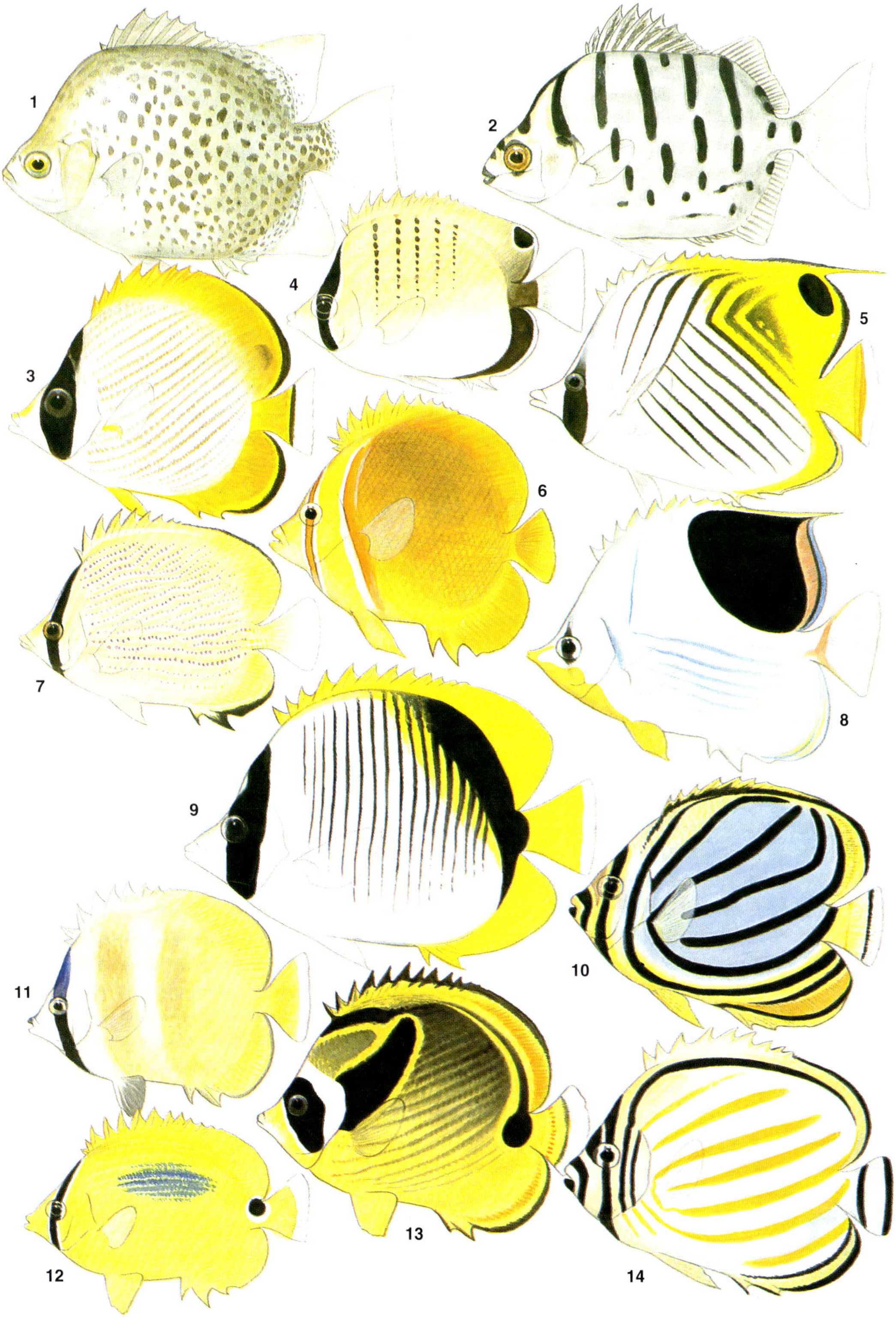
1
2
3
4
5
6
7
8
9
10
11
12
13
14

PLATE 56 : BUTTERFLYFISHES (FAMILY CHAETODONTIDAE)

1 TEARDROP BUTTERFLYFISH
Chaetodon unimaculatus Bloch, 1787 Cairns 10/98
Inhabits coral reefs; distinguished by black spot on white background and yellow fins; N.W. Australia, Great Barrier Reef, and throughout S.E. Asia; Indo-C. Pacific; to 23 cm.

2 SPOT-BANDED BUTTERFLYFISH
Chaetodon punctatofasciatus Cuvier, 1831
Inhabits coral reefs; distinguished by bars on back and spots below; N.W. Australia, Great Barrier Reef, and throughout S.E. Asia; mainly W. Pacific; to 13 cm.

3 OVALSPOT BUTTERFLYFISH
Chaetodon speculum Cuvier, 1831 Cairns 10/98
Inhabits coral reefs; distinguished by round spot on yellow background; N.W. Australia, Great Barrier Reef, and throughout S.E. Asia; mainly W. Pacific; to 15 cm.

4 DOUBLESADDLE BUTTERFLYFISH
Chaetodon ulietensis Cuvier, 1831
Inhabits coral reefs distinguished by 2 broad blackish bars and series of narrow vertical lines on white background; N.W. Australia, Great Barrier Reef, and throughout S.E. Asia; mainly W. Pacific; to 14 cm.

5 CHEVRONED BUTTERFLYFISH
Chaetodon trifascialis Quoy & Gaimard, 1824
Inhabits coral reefs; distinguished by triangular-shaped body and narrow chevron markings; N.W. Australia, Great Barrier Reef, and throughout S.E. Asia; Indo-C. Pacific; to 18 cm.

6 REDFIN BUTTERFLYFISH
Chaetodon trifasciatus Park, 1797
Inhabits coral reefs; distinguished by dark stripes on orange to purple background and reddish anal fin; Indian Ocean ranging to Sumatra, Java, and Bali; *C. lunulatus* (not shown) is a nearly identical species found in Australia and the W. and C. Pacific; to 15 cm.

7 ORANGE-BANDED CORALFISH
Coradion chrysozonus (Cuvier, 1831)
Inhabits coral reefs and trawl grounds; distinguished by orange-brown bars and black pelvic fins; N.W. Australia, Great Barrier Reef, and throughout S.E. Asia; mainly Indo-Australian Archipelago; to 13 cm.

8 MARGINED CORALFISH
Chelmon marginalis Richardson, 1842
Inhabits coral reefs, distinguished by long snout and orange bars; N. Australia only, Shark Bay, W. Australia to Torres Strait, Queensland; to 20 cm.

9 LONG-NOSED BUTTERFLYFISH
Forcipiger flavissimus Jordan & McGregor, 1898
Inhabits coral reefs; distinguished by long snout, black colour on upper half of head and yellow colour of body; N.W. Australia, Great Barrier Reef, and throughout S.E. Asia; Indo-E. Pacific; to 17 cm.

10 PENNANT BANNERFISH
Heniochus chrysostomus Cuvier, 1831
Inhabits coral reefs; distinguished by short "banner", and broad black band across head that is continuous with pelvic fins; N.W. Australia, Great Barrier Reef, and throughout S.E. Asia; mainly W. and C. Pacific; to 17 cm.

11 SINGULAR BANNERFISH
Heniochus singularius Smith & Radcliffe, 1911
Inhabits coral reefs; distinguished by short "banner", slight hump on forehead, and yellow dorsal and caudal fins; N.W. Australia, Great Barrier Reef, and throughout S.E. Asia; mainly W. and C. Pacific; to 25 cm.

12 LONGFIN BANNERFISH Cairns 10/98?
Heniochus acuminatus (Linnaeus, 1758)
Inhabits coral reefs, often in pairs or alone; distinguished by elongated dorsal rays, resembles 13 below, but longer snout and rounded anal fin; found throughout the region; Indo-C. Pacific; to 20 cm.

13 SCHOOLING BANNERFISH
Heniochus diphreutes Jordan, 1903
Inhabits sandy areas around outcrops; usually in schools; distinguished by elongated dorsal rays, resembles 12 above, but has shorter snout and angular anal fin; N.W. Australia, Great Barrier Reef, and throughout S.E. Asia; Indo-C. Pacific; to 20 cm. Cairns 10/98?

14 HUMPHEAD BANNERFISH
Heniochus varius (Cuvier, 1829)
Inhabits coral reefs; distinguished by lack of "banner" and forehead bump and "horns"; offshore reefs of N.W. Australia, Great Barrier Reef, and throughout S.E. Asia; mainly W. and C. Pacific; to 18 cm.

15 OCELLATE CORALFISH
Parachaetodon ocellatus (Cuvier, 1831)
Inhabits silty coral reefs, often seen on sand or silt bottoms near reefs; distinguished by tall triangular dorsal fin and orange-brown bars; found throughout the region; Indo-W. Pacific; to 18 cm.

TWIN SPECIES

Many reef fishes have close relatives or "sister" species that are very similar in overall appearance. These pairs have obviously evolved from a common ancestral population that became fragmented by such barriers as shifting ocean currents, sea temperature changes, and emergent land resulting from lowered sea levels or tectonic processes. In cases where the ancestral stock is widely distributed it may become fragmented into more than two populations evolving into "complexes" of several closely allied species. As the barriers may be only temporary, though persisting for thousands of years (a short span in terms of geological time), members of a pair may once again occur in the same area.

The two similar species of *Heniochus*, 12 and 13, are good examples of this process. Although they now have overlapping distributions they are more or less ecologically separated. The Schooling Bannerfish (13) foms groups which feed on plankton high above the bottom whereas the Longfin Bannerfish (12) lives alone or in small groups which forage on bottom dwelling invertebrates. Other species-pairs continue to exist more or less separately except for a relatively small overlap zone. For example, a number of pairs have one member in the Indian Ocean and another in the Pacific.

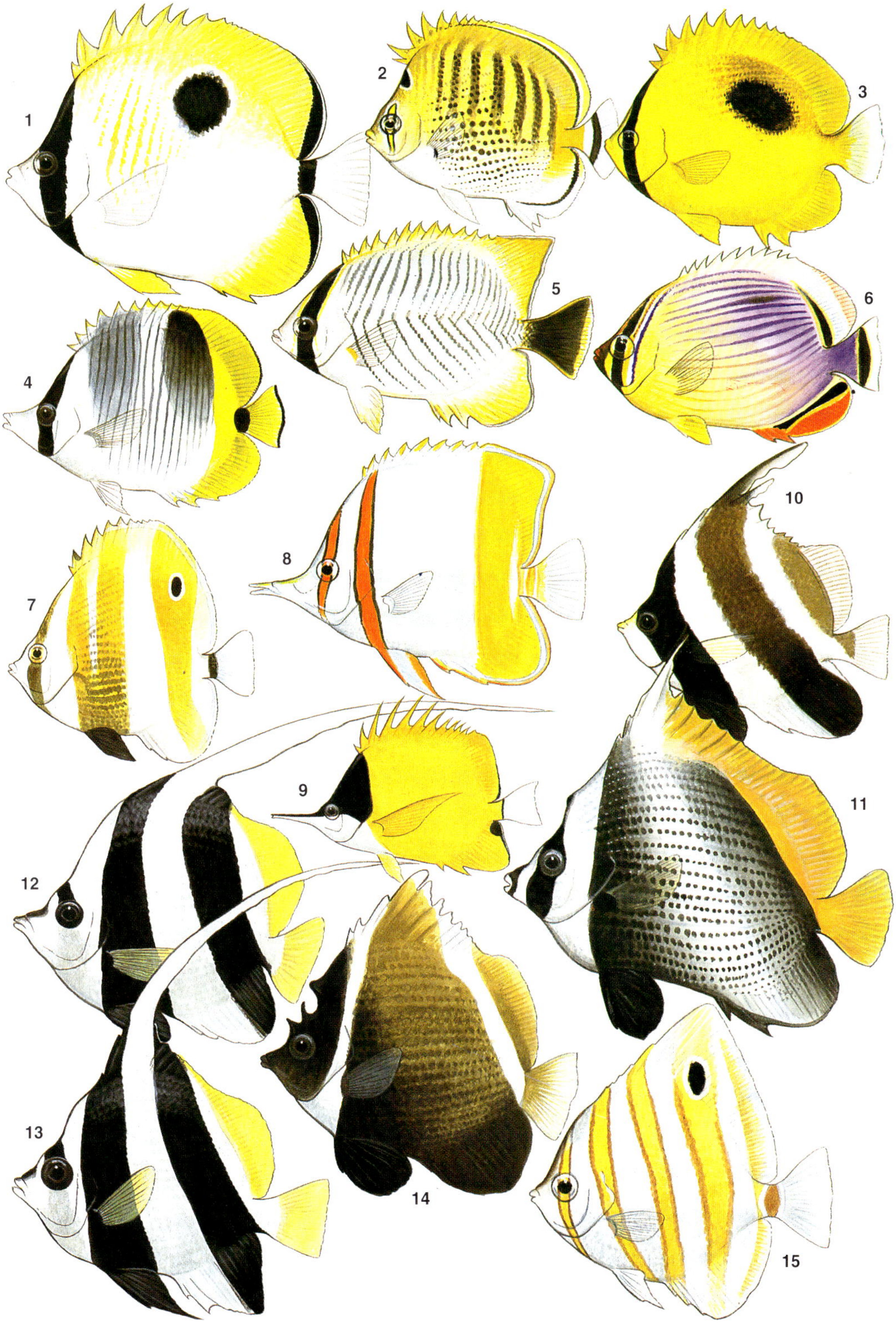
1
2
3
4
5
6
7
8
9
10
11
12
13
14
15

PLATE 57: BUTTERFLYFISHES (FAMILY CHAETODONTIDAE)

1 MERTEN'S BUTTERFLYFISH
Chaetodon mertensii Cuvier, 1831
Inhabits coral reefs; distinguished by orange areas on posterior part of body and tail, and dark chevron markings on side; Great Barrier Reef, New Guinea, and Phillipines; Indo-C. Pacific; to 12 cm; *C. madagaskariensis* of Indian Ocean is a synonym.

2 SPOTTED BUTTERFLYFISH
Chaetodon guttatissimus Bennett, 1835
Inhabits coral reefs, commonly seen on outer slopes; similar to 56-2, but lacks vertical bars; an Indian Oceans species found in our region only at Christmas Island and off the Indian Ocean coast of Sumatra; to 11 cm.

3 LATTICED BUTTERFLYFISH
Chaetodon rafflesi Bennett, 1830
Inhabits coral reefs; distinguished by network pattern on sides and overall yellow colouration with black band on tail and rear part of dorsal fin; Great Barrier Reef, offshore reefs of W.A., and throughout S.E. Asia; E. Indian Ocean and W. and C. Pacific; to 15 cm.

4 DOTTED BUTTERFLYFISH
Chaetodon semion Bleeker, 1855
Inhabits coral reefs, usually seen in pairs on outer slopes; distinguished by overall yellow colour, horizontal rows of small dark spots on side, blue area in front of eyes and on forehead, and broad black band at base of dorsal and anal fins; Great Barrier Reef, offshore reefs of W.A., and throughout S.E. Asia; E. Indian Ocean and W. Pacific; to 23 cm.

5 YELLOW-TAILED BUTTERFLYFISH
Chaetodon xanthurus Bleeker, 1857
Inhabits coral reefs, usually below 15-20 m depth; similar to 1 above, but has reticulated pattern on sides rather than dark chevrons; Indonesia and Philippines north to Japan; to 14 cm.

6 DOT-AND-DASH BUTTERFLYFISH
Chaetodon pelewensis Kner, 1868
Inhabits coral reefs, usually seen on outer reef slopes; distinguished by dark-edged orange bar through eye, diagonal rows of spots and lines on side, and yellow to orange tail base; Great Barrier Reef, offshore reefs of W.A., and throughout S.E. Asia; S.W. and C. Pacific; to 12 cm.

7 BENNETT'S BUTTERFLYFISH
Chaetodon bennetti Cuvier, 1831
Inhabits coral reefs, usually seen on outer reef slopes; distinguished by large pale-edged black spot on upper side, and pair of pale (white to bluish) diagonal bands between gill cover and anal fin; Great Barrier Reef, offshore reefs of W.A., and throughout S.E. Asia; Indo-C. Pacific; to 18 cm.

8 YELLOW-DOTTED BUTTERFLYFISH
Chaetodon selene Bleeker, 1853
Inhabits coral reefs, but generally rare in New Guinea; distinguished by general white colour, diagonal rows of yellow spots on anterodorsal part of body, yellow dorsal and anal fins, and black rim around posterior and dorsal part of body; Indonesia and New Guinea northward to Japan; to 15 cm.

9 SPOTNAPE BUTTERFLYFISH
Chaetodon oxycephalus Bleeker, 1853
Inhabits coral reefs; feeds on coral polyps and anemones; very similar to Pl. 55-9, but has isolated dark patch on forehead; Great Barrier Reef and throughout S.E. Asia; E. Indian Ocean and W. Pacific; to 25 cm.

10 SPOT-TAIL BUTTERFLYFISH
Chaetodon ocellicaudus Cuvier, 1831
Inhabits coral reefs; very similar to 11 below, but has isolated spot on tail base; Great Barrier Reef, offshore reefs of W.A., and throughout S.E. Asia; to 14 cm.

11 BLACK-BACK BUTTERFLYFISH
Chaetodon melannotus Bloch & Schneider, 1801
Inhabits coral reefs; feeds mainly on coral polyps; distinguished by thin eye bar, yellow snout and fins, narrow diagonal black stripes on side, and black upper back; very similar to 10 above, but lacks isolated spot on tail base; Great Barrier Reef, offshore reefs of W.A., and throughout S.E. Asia; Indo-W. Pacific; to 15 cm.

12 VAGABOND BUTTERFLYFISH
Chaetodon vagabundus Linnaeus, 1758
Inhabits coral reefs; distinguished by pattern of narrow diagonal lines on side, blackish band across posterior part of body, and dark bar on middle of tail; Great Barrier Reef, Kimberley coast and offshore reefs of W.A., and throughout S.E. Asia; Indo-C. Pacific; to 18 cm.

13 COLLARE BUTTERFLYFISH
Chaetodon collare Bloch, 1787
Inhabits coral reefs in 3-20 m depth; distinguished by dark overall colour with pale scale centres, brilliant white bar behind eye, and red on basal half of tail; Sumatra, Java, and Bali, mainly on Indian Ocean coasts; mainly Indian Ocean; to 16 cm.

14 RETICULATED BUTTERFLYFISH
Chaetodon reticulatus Cuvier, 1831
Inhabits coral reefs, usually seen in pairs on outer reefs; feeds mainly on corals; distinguished by overall blackish colour with rows of pale spots on side, broad pale band behind head, and pale dorsal fin; Great Barrier Reef, New Guinea, Philippines and N. Sulawesi; W. and C. Pacific; to 16 cm.

HYBRID BUTTERFLYFISHES AND ANGELFISHES

Although hybridisation is common in freshwater fishes, very few marine hybrids are known. Therefore it is surprising that this phenomenon is relatively common among the butterflyfishes (Plates 55-58) and the related angelfishes (Plates 58-60). Approximately 15 cases have been reported for butterflyfishes and another eight for angelfishes. It is interesting to speculate on the circumstances under which this event might occur. Behavioural research reveals that butterflyfishes can be categorised as solitary, pair-forming, or aggregating according to their social disposition. Most of the butterfly and angelfish species involved in hybridisation belong to solitary or pair-forming speices. It seems likely that this phenomenon results when there is a shortage of mates belonging to the same species. Therefore, individuals are forced to spawn with a closely related species. The aberrant colour patterns of hybrid individuals are usually very distinct. They generally possess a mixture of colour pattern features from each of the parent species. A few of the butterflyfish hybrids observed in our region include *C. auriga* x *C. ephippium*, *C. ephippium* x *C. semion*, *C. kleinii* x *C. unimaculatus*, and *C. meyeri* x *C. ornatissimus*.

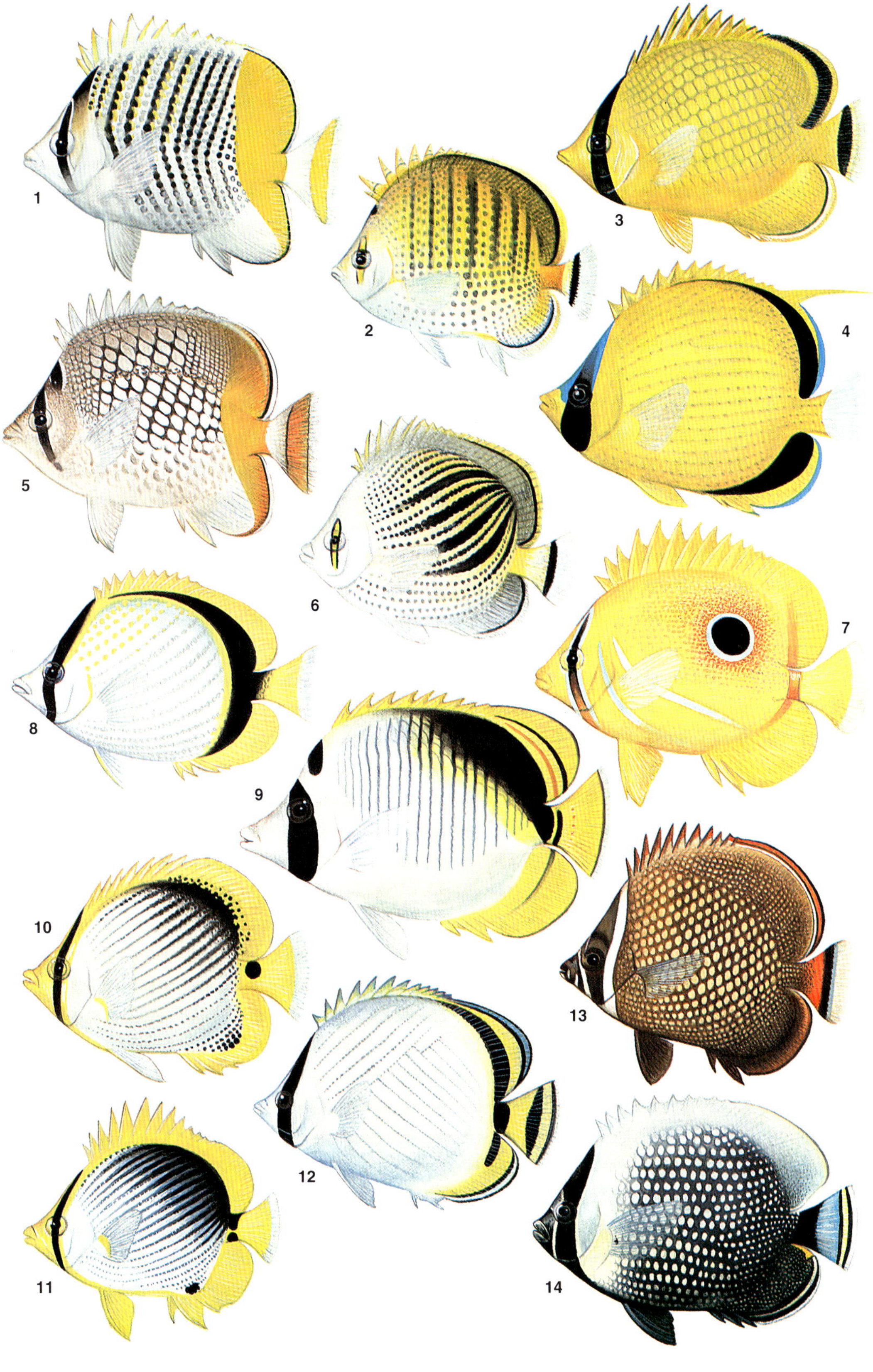
1
2
3
4
5
6
7
8
9
10
11
12
13
14

PLATE 58: BUTTERFLYFISHES AND ANGELFISHES

1 TRIANGULAR BUTTERFLYFISH
Chaetodon baronessa Cuvier, 1831
Inhabits coral reefs, usually seen in rich areas of live coral growth, often amongst branching *Acropora*; feeds on corals; distinguished by elevated shape and dark coloured body with narrow pale chevron markings; Great Barrier Reef, offshore reefs of W.A., and throughout S.E. Asia; E. Indian Ocean and W. Pacific; to 15 cm. (CHAETODONTIDAE)

2 DUSKY BUTTERFLYFISH
Chaetodon flavirostris Günther, 1873
Inhabits coral reefs, often seen in pairs; distinguished by dark body colour and yellow fins; S. Pacific from Great Barrier Reef to Pitcairn Group; to 20 cm. (CHAETODONTIDAE)

3 GÜNTHER'S BUTTERFLYFISH
Chaetodon guentheri Ahl, 1913
Inhabits coral reefs and rocky areas, usually seen in aggregations at depths between 5-40 m; distinguished by spotted pattern and bright yellow dorsal and anal fins; Great BarrierReef (rare) and scattered localities in Indonesia exposed to cool upwelling; edge of W. Pacific; to 14 cm. (CHAETODONTIDAE)

4 RAINFORD'S BUTTERFLYFISH
Chaetodon rainfordi McCulloch, 1923
Inhabits coral reefs; distinguished by orange-margined and inshore areas of Queensland; to 15 cm. (CHAETODONTIDAE)

5 BEAKED CORALFISH
Chelmon rostratus (Linnaeus, 1758)
Inhabits coral reefs, frequently seen in pairs; distinguished by elongate snout and pattern of orange bars with black ocellus on basal part of dorsal fin; Great Barrier Reef and throughout S.E. Asia; Andaman Sea and W. Pacific; to 20 cm. (CHAETODONTIDAE)

6 BURGESS'S BUTTERFLYFISH
Chaetodon burgessi Allen & Stark, 1973
Inhabits coral reefs, usually adjacent to steep dropoffs below 20 m depth; distinguished by diagonal black bar behind head and diagonal demarcation between dark and light areas of body; Indonesia, Philippines, and Caroline Islands; to 14 cm. (CHAETODONTIDAE)

7 HIGHFIN CORALFISH
Chelmon muelleri Richardson, 1842
Inhabits silty coastal reefs and estuaries; similar to **5** above, but bars are brown instead of orange; N. Territory and Queensland; to 18 cm. (CHAETODONTIDAE)

8 HIGHFIN CORALFISH
Coradion altivelis McCulloch, 1916
Inhabits coral reefs, usually in lagoons or on silty coastal reefs; similar to *C. chrysozonus,* but the double bars on the anterior part of the body are narrower and darker, and the ocellus on the dorsal fin is usually faint or absent in adults; Great Barrier Reef, offshore reefs of W.A., and throughout S.E. Asia; W. Pacific; to 15 cm. (CHAETODONTIDAE)

9 TWO-EYED CORALFISH
Coradion melanopus (Cuvier, 1831)
Inhabits coastal, lagoon, and seaward coral reefs; similar to **8** above, but has double bar in front of tail base and pair of ocellated spots; Indonesia, Philippines, and New Guinea; to 15 cm. (CHAETODONTIDAE)

10 PYRAMID BUTTERFLYFISH
Hemitaurichthys polylepis (Bleeker, 1857)
Inhabits coral reefs, seen on outer reef slopes; forms large midwater aggregations that feed on zooplankton; distinguished by a pyramid-shaped white area surrounded by yellow, head is usually brown; Great Barrier Reef, offshore reefs of W.A., and throughout S.E. Asia; W. and C. Pacific; to 18 cm. (CHAETODONTIDAE)

11 BIG LONGNOSE BUTTERFLYFISH
Forcipiger longirostris Jordan & McGregor, 1898
Inhabits coral reefs; distinguished by elongate snout and rich yellow colouration with upper part of head black; very similar to *F. flavissimus,* but has much longer snout and rows of black dots on the chest, it also possesses a rare all black or brown colour phase; Great Barrier Reef, offshore reefs of W.A., and throughout S.E. Asia; Indo-C. Pacific; to 22 cm. (CHAETODONTIDAE)

12 MASKED BANNERFISH
Heniochus monoceros Cuvier, 1831
Inhabits coral reefs, usually seen in pairs that shelter in crevices; similar to *H. singularius,* but dark head bar extends to base of dorsal and pattern of banding on the body is different; Great Barrier Reef, offshore reefs of W.A., and throughout S.E. Asia; Indo-C. Pacific; to 23 cm. (CHAETODONTIDAE)

13 TWO-SPINED ANGELFISH
Centropyge bispinosus (Günther, 1860)
Inhabits coral reefs, common on outer reef slopes; distinguished by blue head and fins, and pattern of blue bars on red-orange background, the number and width of the blue bars is quite variable; Great Barrier Reef, offshore reefs of W.A., and throughout S.E. Asia; Indo-C. Pacific; to 10 cm. (POMACANTHIDAE)

14 FLAME ANGELFISH
Centropyge loriculus (Günther, 1874)
Inhabits coral reefs in lagoons and seaward slope between 2-55 m depth; distinguished by bright red colouration and black bars; Great Barrier Reef and Coral Sea; W. and C. Pacific; to 10 cm. (POMACANTHIDAE)

15 WHITE-TAIL ANGELFISH
Centropyge flavicauda Fraser-Brunner, 1933
Inhabits rubble areas in the vicinity of coral reefs, usually seen below 20 m depth; distinguished by dark blue to blackish colour with contrasted pale caudal fin; Great Barrier Reef, offshore reefs of W.A., and throughout S.E. Asia; W. and C. Pacific; to 7 cm. (POMACANTHIDAE)

LONGNOSE VERSUS LONGNOSE

Longnose butterflyfishes are among the most conspicuous of reef fishes, but their identification is sometimes confusing. The true Longnose Butterflyfish (Plate 56-9) is one of the widest ranging of all coral-reef fishes, ranging from Africa to the Americas. The elongate snout and brilliant yellow colouration are readily diagnostic - except they offer little help when trying to differentiate it from the Big Longnose Butterflyfish (Plate 58-11). Although colour patterns of the two are nearly identical, there is a significant difference in snout length - that of the Big Longnose is nearly one-half to one-third longer. The Big Longnose is further distinguished by several rows of small black dots on the breast. It is also characterised by a relatively rare dark colour variety, which is most often seen around volcanic islands.

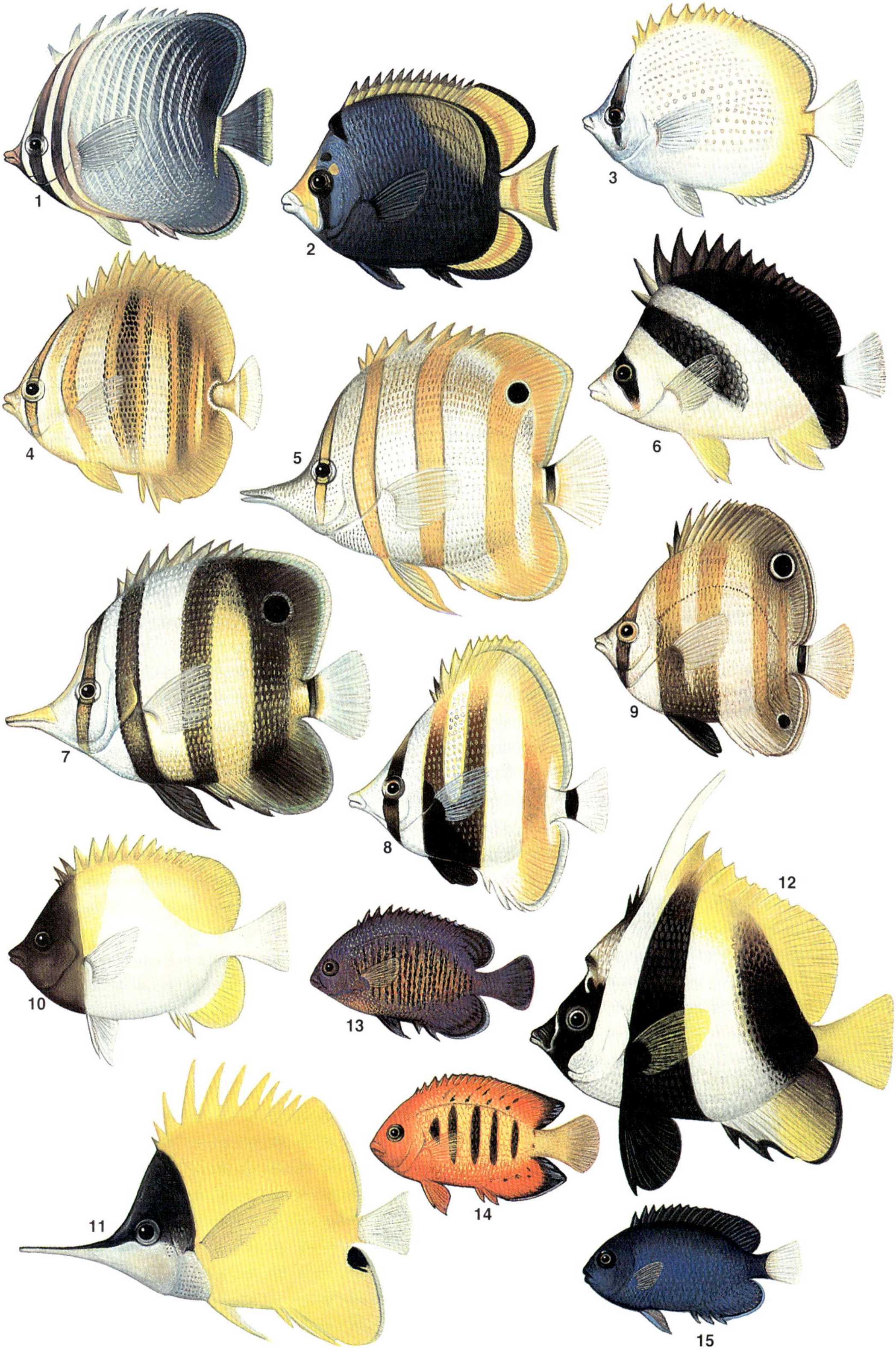
1
2
3
4
5
6
7
8
9
10
11
12
13
14
15

PLATE 59: ANGELFISHES (FAMILY POMACANTHIDAE)

1 THREE-SPOT ANGELFISH
Apolemichthys trimaculatus (Lacepède, 1831)
Inhabits offshore coral reefs, usually below 15 m depth; distinguished by blue lips, black spot on forehead, and broad black margin on anal fin; N.W. Australia, Great Barrier Reef, and throughout S.E. Asia; Indo-W. Pacific; to 20 cm.

2 BICOLOR ANGELFISH
Centropyge bicolor (Bloch, 1787) Cairns 10/98
Inhabits coral reefs; distinguished by blue band through eye, and yellow-blue combination on head and body; N.W. Australia (rare inshore), Great Barrier Reef, and throughout S.E. Asia;; mainly central and W. Pacific; to 15 cm.

3 EIBL'S ANGELFISH
Centropyge eibli Klausewitz, 1963
Inhabits coral reefs; distinguished by black tail and narrow brown or orange bars on side; W. Australia (Ningaloo and offshore reefs) and S. Indonesia (Sumatra to Flores); C. and E. Indian Ocean; to 11 cm.

4 KEYHOLE ANGELFISH
Centropyge tibicen (Cuvier, 1831)
Inhabits coral reefs; distinguished by vertically elongate white spot on middle of side and yellow margin on pelvic and anal fins; N.W. Australia, Great Barrier Reef, and throughout S.E. Asia; mainly W. Pacific; to 14 cm.

5 SCRIBBLED ANGELFISH
Chaetodontoplus duboulayi (Günther, 1867)
Inhabits coastal reefs on rubble, soft bottoms, or open flat bottoms with rock, coral, sponge, and seawhip outcrops; distinguished by dark bar through eye, yellow arc along back, and yellow tail, female (not shown) is dark on sides or has numerous blue or yellowish spots, male has broken wavy blue lines; N. Australia and New Guinea, also reported from Taiwan; to 28 cm.

6 YELLOWTAIL ANGELFISH
Chaetodontoplus personifer (McCulloch, 1914)
Inhabits similar areas to **5** above, usually in the vicinity of reefs; distinguished by blue "mask" with yellow spots on check and above eye, juvenile is black with broad white bar behind eye, and has yellow pelvic fins and tail; N. Australia only between Abrolhos Islands and Gulf of Carpentaria; *C. meredithi* is a nearly identical species from E. Australia; to 35 cm.

7 SIX-BANDED ANGELFISH
Pomacanthus sexstriatus (Cuvier, 1831)
Inhabits coral reefs; distinguished by white bar behind eye, 4-5 dark bars on side, and numerous blue spots on body and fins, juvenile similar to **8** below, but white bars generally broader; N.W. Australia, Great Barrier Reef, and throughout S.E. Asia; Indo-Australian Archipelago; to 45 cm.

8 BLUE-GIRDLED ANGELFISH
Pomacanthus navarchus (Cuvier, 1831)
Inhabits protected lagoon coral reefs; distinguished by broad yellow-orange area on sides with small dark spots, yellow-orange dorsal and tail fins, and bright blue stripe from snout to belly, also narrow blue margins on most fins; offshore reefs of N.W. Australia, Great Barrier Reef, and throughout S.E. Asia; Indo-Australian Archipelago; to 28 cm.

9 EMPEROR ANGELFISH
Pomacanthus imperator (Bloch, 1787)
Inhabits coral reefs; distinguished by blue-edged black band through eye and narrow yellow stripes on side, juveniles dark with narrow white lines, forming concentric circles on rear part of body; N.W. Australia, Great Barrier Reef, and throughout S.E. Asia; Indo-C. Pacific; to 31 cm.

10 BLUE ANGELFISH
Pomacanthus semicirculatus (Cuvier, 1831)
Inhabits coral reefs; distinguished by brown to yellowish colour with overall bluish hue imparted by numerous blue spots on body and fins, also has blue margins on cheek, gill cover, and most fins; juveniles similar to **7** and **8** above, but posterior white bars are strongly curved; N.W. Australia, Great Barrier Reef, and throughout S.E. Asia; Indo-W. Pacific; to 38 cm.

11 REGAL ANGELFISH Cairns 10/98
Pygoplites diacanthus (Boddaert, 1772)
Inhabits coral reefs; distinguished by alternating dark-edged white and yellow-orange cross bars; offshore reefs of N.W. Australia, Great Barrier Reef, and throughout S.E. Asia; Indo-C. Pacific; to 25 cm.

BIOLOGY OF ANGELFISHES

Most angelfishes are greatly dependent on shelter in the form of boulders, caves, and coral crevices. Aside from a few *Chaetodontoplus* species, they are seldom seen in the open. Typically most species are territorial, spending much of the time on the bottom in search of food or dashing back and forth into shelter. Members of the genus *Genicanthus* are usually open water swimmers, often forming aggregations that forage on plankton well above the bottom. Two other basic types of feeding are apparent among the other genera. Many of the smaller species, including those in *Centropyge*, feed mainly on algae, which is grazed from rocky surfaces. Most of the larger angels, particularly those in *Pomacanthus* consume sponges, supplemented with algae, zooantharians, tunicates, gorgonians, hydroids, and seagrasses. Spawning occurs at dusk and usually involves a single pair, although individual males may mate successively with several different females. Males set up territories by driving away other male competitors. They then swim up off the bottom and await the arrival of one or more females. When a prospective mate approaches the male exhibits a courtship display that may include fin erection, rapid back and forth swimming, and body "quivering". Eventually the pair spiral slowly towards the surface, suddenly shed eggs and sperm at the apex of the ascent, then swim back to the bottom. Hatching of the tiny (less than 1 mm) eggs occurs in 15-20 hours. Angelfishes offer a dazzling array of patterns - sparking a great deal of scientific controversy about their possible purpose. Many theories have been proposed, but none have been proved with any certainty. Large angelfishes are famous for their dramatically different juvenile patterns, which gradually change as the fish grows older. According to one theory the striking juvenile livery is actually a type of camouflage, serving to break up the outline of the fish so it effectively blends with the high contrast pattern of bright sunlit patches and shadows that characterise the reef environment.

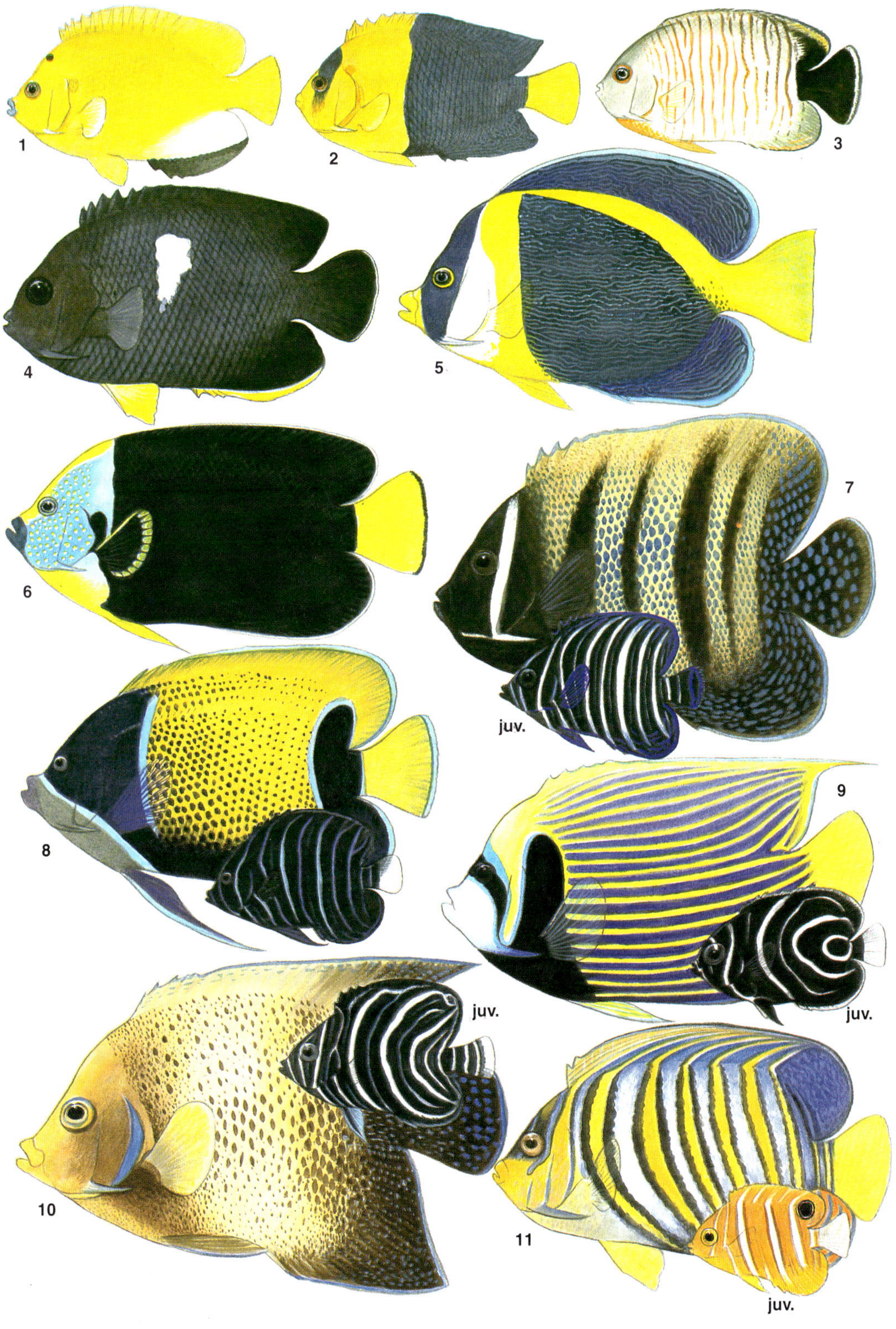
1
2
3
4
5
6
7
juv.
8
9
juv.
juv.
10
11
juv.

PLATE 60: ANGELFISHES (FAMILY POMACANTHIDAE)

1 LEMONPEEL ANGELFISH
Centropyge flavissimus (Cuvier, 1831)
Inhabits coral reefs, common in lagoons of oceanic islands and atolls; distinguished by bright yellow colour with blue cheek spine and blue margin on edge of gill cover followed by thin dark bar; Great Barrier Reef (rare), Christmas and Cocos-Keeling islands; W. and C. Pacific; to 14 cm.

2 HERALD'S ANGELFISH
Centropyge heraldi Wood's & Schultz, 1953
Inhabits coral reefs in 5-20 m depth; similar to **1** above, but yellow colour is duller and has dusky patch behind eye, also lacks blue markings on head; Great Barrier Reef north to Taiwan and eastward to Tuamotu Islands; to 11 cm.

3 MULTI-BARRED ANGELFISH
Centropyge multifasciatus (Smith & Radcliffe, 1911)
Inhabits outer reef slopes, usually seen in caves and crevices of steep slopes below 20 m depth; distinguished by bold pattern of black bars and yellow pelvic and anal fins; Great Barrier Reef and throughout S.E. Asia; W. and C. Pacific; to 10 cm.

4 MIDNIGHT ANGELFISH
Centropyge nox (Bleeker, 1853)
Inhabits coral reefs, but generally rare; distinguished by all black colouration including tail; Great Barrier Reef, offshore reefs of W.A., and throughout S.E. Asia; W. Pacific; to 9 cm.

5 PEARL-SCALED ANGELFISH
Centropyge vroliki (Bleeker, 1853)
Inhabits coral reefs, common in lagoons and on outer reefs; distinguished by grey colour on anterior part of body and dark colouration posteriorly, also has dark streak on upper edge of gill cover; Great Barrier Reef, offshore reefs of W.A., and throughout S.E. Asia; W. and C. Pacific; to 10 cm.

6 QUEENSLAND YELLOWTAIL ANGELFISH
Chaetodontoplus meredithi Kuiter, 1990
Inhabits silty coastal and inner reefs; distinguished by black body, blue head with yellow spots, and yellow tail; similar to *C. personifer* from N.W. Australia; Queensland; to 25 cm.

7 BLACK VELVET ANGELFISH
7 *Chaetodontoplus melanosoma* (Bleeker, 1853)
Inhabits coral and rocky reefs in 10-40 m depth, often on steep dropoffs; distinguished by black body with pale wash at front and yellow margin on dorsal, anal, and caudal fins; Indonesia to New Guinea; W. Pacific; to 7 cm.

8 VERMICULATED ANGELFISH
Chaetodontoplus mesoleucus (Bloch, 1787)
Inhabits coastal coral reefs and sheltered lagoons, usually on silty reefs; distinguished by purplish-brown colour on posterior part of body and on dorsal and anal fins, also has undulating narrow pale lines on side; Indonesia and New Guinea northward to Japan; to 17 cm.

9 LAMARCK'S ANGELFISH
Genicanthus lamarck (Lacepède, 1802)
Inhabits coral reefs, often seen in groups; feeds on plankton; distinguished by pattern of black stripes, black dorsal fin, and prolonged tail filaments; female (not shown) has white pelvic fin and prominent black upper and lower margins on tail; throughout S.E. Asia; Indo-W. Pacific; to 20 cm.

10 BLACKSPOT ANGELFISH
Genicanthus melanospilos (Bleeker, 1857)
Inhabits coral reefs, usually seen on steep outer reef slopes; generally found in groups with one male per several females; feeds on plankton; male distinguished by numerous narrow dark bars on side and black spot on midline of breast, females lack these features, but possess bold black upper and lower margins on the caudal fin; Great Barrier Reef, offshore reefs of W.A., and throughout S.E. Asia; mainly W. Pacific; to 18 cm.

11 WATANABE'S ANGELFISH
Genicanthus watanabei (Yasuda & Tominaga, 1970)
Inhabits coral reefs, usually seen on outer slopes and dropoffs below 20 m depth; male distinguished by blue back and dark stripes on lower side; female is plain light grey with black margins on fins; Queensland eastward to Tuamotus and north to Ryukyu Islands; to 18 cm.

12 YELLOWMASK ANGELFISH
Pomacanthus xanthometopon (Bleeker, 1853)
Inhabits coral reefs, usually seen solitarily on outer reefs; distinguished by blue scribble markings on head, yellow "mask", yellow to orange breast and pectoral fins, black spot at base of posterior part of dorsal fin, and bright yellow tail; juvenile (not shown) is similar to that of *P. sexstriatus*, but has more pale bars (about 15 between eye and tail); Great Barrier Reef, offshore reefs of W.A., and throughout S.E. Asia; Maldive Islands to W. Pacific; to 38 cm.

13 BLUE-RINGED ANGELFISH
Pomacanthus annularis (Bloch, 1787)
Inhabits coral reefs, sometimes seen in silty bays or harbours; distinguished by diagonal blue bands on side, blue ring above upper rear corner of gill cover, and white tail; juveniles are bluish black with narrow white and blue bars; throughout S.E. Asia; E. Indian Ocean and W. Pacific; to 30 cm.

MORE ABOUT ANGELFISHES

Angelfishes of the family Pomacanthidae are close relatives of the butterflyfishes (Plates 55-58) and until recently were considered to belong in the same family. Worldwide there are 82 known species. Most are inhabitants of tropical seas, being found mainly in the vicinity of coral reefs. They occur both as solitary individuals or in aggregations. Many species inhabit shallow water, from only 2-3 m down to 10-15 m depth. Others are restricted to deep water (to at least 75 m depth). Angelfishes are favourite aquarium pets, well known for their brilliant array of colour patterns. Many species exhibit dramatic changes from the juvenile to adult stage. Most angelfishes are dependent on shelter in the form of boulders, caves, or coral crevices. Typically, most species are territorial and spend the daylight hours near the bottom in search of food. The diet varies according to species; some feed almost exclusively on algae, others prefer mainly sponges supplemented by a variety of benthic invertebrates, and a few are mid-water zooplankton feeders. Divers are sometimes startled by the powerful drumming or thumping sound produced by large adults in the genus *Pomacanthus*.

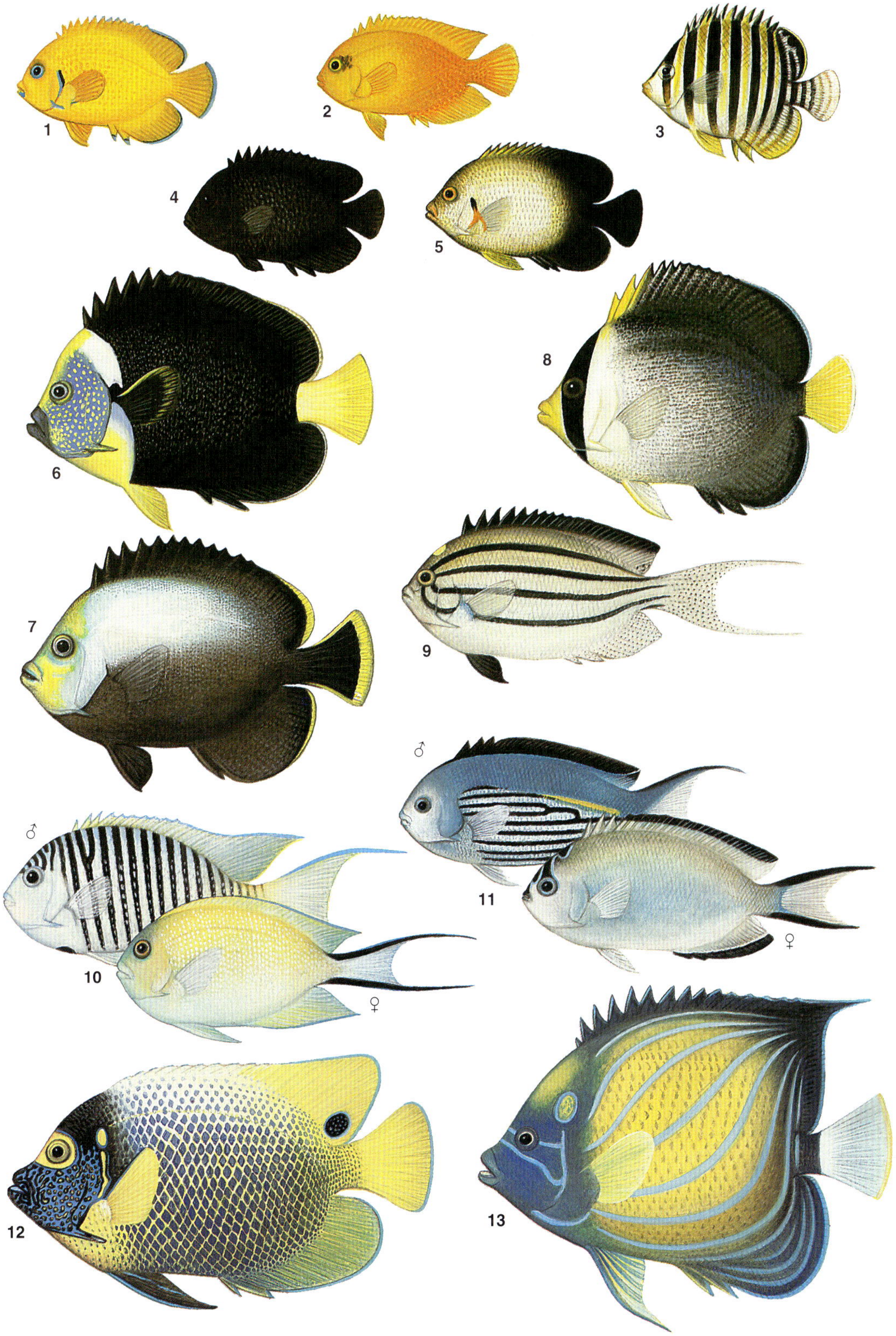
1
2
3
4
5
6
8
7
9
♂
♂
11
10
♀
♀
12
13

PLATE 61: DAMSELFISHES (FAMILY POMACENTRIDAE)

1 BANDED SERGEANT
Abudefduf septemfasciatus (Cuvier, 1830)
Inhabits shallow, wave-swept reefs; distinguished by 7 grey bars; found throughout the region; Indo-C. Pacific; to 22 cm.

2 BLACKSPOT SERGEANT MAJOR
Abudefduf sordidus Forsskål, 1775
Inhabits shallow, wave-swept reefs; similar to **1** above, but has black spot on top of tail base; found throughout the region; Indo-C. Pacific; to 22 cm.

3 SERGEANT MAJOR
Abudefduf vaigiensis (Quoy & Gaimard, 1825)
Inhabits coral, rocky, and weedy reefs; distinguished by 5 dark bars; found throughout the region; Indo-C. Pacific; to 22 cm.

4 NARROW-BANDED SERGEANT MAJOR
Abudefduf bengalensis (Bloch, 1787)
Inhabits coral and weedy reefs, distinguished from other *Abudefduf* by narrower bars and rounded lobes of tail; also known as Bengal Sergeant; found throughout the region; Indo-W. Pacific; to 18 cm.

5 SCISSORTAIL SERGEANT
Abudefduf sexfasciatus (Lacepède, 1802)
Inhabits coral and weedy reefs; similar to **3** above, but has black streaks on tail fin; found throughout the region; Indo-C. Pacific; to 22 cm.

6 STAGHORN DAMSEL
Amblyglyphidodon curacao (Bloch, 1787)
Inhabits coral reefs, often among staghorn corals; distinguished by ovate shape and faint bars; found throughout the region; mainly W. Pacific; to 13 cm.

7 BANDED DAMSEL
Dischistodus darwiniensis (Whitley, 1928)
Inhabits coral and rocky reef near sand; distinguished by brown bands on side and 2 dark spots on dorsal fin; N. Australia only, between Dampier, W. Australia and Gulf of Capentaria; to 15 cm.

8 HONEYHEAD DAMSEL
Dischistodus prosopotaenia (Bleeker, 1852)
Inhabits coral reefs near sand; distinguished by pale band through middle of body; found throughout the region; mainly W. Pacific; to 20 cm.

9 REGAL DEMOISELLE
Neopomacentrus cyanomos (Bleeker, 1856)
Inhabits coral reefs, distinguished by slender shape and white or yellow on rear of dorsal, anal, and tail fins; found throughout the region; Indo-W. Pacific; to 10 cm.

10 BROWN DEMOISELLE
Neopomacentrus filamentosus (Macleay, 1883)
Inhabits coral reefs; distinguished by slender shape, blue border on fins and dark margins on tail; Shark Bay northwards; Indo-Australian Archipelago; to 9 cm.

11 YELLOWTAIL DEMOISELLE
Neopomacentrus azysron (Bleeker, 1877)
Inhabits coral reefs; distinguished by slender shape, dark "ear" spot and yellow tail; found throughout the region; Indo-W. Pacific; to 9 cm.

12 LAGOON DAMSEL
Hemiglyphidodon plagiometopon (Bleeker, 1852)
Inhabits coral reefs, distinguished by plain brown colouration, pointed snout and lack of spines along the margin of the cheek; found throughout the region; mainly W. Pacific; to 22 cm.

13 BLACK DAMSEL
Neoglyphidodon melas (Cuvier, 1830)
Inhabits coral reefs often with soft corals; adults entirely black; found throughout the region; Indo-W. Pacific; to 16 cm.

14 BEHN'S DAMSEL
Neoglyphidodon nigroris (Cuvier, 1830)
Inhabits coral reefs; adults variable - mainly dark brown off N.W. Australia and some parts of W. Indonesia, bright yellow on posterior third of body at other locations; juveniles brilliant yellow with pair of black stripes; found throughout the region; mainly W. Pacific; to 14 cm.

DAMSELS OF THE SEA

The Damselfishes of the family Pomacentridae (Plates 61-67) are one of the most abundant groups of coral reef fishes. Approximately 335 species occur worldwide including about 170 from Australia and S.E. Asia. Most inhabit the tropics, but a number of species live in cooler temperate waters. They display remarkable diversity with regards to habitat preference, feeding habits, and behaviour. Coloration is highly variable ranging from drab hues of brown, grey and black to brilliant combinations of orange, yellow, and neon blue. A number of species have juvenile stages characterised by a yellow body with bright blue stripes crossing the upper head and back Most damselfishes are territorial, particularly algal feeding species such as 12-15 on Plate 66. They zealously defend their small plot against all intruders regardless of size. Damsels exhibit a highly stereo-typed mode of reproduction in which one or both partners clear a next site on the bottom and engage in courtship displays of rapid swimming and fin extension. Males generally guard the eggs which are attached to the bottom by adhesive strands. The eggs hatch within about 2-7 days and the fragile larvae rise to the surface. They are transported by ocean currents for periods which vary between about 10-50 days, depending on the species. Eventually the young fish settle to the bottom and their largely transparent bodies quickly assume the juvenile coloration. The growth rate of juveniles generally ranges from about 5-15 mm per month, gradually tapering off as maturity approaches. There is little reliable data concerning their longevity, but it appears they are capable of living to at least an age of 10 years. An unusual type of sex change has been documented in the anernonefishes (Plates 63 and 64) in which males eventually become females. Damselfishes feed on a wide variety of plant and animal material. Generally, the drab coloured species feed mainly on algae, whereas many of the brightly patterned species and also members of the genus *Chromis* (Plates 64 and 65) obtain their nourishment from current-borne plankton.

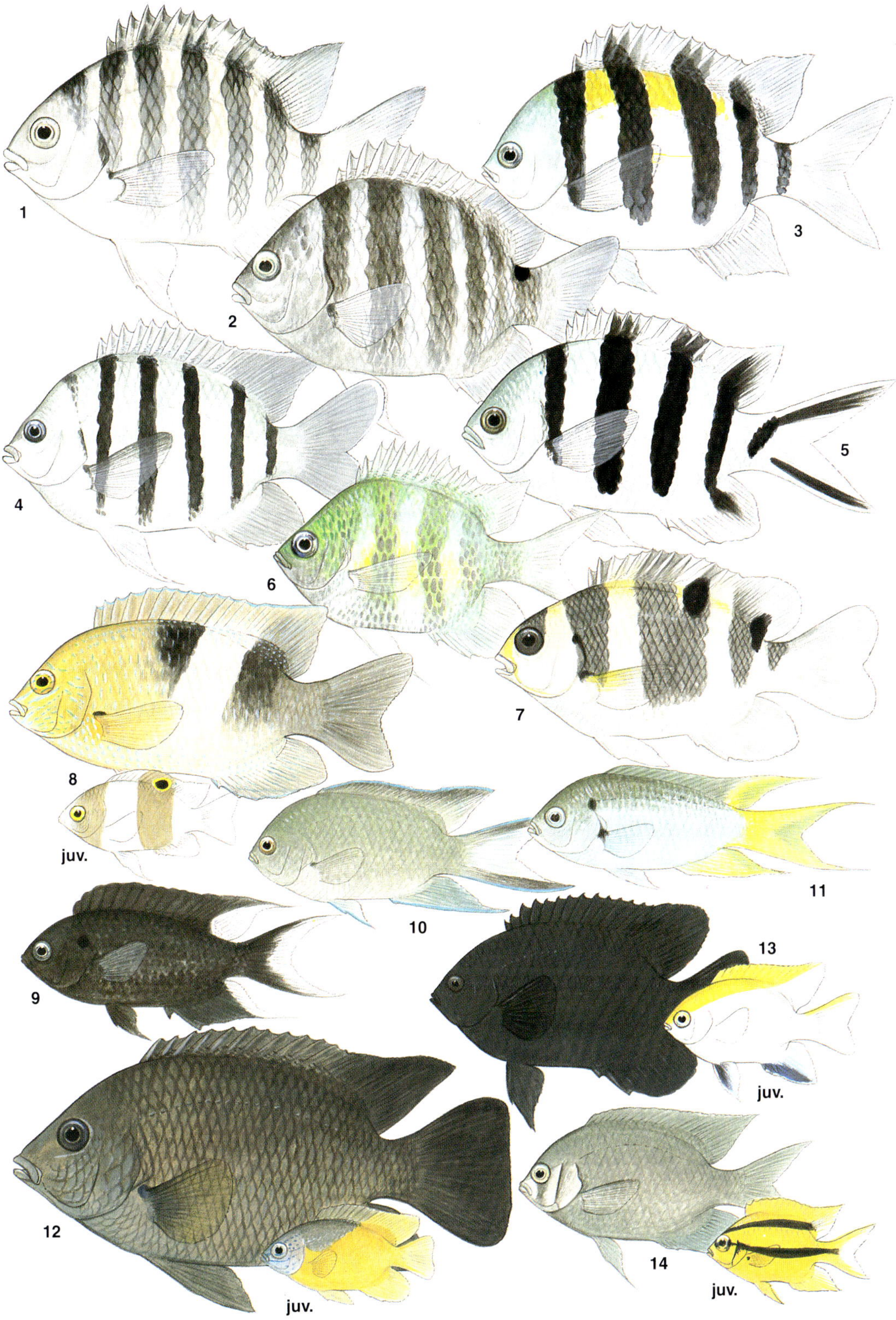
1
2
3
4
5
6
7
8
juv.
9
10
11
12
juv.
13
juv.
14
juv.

PLATE 62: DAMESELFISHES (FAMILY POMACENTRIDAE)

1 BLUE DEVIL
Chrysiptera cyanea (Quoy & Gaimard, 1824)
Inhabits inshore and lagoon coral reefs in 3-10 m depth; feeds mainly on algae and planktonic copepods and amphipods; distinguished by brilliant blue colour with dark stripe on snout; male differs from female in having orange margin on tail (absent at some localities) and female has dark spot at base of posterior part of dorsal fin; Great Barrier Reef, offshore reefs of W. Australia, and throughout S.E. Asia; W. Pacific; to 8.5 cm.

2 GREY DAMSEL
Chrysiptera glauca (Cuvier, 1830)
Inhabits shallow wave-swept reefs; feeds on algae and a variety of benthic and planktonic invertebrates; distinguished by overall light grey colour and black anus; juvenile is blue with neon blue stripe above eye; Great Barrier Reef, offshore reefs of W. Australia, and throughout S.E. Asia; Indo-C. Pacific; to 11 cm.

3 KING DEMOISELLE
Chrysiptera rex (Snyder, 1909)
Inhabits upper edge of outer reef slopes in 1-6 m depth; feeds on algae, fish eggs, and planktonic copepods; distinguished by overall pale yellow to orange colour with bluish tint on head and upper back; Great Barrier Reef, offshore reefs of W. Australia, and throughout S.E. Asia; W. Pacific; to 7 cm.

4 TALBOT'S DEMOISELLE
Chrysiptera talboti (Allen, 1975)
Inhabits coral reefs in 6-35 m depth, usually seen on outer slopes; feeds on algae and plankton; distinguished by yellow head and pelvic fins, purplish body, and black spot at base of dorsal fin; Great Barrier Reef and throughout S.E. Asia; Andaman Sea to Fiji Islands; to 6 cm.

5 ROLLAND'S DEMOISELLE
Chrysiptera rollandi (Whitley, 1961)
Inhabits both inshore and outer reefs in 2-35 m depth; feeds on algae, plankton, and small benthic invertebrates; distinguished by white body with greyish head and back; Great Barrier Reef and throughout S.E. Asia; Thailand to New Caledonia; to 6 cm.

6 TWOSPOT DEMOISELLE
Chrysiptera biocellata (Quoy & Gaimard, 1824)
Inhabits lagoons and coastal coral reefs, usually seen in sandy areas with scattered coral or rock outcrops; feeds on algae, plankton, and benthic invertebrates; distinguished by white bar on middle of side, juvenile is yellow with ocellus at base of middle dorsal rays, and small black spot behind last dorsal ray; Great Barrier Reef, offshore reefs of W. Australia, and throughout S.E. Asia; Indo-C. Pacific; to 10 cm.

7 ONESPOT DEMOISELLE
Chrysiptera unimaculata (Cuvier, 1830)
Inhabits shallow, wave-swept reef flats; feeds on algae, planktonic copepods and amphipods, and a wide variety of benthic invertebrates; similar to 6, but lacks white bar, usually has black spot at base of last few dorsal rays; juvenile mainly yellow with blue stripe on back and ocellus on middle of dorsal fin; Great Barrier Reef and throughout S.E. Asia; Indo-C. Pacific; to 8.5 cm.

8 YELLOWFIN DAMSEL
Chrysiptera flavipinnis (Allen & Robertson, 1974)
Inhabits coral reefs at depths between 3-38 m; distinguished by overall blue colour and yellow upper back and dorsal fin; *C. bleekeri* (not shown) from S.E. Asia is nearly identical; S.W. Pacific, including New Guinea, E. Australia, and Coral Sea; to 8.5 cm.

9 SURGE DEMOISELLE
Chrysiptera leucopoma (Lesson, 1830)
Inhabits shallow reefs exposed to surge; feeds on algae, fish eggs, and small benthic invertebrates; two colour forms are commonly seen: a mainly yellow variety with brilliant blue stripe along the back (shown here) and a dark variety with white bars; Great Barrier Reef, offshore reefs of W. Australia, and throughout S.E. Asia; Indo-C. Pacific; to 8.5 cm.

10 AZURE DEMOISELLE
Chrysiptera parasema (Fowler, 1918)
Inhabits sheltered lagoons in rich coral areas; feeds mainly on algae and zooplankton; distinguished by deep blue body colour and yellow tail; *C. hemicyanea* from offshore reefs of W. Australia and parts of Indonesia is similar, belly, and pelvic/anal fins are entirely yellow; Philippines and Melanesia; to 5.5 cm.

11 SPRINGER'S DEMOISELLE
Chrysiptera springeri (Allen & Lubbock, 1976)
Inhabits sheltered lagoons and inshore coral reefs in depth range of 5-30 m; distinguished by deep blue colour; *C. sinclairi* from New Guinea nearly identical; Indonesia and Philippines; to 5.5 cm.

12 RICHARDSON'S REEF DAMSEL
Pomachromis richardsoni (Snyder, 1909)
Inhabits coral or rocky reefs, frequently in areas exposed to surge; forms aggregations; distinguished by bold black tail margins; Great Barrier Reef, offshore reefs of W. Australia, and scattered localities in S.E. Asia; Indo-W. Pacific; to 8 cm.

13 WHITEPATCH DAMSEL
Dischistodus chrysopoecilus (Schlegel & Muller, 1839)
Inhabits coastal reefs and silty lagoons, sometimes in sea grass beds; feeds on algae and detritus; distinguished by white spot below middle of dorsal fin, also pale band across forehead; Indo-Malay Archipelago; Thailand to Solomon Islands; to 16 cm.

14 BANDED DAMSEL
Dischistodus fasciatus (Cuvier, 1830)
Inhabits lagoon and inshore reefs with silty or sandy bottoms and coral outcrops, often in seagrass areas; feeds on detritus and algae; distinguished by dark body and pale bars; small juveniles white with dark bars; *D. darwiniensis* from N.W. Australia is similar; Indo-Malay Archipelago; to 14 cm.

15 SILVER DEMOISELLE
Neopomacentrus anabatoides (Bleeker, 1847)
Inhabits sandy or silty bottoms around coral or rock outcrops; feeds on zooplankton in huge aggregations; distinguished by overall silvery sheen and black streaks on tail; Indo-Malay Archipelago; to 8 cm.

16 BLACKVENT DAMSEL
Dischistodus melanotus (Bleeker, 1853)
Inhabits lagoon and coastal reefs; feeds on algae and detritus; distinguished by brown head and upper front part of body, and large dark blotch in front of anal fin; Great Barrier Reef and throughout S.E. Asia; edge of W. Pacific; to 16 cm.

17 WHITE DAMSEL
Dischistodus perspicillatus (Cuvier, 1830)
Text on page 164.

18 VIOLET DEMOISELLE
Neopomacentrus violascens (Bleeker, 1848).
Text on page 164.

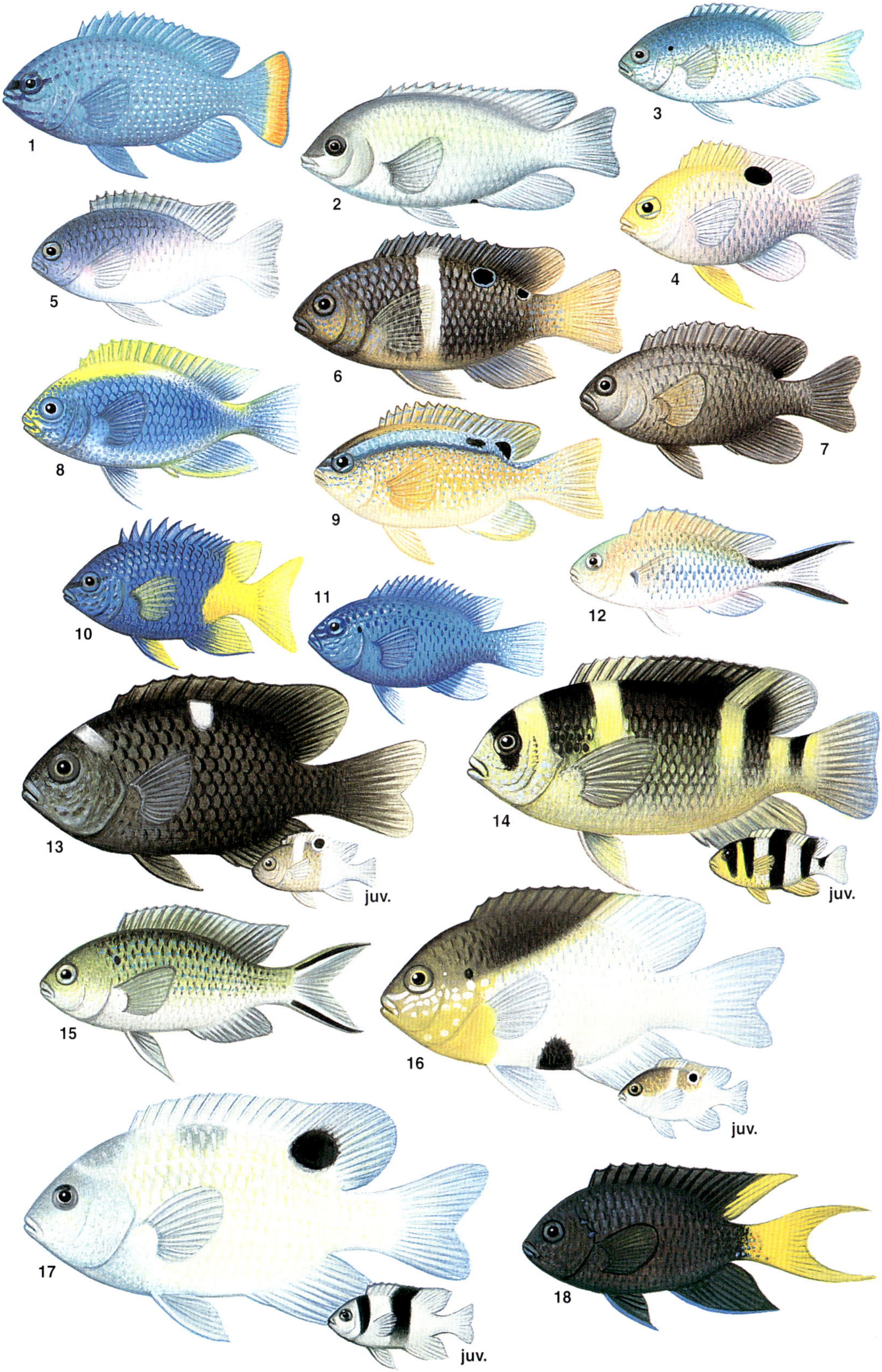
1
2
3
4
5
6
7
8
9
10
11
12
13
juv.
14
juv.
15
16
juv.
17
juv.
18

PLATE 63: DAMSELFISHES (FAMILY POMACENTRIDAE)

1 SPINY CHROMIS
Acanthochromis polyacanthus (Bleeker, 1855)
Inhabits coral reefs from the shallows to 65 m; young stay with parents after hatching, which is unlike the planktonic larval stage of all other damselfishes; feeds on plankton; distinguished by 17 dorsal spines (14 or less in other pomacentrids); colour pattern extremely variable depending on geographic locality (four of these shown here); Great Barrier Reef, Kimberley coast of W.A., and throughout S.E. Asia; Indo-Australian Archipelago; to 14 cm.

2 BLACKTAIL SERGEANT
Abudefduf lorenzi Hensley & Allen, 1977
Inhabits lagoons and sheltered coastal reefs to about 6 m depth, often common immediately adjacent to shore; feeds on algae; distinguished by six black bars and large black patch at tail base; Philippines to Solomon Islands; to 17 cm.

3 WHITLEY'S SERGEANT
Abudefduf whitleyi Allen & Robertson, 1974
Inhabits coral reefs at depths between 1-5 m; distinguished by lime-green colour, 5 narrow dark bars on side, and blackish tail; Grat Barrier Reef, Coral Sea, and New Caledonia; to 17 cm.

4 YELLOW-TAILED SERGEANT
Abudefduf notatus (Day, 1869)
Inhabits rocky inshore areas where there is usually moderate to strong wave action; distinguished by dark grey colouration, 5 narrow white bars, and yellowish tail; found throughout the region, but not yet recorded from Australia; Indo-W. Pacific; to 17 cm.

5 BATUNA'S DAMSEL
Amblyglyphidodon batunai (Allen, 1995)
Inhabits sheltered, silty coastal reefs and lagoons, usually among beds of staghorn coral in 2-10 m depth; feeds mainly on zooplankton and algae; distinguished by overall pale colour becoming greenish on back, pearly lustre of scales on nape, and dark saddle-like mark on upper tail base; *A. ternatensis* (not shown) is similar and lives in the same habitat, but is overall yellowish (sometimes faintly yellow); Kimberley coast of W.A. and Indo-Malay Archipelago; to 10 cm.

6 BLACK-TAILED DASCYLLUS
Dascyllus melanurus Bleeker, 1854
Inhabits sheltered coastal reefs and lagoons, usually associated with isolated coral heads; feeds mainly on zooplankton and algae; similar to *D. aruanus* (Pl. 64-15), but has extra black band across rear part of tail; Great Barrier Reef and throughout S.E. Asia; Indo-Malay Archipelago to W. Micronesia; to 8.5 cm.

7 BLACK ANEMONEFISH
Amphiprion melanopus Bleeker, 1852
Commensal with sea anemones (usually *Entacmaea quadricolor*), in lagoons and on outer reefs; feeds mainly on planktonic copepods and benthic algae; distinguished by single white bar behind the eye, overall blackish colour, and black pelvic and anal fins; young are red with 2-3 white bars; Great Barrier Reef and throughout S.E. Asia W. and C. Pacific; to 12 cm.

8 SADDLEBACK ANEMONEFISH
Amphiprion polymnus (Linnaeus, 1758)
Commensal with sea anemones (usually *Stichodactyla haddoni*), in sandy or silty lagoons and coastal embayments; feeds mainly on zooplankton; distinguished by broad white bar behind eye and large white saddle (variable in length) on back; some individuals have black pelvic and anal fins; amount of yellow colour on head and body is variable (sometimes absent); N. Territory and throughout S.E. Asia; W. Pacific; to 13 cm.

9 BARRIER REEF ANEMONEFISH
Amphiprion akindynos Allen, 1972
Commensal with several sea anemones in lagoons and on outer reefs; feeds mainly on planktonic copepods and benthic algae; distinguished by brown to orange brown body with pair of whitish bars; S.W. Pacific including Great Barrier Reef, N. New South Wales, Coral Sea, New Caledonia, and Loyalty Islands; to 12 cm.

10 RED SADDLEBACK ANEMONEFISH
Amphiprion ephippium (Bloch, 1790)
Commensal with sea anemones (usually *Entacmaea quadricolor* or *Heteractis crispa*), on sheltered, silty inshore reefs; feeds on zooplankton and algae; distinguished by bright red-orange colour and black blotch on side, which becomes larger with increased growth; small juveniles lack black entirely and have a white bar behind the eye; Andaman Sea, Thailand, Malaysia, Sumatra, and Java; to 12 cm.

11 TOMATO ANEMONEFISH
Amphiprion frenatus Brevoort, 1856
Commensal with sea anemones (usually *Entacmaea quadricolor*), in lagoons and on outer reefs; feeds mainly on planktonic copepods and benthic algae; distinguished by single white bar behind the eye overall blackish colour, and red fins; young are red with 2-3 white bars; throughout S.E. Asia; western edge of Pacific to Japan; to 12 cm.

12 ORANGEFIN ANEMONEFISH
Amphiprion chrysopterus Cuvier, 1830
Commensal with several sea anemones, usually in passes or on outer reef slopes; feeds mainly on zooplankton and algae; distinguished by pair of bluish-white bars, orange dorsal fin, and black pelvic and anal fins; young specimens are mainly brown with two white or blue-white bars; Great Barrier Reef (rare) and New Guinea; W. and C. Pacific; to 16 cm.

13 SPINE-CHEEK ANEMONEFISH
Premnas biaculeatus (Bloch, 1790)
Commensal with sea anemones (usually *Entacmaea quadricolor*), both on sheltered inshore reefs and on outer slopes; feeds on zooplankton and algae; distinguished by overall red colour, three pale bars, and enlarged spine below eye; found in pairs, female is usually 2-3 times size of male; female colour less brilliant; Great Barrier Reef, offshore reefs of W. Australia, and throughout S.E. Asia; to 16 cm.

PLATE 62

17 WHITE DAMSEL
Dischistodus perspicillatus (Cuvier, 1830)
Inhabits sandy areas near coral reefs; feeds on algae and detritus; distinguished by mainly white colour with three saddle-like black spots on dorsal part of head and body; Great Barrier Reef, offshore reefs of W. Australia, and throughout S.E. Asia; Andaman Sea to Vanuatu; to 20 cm.

18 VIOLET DEMOISELLE
Neopomacentrus violascens (Bleeker, 1848)
Inhabits inshore reefs on soft bottoms, forming aggregations around coral and rock outcrops, wharf pilings and wreckage; distinguished by dark brown body, yellow tail, and dark "ear" spot; Indo-Malay Archipelago; Indonesia to Vanuatu; to 6 cm.

Papua New Guinea

Coral Sea

N. Great Barrier Reef

N.W. Australia

PLATE 64: DAMSELFISHES (FAMILY POMACENTRIDAE)

1 CLARK'S ANEMONEFISH
Amphiprion clarkii (Bennett, 1830)
Inhabits coral reefs, associated with large sea anemones; distinguished by 2 white bars and and abruptly pale tail; found throughout the region; Indo-W. Pacific; to 13 cm.

2 FALSE CLOWN ANEMONEFISH
Amphiprion ocellaris Cuvier, 1830
Inhabits coral reefs in protected waters; associated with large sea anemones; distinguished by 3 white bars on head and body (middle bar with expanded anterior projection) and black submarginal bands on fins; an all black (except for white bars) occurs off the N. Territory, Australia; N.W. Australia and throughout S.E. Asia; Andaman Sea and Indo-Australian Archipelago, north to Ryukyu Islands; a nearly identical species, *A. percula*, occurs at Queensland and New Guinea; to 7.5 cm.

3 PINK ANEMONEFISH
Amphiprion perideraion Bleeker, 1855
Inhabits coral reefs; associated with large sea anemones; distinguished by a single narrow bar on each side of head; N.W. Australia, Great Barrier Reef, and throughout S.E. Asia; mainly W. and C. Pacific; to 10 cm.

4 ORANGE ANEMONEFISH
Amphiprion sandaracinos Allen, 1972
Inhabits coral reefs; associated with large sea anemones; distinguished by a white stripe along the base of the dorsal fin; N.W. Australia, Great Barrier Reef, and throughout S.E. Asia; mainly W. Pacific; to 13 cm.

5 RED ANEMONEFISH
Amphiprion rubrocinctus Richardson, 1842
Inhabits coral reefs; associated with large sea anemones; distinguished by black to reddish colour and white bar on head (sometimes weakly developed or absent); N. Australia only, between Point Quobba, W. Australia and Gulf of Carpentaria; to 13 cm.

6 BIG-LIP DAMSEL
Cheiloprion labiatus (Day, 1877)
Inhabits coral reefs; usually amongst beds of branching *Acropora* corals; feeds partly on live corals; distinguished by enlarged lips; juveniles are primarily blackish with a neon-blue stripe along the upper back; found throughout the region; Indo-W. Pacific; to 8 cm.

7 BLACK-AXIL CHROMIS
Chromis atripectoralis Welander & Schultz, 1951
Inhabits coral reefs; forms large aggregations above coral bommies; distinguished from **8** below by the black "arm pit" of the pectoral fin; found throughout the region; Indo-C. Pacific; to 10 cm.

8 BLUE-GREEN CHROMIS
Chromis viridis (Cuvier, 1830)
Inhabits coral reefs; forms aggregations that shelter in branching coral; distinguished from **7** above by smaller size and lack of black colour on inside of pectoral fin base; found throughout the region; Indo-C. Pacific; to 8 cm.

9 GREEN CHROMIS
Chromis cinerascens (Cuvier, 1830)
Inhabits coral reefs; distinguished by overall dusky green appearance and faint longitudinal lines along the side of the body; Dampier Archipelago northwards; W. Pacific and E. Indian Ocean; to 10 cm.

10 SMOKEY CHROMIS
Chromis fumea (Tanaka, 1917)
Inhabits coral reefs, usually in deeper water (below 10-15 m); distinguished by dark caudal fin margins and pearly spot behind dorsal fin; W. Australia, E. Malay Peninsula; mainly W. Pacific in subtropical and warm temperate seas; to 9 cm.

11 BICOLOR CHROMIS
Chromis margaritifer Fowler, 1946
Inhabits coral reefs; distinguished by two-tone colour pattern; N.W. Australia, Great Barrier Reef, and throughout S.E. Asia; mainly W. Pacific; to 7.5 cm.

12 WEBER'S CHROMIS
Chromis weberi Fowler & Bean, 1928
Inhabits coral reefs; distinguished by dark scale edges forming a network pattern by the dark tips on the tail; N.W. Australia, Great Barrier Reef, and throughout S.E. Asia; Indo-C. Pacific; to 12 cm.

13 THREE-SPOT DASCYLLUS
Dascyllus trimaculatus (Rüppell, 1828)
Inhabits coral reefs, juveniles sometimes associated with branching corals or large sea anemones; distinguished by white spot on forehead and similar spot on each side of body, which gradually fade with increased size; found throughout the region; Indo-C. Pacific; to 13 cm.

14 RETICULATED DASCYLLUS
Dascyllus reticulatus (Richardson, 1846)
Inhabits coral reefs, occurs in aggregations around small coral formations; distinguished by black bar on front of body; N.W. Australia, Great Barrier Reef, and throughout S.E. Asia; Indo-C. Pacific; to 9 cm.

15 HUMBUG DASCYLLUS
Dascyllus aruanus (Linneaus, 1758)
Inhabits coral reefs, usually seen in aggregations around small coral formations; distinguished by 3 black bars; N.W. Australia, Great Barrier Reef, and throughout S.E. Asia; Indo-C. Pacific; to 8 cm.

SYMBIOTIC DAMSELS

Symbiosis is a biological term that literally means "living together". There are many examples of this phenomenon in nature, but one of the most colourful and well documented involves damselfishes belonging to the genus *Amphiprion*. The genus contains 26 species which occur throughout the tropical Indo-west Pacific region. All of the species are associated with large sea anemones that normally sting fishes which make contact with their tentacles. However, anemonefishes are never stung. There are two factors which contribute to the fishes apparent immunity. Apparently their distinctive swimming behaviour and special chemicals in the external mucus coat of the fish prevent the anemone from firing its nematocysts (stinging cells). Anemonefishes gain security from predators because of their special association and the anemones themselves are protected from coelenterate feeders such as butterflyfishes. The fishes also keep their host free of debris. Experiments reveal that the fishes directly contribute to the anemone's well being by keeping the tentacles in robust condition. Juveniles of the Three-spot Dascyllus (**13**) are also associated with anemones.

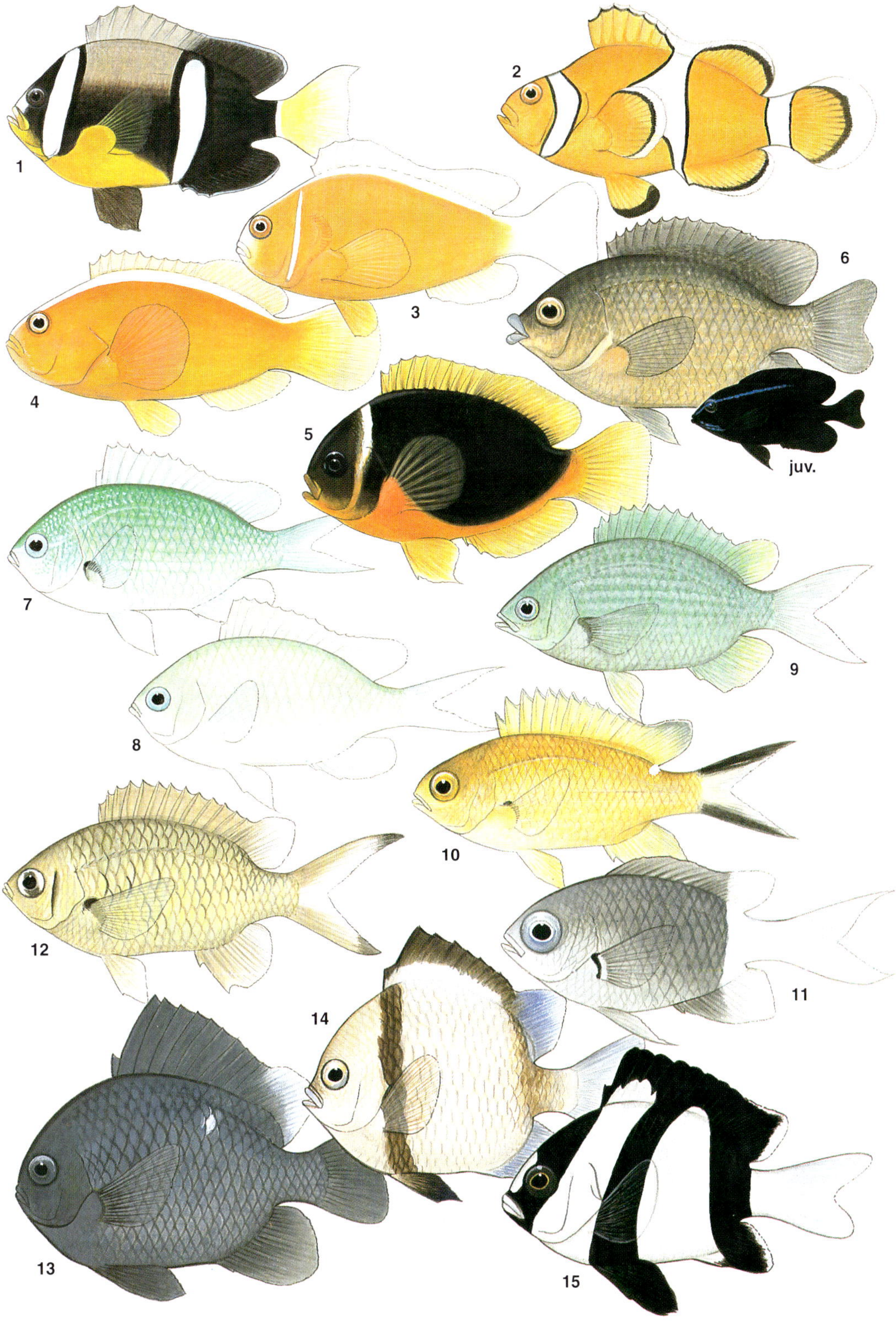
1
2
3
4
5
6
juv.
7
8
9
10
11
12
13
14
15

PLATE 65: DAMSELFISHES (FAMILY POMACENTRIDAE)

1 ALPHA CHROMIS
Chromis alpha Randall, 1987
Inhabits outer reef slopes between 18-95 m depth; distinguished by plain brown colouration with bluish belly and anal fin, sometimes with yellow scale centres on side of body; Great Barrier Reef, offshore reefs of N.W. Australia and throughout S.E. Asia; W. and C. Pacific; to 12 cm.

2 YELLOW CHROMIS
Chromis analis (Cuvier, 1830)
Inhabits coral reefs, usually seen on steep outer slopes between 18-70 m; colouration ranges from yellow-brown with yellow fins to entirely yellow; Great Barrier Reef, offshore reefs of W. Australia, and throughout S.E. Asia; mainly W. Pacific; to 15 cm.

3 AMBON CHROMIS
Chromis amboinensis (Bleeker, 1873)
Inhabits coral reefs, including passages, lagoons, and outer slopes to 65 m depth; distinguished by light golden-brown colour, orange bar at pectoral fin base, and dark upper and lower caudal fin margins; Great Barrier Reef, offshore reefs of W. Australia, and throughout S.E. Asia; W. and C. Pacific; to 8 cm.

4 DARKFIN CHROMIS
Chromis atripes Fowler & Bean, 1928
Inhabits reef passages and outer slopes in 2-35 m depth; distinguished by plain brown colouration, dark areas on soft dorsal and anal fins, and dark marking at base of upper pectoral fin rays; Great Barrier Reef, offshore reefs of W. Australia, and throughout S.E. Asia; mainly W. Pacific; to 7 cm.

5 TWINSPOT CHROMIS
Chromis elerae (Folwer & Bean, 1928)
Inhabits caves, ledges, and thickets of black coral on steep outer slopes at depths between 12-70 m; distinguished by white spot at base of last dorsal and anal rays; Great Barrier Reef, offshore reefs of W. Australia, and throughout S.E. Asia; E. Indian Ocean and W. Pacific; to 6.5 cm.

6 STOUT-BODY CHROMIS
Chromis chrysura (Bliss, 1883)
Inhabits outer coral or rocky reefs at depths between 6-30 m; distinguished by deep-bodied shape, dusky streak on each scale, and abruptly white tail; Great Barrier Reef and Coral Sea; 3 isolated popualtions: (1) Mauritius and Reunion, (2) S. Japan to Taiwan, and (3) E. Australia to Fiji; to 14 cm.

7 HALF AND HALF CHROMIS
Chromis iomelas Jordan & Seale, 1906
Inhabits outer coral reefs in 3-35 m depth; distinguished by bicolour pattern with abrupt demarcation in middle of body between dark and light areas; S.W. Pacific from Great Barrier Reef to Society Islands; to 5.5 cm.

8 SCALY CHROMIS
Chromis lepidolepis Bleeker, 1877
Inhabits coral reefs, both inshore reefs and outer slopes in 2-20 m; feeds on zooplankton; distinguished by black margin on dorsal fin with fine white tips on each spine, and black tips on lobes of tail; Great Barrier Reef, offshore reefs of W. Australia, and throughout S.E. Asia; Indo-C. Pacific; to 8 cm.

9 PHILIPPINES CHROMIS
Chromis scotochilopterus Fowler, 1918
Inhabits coral reefs on sheltered seaward slopes in 5-20 m depth; distinguished by deep-bodied shape, black upper and lower margins on tail, and black band on front of anal fin; Phillipines and Indonesia; to 14 cm.

10 TERNATE CHROMIS
Chromis ternatensis (Bleeker, 1856)
Inhabits lagoons and outer reef slopes in 2-15 m; sometimes forms large aggregations; distinguished by dusky scale margins and dark upper and lower margins on tail; Great Barrier Reef, offshore reefs of W. Australia, and throughout S.E. Asia; Indo-C. Pacific; to 9 cm.

11 LINED CHROMIS
Chromis lineata Fowler & Bean, 1928
Inhabits upper edge of outer reef slopes in 2-10 m; distinguished by horizontal rows of blue spots on side; *C. vanderbilti* (not shown) is similar, but most of anal fin dark and lower lobe of caudal fin with broad black band; throughout Indo-Malay Archipelago to Solomon Islands; to 6 cm.

12 YELLOW-AXIL CHROMIS
Chromis xanthochira (Bleeker, 1851)
Inhabits outer reef slopes in about 10-48 m; feeds on plankton; distinguished by dark rear margin on cheek, yellow pectoral fin base, dark scale outlines, and dark band on each caudal fin lobe; Indo-Malay Archipelago to Great Barrier Reef, New Guinea, and Solomon Islands; to 14 cm.

13 BLACK-BAR CHROMIS
Chromis retrofasciata Weber, 1913
Inhabits coral reef lagoons and outer slopes between depths of 5-65 m; distinguished by yellow-tan colour and prominent black bar across rear of body; Great Barrier Reef and throughout S.E. Asia; Indo-Malay Archipelago to Fiji; to 6 cm.

14 BARRIER REEF CHROMIS
Chromis nitida (Whitley, 1928)
Inhabits coral reefs in 5-25 m depth; distinguished by diagonal demarcation between dark upper body and white lower parts; central and southern Great Barrier Reef south to Sydney and Lord Howe Island; to 8 cm.

15 PALE-TAIL CHROMIS
Chromis xanthura (Bleeker, 1854)
Inhabits outer reef slopes in 10-48 m; feeds on zooplankton; distinguished by black edges on cheek and gill cover, and white tail; Great Barrier Reef, offshore reefs of W. Australia, and throughout S.E. Asia; Indo-C. Pacific; to 15 cm.

16 DUSKY CHROMIS
Chromis caudalis Randall, 1987
Inhabits steep outer reef slopes in 20-50 m depth; distinguished by brownish-grey to nearly blackish body with whitish tail, also has blue spot in axil ("armpit") of pectoral fin; throughout S.E. Asia; E. Indian Ocean and W. Pacific; to 9 cm.

STRENGTH IN NUMBERS

Masses of plankton-feeding Chromis damselfishes form an integral part of the reef community. These fishes occur in prodigious numbers on some reefs, forming an efficient live filtering device that sifts out planktonic organisms from the passing currents. The genus is by far the largest group within the family Pomacentridae, with approximately 80 species; all but 17 inhabit the Indo-west and central Pacific region. Most species dwell on outer reef slopes and passages at depths between five and 40 m.

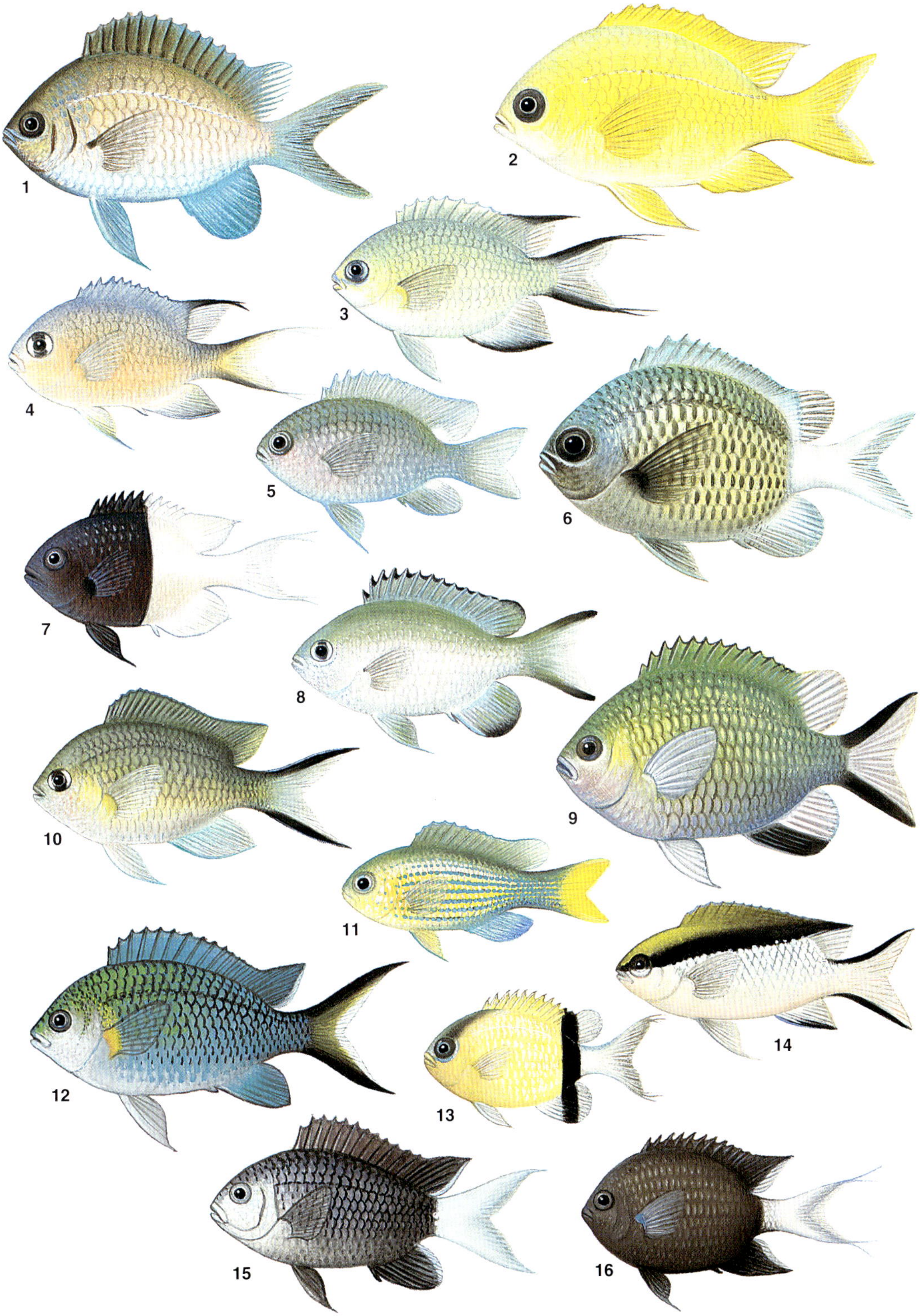
1
2
3
4
5
6
7
8
9
10
11
12
13
14
15
16

PLATE 66: DAMSELFISHES (POMACENTRIDAE)

1 DICK'S DAMSEL
Plectroglyphidodon dickii (Liénard, 1839)
Inhabits coral reefs, frequently associated with *Acropora* coral heads; distinguished by intense blackish bar on rear part of body; N.W. Australia, Great Barrier Reef, and throughout S.E. Asia; Indo-C. Pacific; to 10 cm.

2 JOHNSTON DAMSEL
Plectroglyphidodon johnstonianus Fowler & Ball, 1924
Inhabits coral reefs, frequently associated with *Acropora* or *Pocillopora* coral heads; similar to **1** above, but posterior blackish area broader and more diffuse N.W. Australia, Great Barrier Reef, and throughout S.E. Asia; Indo-C. Pacific; to 9 cm.

3 JEWEL DAMSEL
Plectroglyphidodon lacrymatus (Quoy & Gaimard, 1824)
Inhabits coral reefs, a territorial species that consumes algae; distinguished by numerous bright blue spots on the head and body; found throughout the region; Indo-C. Pacific; to 10 cm.

4 WHITE-BANDED DAMSEL
Plectroglyphidodon leucozona (Bleeker, 1859)
Inhabits coral reefs exposed to wave action; distinguished by broad white bar on middle of sides; N.W. Australia, Great Barrier Reef, and throughout S.E. Asia; Indo-C. Pacific; to 12 cm.

5 ALEXANDER'S DAMSEL
Pomacentrus alexanderae Evermann & Seale, 1907
Inhabits coral reefs; distinguished by light body colour and black pectoral fin base; Indo-Malay Archipelago north to Ryukyu Islands; to 11 cm.

6 AMBON DAMSEL
Pomacentrus amboinensis Bleeker, 1868
Inhabits coral reefs, often in sandy areas; distinguished from **9** below by its preference for open, sandy habitats, less brilliant yellow colour, and slightly larger spot on the upper pectoral fin base; juvenile has spot at rear of dorsal fin; N.W. Australia, Great Barrier Reef, and throughout S.E. Asia; W. Pacific and E. Indian Ocean; to 11 cm.

7 NEON DAMSEL
Pomacentrus coelestis Jordan & Starks, 1901
Inhabits coral reefs, frequently among dead coral rubble; distinguished by bright blue colouration; N.W. Australia, Great Barrier Reef, and throughout S.E. Asia; Indo-C. Pacific; to 9 cm.

8 MILLER'S DAMSEL
Pomacentrus milleri Taylor, 1964
Inhabits coral and rocky reefs, usually near shore; distinguished by plain brownish or grey colour and 14 dorsal spines (other similar species have 12 or 13); juveniles mainly yellow with blue stripes on back; Australia only, between Rottnest Island, W. Australia and Gulf of Carpentaria; to 10 cm.

9 LEMON DAMSEL
Pomacentrus moluccensis Bleeker, 1853
Inhabits coral reefs, usually in the vicinity of live branching corals; distinguished from **6** above by brighter colouration and smaller spot on the upper pectoral fin base; N.W. Australia, Great Barrier Reef, and throughout S.E. Asia; mainly W. Pacific; to 7.5 cm.

10 PRINCESS DAMSEL
Pomacentrus vaiuli Jordan & Seale, 1906
Inhabits coral reefs; distinguished by rows of spots along side of body and spot on rear part of dorsal fin; N.W. Australia, Great Barrier Reef, and throughout S.E. Asia; mainly W. and C. Pacific: to 9 cm.

11 SANDY DAMSEL
Pomacentrus nagasakiensis Tanaka, 1917
Inhabits coral reefs, usually found in sandy areas around coral or rock outcrops; distinguished by black pectoral base, pale tail, and rear part of dorsal and anal fins; juveniles are mainly bluish; N.W. Australia, Great Barrier Reef, and throughout S.E. Asia; W. Pacific and E. Indian Ocean; to 10 cm.

12 PACIFIC GREGORY
Stegastes fasciolatus (Ogilby, 1889)
Inhabits coral reefs, usually in areas exposed to moderate wave action; distinguished by overall brownish colour and lavender spots on cheek and belly; N.W. Australia, Great Barrier Reef, and throughout S.E. Asia; Indo-C. Pacific; to 13 cm.

13 BLUNT-SNOUT GREGORY
Stegastes lividus (Bloch & Schneider, 1801)
Inhabits coral reefs, usually among dead staghorn coral; a very pugnacious fish that continually chases other fishes (and divers) from its territory; distinguished from the similar **12** above by the blunter snout and much wider gap between the eye and mouth; colour is variable, some fish are entirely brown; N.W. Australia, Great Barrier Reef, and throughout S.E. Asia; Indo-C. Pacific; to 15 cm.

14 WESTERN GREGORY
Stegastes obreptus (Whitley, 1948)
Inhabits coral and rocky reefs, sometimes in weedy areas; distinguished by overall dark brown colour; juvenile is bright yellow with a spot on the dorsal fin; *S. apicalis* (not shown) from E. Australia is similar, but has red-orange margin on upper caudal fin lobe and on edge of rear part of dorsal fin; W. and N.W. Australia, Indo-Malay Archipelago; W. Pacific and E. Indian Ocean; to 15 cm.

15 DUSKY GREGORY
Stegastes nigricans (Lacepède, 1802)
Inhabits coral reefs usually among algal covered branches of dead coral; distinguished by blackish spots at base of upper pectoral fin and last dorsal fin ray, nest guarding males assume a very different pattern (15b); N.W. Australia, Great Barrier Reef, and throughout S.E. Asia; Indo-C. Pacific; to 14 cm.

16 GULF DAMSEL
Pristotis obtusirostris (Günther, 1862)
Inhabits flat, sandy bottoms; often caught by trawlers; distinguished by plain colouration, slender shape, forked tail and small black spot on upper pectoral fin base; *P. jerdoni* is a synonym; found throughout the region; Indo-W. Pacifc; to 14 cm.

PLATE 67

1 JAVA DAMSEL
Neoglyphidodon oxyodon (Bleeker, 1857)
Inhabits sheltered reef flats of lagoons and inshore reefs; adults entirely blackish, but lighter on front portion of body; juveniles with neon blue bands and pale bar across front of body; Indo-Malay Archipelago, including Ashmore Reef off N.W. Australia; to 14 cm.

2 BARHEAD DAMSEL
Neoglyphidodon thoracotaeniatus (Fowler & Bean, 1928)
Inhabits outer reef slopes in 15-45 m depth; distinguished by pair of brown bars on head with whitish areas between; Philippines and Indonesia to Solomon Islands; to 12 cm.

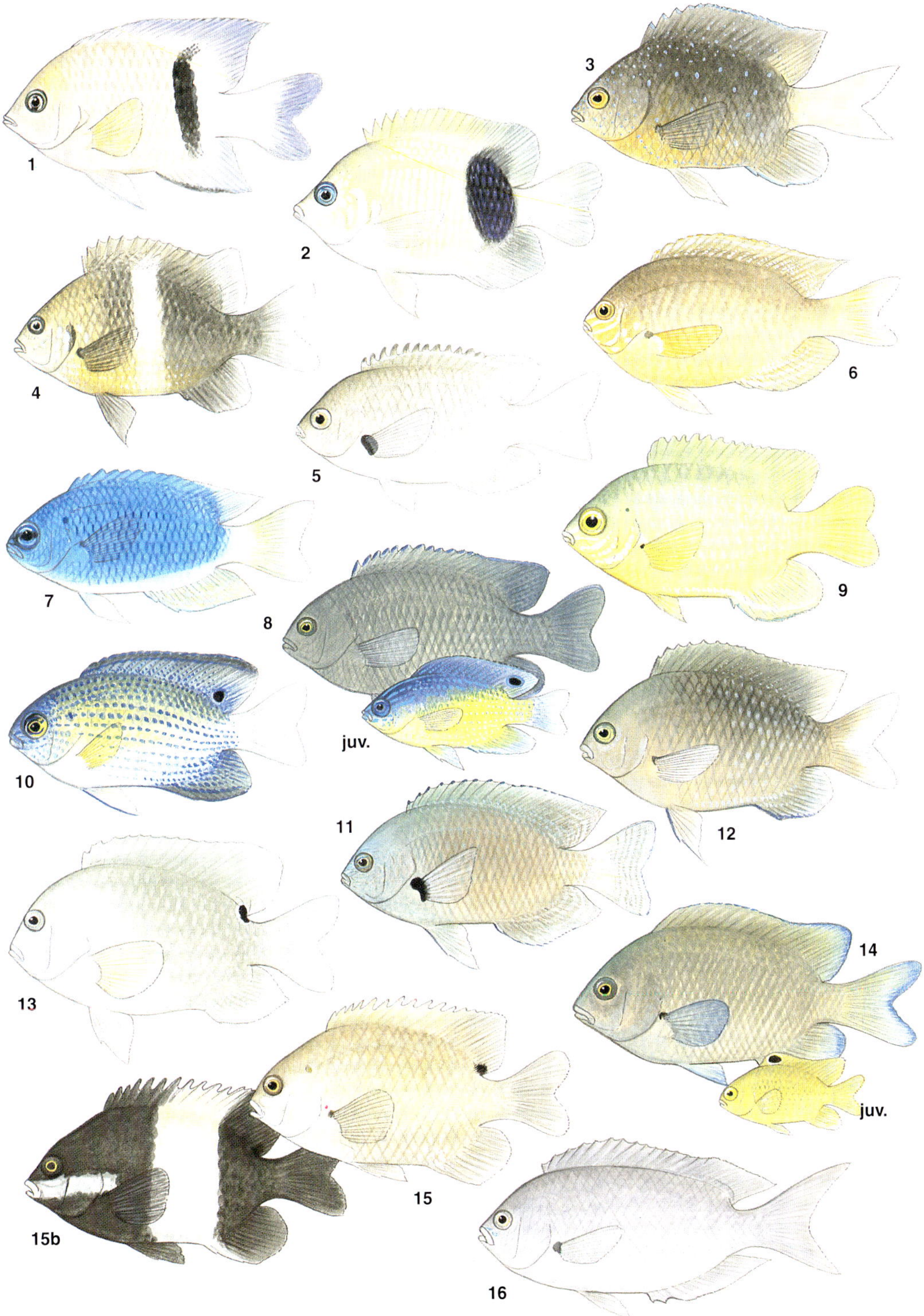
1
2
3
4
5
6
7
8
9
juv.
10
11
12
13
14
juv.
15
15b
16

PLATE 67: DAMSELFISHES (POMACENTRIDAE) AND HAWKFISHES (CIRRHITIDAE)

1 JAVA DAMSEL
Neoglyphidodon oxyodon (Bleeker, 1857).
Text on previous page.

2 BARHEAD DAMSEL
Neoglyphidodon thoracotaeniatus (Fowler & Bean, 1928). Text on previous page.

3 CROSS'S DAMSEL
Neoglyphidodon crossi Allen, 1991
Inhabits rocky areas or coral reefs in sheltered bays and lagoons; feeds on algae and zooplankton; adults are plain brown, juvenile has brilliant orange and blue combination; known only from Indonesia at Sulawesi, Bali, Flores, and Molucca Islands; to 13 cm.

4 GOLDBACK DAMSEL
Pomacentrus nigromanus Weber, 1913
Inhabits lagoons and outer reef slopes in 6-60 m depth; distinguished by black pectoral spot, yellow on rear part of body (absent on fish from N.W. Australia), and broad black margin on anal fin; *P. nigromarginatus* (not shown) is similar, but has fine black border on rear edge of tail; N.W. Australia, Indonesia, Philippines, and New Guinea; Indo-Australian Archipelago; to 9 cm.

5 DUSKY DAMSEL
Pomacentrus adelus Allen, 1991
Inhabits most coral reef habitats in 2-8 m depth; feeds mainly on algae; distinguished by dusky brown colour, often with ocellus on rear part of dorsal fin; northern Great Barrier Reef, offshore reefs of W. Australia, and throughout S.E. Asia; W. Pacific; to 9 cm.

6 SCALY DAMSEL
Pomacentrus lepidogenys Fowler & Bean, 1928
Inhabits lagoons, passes, and outer reefs in 1-12 m depth; feeds on zooplankton and algae; distinguished by pale grey colour, and yellow dorsal and caudal fins; Great Barrier Reef, offshore reefs of W. Australia, and throughout S.E. Asia; Philippines and Indonesia to Fiji; to 9 cm.

7 BLUE DAMSEL
Pomacentrus pavo (Bloch, 1787)
Inhabits sandy areas around coral outcrops or soft coral growths to 16 m; feeds on plankton and algae; distinguished by blue stripes on head, dark "ear" spot, and dusky vertical marks on scales; can "switch" on electric blue-green colour; Great Barrier Reef, offshore reefs of W. Australia, and throughout S.E. Asia; Indo-C. Pacific; to 11 cm.

8 PHILIPPINE DAMSEL
Pomacentrus philippinus Evermann & Seale, 1907
Inhabits passes and outer reef slopes in 2-12 m depth, usually seen near caves and ledges; feeds on zooplankton, algae, and benthic invertebrates; distinguished by black spot on pectoral fin base, dark scale margins, and yellow colour of tail and posterior dorsal and anal rays; Great Barrier Reef, offshore reefs of W. Australia, and throughout S.E. Asia; Indo-W. Pacific; to 11 cm.

9 CHARCOAL DAMSEL
Pomacentrus brachialis Cuvier, 1830
Inhabits passes and outer reef slopes in 6-40 m depth; feeds on plankton and algae; distinguished by dark coloration and black spot covering pectoral fin base; Great Barrier Reef, offshore reefs of W. Australia, and throughout S.E. Asia; W. Pacific; to 11 cm.

10 BLUESPOT DAMSEL
Pomacentrus grammorhynchus Fowler, 1918
Inhabits lagoon and inshore coral reefs, frequently among branching corals in less than 12 m depth; distinguished by general brown colour (often very pale) with brilliant blue spot on upper tail base; Great Barrier Reef, offshore reefs of W. Australia, and throughout S.E. Asia; Indo-Malay Archipelago to Taiwan and Solomon Islands; to 12 cm.

11 BLACKSPOT DAMSEL
Pomacentrus stigma Fowler & Bean, 1928
Inhabits sheltered coastal reefs, usually on fringing reef dropoffs in 2-15 m depth; distinguished by general pale colouration and black blotch on anal fin; Indonesia and Philippines; to 13 cm.

12 BURROUGH'S DAMSEL
Pomacentrus burroughi Fowler, 1918
Inhabits coastal reefs and silty lagoons in 2-16 m depth; feeds mainly on algae; distinguished by one or two pale blotches at base of soft dorsal fin; Philippines to Solomon Islands, but absent from Australia; to 8.5 cm.

13 THREESPOT DAMSEL
Pomacentrus tripunctatus Cuvier, 1830
Inhabits shallow bays and silty coastal reefs to 3 m depth; feeds mainly on algae; distinguished by black spot on upper tail base, young has ocellus on dorsal fin; Great Barrier Reef and throughout S.E. Asia; Sri Lanka to Melanesia; to 10 cm.

14 BLACK-BANDED DAMSEL
Amblypomacentrus breviceps (Schlegel and Müller, 1839)
Inhabits inshore or lagoon areas with sand or silt bottoms around sponge or rock, sometimes in seagrass; distinguished by 3 black bars or saddles on whitish background; Great Barrier Reef and throughout S.E. Asia; Indonesia and Philippines to Solomon Islands; to 6.5 cm.

15 MARBLED HAWKFISH
Cirrhitus pinnulatus (Schneider, 1801)
Inhabits coral reefs, found in the shallow surge zone; feeds on shrimps, crabs, sea urchins, brittle stars and fishes; distinguished by blotchy brown pattern and white spots; Great Barrier Reef, offshore reefs of W. Australia, and throughout S.E. Asia; Indo-C. Pacific; to 28 cm.

16 DWARF HAWKFISH
Cirrhitichthys falco Randall, 1963
Inhabits coral reefs, seen perching on coral heads; feeds on crabs and shrimps; distinguished by series of red-brown saddles that are tapered ventrally and vertical rows of red-brown blotches; Great Barrier Reef and throughout S.E. Asia; Maldives to W. Pacific; to 7 cm.

17 FLAME HAWKFISH
Neocirrhites armatus Castelnau, 1873
Inhabits coral reefs, usually seen hiding among branches of live coral; distinguished by brilliant red colouration and blackened area on back; Great Barrier Reef, Sunda Islands, and Philippines; W. Pacific to Samoa; to 9 cm. (CIRRHITIDAE)

18 LONGNOSE HAWKFISH
Oxycirrhitus typus Bleeker, 1857
Inhabits coral reefs, usually seen on outer slopes below 30 m depth perching on black coral or gorgonians; distinguished by red cross-hatched pattern and long snout; Great Barrier Reef, offshore reefs of N.W. Australia, and throughout S.E. Asia; Indo-E. Pacific; to 13 cm. (CIRRHITIDAE)

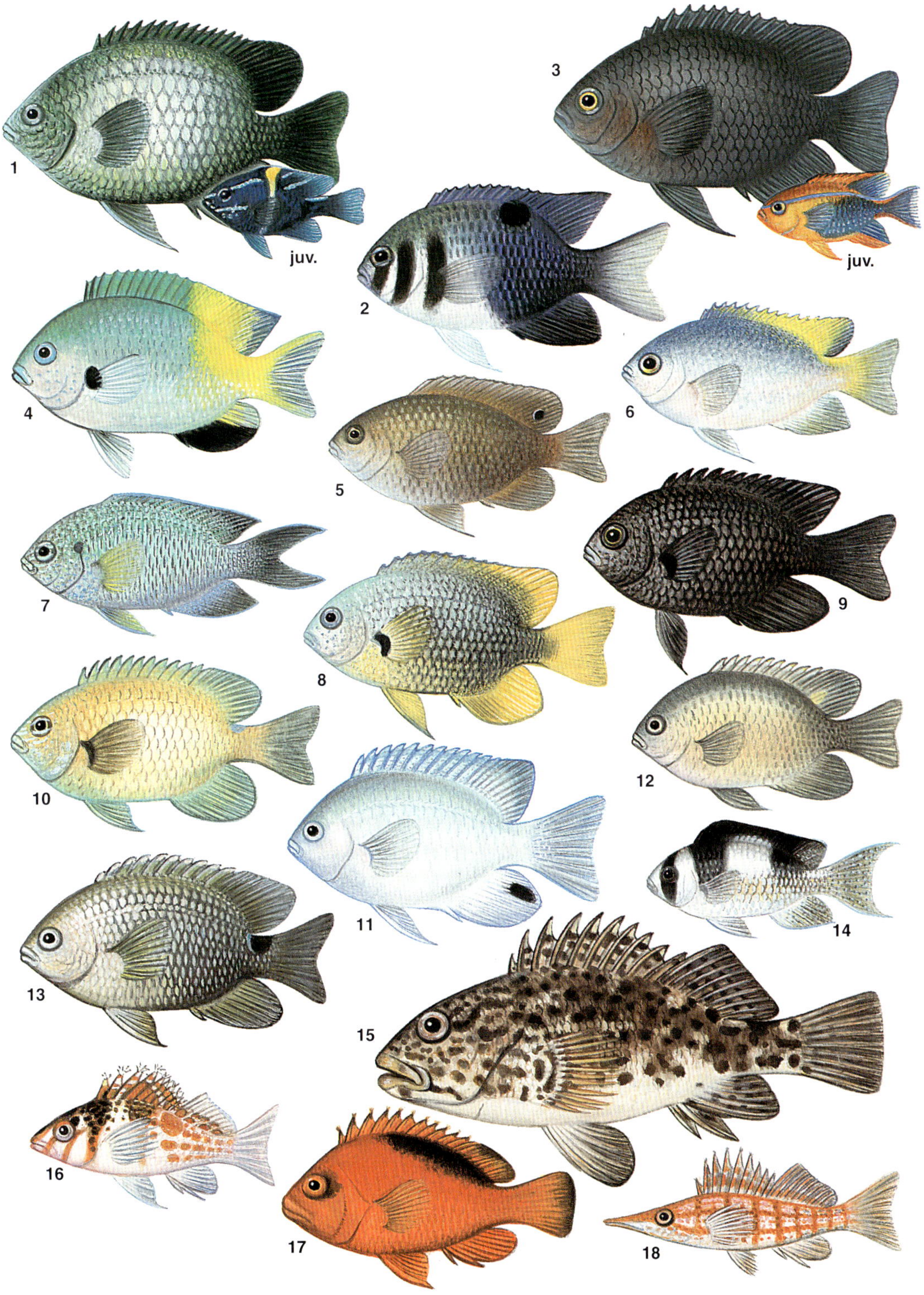
1
juv.
3
juv.
2
4
6
5
7
9
8
10
12
11
14
13
15
16
17
18

PLATE 68: HAWKFISHES AND EEL-BLENNIES

1 THREE-BARRED BOARFISH
Histiopterus typus Temminck & Schlegel, 1844
Inhabits deep reefs on the continental shelf; distinguished by protruding snout, elevated dorsal fin and dark bars; N.W. Australia, Indonesia, and Philippines; mainly W. Pacific; to 35 cm. (PENTACEROTIDAE)

2 TWINSPOT HAWKFISH
Amblycirrhitus bimacula (Jenkins, 1903)
Inhabits coral reefs; distinguished by pale-edged dark spots on gill cover and below rear base of dorsal fin; N.W. Australia, Great Barrier Reef, and throughout S.E. Asia; Indo-C. Pacific; to 8 cm. (CIRRHITIDAE)

3 BLOTCHED HAWKFISH
Cirrhitichthys aprinus (Cuvier, 1829)
Inhabits coral or rocky reefs; distinguished by irregular-shaped brown to red bars on side, brown spot at upper corner of gill cover, and filament at middle of dorsal fin; N.W. Australia, Great Barrier Reef, and throughout S.E. Asia; to 10 cm. (CIRRHITIDAE)

4 SHARP-HEADED HAWKFISH
Cirrhitichthys oxycephalus (Bleeker, 1855)
Inhabits coral or rocky reefs; distinguished by red to brown spots or blotches and prolonged filament at middle of dorsal fin; N.W. Australia, Great Barrier Reef, and throughout S.E. Asia; Indo-E. Pacific; to 9 cm. (CIRRHITIDAE)

5 LYRETAIL HAWKFISH
Cyprinocirrhites polyactis (Bleeker, 1875)
Inhabits outer reefs and passes where currents are strong, often on rubble bottoms; forms small aggregations that swim a short distance above the bottom; distinguished by lack of marks on body, strongly forked tail, and filament at middle of dorsal fin; found throughout the Indo-W. Pacific; to 15 cm. (CIRRHITIDAE)

6 FRECKLED HAWKFISH
Paracirrhites forsteri (Schneider, 1801)
Inhabits coral reefs; distinguished by red to brown spots on head and dark upper half of body; N.W. Australia, Great Barrier Reef, and throughout S.E. Asia; Indo-C. Pacific; to 20 cm. (CIRRHITIDAE)

7 RING-EYED HAWKFISH
Paracirrhites arcatus (Cuvier, 1829)
Inhabits coral reefs; distinguished by orange loop behind eye, 3 orange streaks on lower edge of gill cover, and white streak on sides; N.W. Australia, Great Barrier Reef, and throughout S.E. Asia; Indo-C. Pacific; to 13 cm. (CIRRHITIDAE)

8 ORNATE HAWKFISH
Paracirrhites hemistictus (Günther, 1874)
Inhabits offshore coral reefs; distinguished by dense spotting on upper half of body; offshore reefs of N.W. Australia, Great Barrier Reef, and throughout S.E. Asia; Indo-C. Pacific; to 29 cm. (CIRRHITIDAE)

9 OBTUSE SANDFISH
Squamicreedia obtusata Rendahl, 1921
Inhabits sand bottoms; distinguished by slender body, dorsally positioned eyes, and pattern of small spots; N. Australia only; to 8 cm. (CREEDIIDAE)

10 TOMMYFISH
Limnichthys fasciatus Waite, 1904
Inhabits sand bottoms, frequently buried below surface with only eyes protruding; distinguished by tiny size, pointed snout, dark stripe along middle of side and saddle-like spots on back; found throughout the region; E. Indian Ocean and W. Pacific; to 4.5 cm. (CREEDIIDAE)

11 BANDFISH
Acanthocepola abbreviata (Valenciennes, 1835)
Inhabits sand or silt bottoms; distinguished by elongate, laterally compressed body with pointed tail; found throughout the region; E. Indian Ocean and W. Pacific; to 50 cm. (CEPOLIDAE)

12 OCELLATED EEL-BLENNY
Blennodesmus scapularis Günther, 1872
Inhabits shallow coastal reefs; similar to **13** beow, but pale-edged dark spot above gill cover (not on it) and lacks distinct pale spots arranged in longitudinal rows; N.W. Australia and Queensland only; to 10 cm. (PSEUDOCHROMIDAE)

13 SPINY EEL-BLENNY
Congrogadus spinifer (Borodin, 1933)
Inhabits sand bottoms with weed and sponge; similar to **12** above, but pale-edged dark spot on gill cover (not above it), pale stripe behind eye, and distinct pale spots on side arranged in longitudinal rows; N. Australia only between Exmouth Gulf, W. Australia and Gulf of Carpentaria; to 13 cm. (PSEUDOCHROMIDAE)

14 CARPET EEL-BLENNY
Congrogadus subducens (Richardson, 1843)
Inhabits rock crevices in weedy areas near coral reefs; similar to **13** above, but much larger, bigger lips, less distinct spot on gill cover, and pale spots on side more diffuse; found throughout the region; Andaman Sea and W. Pacific; to 45 cm. (PSEUDOCHROMIDAE)

15 SPOTTED EEL-BLENNY
Notograptus guttatus Günther, 1867
Inhabits turbid inshore reefs; in rocky crevices; distinguished by elongate white body with longitudinal rows of small dark spots; N. Australia and S. New Guinea only; to 14 cm. (NOTOGRAPTIDAE)

16 SHARK BAY EEL-BLENNY
Notograptus gregory Whitley, 1941
Inhabits sand-weed areas; distinguished by general dark colouration and several dark spots on head; a rare species known from only a few examples; known thus far only from Shark Bay, W. Australia; to 10 cm. (NOTOGRAPTIDAE)

HAWKFISHES, ETC.

Hawkfishes of the family Cirrhitidae are benthic inhabitants of coral reefs. Although found in all tropical seas, most of the 35 species are confined to the Indo-Pacific region. Hawkfishes have free projecting rays on the lower part of the pectoral fins adapted for perching on coral branches, gorgonians, etc. They are predators of small fishes and crustaceans. Most are less than 15 cm in length.

Most of the remaining fishes on this Plate are seldom seen due to their cryptic habits. The members of the family Creediidae are small fishes that bury themselves in the sand. The Bandfish (11) of the family Cepolidae occurs over soft bottoms and shelters in burrows. Eel-Blennies (12-14) were formerly classified in a separate family (Congrogadidae), but recent research indicates they belong in Peudochromidae (Plates 28-29). They are usually found among weeds and rocks. The species in the genus *Notograptus* (15-16) are similar in shape, but are classified in a separate family, the Notograptidae.

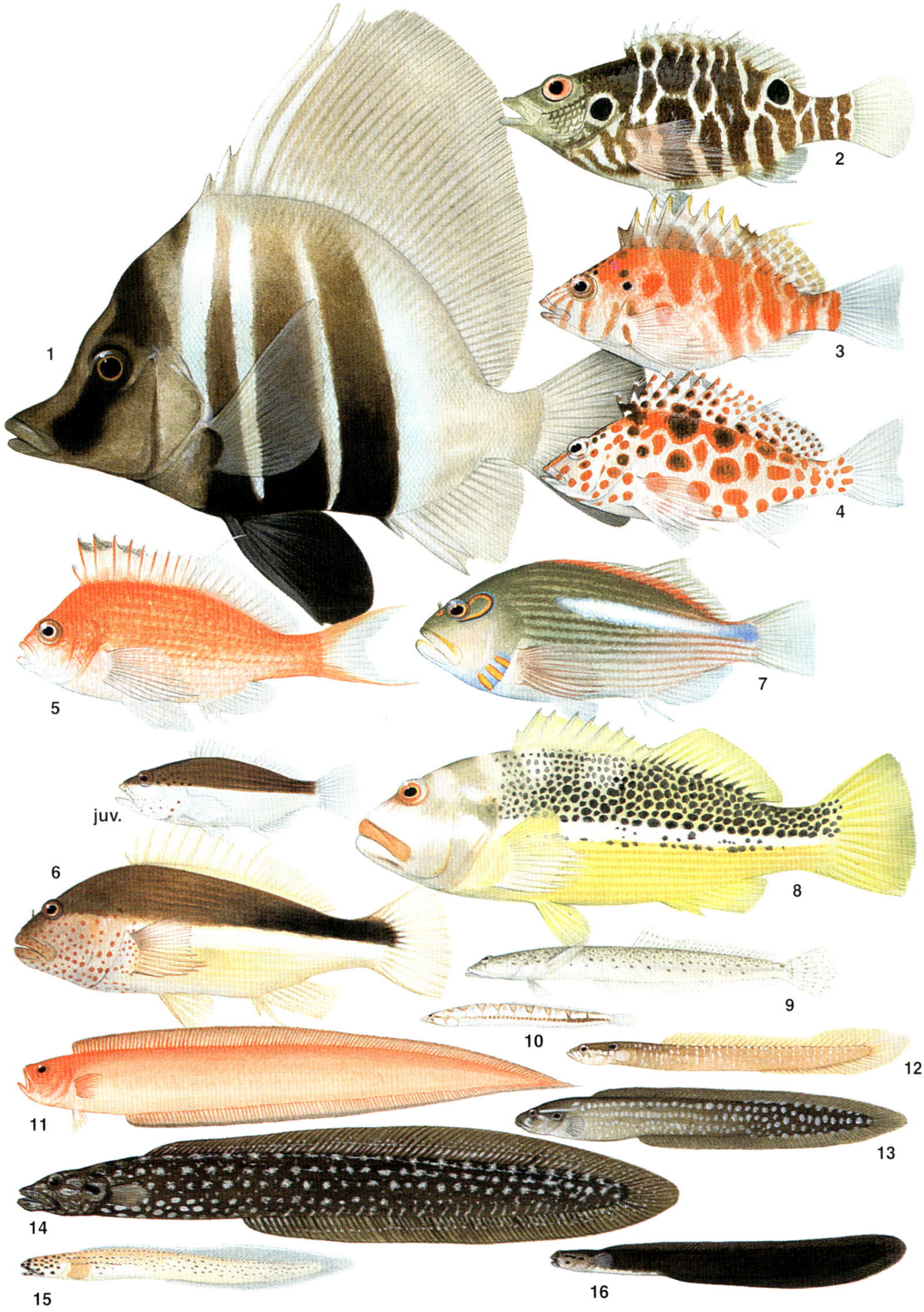

1
2
3
4
5
7
juv.
6
8
9
10
12
11
13
14
15
16

PLATE 69: MULLETS, THREADFINS AND BARRACUDAS

1 FLAT-TAIL MULLET
Liza argentea (Quoy & Gaimard, 1825)
Inhabits coastal waters including estuaries; a plain silvery species without distinguishing marks - has 35-38 scales in lateral line, lacks an enlarged pointed scale at upper pectoral fin base, and has 10 soft (excluding 3 spines) anal fin rays (versus 9 or 10 for other mullets on this page); also known as Jumping mullet and Tiger mullet; mainly confined to southern waters of Australia, but ranges north to Shark Bay, W. Australia; to 32 cm. (MUGILIDAE) ★★★

2 GREENBACK MULLET
Liza subviridis (Valenciennes, 1836)
Inhabits coastal waters including estuaries; distinguished by greenish back, gelatinous membrane partially covering eye, 27-32 scales in lateral line, and lacks an enlarged pointed scale at upper pectoral fin base, also tail narrowly dark-edged; found throughout the region; Indo-C. Pacific; to 30 cm. (MUGILIDAE) ★★★

3 DIAMOND-SCALE MULLET
Liza vaigiensis (Quoy & Gaimard, 1824)
Inhabits coastal waters including estuaries; distinguished by square-shaped tail, scales on upper side with dark blotch giving appearance of stripes, and 24-27 scales in lateral line; juvenile has black pectoral fins; found throughout the region; Indo-C. Pacific; to 55 cm. (MUGILIDAE) ★★★

4 SAND MULLET
Myxus elongatus Günther, 1861
Inhabits coastal waters, frequently off beaches or in estuaries; distinguished by pointed head, black spot at upper pectoral fin base, and 43-46 scales in lateral line; also known as Tallegalane and Lano; mainly confined to subtropical and temperate seas of S.Australia; mainly W. Pacific; to 42 cm. (MUGILIDAE) ★★★

5 SEA MULLET
Mugil cephalus Linneaus, 1758
Inhabits coastal waters, entering estuaries and fresh water; distinguished by flattened head, gelatinous membrane covering eye, an enlarged pointed scale at upper pectoral fin base, 38-42 scales in lateral line, and diffuse stripes on side; found throughout the region; worldwide temperate and tropical seas; to 79 cm. (MUGILIDAE) ★★★

6 BLUE-TAIL MULLET
Valamugil buchanani (Bleeker, 1853)
Inhabits coastal waters including estuaries; distinguished by forked tail, gelatinous membrane around rim of eye, an enlarged pointed scale at upper pectoral fin base, and 32-35 scales in lateral line; found throughout the region; Indo-W. Pacific; to 40 cm. (MUGILIDAE) ★★★

7 NORTHERN THREADFIN
Polydactylus plebius Broussonet, 1782
Inhabits coastal waters, frequently off beaches; distinguished by divided pectoral fin with lower part containing 5 free filamentous rays, and narrow stripes on side; also known as Striped salmon; found throughout the region; Indo-C. Pacific; to 45 cm. (POLYNEMIDAE) ★★★

8 GUNTHER'S THREADFIN
Polydactylus multiradiatus Günther, 1860
Inhabits coastal waters over sand or mud bottoms; distinguished by divided pectoral fin with lower part containing 7 free filamentous rays and most of first dorsal fin blackish; found throughout the region; mainly W. Pacific; to 20 cm. (POLYNEMIDAE) ★★★

9 BLACK-FINNED THREADFIN
Polydactylus nigripinnis Munro, 1964
Inhabits coastal waters over sand or mud bottoms; distinguished by divided pectoral fin with lower part containing 6 free filaments and upper part black; N. Australia and New Guinea only; to 20 cm. (POLYNEMIDAE) ★★

10 GIANT THREADFIN
Eleutheronema tetradactylum (Shaw, 1804)
Inhabits coastal waters over sand or mud bottoms, sometimes in estuaries; distinguished by divided pectoral fin with lower part containing 4 free filaments and lips absent except for small section at rear of lower jaw; also known as Cooktown salmon and Rockhampton kingfish; found throughout the region; Indo-W. Pacific; to 120 cm. (POLYNEMIDAE) ★★★

11 GIANT SEAPIKE
Sphyraena jello Cuvier, 1829
Inhabits coastal waters and offshore reefs; similar to **12** below, but more slender, has smaller scales (see above), and caudal fin yellowish; found throughout the region; Indo-W. Pacific; to 150 cm. (SPHYRAENIDAE) ★

12 BARRACUDA
Sphyraena barracuda (Walbaum, 1792)
Inhabits coastal waters and offshore reefs; distinguished by large size, dark or dusky fins, diffuse dark bars on back, frequently with scattered dark blotches on side (not shown) and truncate tail, **13** below is similar, but more slender and has smaller scales (about 123-135 in lateral line versus 75-90); a curious fish that sometimes follows divers; has been known to attack humans in the Atlantic; found throughout the region; worldwide in tropical seas; to 180 cm. (SPHYRAENIDAE) ★

13 MILITARY SEAPIKE
Sphyraena qenie Klunzinger, 1870
Inhabits coastal waters; distinguished by chevron-shaped markings on side and dark tail; Onslow northwards; Indo-W. Pacific; to at least 90 cm. (SPHYRAENIDAE) ★★

14 STRIPED SEAPIKE   Cairns 10/98
Sphyraena obtusata Cuvier, 1829
Inhabits sand-weed areas, often in the vicinity of rocky or coral reefs, distinguished by pair of dusky yellowish stripes on side and yellow tail; found throughout the region; Indo-C. Pacific; to 55 cm. (SPHYRAENIDAE) ★★

MULLETS, THREADFINS AND BARRACUDAS

Mullets (1-6), Threadfins (7-10), and Barracudas (11-14) are placed together in the suborder Mugiloidei. They are primarily inhabitants of tropical and subtropical seas and estuaries, with a few species also penetrating the lower parts of freshwater streams. Mullets (Mugilidae; worldwide about 75 species) are mainly algal feeders that have very small teeth or may lack them entirely. The largest species is about 90 cm TL, but most are under about 40 cm. Barracudas (Sphyraenidae; worldwide about 18 species) are found near coastal reefs. They have large fang-like teeth and are known to occasionally attack humans. The largest species reaches about 2 m TL. Threadfins (Polynemidae; worldwide about 35 species) inhabit sand or mud bottoms.

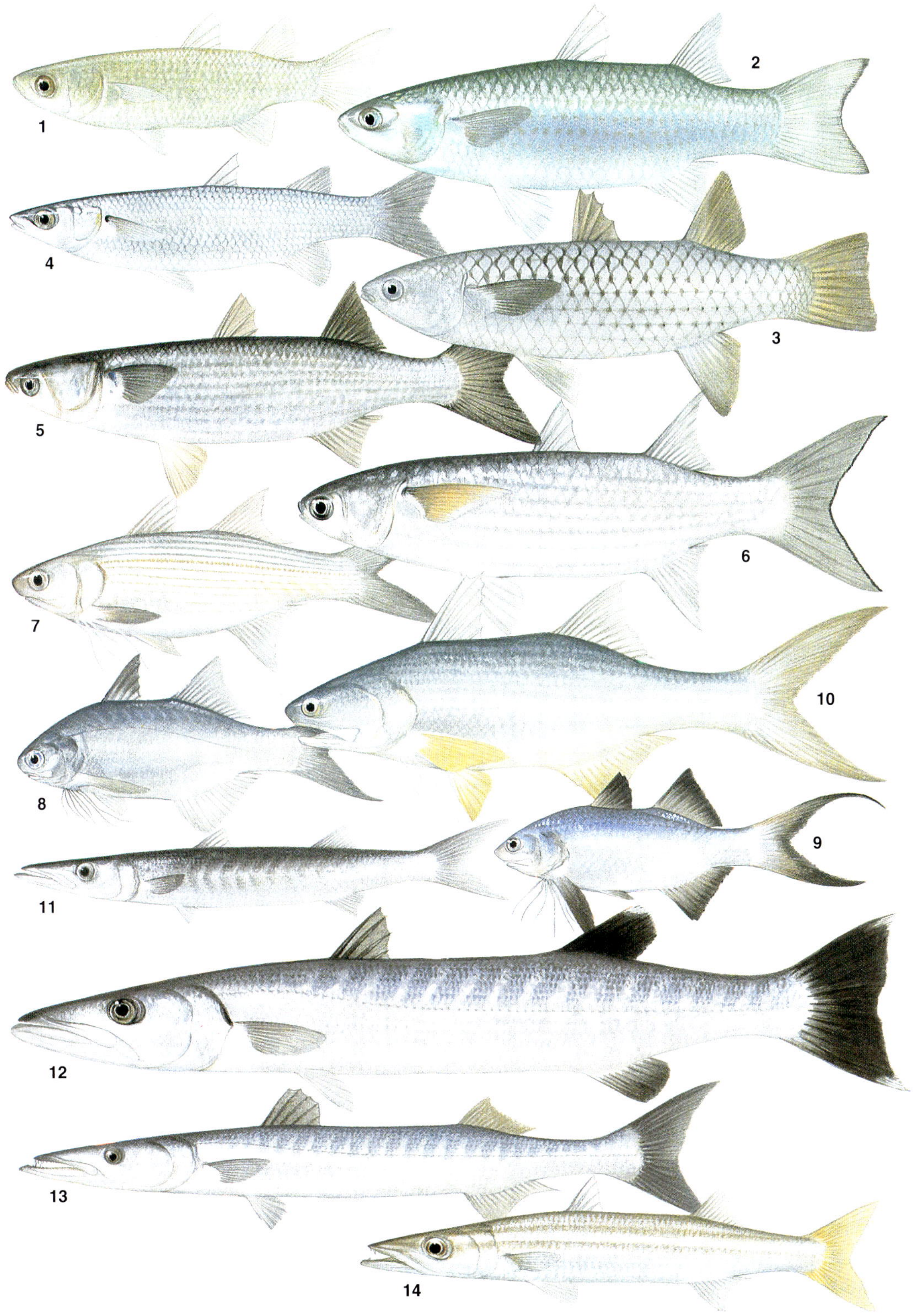
1
2
4
3
5
6
7
10
8
9
11
12
13
14

PLATE 70: WRASSES (FAMILY LABRIDAE)

1 SPOTTED CHISEL-TOOTHED WRASSE
Anampses caeruleopunctatus Rüppell, 1828
Inhabits coral reefs and rocky areas; females distinguished by blue spotting on head and body, male by greenish colour with blue streak on each scale and yellowish bar at level of pectoral fins; N.W. Australia, Great Barrier Reef, and throughout S.E. Asia; Indo-C. Pacific; to 30 cm. ★★

2 SCRIBBLED CHISEL-TOOTHED WRASSE
Anampses geographicus Valenciennes, 1840
Inhabits coastal reefs, frequently in weedy areas; female distinguished by large spot at rear of dorsal and anal fins, male by blue lines on head and numerous blue spots or streaks on sides; found throughout the region; mainly W. Pacific; to 31 cm. ★★

3 BLUE AND YELLOW WRASSE
Anampses lennardi Scott, 1959
Inhabits coral reefs; female distinguished by blue and yellow stripes, male by yellowish patch above pectoral fin and blue streaks on tail; N.W. Australia to Gulf of Carpentaria only; to 28 cm. ★★

4 YELLOWTAIL WRASSE
Anampses meleagrides Valenciennes, 1840
Inhabits coral reefs; female distinguished by white spots and yellow tail, male by pattern on tail which gives appearance of pointed tail lobes; N.W. Australia, Great Barrier Reef, and throughout S.E. Asia; Indo-C. Pacific; to 22 cm. ★★

5 CORAL PIGFISH
Bodianus axillaris (Bennett, 1831)
Inhabits coral reefs; distinguished by large black spot on dorsal and anal fins, also at base of pectoral fin; N.W. Australia, Great Barrier Reef, and throughout S.E. Asia; Indo-C. Pacific; to 20 cm. ★★

6 SADDLEBACK PIGFISH
Bodianus bilunulatus (Lacepède, 1802)
Inhabits coral reefs distinguished by black saddle below rear of dorsal fin, juveniles with large black area at rear of body; W. Australia; scattered locations in Indo-C. Pacific; to 55 cm. ★★★

7 GOLDSPOT PIGFISH
Bodianus perditio (Quoy and Gaimard, 1834)
Inhabits the vicinity of coral and rocky reefs, often over sand or rubble in deeper water; distinguished by yellow patch on middle of back followed by blackish area; N.W. Australia, Great Barrier Reef, and throughout S.E. Asia; Indo-W. Pacfic; to 53 cm and 3 kg. ★★★

8 BLACKSPOT PIGFISH
Bodianus vulpinus (Richardson, 1850)
Inhabits mainly rocky reefs; female distinguished by narrow stripes and sometimes dark blotches on sides, male (see Plate 106) by black blotch at base of middle dorsal spines and reddish margin on upper and lower edges of tail; W. Australia only from Cape Naturaliste northwards to Shark Bay; to 60 cm. ★★★★

9 SCARLET-BREASTED MAORI WRASSE
Cheilinus fasciatus (Bloch, 1791)
Inhabits coral reefs; distinguished by red area at front of body and prominent dark bars; N.W. Australia, Great Barrier Reef, and throughout S.E. Asia; Indo-W. Pacific; to 38 cm. ★★★

10 YELLOW-DOTTED MAORI WRASSE
Cheilinus chlorurus (Bloch, 1791)
Inhabits inshore reefs, in both coral and weed areas; distinguished by mottled pattern with small pale spots on sides; N.W. Australia, Great Barrier Reef, and throughout S.E. Asia; Indo-C. Pacific; to 45 cm. ★★★

11 WHITEBAND MAORI WRASSE
Cheilinus unifasciatus Streets, 1878
Inhabits coral reefs; distinguished by white band across tail base; N.W. Australia, Great Barrier Reef, and throughout S.E. Asia; mainly W. and C. Pacific; to 30 cm. ★★

12 TRIPLETAIL MAORI WRASSE
Cheilinus trilobatus Lacepède, 1802
Inhabits coral reefs; distinguished by numerous pale spots on head, elongated rear part of dorsal and anal fins, and irregular outline of tail; N.W. Australia, Great Barrier Reef, and throughout S.E. Asia; Indo-C. Pacific; to 45 cm. ★★★

INTRODUCING WRASSES

Wrasses (Plates 70-80) are the second largest family of reef fishes - only the goby family has more species. These rainbow-coloured fishes, members of the family Labridae, are one of the most conspicuous groups inhabiting tropical coral reefs. The family is also well represented in warmer temperate seas, for example along the southern coastline of Australia. Worldwide there are an estimated 500 species, including some which are still undescribed. The family is extremely diverse regarding colours, shape, behaviour, and ecological preferences. Most species live over sand, rubble, weed, or coral and rock substrata. Wrasses occur over a wide depth range including shallow tidal pools to depths of at least 100 m. They are diurnally active, feeding on a wide variety of benthic and pelagic invertebrates. At dusk some species bury themselves in the sand where they "sleep" through the night. Razorwrasses (*Xyrichtys*) also dive under the sand when threatened. Female to male sex reversal is common in many species. Dramatic colour changes sometimes occur during growth with juveniles, females and males exhibiting different patterns.

Most species are medium sized (about 20-40 cm), although the Double-headed Maori Wrasse or Giant Wrasse (Plate 71-11) grows to a length of at least 230 cm and weight of 190 kg. In spite of its huge bulk it is a shy fish that is difficult to closely approach. The only thing it is likely to be confused with is the Bumphead Parrotfish (Plate), which reaches 130 cm. However, the parrotfish has a highly distinctive head profile, which is nearly square. The Giant Wrasse is usually seen singly in clear waters of outer reef slopes in 10-100 m depth. It feeds on a wide variety of molluscs, fishes, sea urchins, crustaceans, and other invertebrates. The notorious coral-destroying Crown-of-Thorns Starfish is also occasionally consumed. Large adults are popular display items in public aquaria.

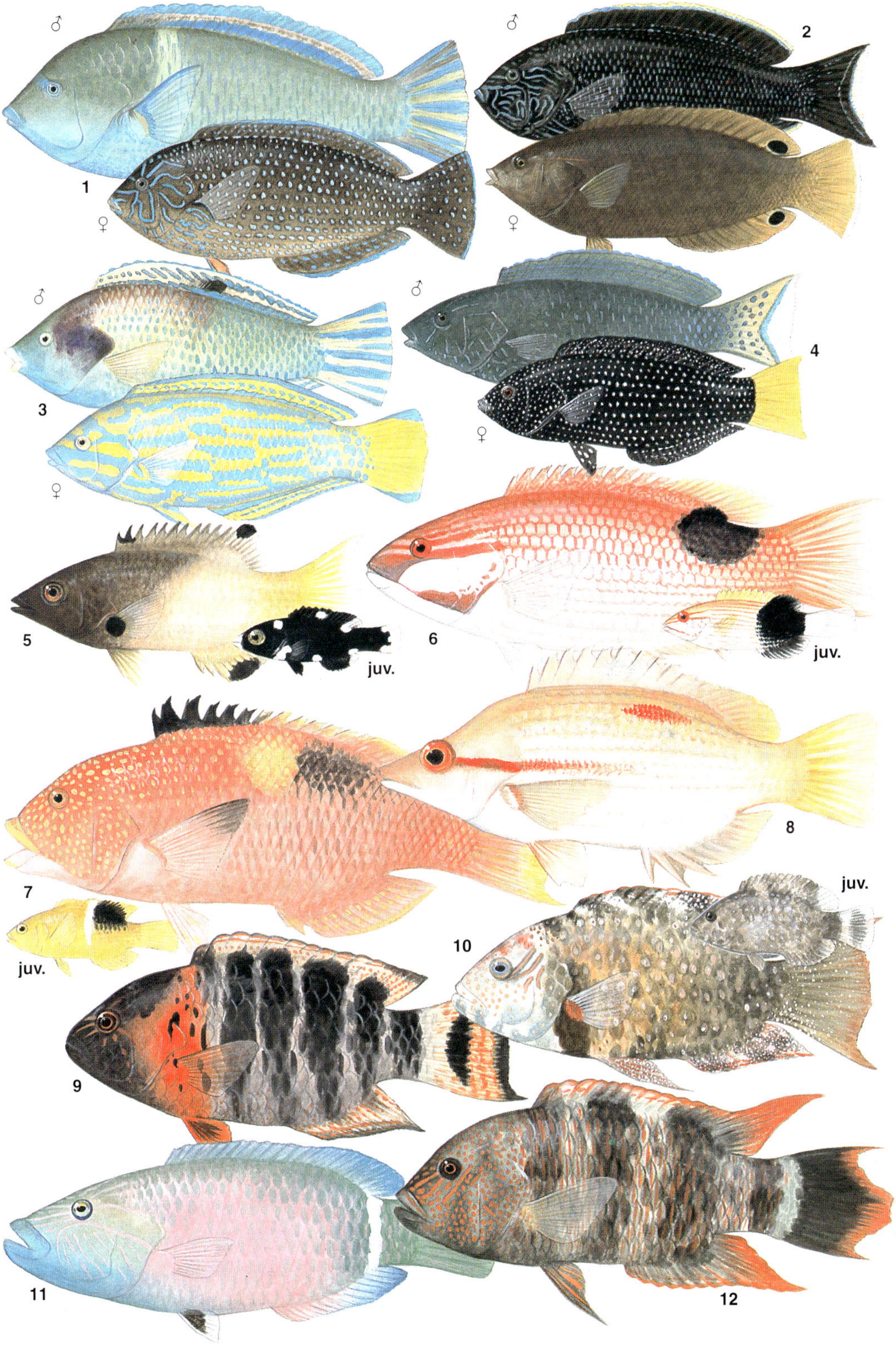

♂
2
♂
1
♀
♀
♂
♂
3
4
♀
♀
5
6
juv.
juv.
8
7
juv.
juv.
10
9
11
12

PLATE 71: WRASSES (FAMILY LABRIDAE)

1 LYRETAIL PIGFISH
Bodianus anthioides (Bennett, 1830)
Inhabits coral reefs, usually seen on outer slopes below 15-20 m depth; distinguished by obliquely demarcated bicolour pattern and lunate or deeply emarginate tail; offshore reefs of N.W. Australia, Great Barrier Reef, and throughout S.E. Asia; Indo-C. Pacific; to 20 cm. ★★

2 TWOSPOT PIGFISH
Bodianus bimaculatus Allen, 1973
Inhabits steep outer reef slopes at depths greater than 30 m; distinguished by small size, slender body, overall reddish-orange colour and black spot just behind head and at base of tail; Great Barrier Reef and throughout S.E. Asia; Indo-W. Pacific; to 10 cm.

3 SPLITLEVEL PIGFISH
Bodianus mesothorax (Bloch & Schneider, 1801)
Inhabits coral reefs; similar to *B. axillaris* (Plate 70-5), but lacks black spot on dorsal and anal fins, also juvenile has yellow instead of white spots; offshore reefs of N.W. Australia, Great Barrier Reef, and throughout S.E. Asia; W. Pacific; to 20 cm. ★★

4 DIANA'S WRASSE
Bodianus diana (Lacepède, 1801)
Inhabits coral reefs; distinguished by purplish head and back, four small pale spots along back, black spot at base of tail, and large red to black blotch on pelvic fins; juvenile with numerous white streaks and spots; offshore reefs of N.W. Australia, Great Barrier Reef, and throughout S.E. Asia; Indo-C. Pacific; to 25 cm. ★★

5 BLACKFIN PIGFISH
Bodianus loxozonus (Snyder, 1908)
Inhabits coral reefs, usually seen on outer slopes; distinguished by broad diagonal black band across rear part of body; Great Barrier Reef and New Guinea; W. and C. Pacific; to 40 cm. ★★

6 TWOSPOT MAORI WRASSE
Oxycheilinus bimaculatus Valenciennes, 1840
Inhabits coral reefs, usually seen amongst rubble or weed; distinguished by small size compared to most other *Cheilinus* and dark stripe on middle of rear half of body; tail rhomboid with prolonged upper lobe in male, rounded in female; offshore reefs of N.W. Australia, Great Barrier Reef, and throughout S.E. Asia; Indo-W. Pacific; to 15 cm.

7 VIOLET-LINED MAORI WRASSE Cairns
Oxycheilinus diagrammus (Lacepède, 1801) 10/98
Inhabits coral reefs; distinguished by dark diagonal lines across lower half of cheek and gill cover, sometimes with dark lateral stripe on side; also known as Cheeklined Maori wrasse; offshore reefs of N.W. Australia, Great Barrier Reef, and throughout S.E. Asia; Indo-C. Pacific; to 30 cm. ★★

8 CELEBES WRASSE
Oxycheilinus celebicus Bleeker, 1853
Inhabits coral reefs; distinguished by relatively slender shape, elongate pointed snout, and diffuse blotches forming midlateral stripe; offshore reefs of N.W. Australia, Great Barrier Reef, and throughout S.E. Asia; mainly W. Pacific; to 20 cm. ★

9 ORIENTAL WRASSE
Oxycheilinus orientalis (Günther, 1862)
Inhabits coral reefs; distinguished by relatively small size, pointed snout, dark diagonal bands on lower edge of cheek, brownish midlateral stripe with white stripe above and below it, and dark spot on each pelvic fin; N. Great Barrier Reef and throughout S.E. Asia; W. Pacific; to 20 cm.

10 POINTED-HEAD MAORI WRASSE
Cheilinus oxycephalus Bleeker, 1853
Inhabits coral reefs, a shy species that stays close to shelter; distinguished by brown to red coloration with scattered pale spots, frequently with 3-4 dark spots in midlateral row on posterior fourth of body and on tail base; also known as Snooty Maori wrasse; Great Barrier Reef and throughout S.E. Asia; Indo-W. Pacific; to 17 cm. ★

11 DOUBLE-HEADED MAORI WRASSE
Cheilinus undulatus Rüppell, 1835 Cairns 13/98
Inhabits coral reefs, large adults usually seen on steep outer reef slopes in 10-100 m depth; distinguished by huge size and hump on forehead; juvenile has pair of dark diagonal stripes through eye; also known as Humphead Maori wrasse, Giant Maori wrasse, and Napolean wrasse; offshore reefs of N.W. Australia, Great Barrier Reef, and throughout S.E. Asia; Indo-C. Pacific; the largest species in the family, to 229 cm and 190 kg. ★★★

12 ANCHOR TUSKFISH
Choerodon anchorago (Bloch, 1791)
Inhabits coral reefs, frequently seen on silty inshore reefs; distinguished by pale diagonal bar at level of pectoral fin, black area on middle of upper side, and large white saddle behind dorsal fin, also has white chin and belly; also known as Yellow-Cheeked tuskfish; Great Barrier Reef, and throughout S.E. Asia; E. Indian Ocean and W. Pacific; to 38 cm. ★★★

13 GRASS TUSKFISH
Choerodon cephalotes (Castelnau, 1875)
Inhabits coral reefs, often in nearby seagrass beds; distinguished by pale spots on cheek, blue markings on head, white blotch above pectoral fin, and wavy yellow bands on fins; also known as Purple tuskfish; Queensland; S. New Guinea, and Indonesia; to 38 cm. ★★★

14 HARLEQUIN TUSKFISH
Choerodon fasciatus (Günther, 1867)
Inhabits coral reefs; distinguished by brilliant red bars; Great Barrier Reef to New Caledonia, and Ryukyu Islands to Taiwan; to 30 cm. ★★★

15 GRAPHIC TUSKFISH
Choerodon graphicus (De Vis, 1885)
Inhabits sand-rubble bottoms with small coral heads in the vicinity of reefs; distinguished by green colour with dark bands through eye and irregular dark bars on side; *C. transversalis* is a synonym; Queensland and New Caledonia; to 46 cm. ★★★

16 VENUS TUSKFISH
Choerodon venustus (De Vis, 1884)
Inhabits vicinity of reefs, often on sand-rubble or weed bottoms; distinguished by greenish to red background colour with numerous pale spots on side; Australia only, N. New So. Wales and E. Queensland; to 65 cm. ★★★

17 ZOSTER WRASSE
Choerodon zosterophorus (Bleeker, 1868)
Inhabits sand-rubble bottoms in the vicinity of coral reefs; similar to *C. jordani* (Pl. 74-6), but black and white markings on side of body are different; Indo-Malay Archipelago, including Philippines; to 20 cm.

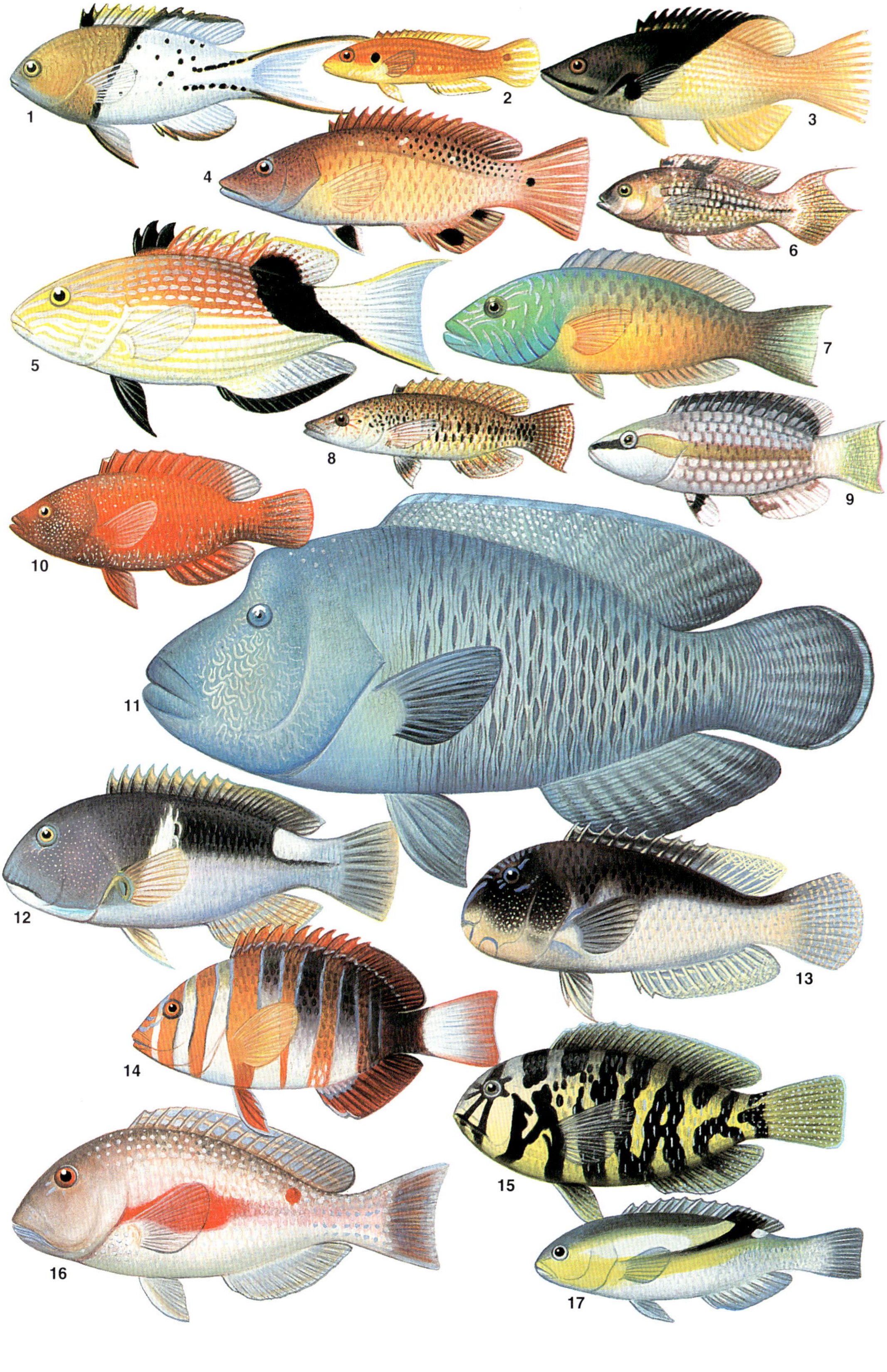
1
2
3
4
6
5
7
8
9
10
11
12
13
14
15
16
17

PLATE 72: WRASSES (FAMILY LABRIDAE)

1 BLUESIDE WRASSE
Cirrhilabrus cyanopleura (Bleeker, 1851)
Inhabits coral reefs over rubble bottoms; distinguished by large dark blue patch behind head; female (not shown) is light red; Great Barrier Reef and throughout S.E. Asia; E. Indian Ocean and W. Pacific; to 15 cm.

2 MAGENTA-STREAKED WRASSE
Cirrhilabrus laboutei Randall & Lubbock, 1982
Inhabits coral reefs over rubble bottoms; distinguished by magenta stripes edged in yellow; Great Barrier Reef to New Caledonia and Loyalty Islands; to 12 cm.

3 EXQUISITE WRASSE Cairns 10/98
Cirrhilabrus exquisitus Smith, 1957
Inhabits outer reefs and passes, usually seen over rubble bottoms in less than 10 m depth; distinguished by brilliant red margin on fins, blue spots on tail, and black spot on tail base, female mainly reddish with white patch on snout; Great Barrier Reef, offshore reefs of W. Australia and throughout S.E. Asia; Indo-C. Pacific; to 12 cm.

4 LUBBOCK'S WRASSE
Cirrhilabrus lubbocki Randall & Carpenter, 1980
Inhabits coral reefs over rubble and coral patches in 5-50 m depth; distinguished by bright yellow head and back with yellow dorsal fin; female (not shown) is red with small dark spot on upper tail base; Indonesia and Philippines; to 8 cm.

5 PURPLE-LINED WRASSE
Cirrhilabrus lineatus Randall and Lubbock, 1982
Inhabits coral reefs over rubble bottoms; distinguished by bold blue stripes on head that break up into spots on body, blue stripe along upper back and lowermost part of body, and red tail with blue cross-bands; Great Barrier Reef to New Caledonia and Loyalty Islands; to 12 cm.

6 DOTTED WRASSE
Cirrhilabrus punctatus Randall & Kuiter, 1989
Inhabits coral reefs over rubble bottoms; distinguished by numerous pale dots on upper three-fourths of body and black bar across pectoral-fin base, female (not shown) mainly red; New South Wales northward to Great Barrier Reef and New Guinea, and east to Fiji; to 13 cm.

7 SCOTT'S WRASSE Cairns 10/98
Cirrhilabrus scottorum Randall & Pyle, 1988
Inhabits coral reefs over rubble bottoms; distinguished by overall dark appearance with shades of red, and red dorsal and anal fins with deep purple outer margin; Great Barrier Reef and across tropical South Pacific to Pitcairn Group; to 13 cm.

8 RANDALL'S WRASSE
Cirrhilabrus randalli Allen, 1995
Inhabits rubble bottoms of lagoons and outer reefs in 10-40 m depth; distinguished by broad orange stripe on middle of side; offshore reefs of W. Australia including Rowley Shoals and Scott, Seringapatam, Ashmore, Cartier, and Hibernia reefs; to 11 cm.

9 THREADFIN WRASSE
Cirrhilabrus filamentosus (Klausewitz, 1976)
Inhabits rubble bottoms of sheltered reefs in 10-40 m depth; male has long filamentous extension on middle of dorsal fin, an abruptly white belly, long pelvic fins, and fan-shaped tail; female mainly red; Indo-Malay Archipelago; to 7 cm.

10 FILAMENTOUS WRASSE
Paracheilinus filamentosus Allen, 1974
Inhabits rubble areas in passages and on the outer reef slope, occasionally seen in lagoons; distinguished by reddish, striped body, filamentous dorsal rays, and lunate caudal fin; male is brighter than female, has longer dorsal fin filaments and a more lunate tail; Philippines and Indonesia to Solomon Islands; to 9 cm.

11 MCCOSKER'S WRASSE
Paracheilinus mccoskeri Randall & Harmelin-Vivien, 1977
Inhabits outer reef slopes over rubble bottoms in 12-40 m depth; distinguished by filamentous extension in middle of dorsal fin and bright blue stripe on lower half of body; female (not shown) is plain; offshore reefs of W. Australia and W. Indonesia; widely scattered localities mostly in Indian Ocean; to 7 cm.

12 SPOTFIN WRASSE
Coris dorsomacula (Fowler, 1908)
Inhabits areas of mixed sand, weed, and rubble near coral reefs; distinguished by narrow pale bars on side, ocellated "ear" spot and small ocellated spot at base of rear part of dorsal fin; Great Barrier Reef and throughout S.E. Asia; W. Pacific; to 20 cm.

13 VARIEGATED WRASSE
Coris batuensis (Bleeker, 1856)
Inhabits sand-rubble areas near coral reefs; distinguished by variegated pattern with short blackish bars on back; usually misidentified as *C. variegata* (a Red Sea species) or *C. schroederi*; Great Barrier Reef, offshore reefs of W. Australia, and throughout S.E. Asia Indo-W. Pacific; to 17 cm.

14 MIDGET WRASSE
Pseudocheilinops ataenia Schultz, 1960
Inhabits coral reefs, usually seen among rubble; a secretive fish that rarely exposes itself; distinguished by small size, "stubby" body shape, reddish colour and dark spot on each pelvic fin; Indo-Malay Archipelago; to 5 cm.

15 CONNIE'S WRASSE
Conniella apterygia Allen, 1983
Inhabits rubble bottom with scattered coral outcrops on outer reefs in 25-50 m depth; distinguished by candy-striped pattern and lack of pelvic fins; offshore reefs of W. Australia including Rowley Shoals and Scott and Seringapatam Reefs; to 7 cm.

16 EIGHT-LINED WRASSE
Pseudocheilinus octotaenia Jenkins, 1900
Inhabits rubble and live coral areas on seaward reefs to depths of 50 m; distinguished by overall reddish colour and eight dark stripes on side; Great Barrier Reef, offshore reefs of W. Australia, and throughout S.E. Asia; Indo- C. Pacific; to 13.5 cm.

17 DISAPPEARING WRASSE
Pseudocheilinus evanidus Jordan & Evermann, 1903
Inhabits coral reef crevices, a very secretive fish; distinguished by bluish-white streak on cheek, general red colouration, and thin white stripes on side; Great Barrier Reef and throughout S.E. Asia; Indo- C. Pacific; to 8 cm.

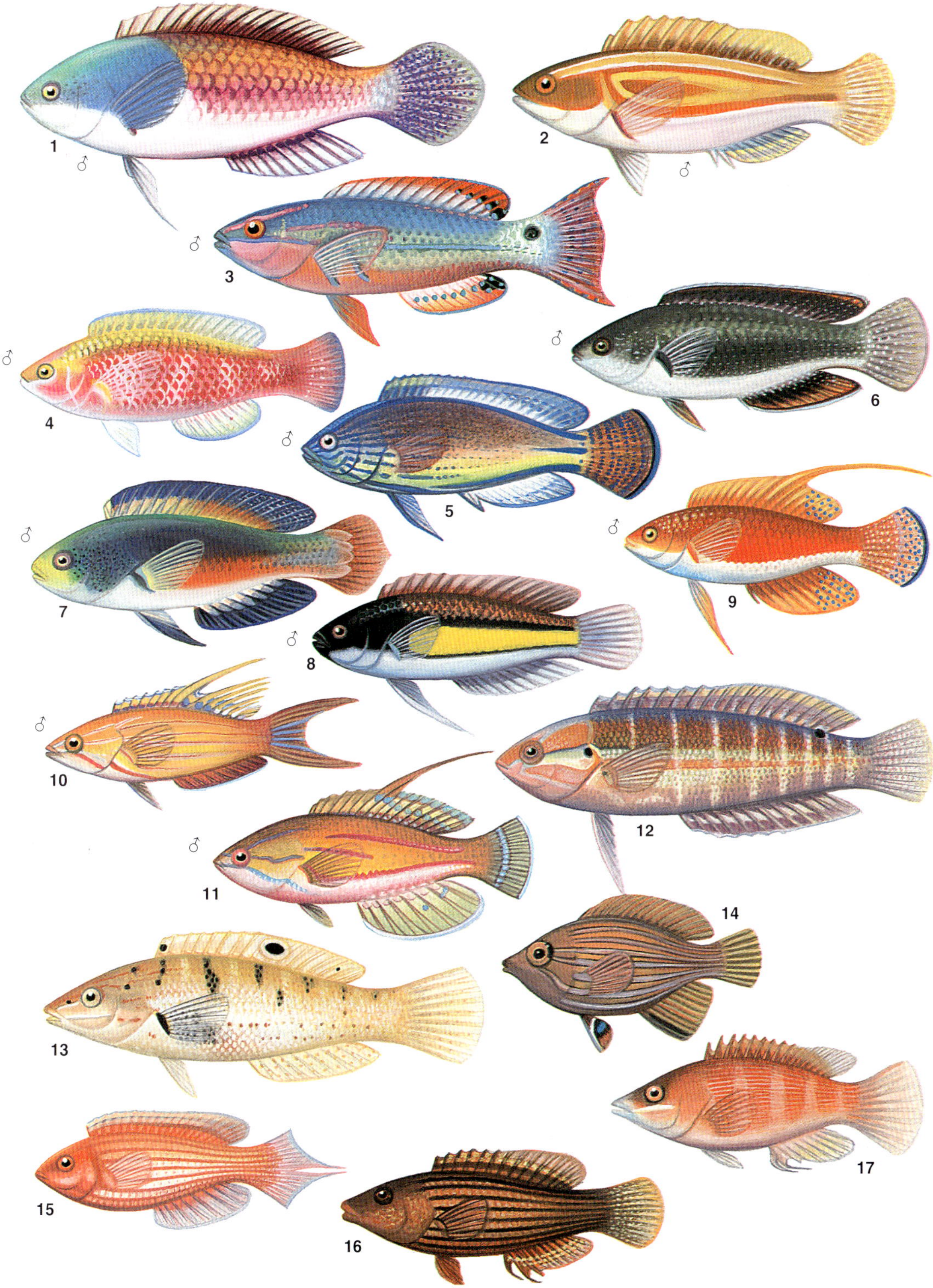
1
♂
2
♂
♂
3
♂
4
6
♂
5
♂
♂
7
9
♂
8
♂
10
12
♂
11
14
13
17
15
16

PLATE 73: WRASSES (LABRIDAE)

1 FEMININE WRASSE
Anampses femininus Randall, 1972
Inhabits coral and rocky reefs; female recogised by brilliant orange colour with narrow blue stripes, becoming entirely blue on tail and rear part of body; male (not shown) similar to male of *A. caeruleopunctatus* (Pl. 70-1), but has broad blue stripes on head; New So. Wales and S. Great Barrier Reef across S. Pacific to Easter Island; to 24 cm.

2 BLACKTAIL WRASSE
Anampses melanurus Bleeker, 1857
Inhabits coral reefs; distinguished by white spotting and black tail with pale yellow bar across basal part; offshore reefs of N.W. Australia and throughout S.E. Asia; Indo-C. Pacific; to 12 cm.

3 NEW GUINEA WRASSE
Anampses neoguinaicus Bleeker, 1878
Inhabits coral reefs; distinguished by blackish back and white colour on most of head and body, males and females are similar; Great Barrier Reef, offshore reefs of N.W. Australia, and throughout S.E. Asia; W. Pacific; to 17 cm.

4 YELLOW-BREASTED WRASSE
Anampses twistii Bleeker, 1856
Inhabits coral reefs; distinguished by yellow region on breast and lower part of head, also by pattern of dark-edged pale spots and ocellus at rear of dorsal and anal fins; Great Barrier Reef, offshore reefs of N.W. Australia, and throughout S.E. Asia; Indo-C. Pacific; to 18 cm.

5 YELLOWTAIL TUBELIP
Diproctacanthus xanthurus (Bleeker, 1856)
Inhabits rich coral areas, usually seen in shallow, protected lagoons; feeds mainly on coral polyps, juveniles "clean" other fishes; distinguished by black and white stripes, and yellow tail; Great Barrier Reef, Kimberley coast of N.W. Australia, and throughout S.E. Asia; Indo-Australian Archipelago; to 10 cm.

6 PASTEL RINGWRASSE
Hologymnosus doliatus (Lacepède, 1801)
Inhabits coral reefs; distinguished by elongate shape, pointed snout, and series of thin bars on side; males are predominately bluish with a broad blue-margined pale zone just behind pectoral fin base; Great Barrier Reef, offshore reefs of N.W. Australia, and throughout S.E. Asia; Indo-W. Pacific; to 50 cm.

7 BREASTSPOT CLEANERFISH
Labroides pectoralis Randall & Springer, 1975
Inhabits coral reefs; "cleans" parasites from other fishes; similar to *L. dimidiatus* (Pl. 77-3), but has large black spot at lower edge of pectoral fin base; Great Barrier Reef, offshore reefs of N.W. Australia, and throughout S.E. Asia; mainly W. Pacific; to 8 cm.

8 TAILBLOTCH TUBELIP
Labropsis manabei Schmidt, 1930
Inhabits coral reefs, often seen on steep slopes in 10-30 m depth; male distinguished by yellow blotch on anterior body region and another on basal half of tail; female (not shown is similar, but lacks yellow markings; Indo-Malay and Melanesian archipelagos; W. Pacific; to 13 cm.

9 ALLEN'S TUBELIP
Labropsis alleni Randall, 1981
Inhabits coral reefs, usually seen on outer reef slopes to 52 m depth; juveniles "clean" other fishes, adults probably feed on coral polyps; distinguished by dark head, greenish body, white tail, and prominent pale-edged black spot at base of pectoral fin; Philippines, Indonesia, Micronesia, and Melanesia; to 10 cm.

10 YELLOWBACK TUBELIP
Labropsis xanthonota Randall, 1981
Inhabits coral reefs, most often seen on outer reef slopes in 10-50 m depth; feeds on coral polyps, juveniles "clean" other fishes; female distinguished by yellow back and dorsal fin; male by dark caudal fin lobes and orange streak on rear edge of gill cover; Great Barrier Reef, offshore reefs of N.W. Australia, and throughout S.E. Asia; Indo-W. Pacific; to 13 cm.

11 SOUTHERN TUBELIP
Labropsis australis Randall, 1981
Inhabits coral reefs, usually seen on outer slopes to 55 m depth; feeds on coral polyps; juveniles (black and white striped) "clean" other fishes; female with grey head and three faint brown stripes, male as illustrated; Melanesia and Great Barrier Reef to Samoa; to 10 cm.

12 SHOULDERSPOT WRASSE
Leptojulis cyanopleura (Bleeker, 1853)
Inhabits coral reefs or at least in the vicinity of reefs, sometimes on sand-rubble bottoms; occurs in aggregations with females greatly outnumbering males; female distinguished by whitish background and dark mid-lateral stripe; male (not shown) is bluish grey with blue-edged orange stripes and blotches on head, body, and fins; Great Barrier Reef and throughout S.E. Asia; Indo-W. Pacific; to 13 cm.

13 CHOAT'S WRASSE
Macropharygodon choati Randall, 1978
Inhabits coral reefs between 1 and 30 m depth; distinguished by white background with irregular longitudinal red-orange bands and blotches; S. Great Barrier Reef; to 10 cm.

14 KUITER'S WRASSE
Macropharygodon kuiteri Randall, 1978
Inhabits coral and rocky reefs in depth range of 5-55 m; distinguished by yellow orange colour with prominent blue-rimmed black spot on upper part of gill cover; New So. Wales, S. Great Barrier Reef, and New Caledonia; to 10 cm.

15 RETICULATED WRASSE
Macropharyngodon meleagris (Valenciennes, 1839)
Inhabits coral reefs; female distinguished by leopard-like spotting and male (not shown) by dark-edged green spot on each scale of body, also has green spots and short bands on head; Great Barrier Reef, offshore reefs of N.W. Australia, and throughout S.E. Asia; mainly W. and C. Pacific; to 15 cm.

TUBELIPS

Wrasses belonging to the genus *Labropsis* are closely associated with areas of rich coral development. They are called Tubelips because of the rounded shape of their mouth and swollen lips. The latter feature is possibly an adaptation for feeding on live corals. Juveniles are often seen picking at the bodies of other reef fishes, presumably feeding on external parasites in the same manner as the cleaner wrasses (see Plate 77). Unlike the latter fishes, which are territorial, tubelips range over a larger area.

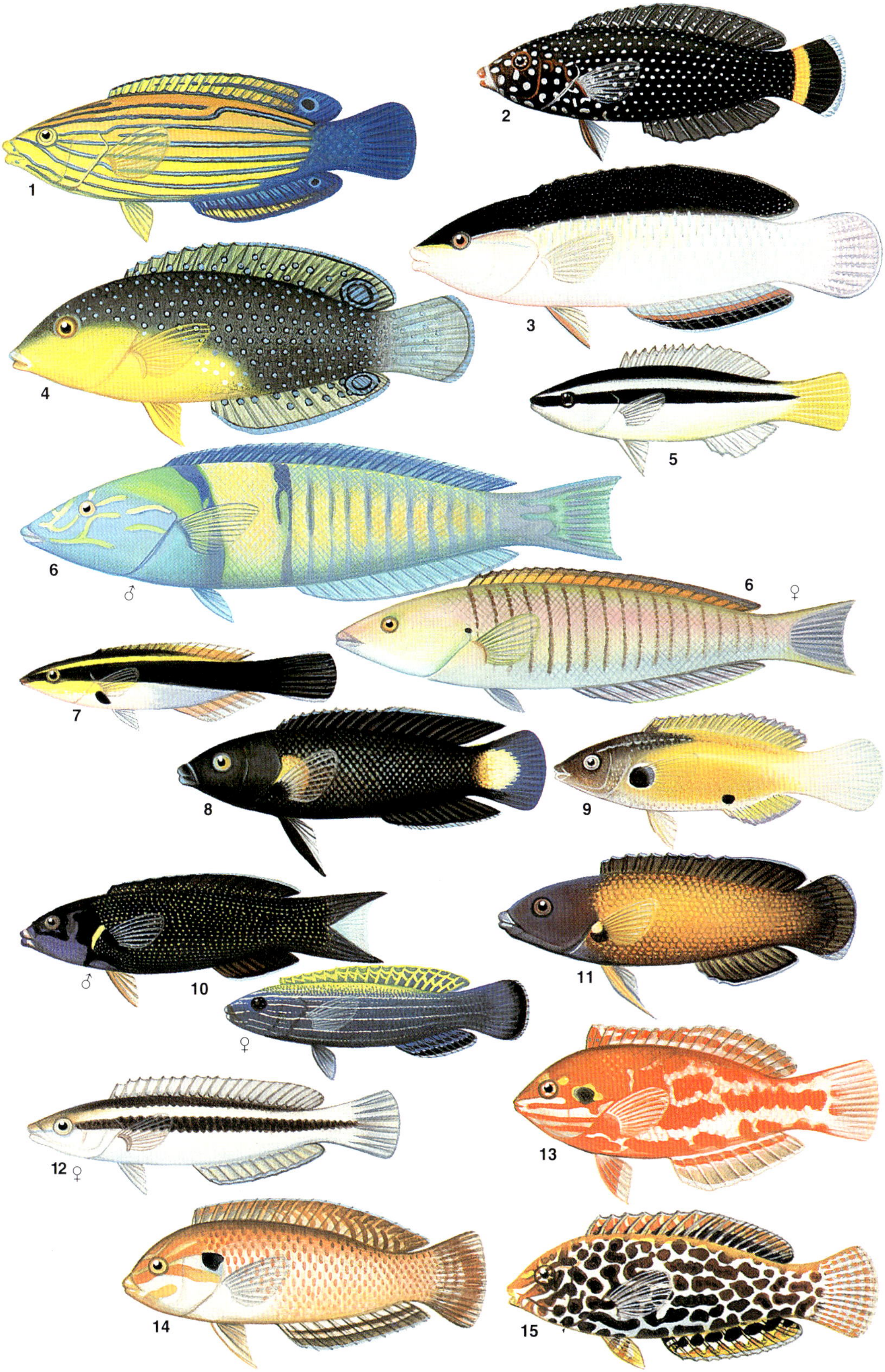
1
2
3
4
5
6
♂
6 ♀
7
8
9
10
♂
♀
11
12 ♀
13
14
15

PLATE 74: WRASSES (FAMILY LABRIDAE)

1 SHARP-NOSED WRASSE
Cheilio inermis (Forsskål, 1775)
Inhabits weed beds; distinguished by elongate shape - males have blackish patch behind pectoral fins, females frequently display a narrow dark stripe along middle of sides; found throughout the region; Indo-C. Pacific; to 50 cm. ★

2 PEACOCK WRASSE
Cirrhilabrus temmincki Bleeker, 1853
Inhabits outer coral reefs, usually where there is loose rubble; male distinguished by red colour on upper half of head and body and very elongate pelvic fin filaments, females less vivid and lack pelvic filaments; W. Australia (Abrolhos to North West Cape) and scattered locations in S.E. Asia; mainly W. Pacific; to 10 cm.

3 BLUE TUSKFISH
Choerodon cyanodus (Richardson, 1843)
Inhabits coral reefs and flat bottoms, distinguished by white chin, white spot on middle of back, and scribble markings on tail; N. Australia only; to 60 cm and 7 kg. ★★★

4 PURPLE TUSKFISH
Choerodon cephalotes (Castelnau, 1875)
Inhabits reefs and flat bottoms; distinguished by dark colour on upper half and light colour on lower half with whitish patch above pectoral fin; N. Australia only; to 30 cm. ★★★

5 BALDCHIN GROPER
Choerodon rubescens (Günther, 1862)
Inhabits coral reefs and rock-weed areas; distinguished by abruptly pale chin and pale area at base of pectoral fins, the head profile becomes increasingly steep with growth; W. Australia only, between Geographe Bay and Coral Bay; to 90 cm and 7 kg. ★★★★

6 JORDAN'S WRASSE
Choerodon jordani (Snyder, 1908)
Inhabits sandy areas adjacent to reefs; distinguished by blackish wedge on rear part of body and large white spot below end of dorsal fin; Great Barrier Reef, W. Australia, and scattered locations in S.E. Asia; mainly W. Pacific; to 13 cm.

7 REDSTRIPE TUSKFISH
Choerodon vitta Ogilby, 1910
Inhabits flat sandy or weedy areas; distinguished by reddish stripe along middle of sides; N. Australia only; to 20 cm. ★★

8 BLACKSPOT TUSKFISH
Choerodon schoenleinii (Valenciennes, 1839)
Inhabits sand and weed areas adjacent to coral reefs; distinguished by overall bluish colour and black spot at base of middle of dorsal fin; found throughout the region; mainly W. Pacific; to 80 cm and 9 kg. ★★★

9 BLUESPOTTED TUSKFISH
Choerodon cauteroma Gomon & Allen, 1987
Inhabits sand and weed areas adjacent to coral reefs; distinguished by yellow colouration and dark streak below middle of spiny dorsal fin; N.W. Australia only; to 36 cm. ★★

10 WEDGE-TAILED TUSKFISH
Choerodon sugillatum Gomon, 1987
Inhabits flat sandy or weedy areas; distinguished by blue band at pectoral fin base and blue streak above pectoral fin; N. Australia only; to 24 cm. ★★

11 DARKSPOT TUSKFISH
Choerodon monostigma (Ogilby, 1910)
Inhabits flat sandy or weedy areas; distinguished by prominent black spot at middle of dorsal fin and striped tail; N. Australia only; to 25 cm. ★★

12 ZAMBOANGA TUSKFISH
Choerodon zamboangae (Seale & Bean, 1907)
Inhabits flat sandy or weedy areas; distinguished by dark back and wedge-shaped reddish mark; N.W. Australia and throughout S.E. Asia; mainly W. Pacific; to 25 cm. ★★

13 RED-FINNED RAINBOWFISH
Coris gaimardi (Quoy & Gaimard, 1824)
Inhabits sandy areas adjacent to coral reefs; distinguished by bright colour pattern and elongate spines at front of dorsal fin, juveniles bright red with white saddles; Great Barrier Reef, N.W. Australia, and throughout S.E. Asia; Indo-C. Pacific; to 30 cm.

14 HUMP-HEADED WRASSE
Coris aygula Lacepède, 1802
Inhabits rubble weed and sandy areas adjacent to reefs; male distinguished by hump on forehead and ragged tail margin, females by bicolour pattern with spots on the head and fins, juveniles with twin "eye-spots"; found throughout the region; Indo-C. Pacific; to 100 cm.

15 BLACK-STRIPED WRASSE
Coris pictoides Randall & Kuiter, 1982
Inhabits rubble, sand, or weedy areas adjacent to coral reefs; distinguished by black stripes on upper sides; Dampier Archipelago northwards; New South Wales, Great Barrier Reef, N.W. Australia, and throughout S.E. Asia; W. Pacific and E. Indian Oceans; to 11 cm.

16 SPOTTED-TAIL WRASSE
Coris caudimacula (Quoy & Gaimard, 1834)
Inhabits sand or weedy areas adjacent to coral reefs; distinguished by dark band or blotches along upper side, broken bands or spots on tail, and usually a prominent "ear-spot" at rear edge of gill cover; N.W. Australia, Indonesia, and Malaysia; Indian Ocean; to 20 cm.

LIVING RAINBOWS

The striking Red-finned Rainbow Wrasse (**13**) frequents sand and rubble patches intermingled with coral where it feeds on small invertebrates. Juveniles are brilliant red with four white saddles along the top of the head and back. This pattern gradually disappears with increased growth and is replaced by the adult colours shown here. Males and females of this species are very similar in appearance, but in many other wrasses the terminal male phase exhibits a gaudy colouration that is significantly different from that of the female. Like most wrasses, the Red-finned Rainbow begins its adult life as a female and can later change sex. In some species the initial adult stage contains both males and females, which tend to be relatively dull in colour. These initial phase fish tend to spawn in aggregations. In contrast, the bright coloured terminal male phase usually engages in pair spawning with a single parner, which appears to be the reproductive pattern of the Red-finned Rainbow Wrasse. At night this species buries itself in the sand.

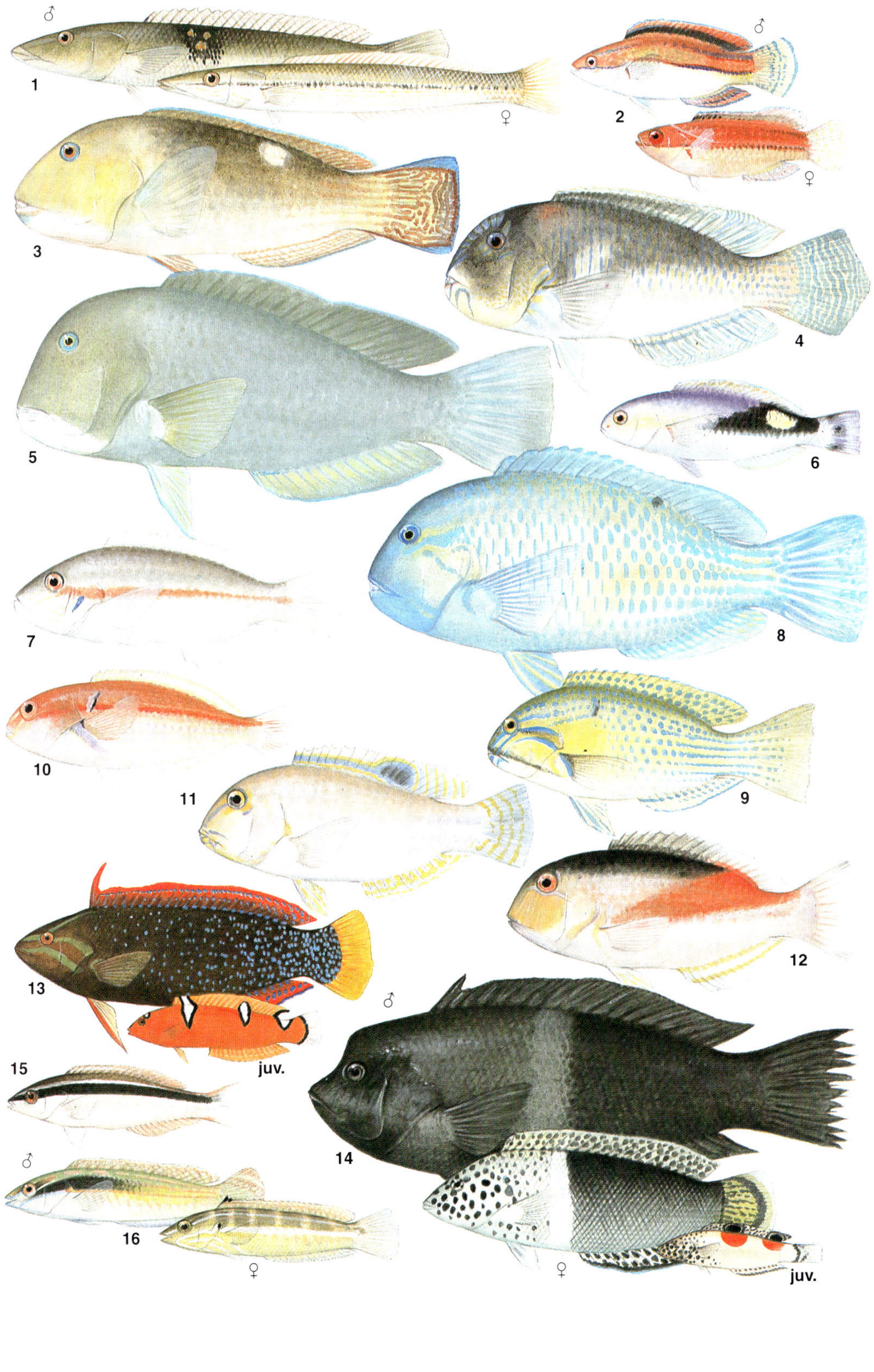
♂
1
♀
♂
2
♀
3
4
5
6
7
8
10
11
9
12
13
♂
juv.
15
14
♂
16
♀
♀
juv.

PLATE 75: WRASSES (FAMILY LABRIDAE)

1 SLINGJAW WRASSE
Epibulus insidiator (Pallas, 1770)
Inhabits coral reefs; distinguished by its highly protrusible jaw - two colour varieties are commonly encountered, one that is largely bright yellow and a dark variety; Great Barrier Reef, N.W. Australia, and throughout S.E. Asia; Indo-C. Pacific; to 35 cm. ★★

2 CLUBNOSED WRASSE Cairns 10/98
Gomphosus varius (Lacepède, 1801)
Inhabits coral reefs; distinguished by elongate snout - females are generally lighter in colour than males; Great Barrier Reef, N.W. Australia, and throughout S.E. Asia; Indo-C. Pacific; to 28 cm.

3 FOURSPOT WRASSE
Halichoeres hortulanus (Lacepède, 1802)
Inhabits coral reefs; female distinguished by blackish area below spiny dorsal fin, and ocellus on middle of dorsal fin, male by yellow spot below front of dorsal fin and broad pale bar behind head; Great Barrier Reef, N.W. Australia, and throughout S.E. Asia; Indo-C. Pacific; to 26 cm.

4 RED-LINED WRASSE
Halichoeres biocellatus Schultz, 1960
Inhabits coral reefs; distinguished by series of red streaks and lines on head and sides; Great Barrier Reef, N.W. Australia, and throughout S.E. Asia; mainly W. Pacific; to 15 cm.

5 SPECKLED RAINBOWFISH
Halichoeres marginatus Rüppell, 1835
Inhabits coral reefs; distinguished by dark green-brown colour with darker stripes (most apparent anteriorly) and yellow margin on tail, young have striped pattern with an ocellus on the dorsal fin; Great Barrier Reef, N.W. Australia, and throughout S.E. Asia; Indo-C. Pacific; to 18 cm.

6 THREESPOT WRASSE
Halichoeres trimaculatus (Griffith, 1834)
Inhabits sand and rubble flats adjacent to coral reefs; female distinguished by overall pale colour and spot on upper tail base, male is more ornate, but also has prominent spot on upper tail base; found throughout the region; Indo-C. Pacific; to 25 cm.

7 PURPLE WRASSE
Halichoeres melanochir Fowler & Bean, 1928
Inhabits coral reefs; distinguished by overall purple-brown colour and yellow-orange pelvic fins, young specimens have 2 spots on the dorsal fin and a third spot on the upper tail base; N.W. Australia and throughout S.E. Asia; mainly W. Pacific, to 10 cm.

8 DIAMOND WRASSE
Halichoeres dussumieri (Valenciennes, 1839)
Inhabits rubble and weed areas near coral reefs; distinguidhed by 4-5 diffuse, broad dark bars on upper two-thirds of side with yellow areas in between and dark spot at pectoral fin base; N.W. Australia and throughout S.E. Asia; Indo-Australian Archipelago; to 13 cm.

9 SADDLED RAINBOWFISH
Halichoeres margaritaceous (Valenciennes, 1839)
Inhabits rubble and weed areas near coral reefs; similar to **10** below, but the bands running across cheek rise posteriorly (/) and usually has 13 pectoral-fin rays; Great Barrier Reef, and throughout S.E. Asia; Indo-C. Pacific; to 12 cm.

10 NEBULOUS WRASSE
Halichoeres nebulosus (Valenciennes, 1839)
Inhabits rubble and weed near coral reefs; similar to **9** above, but the bands running across cheek descend posteriorly (\) and usually has 14 pectoral-fin rays; Great Barrier Reef, N.W. Australia, and throughout S.E. Asia; IndoW. Pacific; to 12 cm.

11 FIVE-BANDED WRASSE Cairns 10/98
Hemigymnus fasciatus (Bloch, 1792)
Inhabits coral reefs; distinguished by fleshy lips and broad dark bands on sides; Great Barrier Reef, N.W. Australia, and throughout S.E. Asia; Indo-C. Pacific; to 80 cm.

12 THICK-LIPPED WRASSE
Hemigymnus melapterus (Bloch, 1791)
Inhabits coral reefs; distinguished by large size and fleshy lips, smaller individuals (up to about 40-50 cm) have characteristic bicolour pattern (pale anteriorly and dark posteriorly); Great Barrier Reef, N.W. Australia, and throughout S.E. Asia; Indo-C. Pacific; to 90 cm.

THE SLINGJAW

The colourful Slingjaw Wrasse (**1**) is equipped with a unique feeding apparatus. It slowly and deliberately stalks its prey - mainly small fishes, crabs, and prawns. When the victim is within range it shoots its distensible jaws forward with lightning speed, forming a peculiar tube-like structure that efficiently sucks in the meal. When not in use the jaws are neatly tucked away, giving the fish a normal wrasse profile.

Three distinct colour patterns are evident in this species. Females are either bright yellow or entirely dark brown. Terminal phase males are brown with bright green scale edges with a diffuse yellow bar on the sides, just behind the pectoral fin. They also have a light grey head with dark lines radiating from the eye. The species ranges widely in the Indo-Pacific region, from the Red Sea and Africa eastward to the Hawaiian Islands, and from Australia north to Japan.

WRASSE POWER

Wrasses are extremely abundant on coral reefs. As far as species total is concerned they are surpassed only by gobies. These colourful fishes, which range in size from 4 centimetre-long dwarfs to over two metres, have successfully invaded every reef habitat. The genus *Halichoeres*, of which several representatives are illustrated on the opposite plate, is the largest genus in the family, with approximately 50 species occurring in the Indo-Pacific region and several others in the tropical Atlantic. They are active fishes, constantly foraging for small crabs, shrimps, worms, and other small bottom-living invertebrates. They will quickly congregate to feed on the encrusting growth when a diver turns over a dead coral slab.

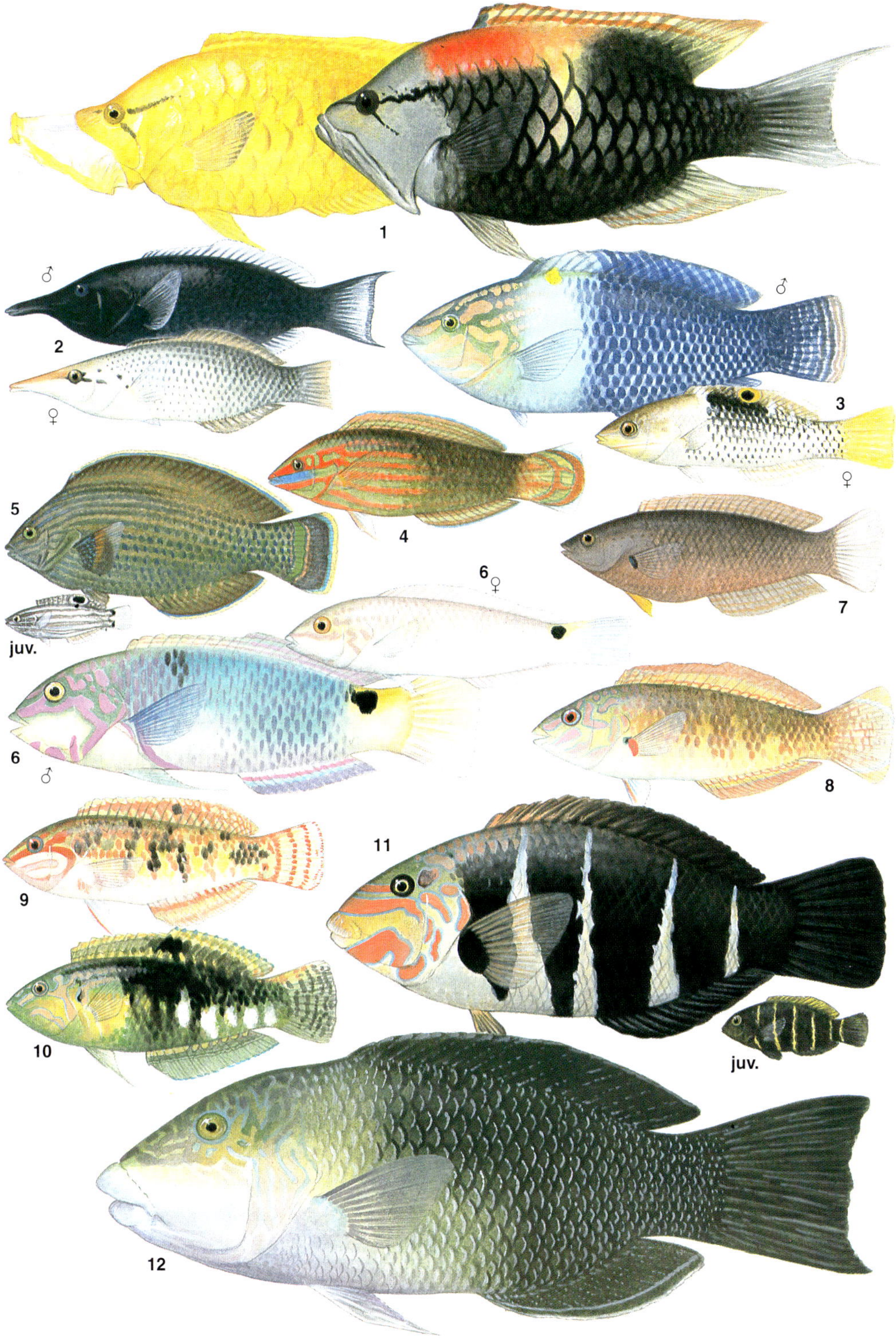

1
♂
2
♀
♂
3
♀
5
4
juv.
6♀
7
6
♂
8
9
11
10
juv.
12

PLATE 76: WRASSES (LABRIDAE)

1 GREEN-SPOTTED WRASSE
Halichoeres chloropterus (Bloch, 1791)
Inhabits protected lagoons and silty inshore reefs; female (shown here) distinguished by lime colour and small dark spots on side and frequently has dark smudge on middle of body, male with lavender-pink stripes on head and spots of same colour on side; also known as Pastel-Green wrasse; Great Barrier Reef and throughout S.E. Asia; W. Pacific; to 19 cm.

2 GOLDEN WRASSE
Halichoeres chrysus Randall, 1981
Inhabits vicinity of coral reefs, usually seen around small bommies in sand or rubble areas to 50 m depth; distinguished by bright yellow colour and black spot at front of dorsal fin; offshore reefs of W. Australia, Great Barrier Reef and throughout S.E. Asia; W. Pacific; to 19 cm.

3 REDHEAD WRASSE
Halichoeres rubricephalus Kuiter & Randall, 1994
Inhabits protected inner reefs between 15-35 m depth; distinguished by dark body and contrasting red colour of head; known thus far only from Maumere Bay on the Indonesian island of Flores; to 10 cm.

4 GOLDSTRIPE WRASSE   Cairns 10/98
Halichoeres hartzfeldi Bleeker, 1852
Inhabits open sand or sand-rubble bottoms, sometimes near isolated coral heads; female distinguished by broad yellow stripe on middle of side, male (shown here) by similar stripe, but with jagged edge and dark blotch on side beneath outer part of pectoral fin; Great Barrier Reef, Kimberley coast of N.W. Australia, and throughout S.E. Asia; Indo-W. Pacific; to 20 cm.

5 CHAIN-LINED WRASSE
Halichoeres leucurus (Walbaum, 1792)
Inhabits coastal coral reefs, frequently in silty areas; distinguished by greenish body, dark head, and blue stripes on side; *H. richmondi* is a synonym; Kimberley coast of N.W. Australia and throughout S.E. Asia; mainly W. Pacific; to 14 cm.

6 THREE-EYED WRASSE   Cairns 10/98
Halichoeres melanurus (Bleeker, 1851)
Inhabits shallow coral reefs; female distinguished by blue stripes and small ocellus on upper base of tail, male by red stripes and yellow pectoral fin base; *H. hoeveni* is a synonym; also known as Yellow-Tailed Wrasse and Tailspot Wrasse; N.W. Australia, Great Barrier Reef, and throughout S.E. Asia; mainly W. Pacific; to 10 cm.

7 SILTY WRASSE
Halichoeres purpurescens (Bloch & Schneider, 1801)
Inhabits silty inshore reefs; female similar to female of **5**, but stripes less distinct, male distinguished by blue stripes on head, yellowish-brown spots on body, and blue submarginal band on tail; Kimberley coast and Indo-Malay Archipelago; to 12 cm.

8 BREASTSPOT WRASSE
Halichoeres podostigma (Bleeker, 1854)
Inhabits coastal coral reefs; distinguished by dark, pale-edged scales over most of body, pale head and tail base, and prominent black spot on pectoral fin base, juveniles have dark spot on pelvic fins; Indonesia and Philippines; to 18 cm.

9 OCELLATED WRASSE
Halichoeres melasmapomus Randall, 1980
Inhabits steep outer reef slopes in 20-55 m depth; distinguished by yellow bands on head and blue-edged ocellus behind eye; also known as Cheekspot Wrasse; Great Barrier Reef, offshore reefs of W. Australia, Christmas and Cocos-Keeling Is., and throughout S.E. Asia; E. Indian Ocean and W. and C. Pacific; to 14 cm.

10 CIRCLE-CHEEK WRASSE
10 *Halichoeres miniatus* (Valenciennes, 1839)
Inhabits shallow coral reefs; male distinguished by irregular dusky bars on lower side, dark blotch on middle of dorsal fin, and pinkish bands on head; female (not shown) has series of close-set dark stripes on side; *H. nebulosus* (Pl. 75-10) and *H. margarataceous* (Pl. 75-9) are very similar species, particularly males, but the dominant pink markings on the cheek tends to form a diagonal band rather than a partial or fully formed ring; N. Queenland and throughout S.E. Asia; W. edge of Pacific; to 10 cm.

11 SCHWARTZ'S WRASSE
Halichoeres schwartzi (Bleeker, 1849)
Inhabits shallow coral reefs, frequently in weedy areas; distinguished by irregular bar pattern on side, pinkish bands on head, and blackish outer edge of tail; Indo-Malay Archipelago; to 12 cm.

12 ARGUS WRASSE
Halichoeres argus (Bloch & Schneider, 1801)
Inhabits shallow coral reefs, frequently in sand-weed areas or seagrass beds; distinguished by strongly reticulated pattern composed of pale scale centres surrounded by thick dark margins; Indo-Malay and Melanesian archipelagos; to 11 cm.

13 ORNAMENTAL WRASSE
Halichoeres ornatissimus (Garrett, 1863)
Inhabits coral reefs of lagoons and outer slopes; distinguished by irregular pink stripes on sides, a row of iridescent green spots on back, and small black spot behind eye; females less than about 10 cm with ocellated black spot on middle of dorsal fin; Great Barrier Reef, offshore reefs of W. Australia. and throughout S.E. Asia.; island areas of W. and C. Pacific.

14 TWOTONE WRASSE
Halichoeres prosopeion (Bleeker, 1853)
Inhabits coral reefs, frequently seen on outer slopes; distinguished by purplish head and anterior part of body, and yellowish colour posteriorly; juvenile has black and white stripes; Great Barrier Reef, and throughout S.E. Asia; W. Pacific; to 13 cm.

15 ZIGZAG WRASSE
Halichoeres scapularis (Bennett, 1831)
Inhabits shallow reefs, usually in lagoons or bays on sand, rubble, or sea grass bottoms; distinguished by general pale coloration with dark "zipper-like" stripe on upper side, male is generally greenish with lavender bands on head and similar zigzag stripe (female shown here); Great Barrier Reef, offshore reefs of W. Australia, and throughout S.E. Asia; Indo-W. Pacific; to 20 cm.

16 GREEN WRASSE
Halichoeres solorensis (Bleeker, 1853)
Inhabits coral reefs; distinguished by overall greenish colour with pink stripes on head and pale spots on scales of body; Indo-Malay Archipelago; to 18 cm.

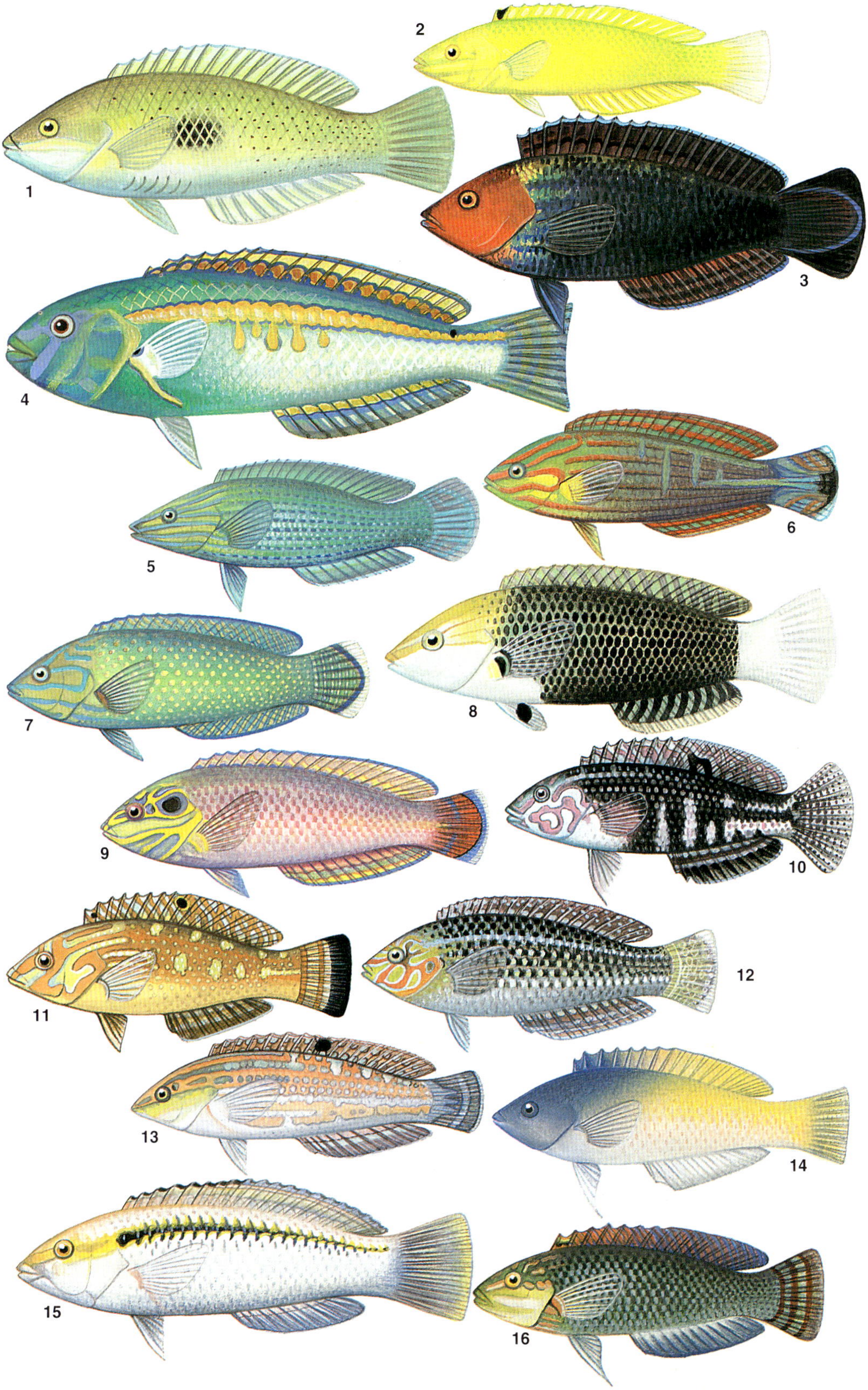
1
2
3
4
5
6
7
8
9
10
11
12
13
14
15
16

PLATE 77: WRASSES (FAMILY LABRIDAE)

1 RINGED SLENDER WRASSE
Hologymnosus annulatus (Lacepède, 1801)
Inhabits coral reefs, often seen over sand or rubble areas; male distinguished by series of narrow dark bars on side, these also visible in female, which has a more slender body; juveniles black except golden yellow on back and top of head; Great Barrier Reef, N.W. Australia and throughout S.E. Asia; Indo-W. Pacific; to 40 cm.

2 BICOLOR CLEANERFISH
Labroides bicolor Fowler & Bean, 1928
Inhabits coral reefs, feeds on ectoparasites from other fishes; distinguished by light coloured tail region; Great Barrier Reef, N.W. Australia and throughout S.E. Asia; Indo-C. Pacific; to 14 cm.

3 CLEANERFISH
Labroides dimidiatus (Valenciennes, 1839)
Inhabits coral reefs, feeds on ectoparasites from other fishes; similar to *L. pectoralis* (Pl. 73-7), but lacks dark spot on pectoral fin base; found throughout the region; Indo-C. Pacific; to 12 cm.

4 ONE-LINED WRASSE
Labrichthys unilineatus (Guichenot, 1847)
Inhabits coral reefs, frequently seen amongst branching *Acropora* coral; distinguished by thin blue lines on side - male has pale bar behind head and juveniles have single pale stripe along middle side; Great Barrier Reef, N.W. Australia and throughout S.E. Asia; Indo-W. Pacific; to 17 cm.

5 SHOULDERSPOT WRASSE
Leptojulis cyanopleura (Bleeker, 1853)
Inhabits coral reefs or at least in the vicinity of reefs, sometimes on sand-rubble bottoms; occurs in aggregations with females greatly outnumbering males; female distinguished by whitish background and dark mid-lateral stripe; male is bluish grey with blue-edged orange stripes and blotches on head, body, and fins; Great Barrier Reef and throughout S.E. Asia; Indo-W. Pacific; to 13 cm. (See also Pl. 72-12)

6 SIX-LINED WRASSE
Pseudocheilinus hexataenia (Bleeker, 1857)
Inhabits coral crevices; distinguished by 6 red-orange stripes on side and small black spot on upper edge of tail base; Great Barrier Reef, N.W. Australia and throughout S.E. Asia; Indo-C. Pacific; to 8 cm.

7 FLAGFIN WRASSE
Pteragogus amboinensis (Bleeker, 1856)
Inhabits weedy patches on coral reefs; distinguished by free filamentous spines at front of dorsal fin and dark spot on gill cover; Great Barrier Reef, N.W. Australia and throughout S.E. Asia; mainly W. Pacific; to 20 cm.

8 ORNATE WRASSE
Macropharyngodon ornatus Randall, 1978
Inhabits coral reefs; female distinguished by pale spots in rows on side and reddish fins with light markings, males similar, but generally darker; N.W. Australia and throughout S.E. Asia; E. Indian Ocean and W. Pacific; to 11 cm.

9 BLACK LEOPARD WRASSE
Macropharyngodon negrosensis Herre, 1932
Inhabits coral reefs; female distinguished by blackish colour with blue-green scale margins, also dark streaks on upper and lower edges of tail; Great Barrier Reef, N.W. Australia and throughout S.E. Asia; mainly W. Pacific; to 12 cm.

10 REDSPOT WRASSE
Stethojulis bandanensis (Bleeker, 1851)
Inhabits coral and rocky reefs, often in weed or rubble areas; distinguished by red patch above pectoral fin base - male has curved blue stripe on cheek and female with pale spotting on back; found throughout the region; Indo-C. Pacific; to 15 cm.

11 SILVER-STREAKED WRASSE
Stethojulis strigiventer (Bennett, 1832)
Inhabits weedy areas in the vicinity of coral and rocky reefs; female distinguished by narrow stripes on ventral half of body, male similar to **10** above, but lacks blue stripe on the cheek; found throughout the region; Indo-W. Pacific; to 15 cm.

12 SOELA WRASSE
Suezichthys soelae Russell, 1985
Inhabits sandy areas in 50-80 m; distinguished by dark spots on tail; N.W. Australia only; to 10 cm.

13 PEARLY RAINBOWFISH
Xenojulis margaritaceous (Macleay, 1884)
Inhabits weedy areas in the vicinity of coral reefs; distinguished by dorsal fin shape (short in front and elevated posteriorly) and mottled pattern with pearly or pinkish spots on side; N.W. Australia and throughout S.E. Asia; mainly W. Pacific; to 15 cm.

FISH SERVICE STATIONS

The Cleaner wrasses (2 and 3 above, and Pl. 72-7) are small, colourful reef inhabitants which render an important service to other fishes. They remove parasites from the body, mouth cavity, and gill chamber of numerous species spanning a considerable size range. Cleaning "stations" occur at regular intervals on coral reefs. Research shows they are an integral component for maintaining good health within the local fish community. Each station is occupied by one or more cleaner wrasses. The demand for their services is readily apparent to even casual observers. Several fishes often queue while patiently waiting their turn to be inspected for parasites. Cleaner wrasses are not timid. They even enter the mouth of large voracious predators such as moray eels and gropers. The False Cleanerfish (Pl.84-2) is a member of the blenny family that is cleverly disguised as a cleaner wrasse. Not only is the colour pattern identical, but it also swims with an exaggerated undulating motion, exactly like the wrasse. Its uses this disguise to approach unsuspecting victims. But instead of having their parasites removed the blenny dashes in and rips off a chunk of flesh, scales, or fin with its vampire-like fangs.

Like other wrasses, the Cleanerfish (3) exhibits sex change, but with an interesting twist. Each cleaner station is composed of a single dominant male and several females. Sex change in this species is dependent on behavioural interactions between the male and its harem of females. The normal aggressive behaviour of the male prevents its female companions from changing sex. However, if the male is experimentally removed the dominant female assumes command within a few hours and after several days it can sexually function as a male.

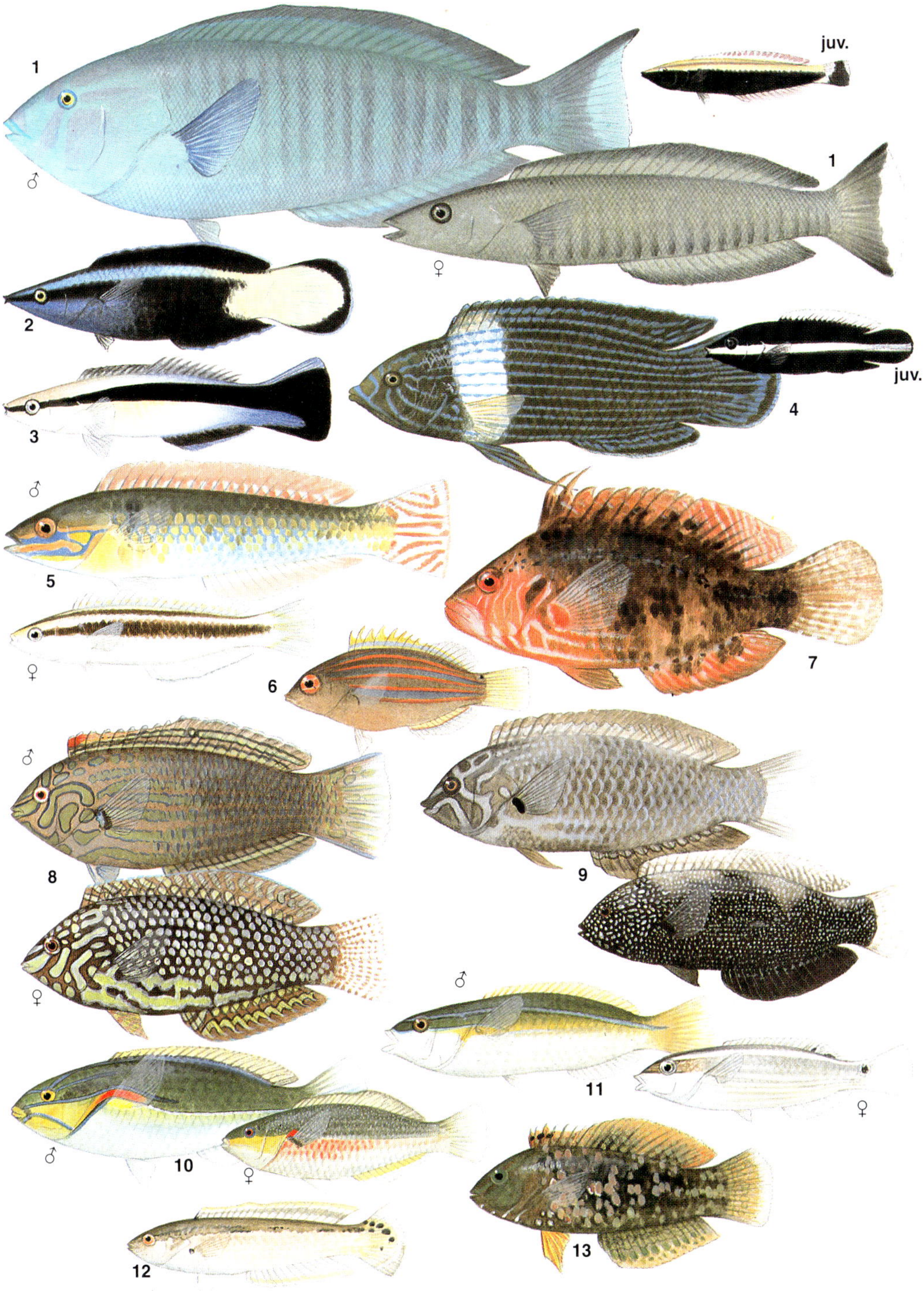
1
♂
juv.
1
♀
2
3
4
juv.
♂
5
♀
6
7
♂
8
9
♀
♂
11
♀
♂
10
♀
12
13

PLATE 78: WRASSES (FAMILY LABRIDAE)

1 RED AND GREEN WRASSE
Thalassoma purpureum (Forsskål, 1775)
Inhabits rocky reefs and inshore coral reefs, usually where there is wave action; male distinguished by ornate pattern highlighted by reddish pink to purple stripes, female and juvenile greenish with double row of elongate red to brown blotches or stripes on middle and lower side; also known as Surge wrasse; Great Barrier Reef, N.W. Australia, and throughout S.E. Asia; Indo-C. Pacific; to 40 cm. ★★

2 BLUE-HEADED WRASSE
Thalassoma amblycephalum (Bleeker, 1856)
Inhabits coral and rocky reefs, usually several females seen with each male; male distinguished by bluish head with broad pale band just behind head, female has dark stripe or is overall brownish on upper half of body and white below; Great Barrier Reef, N.W. Australia, and throughout S.E. Asia; Indo-W. Pacific; to 15 cm.

3 SEVEN-BANDED WRASSE
Thalassoma septemfasciata Scott, 1959
Inhabits rocky and weedy reefs; male distinguished by uniform dark body and yellow pectoral fins, female by broad greenish bars on side and yellow colour on pectoral fin base; W. Australia only, between Rottnest Island and Coral Bay; to 31 cm. ★★

4 MOON WRASSE
Thalassoma lunare (Linnaeus, 1758)
Inhabits coral and rocky reefs; distinguished from **5** below by magenta central portion of pectoral fin; juveniles (not shown) have a bluish belly and large black spot at base of tail; Great Barrier Reef, N.W. Australia, and throughout S.E. Asia; Indo-W. Pacific; to 30 cm. ★★

5 GREEN MOON WRASSE
Thalassoma lutescens (Lay & Bennett, 1839)
Inhabits coral and rocky reefs; male similar to **4** above, but with blue-edged pectoral fin and lighter body colour, juvenile (not shown) with black midlateral stripe; Great Barrier Reef, N.W. Australia, and throughout S.E. Asia; Indo-C. Pacific; to 30 cm.

6 SIX-BANDED WRASSE
Thalassoma hardwickei (Bennett, 1828)
Inhabits coral reefs; distinguished from **7** below by narrower bars, some of which extend to belly region; Great Barrier Reef, N.W. Australia, and throughout S.E. Asia; Indo-C. Pacific; to 18 cm.

7 JANSEN'S WRASSE
Thalassoma jansenii (Bleeker, 1856)
Inhabits coral reefs; similar to **6** above, but has broader bars that do not extend onto belly; Great Barrier Reef, N.W. Australia, and throughout S.E. Asia; Indo-W. Pacific; to 18 cm.

8 LONG GREEN WRASSE
Pseudojuloides elongatus Ayling & Russell, 1977
Inhabits weed beds; distinguished by elongate shape, male has orange blotch at pectoral base and blue spots on upper side female uniform greenish; W. Australia (between Abrolhos Islands and Dampier Archipelago), New So. Wales, New Zealand, and Japan; to 15 cm.

9 BLUE-TOOTHED TUSKFISH
Xiphocheilus typus Bleeker, 1856
Inhabits flat sandy bottoms or rubble; distinguished by blue and yellow stripes in front of eye and yellowish fins; N. Australia (North West Shelf to Cape York), and throughout S.E. Asia; Indo-Australian Archipelago; to 14 cm.

10 PAVO RAZORFISH
Xyrichtys pavo Valenciennes, 1839
Inhabits sand bottoms, uses keeled forehead to bury into the sand when threatened; distinguished by broad brown bars and antenna-like first dorsal fin; Great Barrier Reef, N.W. Australia, and throughout S.E. Asia; Indo-C. Pacific; to 35 cm.

11 BLACKSPOT RAZORFISH
Xyrichtys dea Temminck & Schlegel, 1846
Inhabits sand bottoms; buries in sand when threatened; distinguished by antenna-like first dorsal fin, red or pinkish colour with vague bars, and small black spot below front of dorsal fin; juveniles variable, black to pale in colour; North West Shelf; mainly W. Pacific; to 35 cm.

12 CARPET WRASSE
Novaculichthys taeniurus (Lacepède, 1802)
Inhabits rubble and weedy areas near coral reefs; distinguished by diagonal stripes on cheek, speckled fins and white bar across base of tail; juvenile has elongate spines at front of dorsal fin; Great Barrier Reef, N.W. Australia, and throughout S.E. Asia; Indo-C. Pacific; to 25 cm.

OF RAZORS AND KNIVES

Razorfishes (Plate 78-10 and 11, Plate 79-14-15) and knifefishes (Plate 79-13 and 16) are peculiar wrasses that usually live on beds of clean white or black sand in 10-40 metres depth. Their head and body is specially adapted for burrowing into the bottom - they are very compressed from side to side (ie thin), and the forehead forms a relatively sharpened keel (which is responsible for their common names). Reef-dwelling wrasses rely on the shelter afforded by the myriad of crevices and fissures. In their desert-like surroundings, which are absolutely devoid of hiding places, razorfishes and knifefishes have evolved an ingenious escape mechanism when threatened by predators or passing divers. They quickly vanish by plunging headlong into the sand. On several occasions when the author attempted to dig them out, it was obvious these fishes do not just simply bury themselves, but are adept at moving at a fairly rapid rate beneath the sandy surface. These fishes can sometimes be taken with a baited hook, but extreme care must be exercised when handling them. The powerful jaws are equipped with large tusk-like teeth that can inflict a nasty wound. Normally they are used for feeding on sand-dwelling invertebrates such as shellfish, crabs, and shrimps.

Like most other wrasses, razorfishes have different colour patterns according to sex and growth stage. Juveniles appear to mimic drifting debris, leaning over to one side and swimming with a peculiar swaying motion. The small young of the Pavo Razorfish (**10**) and a few other species have a very elongate antenna on top of the head, which is actually part of the dorsal fin.

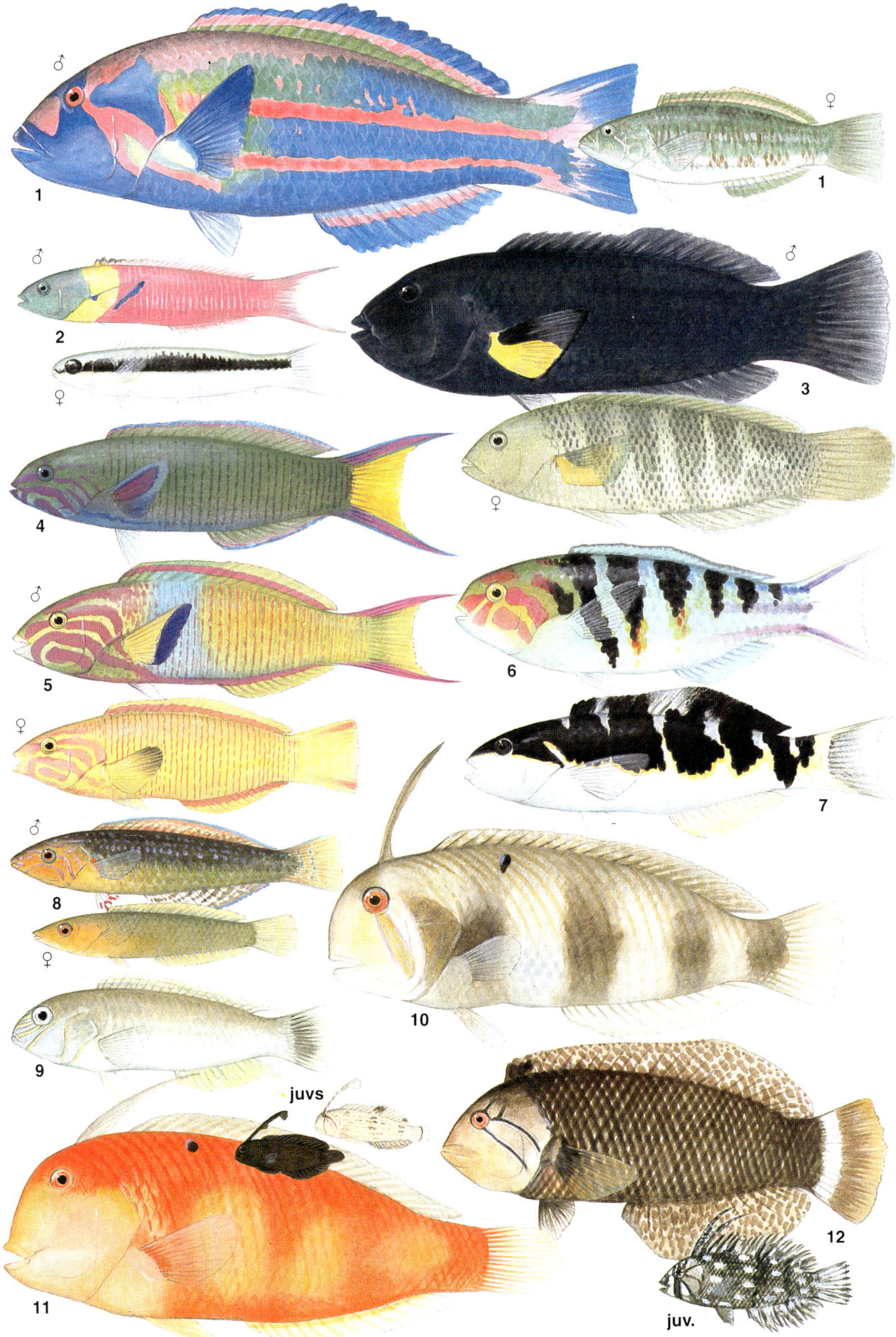

♂
1
♀
1
♂
2
♀
♂
3
4
♀
♂
5
6
♀
7
♂
8
♀
10
9
juvs
11
12
juv.

PLATE 79 - WRASSES (LABRIDAE)

1 JAPANESE WRASSE
Pseudocoris yamashiroi (Schmidt, 1930)
Inhabits coral reefs, usually seen in aggregations on outer slopes; females greatly outnumber males; feeds on zooplankton; male greyish on upper two-thirds and white below, further distinguished by dark-edged caudal fin lobes and prolonged spines at front of dorsal fin; female (not shown) is mainly pink to reddish; Great Barrier Reef, offshore reefs of N.W. Australia, and throughout S.E. Asia; W. and C. Pacific; to 15 cm.

2 ZEBRA WRASSE
Pseudocoris heteroptera (Bleeker, 1857)
Inhabits coral reefs, frequently seen on outer slopes; distinguished by brilliant male colour pattern of black and yellow bars on rear half of body; Indo-Malay Archipelago; to 24 cm.

3 CUTRIBBON WRASSE
Stethojulis interrupta (Bleeker, 1851)
Inhabits coral reefs in areas with scattered sand pockets; male distinguished by interrupted blue stripe on middle of sides; female (not shown) is brownish above and white below with black dots on lower half; Great Barrier Reef and throughout S.E. Asia; Indo-W. Pacific; to 13 cm.

4 CRYPTIC WRASSE
Pterogogus cryptus (Randall, 1981)
Inhabits reefs with abundant hard or soft corals between 4-70 m; a secretive fish that never exposes itself for more than a few seconds at a time; distinguished by white stripe on snout, continuing above eye to back of head, and vertically elongate dark spot on gill cover; Great Barrier Reef, offshore reefs of W. Australia, and throughout S.E. Asia; Indo-W. Pacific; to 7 cm.

5 RUST-BANDED WRASSE
Pseudocoris aurantifasciata Fourmanoir, 1971
Inhabits coral reefs, most frequently on steep dropoffs, occurring in small aggregations; male distinguished by elongate first dorsal fin, relatively deep body with several dark bars on side and single pale bar on posterior half of body; Indo-Malay Archipelago; W. and C. Pacific to Tuamotus; to 24 cm.

6 PHILIPPINES WRASSE
Pseudocoris philippina (Fowler & Bean, 1928)
Inhabits coral reefs; male distinguished by black first dorsal spine, yellow botch on middle of side, and dark margins on tail; female is light brown with brown band on snout and dark blotches on end of gill cover and upper tail base; Philippines to Ryukyu Islands; to 15 cm.

7 CHISELTOOTH WRASSE
Pseudodax moluccanus (Valenciennes, 1839)
Inhabits coral reefs, usually seen on outer slopes; juvenile pick parasites from other fishes; distinguished by orange-red wash on nape and dorsal fin, and pair of large spatulate teeth at front of jaws; juvenile resembles *Labroides* and has pair of blue stripes on side; Great Barrier Reef, offshore reefs of W. Australia, and throughout S.E. Asia; Indo-C. Pacific; to 25 cm.

8 THREE-LINED WRASSE
Stethojulis trilineata (Bloch & Schneider, 1801)
Inhabits shallow reefs exposed to mild or moderate wave action; a very fast swimming fish; male distinguished by three bright blue stripes running full length of body and shorter stripe from snout to pectoral region; female (not shown) is dark above and white below with numerous small white spots on upper half of body; Great Barrier Reef, offshore reefs of W. Australia, and throughout S.E. Asia; E. Indian Ocean and W. Pacific; to 15 cm.

9 FIVESTRIPE WRASSE
Thalassoma quinquevittatum (Lay & Bennett, 1839)
Inhabits shallow coral reefs from shallow surge areas down to 30 m depth; male distinguished by green and purplish bands on head, pink to purplish midlateral stripe, and horizontal row of vertically elongate green spots on upper side; female relatively dull greenish with small black spot at front of dorsal fin; Great Barrier Reef, offshore reefs of N.W. Australia, and throughout S.E. Asia; Indo-C. Pacific; to 17 cm.

10 LADDER WRASSE
Thalassoma trilobatum (Lacepède, 1801)
Inhabits shallow coral reefs, usually seen in areas exposed to surge; male distinguished by broad red midlateral stripe with double row of vertically elongate rectangular patches of green; female is very similar to that of *T. purpureum*; Great Barrier Reef, offshore reefs of N.W. Australia, and throughout S.E. Asia; Indo-C. Pacific; to 30 cm.

11 DIAGONAL-LINED WRASSE
Wetmorella albofasciata Schultz & Marshall, 1954
Inhabits reef crevices and caves, rarely seen; distinguished by narrow white bars on body, ocellated spots on dorsal and anal fins, and black spot on pelvic fins; Great Barrier Reef, offshore reefs of N.W. Australia, and throughout S.E. Asia; Indo-C. Pacific; to 5.5 cm.

12 SHARPNOSE WRASSE
Wetmorella nigropinnata (Seale, 1901)
Inhabits reef crevices and caves, rarely seen; distinguished by yellow bars behind eye and across tail base, also ocellated spot on dorsal and anal fins, and black spot on pelvic fin; Great Barrier Reef, offshore reefs of N.W. Australia, and throughout S.E. Asia; Indo- C. Pacific; to 8 cm.

13 KNIFEFISH
Cymolutes praetextatus (Quoy & Gaimard, 1834)
Inhabits clean sandy areas near coral reefs; dives under sand when disturbed; a slender, highly compressed pale-coloured fish without distinguishing marks; Great Barrier Reef, offshore reefs of N.W. Australia, and throughout S.E. Asia; Indo-C. Pacific; to 12 cm.

14 WHITEPATCH RAZORFISH
Xyrichtys aneitensis (Günther, 1862)
Inhabits clean sandy areas near coral reefs; dives under sand when disturbed; distinguished by large white patch above belly; also known as Pale Razorfish; Great Barrier Reef and throughout S.E. Asia; Indo-W. Pacific; to 20 cm.

15 FIVEFINGER RAZORFISH
Xyrichtys pentadactylus (Linnaeus, 1758)
Inhabits clean sandy areas near coral reefs; dives under sand when disturbed; male distinguished by several black spots behind eye, female (not shown) has blackish spot on side near tip of pectoral fin; Great Barrier Reef and throughout S.E. Asia; Indo-W. Pacific; to 25 cm.

16 COLLARED KNIFEFISH
Cymolutes torquatus Valenciennes, 1840. Text on page 198.

17 SEAGRASS WRASSE
Novaculichthys macrolepidotus (Bloch, 1791). Text on page 198.

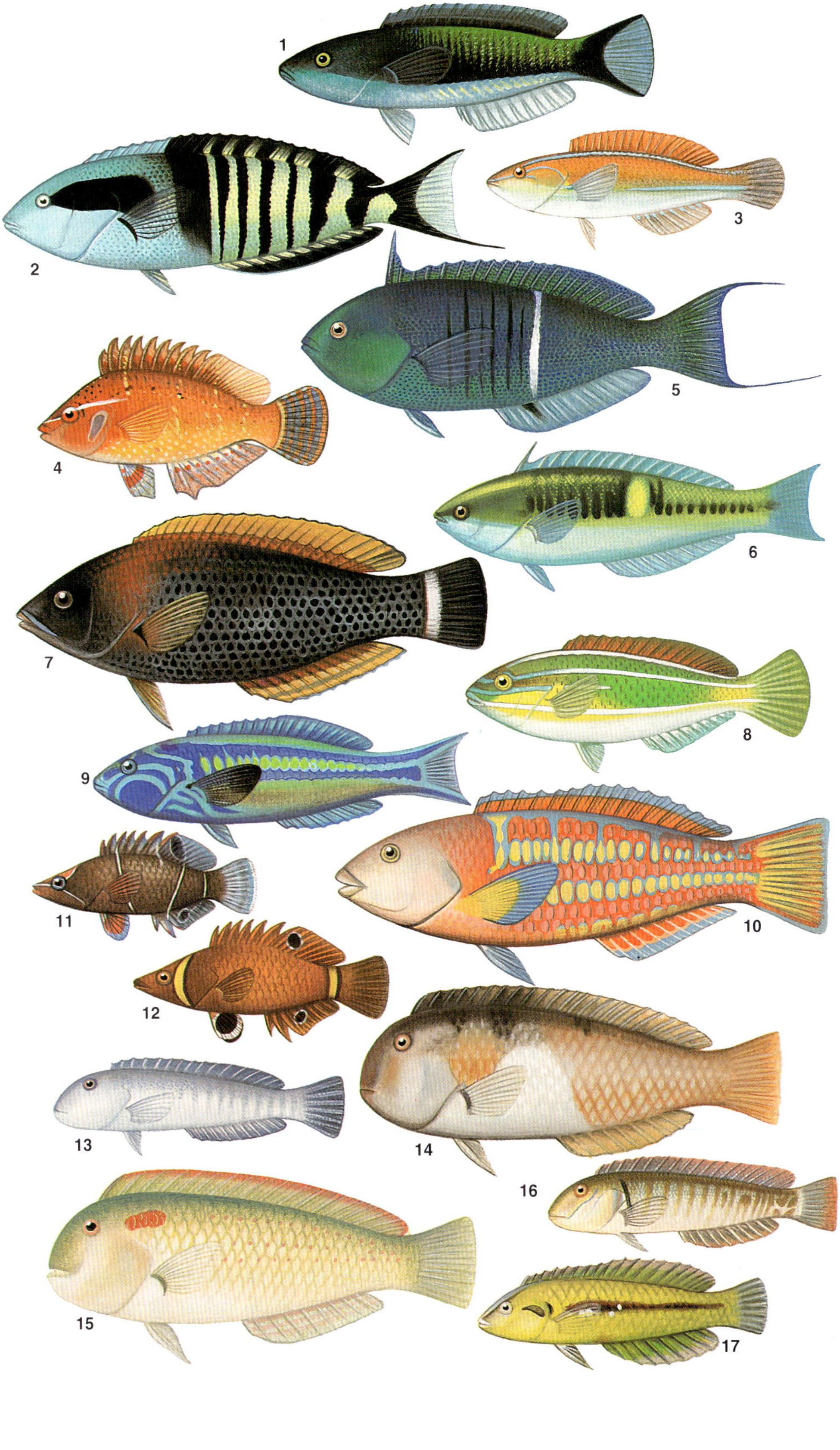
1
2
3
4
5
6
7
8
9
10
11
12
13
14
15
16
17

PLATE 80: PARROTFISHES (FAMILY SCARIDAE)

1 DOUBLE-HEADED PARROTFISH
Bolbometopon muricatum (Valenciennes, 1840)
Inhabits coral reefs; distinguished by large size and hump on forehead; the largest of all parrotfishes; Great Barrier Reef, N.W. Australia, and throughout S.E. Asia; Indo-C. Pacific; to 130 cm. ★★★

2 RED-SPECKLED PARROTFISH
Cetoscarus bicolor (Rüppell, 1828)
Inhabits coral reefs, often in pairs; male distinguished by orange-red to pink scale margins and spots on head and front part of body, and pale stripe from mouth to anal fin, females by dense black spotting on lower two-thirds of side; juvenile mainly white with broad orange bar on head; Great Barrier Reef, N.W. Australia, and throughout S.E. Asia; Indo-C. Pacific; to 50 cm. ★★★

3 SPINYTOOTH PARROTFISH
Calotomus spinidens (Quoy & Gaimard, 1824)
Inhabits weed and seagrass beds; distinguished by marbled pattern of green and brown, differs from most other parrotfishes in having separate teeth (not fused to beak-like structure); can change colour rapidly to blend with surroundings; Great Barrier Reef, N.W. Australia, and throughout S.E. Asia; Indo-C. Pacific; to 19 cm.

4 BLUE-SPOTTED PARROTFISH
Leptoscarus vaigiensis (Quoy & Gaimard, 1824)
Inhabits weed and seagrass beds; similar to **3** above, but has fused teeth, male has pale stripe along sides and small blue spots on head, female mottled and spotted with whitish and dark brown; Great Barrier Reef, N.W. Australia, and throughout S.E. Asia; Indo-C. Pacific; to 38 cm.

5 LONG-NOSED PARROTFISH
Hipposcarus longiceps (Bleeker, 1862)
Inhabits coral reefs, sometimes seen over sand or rubble bottoms, often forms schools; distinguished by pointed head, female has yellowish fins and is overall pale, male bluish on sides and on snout with blue margins on most fins; Great Barrier Reef, N.W. Australia, and throughout S.E. Asia; Indo-C. Pacific; to 45 cm. ★★★

6 SIX-BANDED PARROTFISH
Scarus frenatus Lacepède, 1802
Inhabits coral reefs, usually occurring in small groups; male distinguished by abruptly lighter areas on lower half of head and posterior part of body female by series of 6-7 dark stripes on side and reddish fins; Great Barrier Reef, N.W. Australia, and throughout S.E. Asia; Indo-C. Pacific; to 40 cm. ★★★

7 SADDLED PARROTFISH
Scarus dimidiatus Bleeker, 1859
Inhabits coral reefs, usually alone or in small groups; male distinguished by dusky band from eye across gill cover with white band below, female by 3 saddle-like bars on back; Great Barrier Reef, N.W. Australia, and throughout S.E. Asia; Indo-C. Pacific; to 34 cm. ★★★

8 BLUE-BARRED PARROTFISH
Scarus ghobban Forsskål, 1775
Inhabits coral reefs, seagrass and weed areas, and deeper offshore trawling grounds; male distinguished by broad blue scale margins, 3 dark streaks behind and below eye, and short blue stripe on chin, female by overall yellowish colour with diffuse blue bars and spots on side; found throughout the region; Indo-E. Pacific; to 100 cm and 6.5 kg. ★★★

9 CHAMELEON PARROTFISH
Scarus chameleon Choat & Randall, 1986
Inhabits coral reefs; male distinguished by dark stripe from snout, passing below eye across gill cover, another short stripe behind eye and vertical band above eye, also usually has large oval-shaped pale area occupying most of lower side, female generally dusky (may also have pale blotches) on upper half of side and abruptly pale below; Great Barrier Reef, N.W. Australia, and throughout S.E. Asia; mainly W. Pacific; to 28 cm. ★★

PLATE 79

16 COLLARED KNIFEFISH
Cymolutes torquatus Valenciennes, 1840
Inhabits clean sandy areas near coral reefs; dives under sand when disturbed; similar to **12** above, but has faint bars on side and dark diagonal bar above pectoral fin; Great Barrier Reef and throughout S.E. Asia; Indo-W. Pacific; to 12 cm.

17 SEAGRASS WRASSE
Novaculichthys macrolepidotus (Bloch, 1791)
Inhabits seagrass beds, usually in small groups; distinguished by green colouration and dark stripe on middle of side; Queensland and S.E. Asia; Indo-W. Pacific; to 15 cm.

PARROTFISHES

Parrotfishes of the family Scaridae (Plates 80-82) are closely related to the wrasses (Plates 70-79), but rather than having individual teeth in the jaws, the dental plates are usually fused to form a distinctive beak-like structure. This structure is well adapted for scraping algal food from the surface of the reef. Parrotfishes also ingest large amounts of coral rock and sand along with the algae. This material is ground to a fine powder by special teeth at the back of the throat and is passed out with the faeces, contributing significantly to the bottom sediment.

Like the closely related wrasses, parrotfishes undergo female to male sex change and display different colour patterns according to growth stage or sex. Three decades ago, this discovery was considered a unique revelation, but detailed studies indicate that sex reversal is extremely common for a large portion of reef fishes belonging to many different families. In many species, the male and female colour patterns are so different they have long been regarded as different species. Diving scientists have been able to properly link most male-female pairings by observing courtship and spawning.

The estimated 75 species occurring worldwide are mainly inhabitants of coral reefs, although a few are found in weedy areas. They occur both individually or graze over the reef in large schools. At dusk parrotfishes retire to shelter in the form of crevices and ledges. Many of the species exude a strange cocoon-like mucus envelope which they remain inside during the night. This appears to be an adaptation which masks their scent, thus preventing predation by fishes such as moray eels that rely heavily on their sense of smell in locating prey. The largest parrotfishes are slightly over 1 m in length, but most species are under 50 cm.

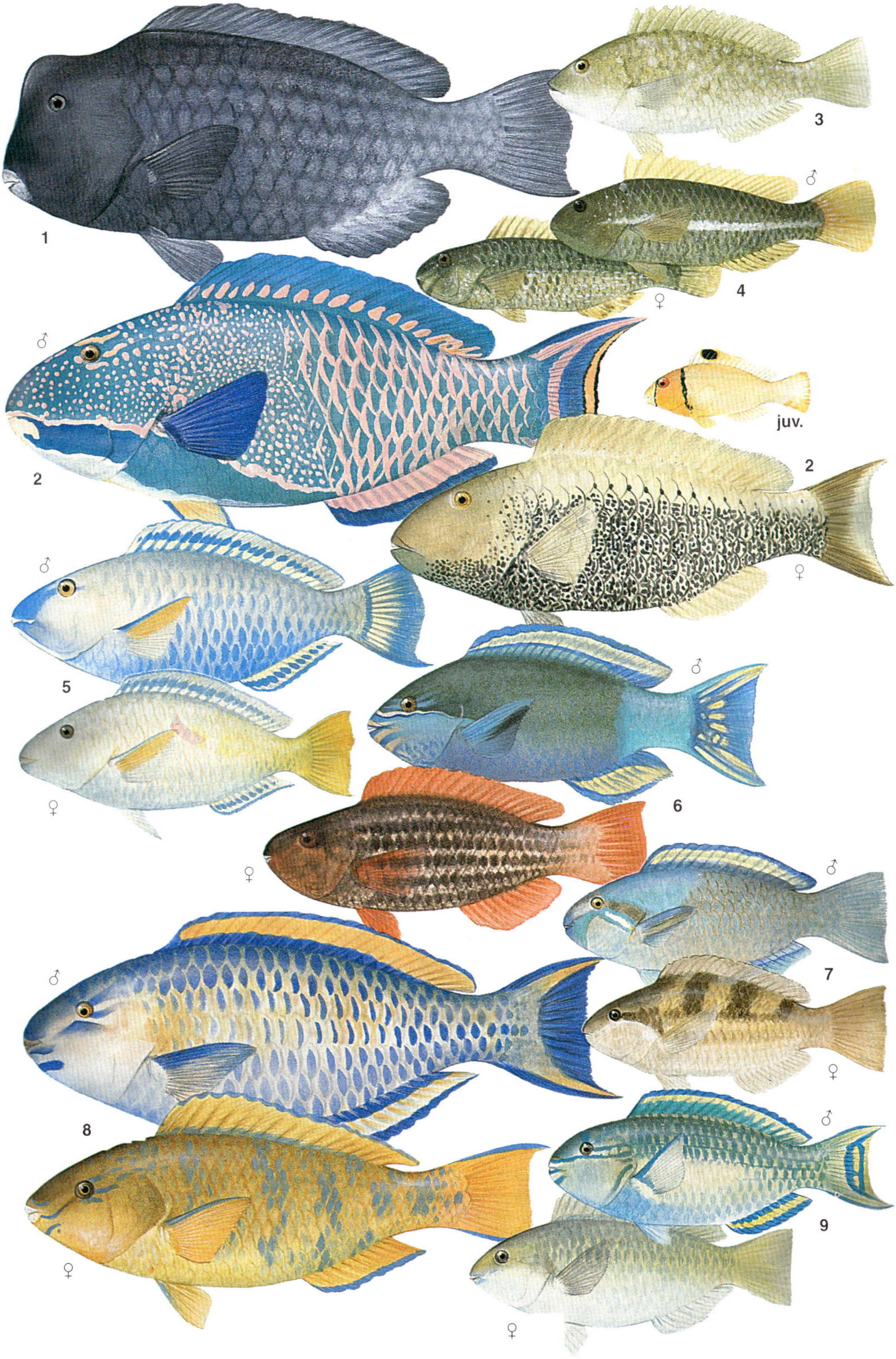

1
3
♂
4
♀
♂
juv.
2
2
♀
♂
5
♂
♀
6
♀
♂
7
♂
8
♀
♂
9
♀
♀

PLATE 81: PARROTFISHES (FAMILY SCARIDAE)

1 STEEPHEAD PARROTFISH
Scarus microhinos (Bleeker, 1854)
Inhabits coral reefs and southern rocky reefs; distinguished by steep, blunt snout profile, becoming steeper with increased size, male and female similar but 2 colour phases are exhibited, a more common green one and a red one, large terminal male sometimes mainly purple with blue area ventrally; juvenile (not shown) is dark brown with 3 white stripes on side; Great Barrier Reef, N.W. Australia, and throughout S.E. Asia; Indo-C. Pacific; to 50 cm. ★★★

2 VIOLET-LINED PARROTFISH
Scarus globiceps Valenciennes, 1840
Inhabits coral reefs; male distinguished by "maze" pattern of spots and dashes on upper back and top of head, female difficult to separate from that of **9** below, but often seen in company with male; Great Barrier Reef, N.W. Australia, and throughout S.E. Asia; Indo-C. Pacific; to 30 cm.★★★

3 BLUE PARROTFISH
Scarus oviceps Valenciennes, 1839
Inhabits coral reefs; male distinguished by abruptly dark area on upper part of head and adjacent portion of back, female by similar darkened area followed by a pair of light and dark patches below dorsal fin; Great Barrier Reef, N.W. Australia, and throughout S.E. Asia; mainly W. and C. Pacific, to 35 cm. ★★★

4 DUSKY PARROTFISH
Scarus prasiognathus Valenciennes, 1839
Inhabits coral reefs; male distinguished by blue-green colour on head below eyes and orange colour above, female by blackish or dark brown colour, numerous small white spots on side, and reddish fins; N.W. Australia, and throughout S.E. Asia; W. Pacific and E. Indian Ocean; to 70 cm. ★★★

5 PALENOSE PARROTFISH
Scarus psittacus Forsskål, 1775
Inhabits coral reefs; male distinguished by 2-3 blue bands behind eye and lavender or bluish-grey colour on snout and forehead, female by overall dark colour, red pelvic fins, and pale snout; Great Barrier Reef, N.W. Australia, and throughout S.E. Asia; mainly W. and C. Pacific; to 30 cm. ★★★

6 EMBER PARROTFISH
Scarus rubroviolaceus Bleeker, 1849
Inhabits coral reefs; male distinguished by blunt snout and bicolour pattern, female by red colour with irregular dark stripes; found throughout the region; Indo-E. Pacific; to 65 cm. ★★★

7 SCHLEGEL'S PARROTFISH
Scarus schlegeli (Bleeker, 1861)
Inhabits coral reefs; male distinguished by 2 yellow patches below dorsal fin, the rear one forming a narrow bar between the dorsal and anal fin, female by overall dark colour and 4-5 narrow pale bars (often faint, sometimes absent) on side; found throughout the region; mainly W. and C. Pacific; to 35 cm.. ★★★

8 GREEN-FINNED PARROTFISH
Chlorurus sordidus (Forsskål, 1775)
Inhabits coral reefs; male distinguished by overall green colour with pale tail base and bluish cheeks, throat and breast, female by white tail with round black spot at base; Great Barrier Reef, N.W. Australia, and throughout S.E. Asia; Indo-C. Pacific; to 45 cm. ★★★

9 SURF PARROTFISH
Scarus rivulatus Valenciennes, 1840
Inhabits coral reefs; male distinguished by orange patch on cheek, wavy lines on snout, and light yellow-green colour of pectoral fins, females are variable, but generally very pale; found throughout the region; mainly W. Pacific; to 45 cm. ★★★

GOBIES
(See Plates 88-92)

Gobies are by far the largest family of fishes in the region. There are no accurate estimates available, but in the area covered by this book there is probably well in excess of 800 species. Worldwide it is the largest family of marine fishes with an estimated 220 genera and about 2,000 species - almost one out of every 10 fishes in the world, including both freshwater and saltwater species, belongs to this massive family. Although the family is large, many of its members are definitely not. Pygmy gobies belonging to the genus *Trimmatom* and *Pandaka* are full grown at about 1 cm - making them the smallest known vertebrate animals. It's difficult to define a typical goby, but they are usually small (less than 15 cm), bottom-resting fishes, that depend on shelter in the form of rocks, coral, or sandy burrows. The majority live in the sea on reefs, but there are numerous freshwater species as well (particularly in the subfamily Eleotridinae). Most of the marine species have the pelvic fins completely or partially fused, forming a disk-like apparatus that is used as a "tripod" when resting. Although gobies exhibit a wide range of feeding habits, most consume a variety of small invertebrates.

Reef gobies offer some of the best examples of "symbiosis" - the association of two unrelated organisms. Some live among the branches of particular corals, others are found on the surface of sponges, soft corals, and tunicates, but the most fascinating examples involve associations with alpheid shrimps. The fish (often in pairs and usually in the genera *Amblyeleotris, Cryptocentrus, Ctenogobiops, Stonogobiops,* and *Vanderhorstia*) stands sentry duty at the entrance of the burrow, its body or tail constantly maintaining contact with the long antennae of the shrimp. If the coast is clear of intruders, the fish signals its shrimp partner with a little wiggle. The shrimp then emerges with a load of sediment, carefully balanced on its large claw. It dumps the load a short distance from the entrance, then re-enters the burrow to repeat the process. Burrows are generally constructed in areas of very soft sand or silt, and therefore require constant maintenance. Both fish and shrimp benefit greatly from their association and it is doubtful they could survive without it. The shrimp provides the fish with a home and also helps to unearth food. The fish performs a "watchdog" role for the shrimp, allowing it to dig a burrow and forage (on algae, detritus, and bacteria) without fear of predators.

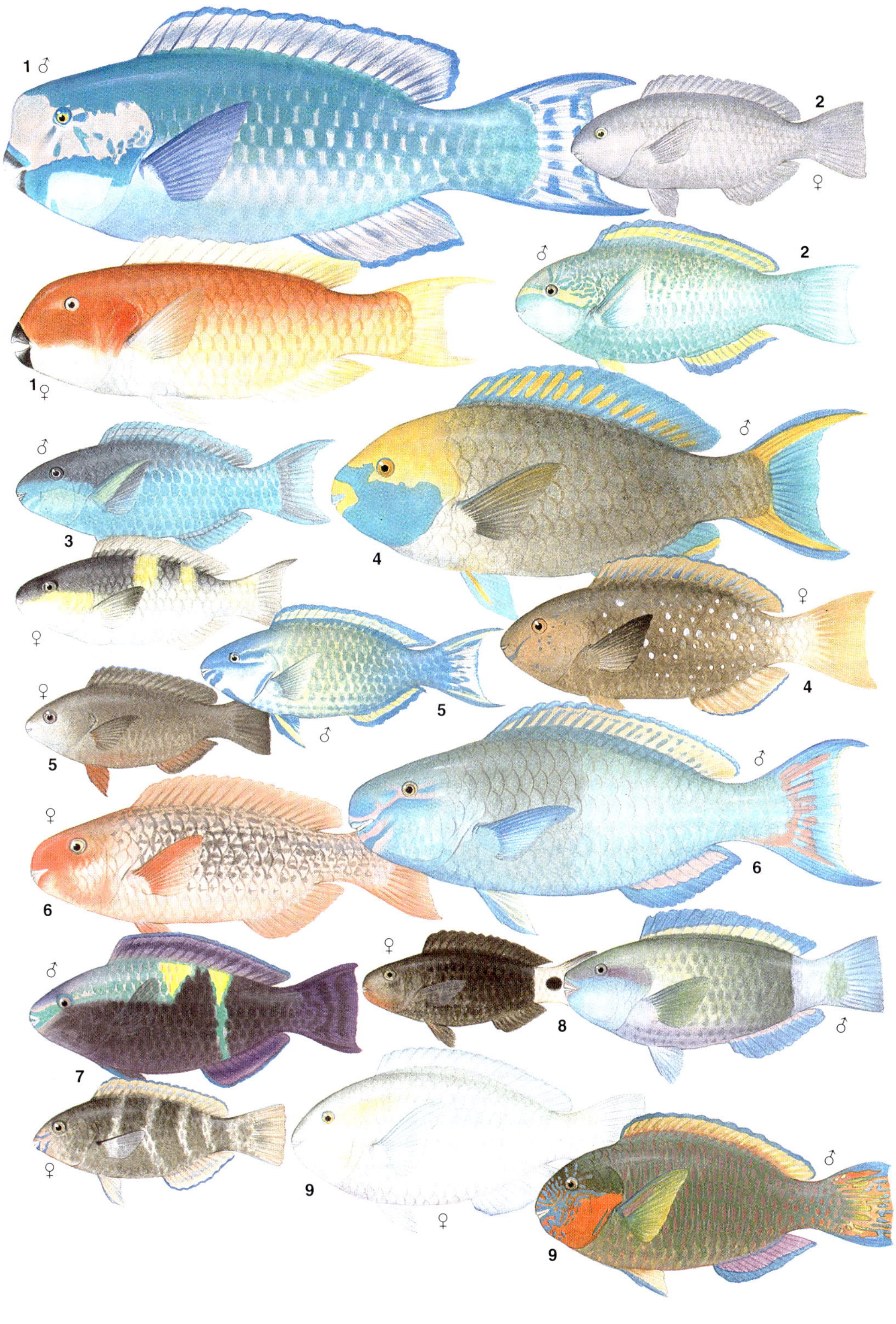
1 ♂
2
♀
♂
2
1 ♀
♂
♂
3
4
♀
♀
5
♂
4
♀
5
♀
♂
6
6
♂
♀
8
♂
7
♀
9
♀
♂
9

PLATE 82: PARROTFISHES (FAMILY SCARIDAE)

1 MINIFIN PARROTFISH
Scarus altipinnis (Steindachner, 1879)
Inhabits coral reefs; differs from other parrotfishes in having the middle part of the dorsal fin distinctly taller and forming short filaments; female (not shown) also has elevated rays in the middle of the dorsal fin and is reddish brown with 4-5 vertical series of small whitish spots on side; Great Barrier Reef and islands of the tropical W. and C. Pacific Ocean; to 60 cm. ★★★

2 BLEEKER'S PARROTFISH
Chlorurus bleekeri (de Beaufort, 1940)
Inhabits coral reefs; male distinguished by large, squarish patch on cheek bordered by green; female is dark brown, usually with about four pale bars on side; Great Barrier Reef, offshore reefs of N.W. Australia, and throughout S.E. Asia; W. Pacific; to 30 cm. ★★★

3 YELLOWFIN PARROTFISH Cairns 10/98
Scarus flavipectoralis Schultz, 1958
Inhabits sheltered coral reefs; distinguished by yellowish pectoral fin, male has broad green band from snout to region above pectoral fin, also a lime-green patch often evident on middle of tail base; female (not shown) pale greyish to brown on back and head, sometimes yellowish ventrally with whitish stripes on belly; Great Barrier Reef, offshore reefs of N.W. Australia, and throughout S.E. Asia; W. Pacific; to 30 cm. ★★★

4 WHITESPOT PARROTFISH
Scarus forsteni (Bleeker, 1861) Cairns 10/98
Inhabits coral reefs; male distinguished by green "moustache" that continues as green stripe below eye, bluish belly, darkened area on upper half of head, and adjacent part of body, and large diffuse pale patch on side; female has blue to yellowish streak or patch under pectoral fin and white spot on upper side; Great Barrier Reef, offshore reefs of N.W. Australia, and throughout S.E. Asia; W. and C. Pacific; to 40 cm. ★★★

5 REEFCREST PARROTFISH
Chlorurus frontalis (Valenciennes, 1840)
Inhabits coral reefs, usually seen in small schools in shallow water; differs from most other parrotfishes in not having distinct male and female colour patterns - generally geen with salmon-pink bar on each scale of body and irregular light salmon-pink to lavender bands in front of and above eye and on chin; Great Barrier Reef and scattered localities in SE Asia; W. and C. Pacific; to 50 cm. ★★★

6 HIGHFIN PARROTFISH
Scarus longipinnis Randall & Choat, 1980
Inhabits coral reefs, usually between 20-55 m depth; distinguished by unusual shape of dorsal fin, which is more elevated than in other parrotfishes (similar to that of wrasses in genus *Cirrhilabrus*); female is light brownish orange, often with dark brown bars, and has trio of blue-green stripes ventrally on the side and head markings similar to the male (shown here); Great Barrier Reef and Coral Sea across the S. Pacific to Pitcairn Island; to 40 cm. ★★★

7 SWARTHY PARROTFISH
Scarus niger Forsskål, 1775
Inhabits coral reefs; distinguished by overall dark colouration (female reddish brown) with pale mark just above upper corner of gill cover; Great Barrier Reef, offshore reefs of N.W. Australia, and throughout S.E. Asia; Indo-W. Pacific; to 35 cm. ★★★

8 REDTAIL PARROTFISH Cairns 10/98
Scarus pyrrhurus (Jordan & Seale, 1906)
Inhabits coral reefs; male distinguished by broad orange band on dorsal and anal fins, large yellowish or tan area on posterior part of side, and pale cheek; female (not shown) is overall dark brown with reddish tail; Great Barrier Reef, E. Indonesia, New Guinea, and Philippines; W. Pacific; to 30 cm. ★★★

9 GREENSNOUT PARROTFISH
Scarus spinus (Kner, 1868)
Inhabits coral reefs, usually seen on outer slopes; male distinguished by greenish snout and broad yellowish area on head; female (not shown) is dark brown, often with 4-5 indistinct pale bars on side; Great Barrier Reef, offshore reefs of N.W. Australia, and throughout S.E. Asia; W. and C. Pacific; to 30 cm. ★★★

10 QUOY'S PARROTFISH
Scarus quoyi Valenciennes, 1840
Inhabits coral reefs, usually in sheltered areas; distinguished by mainly orange dorsal fin and lime green saddle on base of tail; throughout S.E. Asia; E. Indian Ocean and W. Pacific; to 30 cm. ★★★

11 TRICOLOR PARROTFISH
Scarus tricolor Bleeker, 1847
Inhabits coral reefs, usually seen on outer slopes; female distinguished by bright red fins and prominent reticulated pattern due to dark scale edges; male has blue or green stripes above and below eye, on chin, and on margins of all fins except pelvics, also has exceptionally long pointed lobes on tail; Great Barrier Reef and throughout S.E. Asia; Indo-C. Pacific; to 45 cm. ★★★

PARROTFISH REPRODUCTION

Parrotfishes have a reproductive mode very similar to that of wrasses (Labridae). In both groups the male and female usually exhibit dramatically different colour patterns. The first mature phase in the life cycle is termed the initial phase and may include both female and male individuals. Females of most species in this phase can change to the male sex and alter their colouration to that of the more gaudy terminal phase. The initial phase is frequently drab brownish or grey compared to a terminal phase that is often dominated by green. Two styles of reproduction are found in many species: group spawning of initial phase fish in which males greatly outnumber females, and pair spawning of an initial phase female and a terminal male. In both cases eggs and sperm are released at the apex of a rapid rush toward the surface. Terminal males tend to establish territories and maintain harems consisting of a number of females. Parrotfish eggs are spherical, ranging in diameter between 0.6 to 1.1 mm. They have a single yellow or orange oil droplet and are positively buoyant. The eggs hatch in about 25 hours at 26 degrees C and the fry are slightly under 2.0 mm in length. The larvae begin feeding after about three days and are planktonic for an undetermined period, probably at least several weeks.

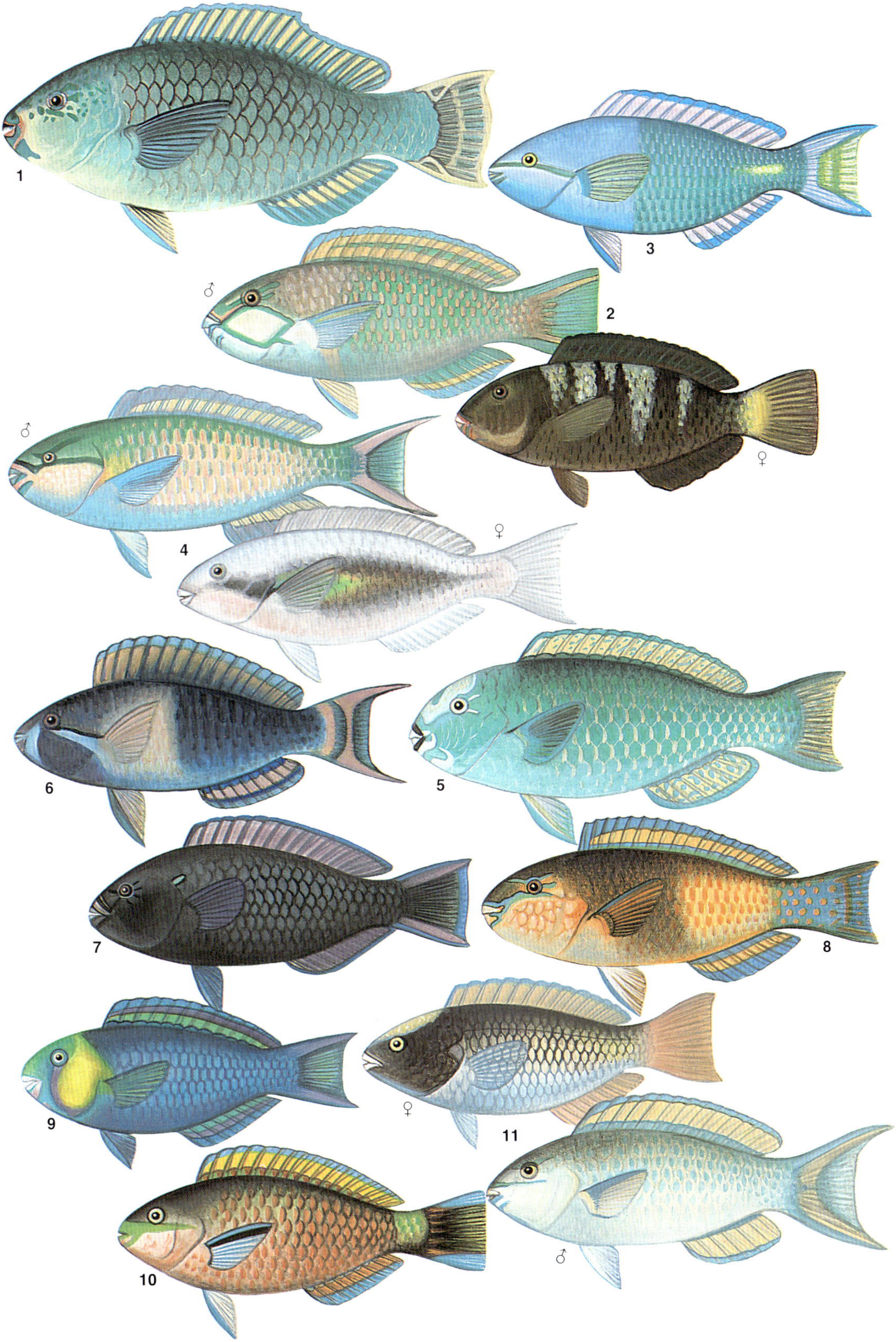
1
3
♂
2
♂
♀
♀
4
6
5
7
8
9
♀
11
10
♂

PLATE 83: GRUBFISHES, JAWFISHES AND STARGAZERS

1 BLUE-NOSED GRUBFISH
Parapercis alboguttata (Günther, 1872)
Inhabits trawling grounds; distinguished by diffuse dark bars or blotches on side and bluish snout; N.W. Australia and S.E. Asia; N. Indian Ocean and W. Pacific; to 22 cm. (PINGUIPEDIDAE)

2 DOUBLESPOT GRUBFISH
Parapercis diplospilus Gomon, 1980
Inhabits trawling grounds; distinguished by diffuse dark spots or blotches on side and 2 spots on tail base; Exmouth Gulf northwards; Indo-Australian Archipelago; to 9 cm. (PINGUIPEDIDAE)

3 SPOTHEAD GRUBFISH
Parapercis clathrata Ogilby, 1911
Inhabits coral reefs; distinguished by row of enlarged spots along lower side, these usually connected by a thin stripe, also frequently with pale-edged black spot above gill cover; Great Barrier Reef, N.W. Australia, and throughout S.E. Asia; Indo-C. Pacific; to 17 cm. (PINGUIPEDIDAE)

4 NARROW-BARRED GRUBFISH
Parapercis macrophthalma (Pietschmann, 1911)
Inhabits trawling grounds; distinguished by 5 narrow bars on side and dark spot on upper tail base; N.W. Australia and S.E. Asia; mainly W. Pacific; to 14 cm. (PINGUIPEDIDAE)

5 ROSY GRUBFISH
Parapercis gushikeni Yoshino, 1975
Inhabits trawling grounds; distinguished by narrow lines on upper side and spotted tail with pointed upper edge; N.W. Australia and S.E. Asia; mainly W. Pacific; to 30 cm. (PINGUIPEDIDAE)

6 BANDED GRUBFISH
Parapercis mimaseana (Kamohara, 1937)
Inhabits trawling grounds; distinguished by black spot at front of dorsal fin, wavy stripes on upper side, and narrow bars on tail; N.W. Australia and S.E. Asia; mainly W. Pacific; to 20 cm. (PINGUIPEDIDAE)

7 RED-BANDED GRUBFISH
Parapercis multiplacata Randall, 1984
Inhabits offshore coral reefs; distinguished by black spiny dorsal fin and series of red bars on side; Great Barrier Reef, N.W. Australia, and Indonesia; mainly W. Pacific; to 10 cm. (PINGUIPEDIDAE)

8 RED-BARRED GRUBFISH
Parapercis nebulosa (Quoy & Gaimard, 1824)
Inhabits trawling grounds; similar to **7** above, but cross bars wider and more diffuse, also lacks row of dark spots along lower side; Australia only, New So. Wales northward to Great Barrier Reef and across far north to W. Australia; to 20 cm. (PINGUIPEDIDAE)

9 ABROLHOS JAWFISH
Opistognathus sp.
Inhabits rubble and sand bottoms below 20 m depth; distinguished by small size and irregular dark longitudinal band on sides; Abrolhos to Monte Bello Islands, Western Australia; to 10 cm. (OPISTOGNATHIDAE)

10 DARWIN JAWFISH
Opistognathus darwiniensis Macleay, 1878
Inhabits shallow reefs, usually in sandy or rubble areas; distinguished by yellowish colour of fins, dense spotting on head and sides, and prominent banding on dorsal, anal and tail fins; N. Australia only, between Ningaloo Reef, W. Australia and Gulf of Carpentaria; to 50 cm. (OPISTOGNATHIDAE)

11 BLOTCHED JAWFISH
Opistognathus latitabundus (Whitley, 1937)
Inhabits coastal waters, usually on rubble or soft bottoms; distinguished by large dark blotches on back and base of dorsal fin; also known as Spotted pug; N. Australia only, between Broome, W. Australia and Gulf of Carpentaria; to 30 cm. (OPISTOGNATHIDAE)

12 BLACK JAWFISH
Opistognathus inornatus Ramsay & Ogilby, 1887
Inhabits coastal waters, usually on rubble or soft bottoms; distinguished by overall dark colour without markings; N. Australia only, between Exmouth Gulf, W. Australia and Gulf of Carpentaria; to 55 cm. (OPISTOGNATHIDAE)

13 LEOPARD JAWFISH
Opistognathus reticulatus McKay, 1969
Inhabits coastal waters, usually on rubble or soft bottoms; distinguished by large black spots; has bitten people wading in shallow water, but usually harmless; N. Australia only, between Exmouth Gulf, W. Australia and Gulf of Carpentaria; to 50 cm. (OPISTOGNATHIDAE)

14 BANDED STARGAZER
Ichthyscopus fasciatus Haysom, 1957
Inhabits trawling grounds; distinguished by 5 dark bars on upper side; N. Australia only; to 25 cm. (URANOSCOPIDAE)

15 DOUBLE-BANDED STARGAZER
Ichthyscopus insperatus Mees, 1960
Inhabits trawling grounds; distinguished by narrow double-bars on back and barred tail; N. Australia only; to 30 cm. (URANOSCOPIDAE)

16 MARBLED STARGAZER
Uranoscopus bicinctus Temminck & Schlegel, 1850
Inhabits trawling grounds; distinguished by white patches on back and diffuse broad bar below each dorsal fin, has stout spine behind upper edge of gill cover; N.W. Australia and S.E. Asia; mainly W. Pacific; to 20 cm. (URANOSCOPIDAE)

17 YELLOWTAIL STARGAZER
Uranoscopus cognatus Cantor, 1849
Inhabits trawling grounds; distinguished by overall brown colour except for black spot on first dorsal fin and lighter tail, has stout spine behind upper edge of gill cover; N.W. Australia and S.E. Asia; Indo-Australian Archipelago; to 22 cm. (URANOSCOPIDAE)

18 KAI STARGAZER
Uranoscopus kaianus Günther, 1880
Inhabits trawling grounds; similar to **17** above, but has variegated pattern on upper half of body; N.W. Australia and S.E. Asia; Indo-Australian Archipelago; to 22 cm. (URANOSCOPIDAE)

19 WHITE-SPOTTED STARGAZER
Uranoscopus sp.
Inhabits trawling grounds; distinguished by small white spots on upper part of head and body, has stout spines behind upper edge of gill cover; possibly N. Australia only; to 15 cm. (URANOSCOPIDAE)

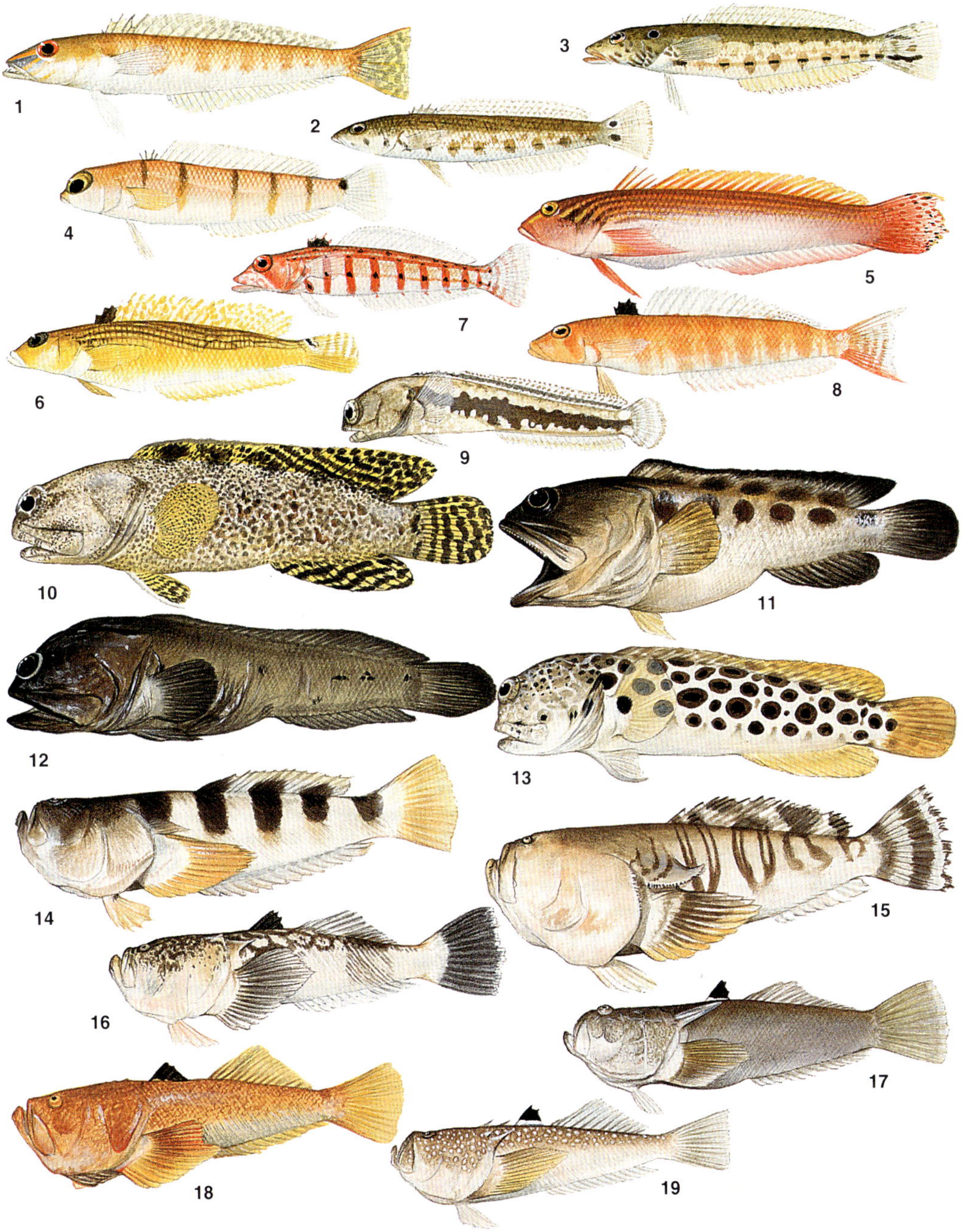
1
2
3
4
5
6
7
8
9
10
11
12
13
14
15
16
17
18
19

PLATE 84: BLENNIES (FAMILY BLENNIIDAE)

1 SLENDER SABRETOOTH BLENNY
Aspidontus dussumieri (Valenciennes, 1836)
Inhabits coral reefs and southern rocky reefs; similar to **2** below, but has yellow fins; Great Barrier Reef, W. Australia, and throughout S.E. Asia; Indo-C Pacific; to 14 cm.

2 FALSE CLEANERFISH
Aspidontus taeniatus Quoy & Gaimard, 1834
Inhabits coral reefs; nearly identical in colour to the Cleanerfish (Pl. 77-3), but has more pointed snout, longer dorsal fin base, and enlarged fangs at back of lower jaw; uses its "disguise" to attack other fishes, taking bites of skin and scales; Great Barrier Reef, W. Australia, and throughout S.E. Asia; Indo-C. Pacific; to 12 cm.

3 BROWN CORAL BLENNY
Atrosalarias fuscus (Rüppell, 1835)
Inhabits coral reefs; distinguished by blackish to brown body and relatively tall dorsal and anal fins with abruptly pale tail; an entirely yellow colour phase is sometimes seen; Great Barrier Reef, N.W. Australia, and throughout S.E. Asia; Indo-C. Pacific; to 10 cm.

4 MANY-SPOTTED BLENNY
Laiphognathus multimaculatus Smith, 1955
Inhabits coral reefs, distinguished by numerous small red spots and 1-2 rows of white spots just below base of dorsal fin; found throughout the region; W. Australia and S.E. Asia; W. Pacific and E. Indian Ocean; to 5.5 cm.

5 MIMIC BLENNY
Mimoblennius atrocinctus (Regan, 1909)
Inhabits coral reefs; distinguished by faint bars and row of double-spots along midside, underside of head is dusky to black; W. Australia and S.E. Asia; W. Pacific and E. Indian Ocean; to 4 cm.

6 TALBOT'S BLENNY
Stanulus talboti Springer, 1968
Inhabits coral reefs; distinguished by small white spots on cheek and lower part of head, larger white spots or blotches on body, and dark irregular blotches above pectoral fin; Great Barrier Reef, N.W. Australia, and throughout S.E. Asia; Indo-W. Pacific; to 7 cm

7 LEOPARD BLENNY
Exallias brevis (Kner, 1868)
Inhabits coral reefs, usually seen among branches of *Pocillopora* and other corals; feeds on coral polyps; distinguished by dense network of leopard-like spotting; Great Barrier Reef, N.W. Australia, and throughout S.E. Asia; Indo-C Pacific; to 12 cm.

8 SHORT-HEADED SABRETOOTH BLENNY
Petroscirtes breviceps (Valenciennes, 1836)
Inhabits weed-sand areas in the vicinity of reefs; distinguished by mottled colour, usually with dark stripe or broken band along middle of side; also known as Weed blenny; W. Australia and throughout S.E. Asia; Indo-W. Pacific; to 15 cm.

9 HIGH-FINNED BLENNY
Petroscirtes mitratus Rüppell, 1830
Inhabits weed-sand areas in the vicinity of reefs; distinguished by tall "mast" at front of dorsal fin, juvenile similar to **8** above, but slightly deeper-bodied; Great Barrier Reef, W. Australia, and throughout S.E. Asia; Indo-W Pacific; to 7 cm.

10 BLACK-BANDED BLENNY
Meiacanthus grammistes (Valenciennes, 1836)
Inhabits sand-weed areas and coral reefs; distinguished by black stripes that become broken into spots on rear of body and tail; Great Barrier Reef, N.W. Australia, and throughout S.E. Asia; mainly W. Pacific; to 10 cm.

11 GERMAIN'S BLENNY
Omobranchus germaini (Sauvage, 1883)
Inhabits shallow reefs, usually in crevices just below the level of low tide; distinguished by "ear" spot and series of narrow white to blue lines on side; Great Barrier Reef, N.W. Australia, and S.E. Asia; to 8 cm.

12 ROTUND BLENNY
Omobranchus ferox (Herre, 1927)
Inhabits mangrove bays or estuaries; distinguished by semi-transparent body and white streak behind eye; N.W. Australia and S.E. Asia; Indian Ocean and Indo-Australian Archipelago; to 7 cm.

13 ROUND HEADED BLENNY
Omobranchus lineolatus (Kner, 1868)
Inhabits mangrove bays and estuaries; distinguished by "ear" spot, pronounced dark bars on head, and fainter irregular bars on side; found throughout the region; N. Australia and S. New Guinea; to 9 cm.

14 MUZZLED BLENNY
Omobranchus punctatus (Valenciennes, 1836)
Inhabits inshore reefs and estuaries; distinguished by vertical bars on head and broken thin stripes on side; Great Barrier Reef, W. Australia, and S.E. Asia; Indo-W. Pacific; to 9 cm.

15 YELLOW SABRETOOTH BLENNY
Plagiotremus tapeinosoma (Bleeker, 1857)
Inhabits coral and rocky reefs; distinguished by broad black stripe which becomes broken (or has wavy margin) towards rear of body, and broad dark margins on dorsal and anal fins; this fish and **16** below have pair of enlarged fangs and sometimes bite divers; Great Barrier Reef, N.W. Australia, and throughout S.E. Asia; Indo-C. Pacific; to 13cm.

16 BLUE-LINED SABRETOOTH BLENNY
Plagiotremus rhinorhynchos (Bleeker, 1852)
Inhabits coral and rocky reefs, distinguished by narrower blue stripes on each side of broad brownish to orange stripe, and has pale fins; Great Barrier Reef, N.W. Australia, and throughout S.E. Asia; Indo-C. Pacific; to 12 cm.

17 HAIR-TAIL BLENNY
Xiphasia setifer Swainson, 1839
Inhabits sand-silt bottoms in the vicinity of coral reefs, occupies a burrow; distinguished by extremely elongate, tapering body; Great Barrier Reef, N.W. Australia, and throughout S.E. Asia; Indo-W. Pacific; to 56 cm

---

FANGED FISHES

Blennies of the family Blennidae (Plates 84-86) are small, elongate, scaleless fishes that are common on shallow reefs, mainly in tropical seas. Most of the estimated 300 species occurring worldwide are under 10-15 cm. They feed mainly on small invertebrates, algae, and bottom detritus. Several of the species shown on Plate 84 (1-3, 8-10,15-17) are commonly referred to as saber-toothed blennies. The name refers to the large teeth of the lower jaw which appear to be used mainly for defense. Scale and fin predation is a characteristic feeding habit of *Aspidontus* (1-2) and *Plagiotremus* (15-16).

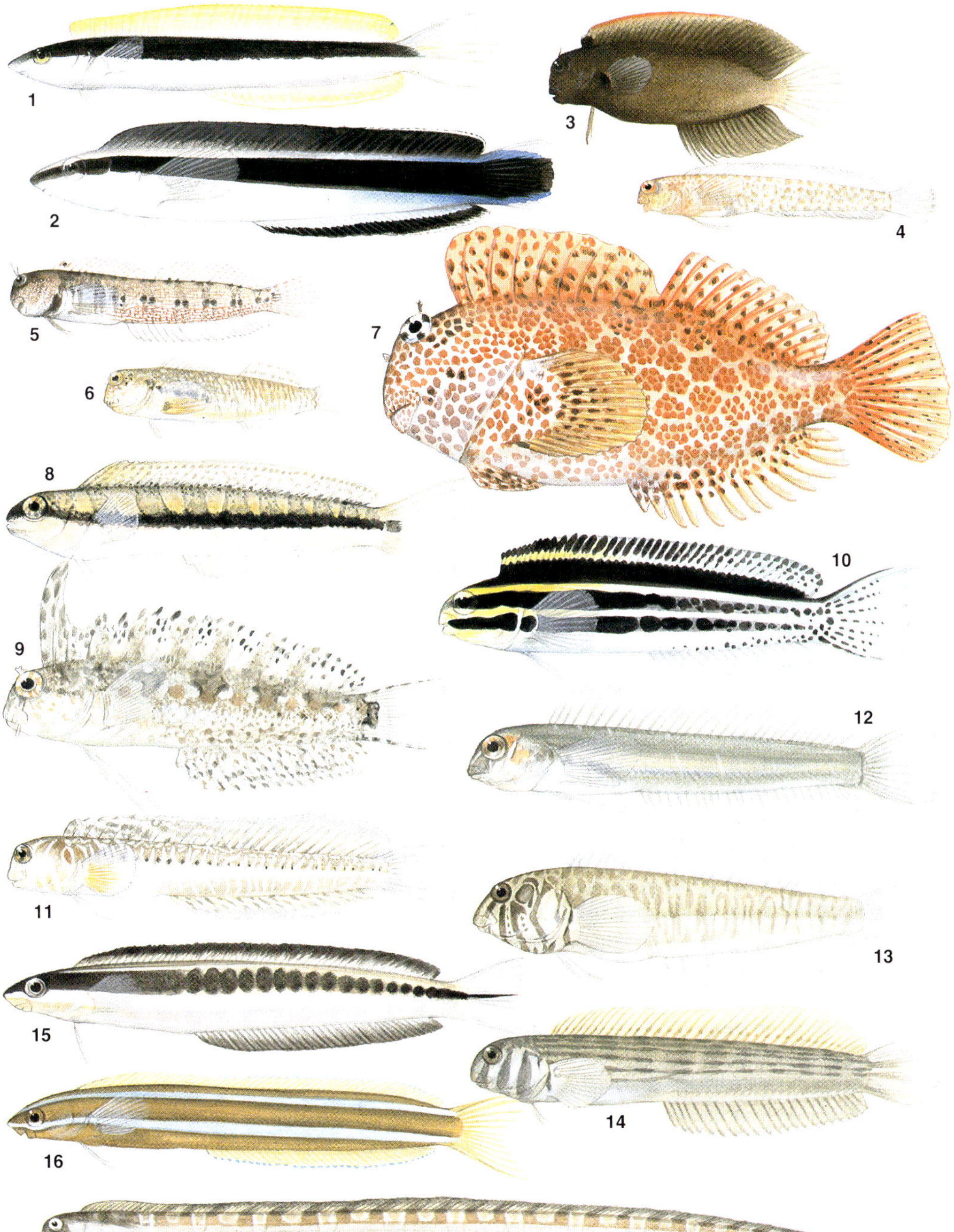
1
2
3
4
5
6
7
8
9
10
11
12
13
14
15
16
17

PLATE 85: BLENNIES (FAMILY BLENNIIDAE)

1 DUSKY BLENNY
Cirripectes filamentosus (Alleyne & Macleay, 1877)
Inhabits inshore reef crevices, frequently where there is wave action; distinguished by fringe of tentacles on neck, overall dusky colour and elevated fin rays at front of dorsal fin; found throughout the region; Indo-W. Pacific; to 10 cm.

2 SPOTTED-CHIN BLENNY
Cirripectes castaneus (Valenciennes, 1836)
Inhabits inshore reef crevices exposed to surge; distinguished by fringe of tentacles on neck, pale spots on chin, and series of narrow dark bars on side (sometimes barely visible); found throughout the region; Indo-W. Pacific; to 13 cm.

3 BICOLOR BLENNY
Ecsenius bicolor (Day, 1888)
Inhabits coral reef crevices; 2 colour varieties shown here are common: one is bright orange posteriorly and the other is overall dark brown; C. Indian Ocean to W. Pacific; to 8 cm.

4 LINED BLENNY
Ecsenius lineatus Klausewitz, 1962
Inhabits coral reefs, distinguished by black stripe on upper side with brown area above and abruptly whitish on lower half; W. Australia, W. Indonesia, and Philippines; Indo-W. Pacific; to 7 cm.

5 OCULAR BLENNY
Ecsenius oculatus Springer, 1988
Inhabits coral reef crevices distinguished by series of pale-edged black spots connected by pale lines on side; W. Australia and Christmas Island (Indian Ocean); mainly W. Pacific; to 6.5 cm. *E. paroculus* is a similar species occuring in W. Indonesia and W. coast of Malay peninsula.

6 CORAL BLENNY
Ecsenius yaeyamensis (Aoyagi, 1954)
Inhabits coral reef crevices; distinguished by dark stripe behind eye, pale spots on side and pair of thin lines across base of pectoral fin; throughout S.E. Asia and N.W. Australia to Gulf of Carpentaria; mainly W. Pacific; to 6.5 cm. *E. stictus*, a nearly identical species, occurs on the Great Barrier Reef.

7 TWIN SPOTS BLENNY
Entomacrodus thalassinus (Jordan & Seale, 1906)
Inhabits shallow coral reefs; distinguished by dark streak behind eye, row of large spots arranged in pairs along middle of side and spotted dorsal and tail fins; found throughout the region; Indo-W. Pacific; to 6.5 cm.

8 WAVY-LINED BLENNY
Entomacrodus decussatus (Bleeker, 1858)
Inhabits shallow reefs exposed to wave action; distinguished by wavy lines and irregular light and dark blotches on sides; Great Barrier Reef, N.W. Australia and throughout S.E. Asia; Indo-C. Pacific; to 18 cm.

9 STRIATED BLENNY
Entomacrodus striatus (Quoy & Gaimard, 1836)
Inhabits shallow reefs exposed to wave action; distinguished by clusters of dark spots on back superimposed on 4-5 broad diffuse dark bars; Great Barrier Reef, N.W. Australia, and throughout S.E. Asia; Indo-C. Pacific; to 10 cm.

10 RED-SPOTTED BLENNY
Blenniella chrysospilos (Bleeker, 1857)
Inhabits shallow reefs and tide pools exposed to wave action; distinguished by small red spots on head and body, and series of dark vertical streaks arranged in pairs on side; Great Barrier Reef, N.W. Australia, and throughout S.E. Asia; Indo-C. Pacific; to 13 cm.

11 RIPPLED BLENNY
Istiblennius edentulus (Schneider, 1801)
Inhabits shallow reefs and tide pools exposed to wave action; male has skin flap on top of head, and dark bars (usually in pairs) and pale streaks on side; female has lighter bars on sides, spots or lines on dorsal and anal fins, and often with numerous red to brown spots on rear part of body; Great Barrier Reef, N.W. Australia, and throughout S.E. Asia; Indo-C. Pacific; to 13 cm.

12 SPOTTED BLENNY
Istiblennius meleagris (Valenciennes, 1836)
Inhabits shallow reefs and tide pools exposed to wave action; male with low skin flap on top of head and rows of pale spots on side, female lacks skin flap and has faint forward-slanting bars and scattered pale spots on side; Australia only, occurs around entire coastline, except extreme south; to 15 cm.

13 TIDEPOOL BLENNY
Istiblennius lineatus (Valenciennes, 1836)
Inhabits shallow reefs and tide pools exposed to wave action; distinguished by skin flap on head and general pale colouration with narrow longitudmal lines on side; Great Barrier Reef, N.W. Australia, and throughout S.E. Asia; C. Indian Ocean to C. Pacific; to 20 cm.

14 BLUE-STREAKED BLENNY
Blenniella periophthalmus (Valenciennes, 1836)
Inhabits shallow reef and tide pools exposed to wave action; distinguished by H-shaped bars on side and double row of silvery-blue streaks or spots on middle of side; found throughout the region; Indo-W. Pacific; to 14 cm. *B. paula*, a nearly identical species, occurs on the Great Barrier Reef.

15 BANDED BLENNY
Salarias fasciatus (Bloch, 1786)
Inhabits inshore coral reefs, often in weeds; distinguished by light and dark bars with overlay of narrow dark lines; Great Barrier Reef, N.W. Australia, and throughout S.E. Asia; Indo-W. Pacific; to 13 cm.

16 STARRY BLENNY
Salarias ramosus Bath, 1992
Inhabits sand-weed areas on rocky outcrops; distinguished by frilly tentacles above eye and numerous small white spots; N.W. Australia, Indonesia, and Philippines; Indo-Australian Archipelago; to 9 cm.

17 SPALDING'S BLENNY
Salarias spaldingi Macleay, 1878
Inhabits coral reefs; distinguished by narrow dark bars on side and "pepper-spotting" on upper side; N.Australia only, between Exmouth Gulf, Western Australia and Gulf of Carpentaria; to 9 cm.

TIDEPOOL BLENNIES

Most of the blennies shown on this plate, particularly 7-17, are common inhabitants of shallow reef flats, the intertidal zone and splash pools along rocky shores. They are one of the most frequently observed groups of fishes encountered by beachcombers. A few of the species are commonly called rock-skippers because when disturbed they use their body musculature and stout pelvic fins to skip over rocks from one pool to another.

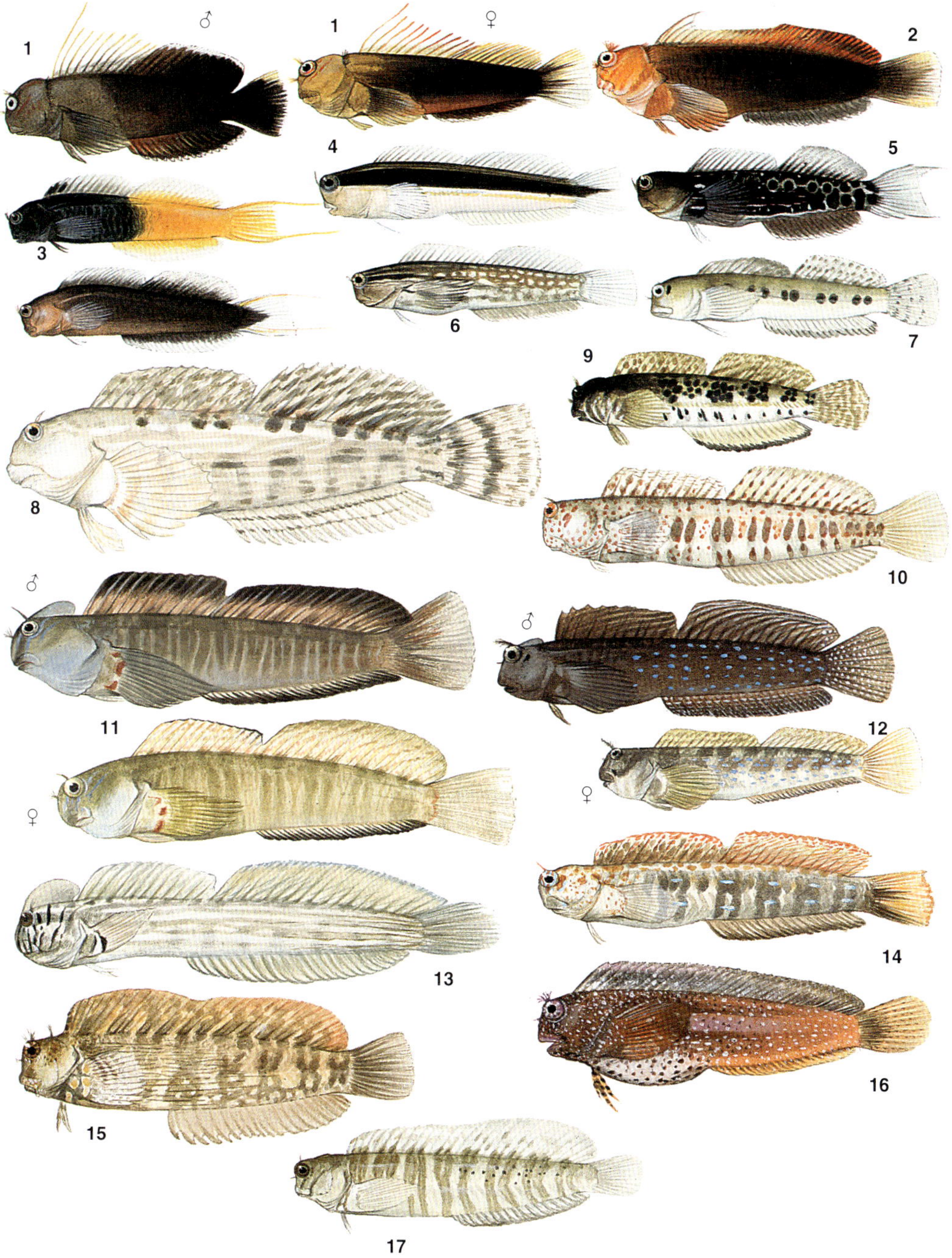
1 ♂
1 ♀
2
3
4
5
6
7
8
9
10
♂
11
♀
♂
12
♀
13
14
15
16
17

PLATE 86: BLENNIES (FAMILY BLENNIIDAE)

1 YELLOWTAIL FANGBLENNY Cairns 1•/98
Meiacanthus atrodorsalis (Günther, 1877)
Inhabits coral reefs, swims close to bottom with rapid darting motion interspersed with stationary hovering; nearly identical to **5** below, but is not as slender and has diagonal dark band through eye; Great Barrier Reef, offshore reefs of W. Australia, and E. Indonesia; mainly W. Pacific; to 11 cm.

2 DOUBLEPORE FANGBLENNY
Meiacanthus ditrema Smith-Vaniz, 1976
Inhabits coral reefs, behaviour as for **1** above; distinguished by pair of black stripes on lower part of head that join into one stripe at pectoral fin base; Great Barrier Reef and Indonesia; W. Pacific; to 6.5 cm.

3 SMITH'S FANGBLENNY
Meiacanthus smithi Klausewitz, 1961
Inhabits coral reefs, behaviour as for **1** above; distinguished by broad black stripe on dorsal fin and oblique dark band above eye; W. Indonesia and W. coast of Malay Peninsula; C. Indian Ocean to Java; Great Barrier Reef and Indonesia; W. Pacific; to 7 cm.

4 YELLOW FANGBLENNY
Meiacanthus luteus Smith-Vaniz, 1987
Inhabits coral reefs, behaviour as for **1** above; distinguished by broad black stripe on middle of side and much narrower one at base of dorsal fin with bright yellow between; is mimicked by *Petroscirtes fallax* (not shown); N. Australia only; to 10 cm.

5 BICOLOR FANGBLENNY Cairns
Plagiotremus laudanus (Whitley, 1961) 10/98
Inhabits coral reefs; it mimics the nearly identical **1** above, but is more slender and lacks diagonal dark band through eye; Great Barrier Reef, offshore reefs of W. Australia, and E. Indonesia; mainly W. Pacific; to 7.5 cm.

6 RETICULATED BLENNY
Cirripectes stigmaticus Strasburg & Schultz, 1953
Inhabits coral reefs; distinguished by scarlet (male) to rust-coloured (female) spots and lines on side; Great Barrier Reef, offshore reefs of W. Australia, E. Indonesia, and Melanesia; Indo-W. Pacific; to 12 cm.

7 KIMBERLEY BLENNY
Cirripectes alleni Williams, 1993
Inhabits coral reefs in turbid water; distinguished by frill of skin flaps on nape and whitish colour with broad dark stripe on middle of side; Kimberley district of N.W. Australia only; to 6.5 cm.

8 TRIPLESPOT BLENNY
Crossosalarias macrospilus Smith-Vaniz & Springer, 1971
Inhabits coral reefs, usually in less than 10 m depth; distinguished by mottled, pale-spotted pattern, low fleshy flap in front of dorsal fin, and pair of dark spots on each side of throat; Great Barrier Reef and S.E. Asia; mainly W. Pacific; to 8 cm.

9 MIDAS BLENNY
Ecsenius midas Starck, 1969
Inhabits coral reefs, usually on outer slopes; mimics *Pseudanthias squamipinnis* (Pl. 27-4); colour variable, usually orange-yellow or dark slate blue, also has small dark spot in front of anus; Great Barrier Reef and throughout S.E. Asia; Indo-W. Pacific; to 13 cm.

10 ALLEN'S BLENNY
Ecsenius alleni Springer, 1988
Inhabits coral reefs; distinguished by black spot on pectoral-fin base and combination of white botches and black bars on side; offshore reefs of W. Australia only; to 4 cm.

11 AUSTRALIAN BLENNY
Ecsenius australianus Springer, 1988
Inhabits coral reefs; distinguished by black stripe behind eye and double row of rectangular white spots on side; N. Great Barrier Reef only; to 4 cm.

12 BATH'S BLENNY
Ecsenius bathi Springer, 1988
Inhabits coral reefs; has two distinct patterns: one with double stripes shown here and another that is similar to **11** above; E. Indonesia only; to 4 cm.

13 AXELROD'S BLENNY
Ecsenius axelrodi Springer, 1988
Inhabits coral reefs; distinguished by black "ear" spot, white stripe on middle of sides crossed by 6-7 black bars; *E. tigris* from the Coral Sea and *E. dilemma* from the Philippines are similar, but lack "ear" spot; Sulawesi to Solomon Islands; to 4 cm.

14 BANDA BLENNY
Ecsenius bandanus Springer, 1971
Inhabits coral reefs; distinguished by pearly stripe with black stripe immediately above between eye and upper edge of gill cover, with brown colour above and light colour below stripes; Indonesia only, between Java and N. Irian Jaya; to 4 cm.

15 BLACKASS BLENNY
Ecsenius lividanalis Chapman & Schultz, 1952
Inhabits coral reefs; entirely yellow to mainly slate grey with yellow fins, all varieties with black spot covering anus; N.W. Australia (Kimberley district to Darwin) and Bali, Indonesia to Solomon Islands; to 4 cm.

16 YELLOWEYE BLENNY
Ecsenius melarchus McKinney & Springer, 1971
Inhabits coral reefs; distinguished by bluish colour on lower part of head, small black spot behind eye, and black spot covering anus; 2 varieties seen: "bicolour" form as shown and one which is uniformly brown on body; E. Indonesia and Philippines; to 4 cm.

17 NAMIYE'S BLENNY
Ecsenius namiyei (Jordan & Evermann, 1903)
Inhabits coral reefs; head and body dark brown to blackish; tail either greyish white or bright yellow; Philippines, E. Indonesia, and offshore reefs of N.W. Australia; W. Pacific; to 9 cm.

18 WHITE-LINED BLENNY
Ecsenius pictus McKinney & Springer, 1976
Inhabits coral reefs; distinguished by dark brown body with thin white lines and spots; E. Indonesia to Solomon Islands; to 4.5 cm.

19 TWINSPOT BLENNY
Salarias segmentatus Bath & Randall, 1991
Inhabits sheltered coral reefs, often in silty areas; distinguished by broken brown bars on side and large dark-rimmed pale spot below pectoral-fin base; E. Indonesia to Solomon Islands; to 7.5 cm.

20 PATZNER'S BLENNY
Salarias patzneri Bath, 1992. Text on page 212.

21 AMBON BLENNY
Paralticus amboinensis (Bleeker, 1857). Text on page 212.

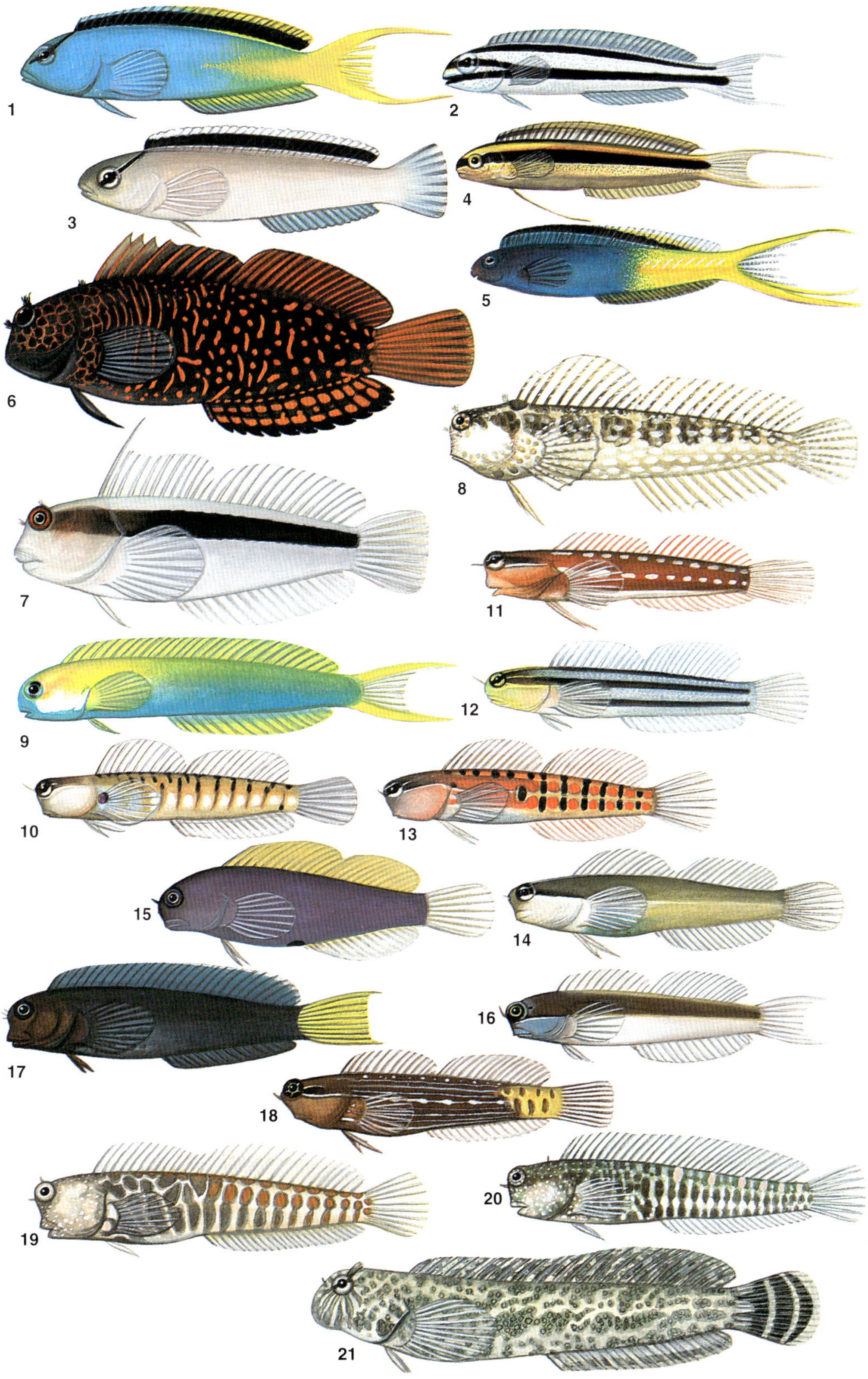
1
2
3
4
5
6
7
8
9
10
11
12
13
14
15
16
17
18
19
20
21

PLATE 87: THREEFINS (TRIPTERYGIIDAE) AND DRAGONETS (CALLIONYMIDAE)

1 BLACK-THROATED THREEFIN
Helcogramma decurrens McCulloch & Waite, 1918
Inhabits inshore reefs; male distinguished by black on lower half of head and body, female is variably coloured, either red, green, or brown; Australia only, southward of N.W. Cape, W. Australia, ranging to S.Australia; to 7 cm.

2 NEON THREEFIN
Helcogramma striata Hansen, 1986
Inhabits coral reef crevices; distinguished by reddish colour with 3 narrow golden stripes on side; found throughout the region; E. Indian Ocean to C. Pacific; to 3.5 cm.

3 SCALY-HEAD THREEFIN
Norfolkia brachylepis (Schultz, 1960).
Inhabits inshore coral reefs; distinguished by presence of scales on cheek and gill cover, a low fleshy flap above eye, and diagonal cross bands on anal fin; N.W. Australia, Indonesia, and Philippines; Indo-W. Pacific; to 6 cm.

4 GOODLAD'S STINKFISH
Callionymus goodladi Whitley, 1944
Inhabits trawling grounds, generally on sand bottoms; distinguished by overall pale colouration with brown mottling on back and spotted tail, male has 2 filamentous spines at front of first dorsal fin and female (not shown) has one filament; W.Australia only; to 22 cm.

5 QUEENSLAND STINKFISH
Callionymus moretonensis Johnson, 1971
Inhabits trawling grounds, generally on sand bottoms; similar to **6** and **7** below, but has shorter tail and stout spine on lower edge of cheek has 2 large teeth on its upper surface (versus many small teeth); N. Australia, New Caledonia and New Ireland; to 12 cm.

6 AUSTRALIAN STINKFISH
Callionymus margaretae australis Fricke, 1983
Inhabits trawling grounds, generally on sand bottoms; similar to **5** above, but has 5-7 small teeth on upper surface of cheek spine (versus 2 large teeth); also resembles **7** below, but has only outer tips of anal fin rays blackish (versus outer half of entire fin); N.W. Australia only, but another subspecies *C. margaretae margaretae* widespread in N. Indian Ocean; to 18 cm.

7 JAPANESE STINKFISH
Callionymus japonicus Houttyn, 1782
Inhabits trawling grounds, generally on sand bottoms; similar to **5** and **6** above, but has wider black margin on anal fin and tail often wider and more elongate; N. Australia, New Guinea, and Philippines; mainly W. Pacific: to 30 cm.

8 GROSS'S STINKFISH
Callionymus grossi Ogilby, 1910
Inhabits trawling grounds, generally on sand bottoms; similar to **4** above, but has larger, differently shaped dorsal fin; N. Australia only; to 20 cm.

9 ROSY DRAGONET
Synchiropus altivelis (Temminck & Schlegel, 1850)
Inhabits trawling grounds in deep (70-600 m) water; distinguished by overall reddish or pink colour and relatively short first dorsal fin (sometimes with filament); N.W. Australia and Philippines to Japan; to 20 cm.

10 MORRISON'S DRAGONET
Synchiropus morrisoni Schultz, 1960
Inhabits coral reefs and rocky areas usually below 25 m depth; male distinguished by tall flat-topped first dorsal fin with dark vertical streaks and broad dark submarginal band on anal fin, female by large dark blotch covering pectoral fin base and diagonal dark streaks on anal fin; Great Barrier Reef, N.W. Australia and Indonesia; mainly W. Pacific; to 6.5 cm.

11 HIGH-FINNED DRAGONET
Synchiropus rameus (McCulloch, 1926)
Inhabits trawling grounds, generally on sand or rubble bottoms; distinguished by huge sail-like dorsal fin and blue-spotted anal fin; N. Australia and New Caledonia; to 15 cm.

12 WESTERN DRAGONET
Synchiropus picturatus occidentalis Fricke, 1983
Inhabits coral reefs; distinguished by robust shape and ornate pattern of ocellated spots and blotches; known only from N.W. Australia, another subspecies *S. picturatus picturatus* occurs in the Indo-Malay Archipelago; to 5.5 cm.

13 NORTHERN DRAGONET
Diplogrammus xenicus (Jordan & Thompson, 1914)
Inhabits sand-rubble areas near reefs; male distinguished by relatively short filament on first dorsal fin, blue spots and lines on head, dusky anal fin and fan-like, dark-edged pelvic fins, female differs from male in size, shape, and colour of first doral and pelvic fins, also the anal fin is clear instead of dusky; W. Australia and Okinawa to Japan; to 7 cm.

14 FINGERED DRAGONET
Dactylopus dactylopus (Valenciennes, 1837)
Inhabits sand-weed bottoms; distinguished by detached first ray on large fan-like pelvic fin; Great Barrier Reef, W. Australia, and S.E. Asia; E. Indian Ocean and W. Pacific; to 30 cm.

PLATE 86

20 PATZNER'S BLENNY
Salarias patzneri Bath, 1992
Inhabits coral reefs; distinguished by mottled pattern with numerous small white spots on head and body; Indonesia and Philippines; to 5.5 cm.

21 AMBON BLENNY
Paralticus amboinensis (Bleeker, 1857)
Inhabits protected shores, sometimes among mangroves; distinguished by strongly mottled pattern and bushy flap above each eye; Sabah and E. Indonesia; to 13 cm.

TRIPLEFINS AND STINKFISHES

The Threefins of the family Tripterygiidae (1-3), as their name suggests, are characterised by three separate dorsal fins. They are mainly tiny fishes with well over 100 species occurring worldwide. Due to their small size and numerous species the identification is usually difficult. They are bottom dwellers found on the surface of rocks or corals.

Stinkfishes of the family Callionymidae (4-14) occur in tropical and temperate seas where they generally are found on sand or mud bottoms in the vicinity of reefs. Some species occur on the continental shelf and slope to depths of about 900 m. Nearly all of the estimated 135 species are small in size (usually under 20 cm TL) and most are inhabitants of the Indo-Pacfic region.

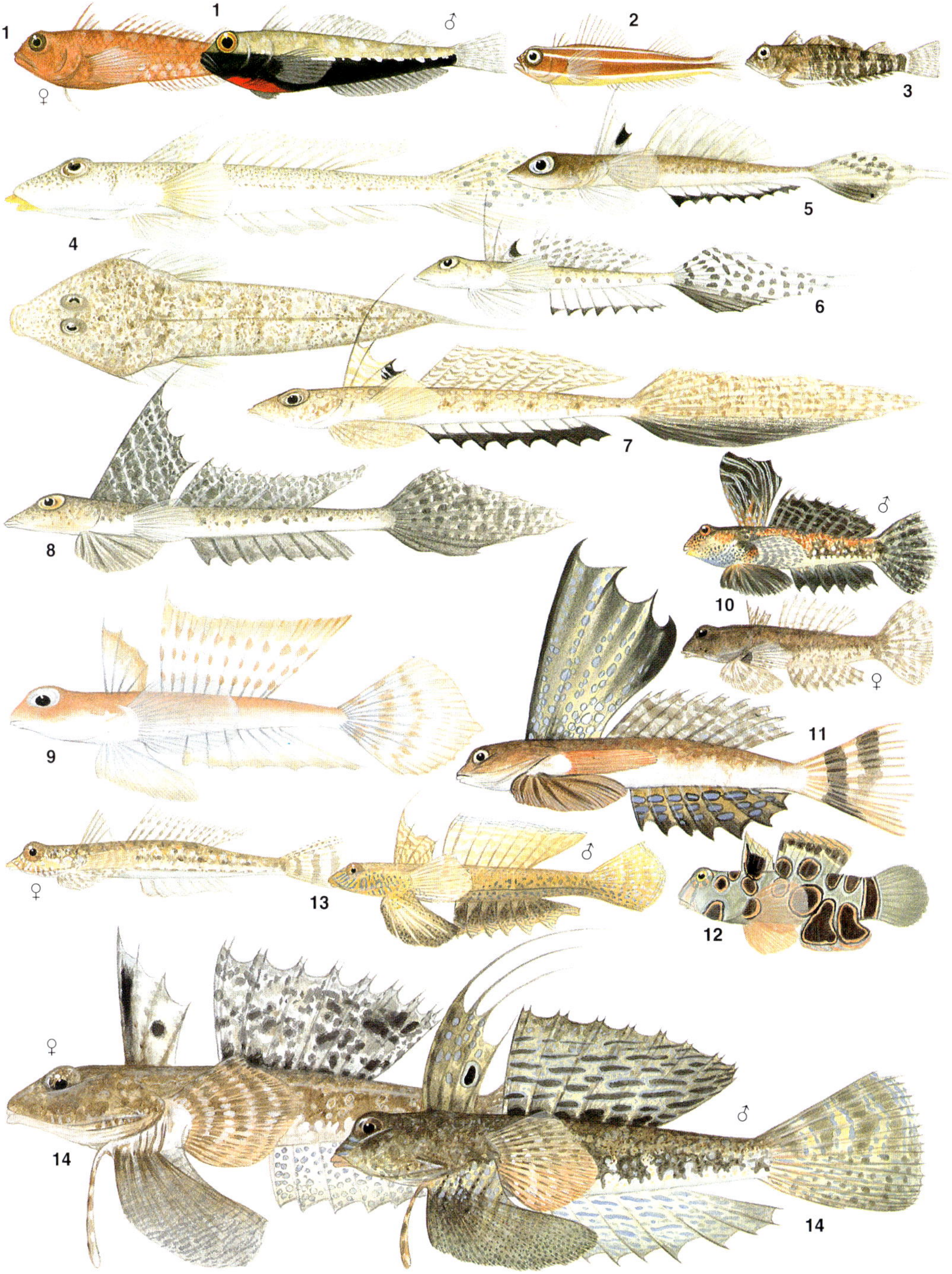
1
♀
1
♂
2
3
4
5
6
7
8
10
♂
9
11
♀
13
♂
♀
12
♀
14
♂
14

PLATE 88: DRAGONETS (CALLIONYMIDAE) AND GOBIES (GOBIIDAE)

1 MANDARINFISH
Synchiropus splendidus (Herre, 1927)
Inhabits shallow reefs; distinguished by green "maze" pattern on orange or orange-brown background; Great Barrier Reef, offshore reefs of N.W. Australia, and S.E. Asia; W. Pacific; to 6 cm.

2 OCELLATED DRAGONET
Synchiropus ocellatus (Pallas, 1770)
Inhabits rubble bottoms; distinguished by blue markings on head and sail-like first dorsal fin; N. Great Barrier Reef, E. Indonesia, and Philippines; W. and C. Pacific; to 7 cm.

3 WEEDY DRAGONET
Anaora tentaculata Gray, 1835
Inhabits weedy areas; has excellent camouflage colour, distinguished by numerous small skin flaps on body; Borneo and Philippines to New Guinea; to 5.5 cm.

4 SPOTTED SHRIMPGOBY
Amblyeleotris guttata (Fowler, 1938)
Inhabits sand-rubble; shares burrow with alpheid shrimp; distinguished by orange spots on head and body, and black pelvics and belly region; Great Barrier Reef and S.E. Asia; W. Pacific; to 8 cm.

5 STEINITZ'S SHRIMPGOBY
Amblyeleotris steinitzi (Klausewitz, 1974)
Inhabits sand-rubble fringe of coral reefs; shares burrow with alpheid shrimp; distinguished by whitish body and five dull brown bars; Great Barrier Reef, offshore reefs of N.W. Australia, and S.E. Asia; Indo-W. Pacific; to 8 cm.

6 PERIOPHTHALMA SHRIMPGOBY
Amblyeleotris periophthalma (Bleeker, 1853)
Inhabits sand-rubble; shares burrow with alpheid shrimp; distinguished by diffuse brown bars and horizontally elongate blotches on side; Great Barrier Reef and S.E. Asia; W. Pacific; to 8 cm.

7 RANDALL'S SHRIMPGOBY
Amblyeleotris randalli Hoese & Steene, 1978
Inhabits ledges on outer reef slopes; shares burrow with alpheid shrimp; distinguished by thin yellow-orange bars and large first dorsal fin; Great Barrier Reef, E. Indonesia, Phillipines, New Guinea, and Solomon Islands; to 8 cm.

8 WIDE-BARRED SHRIMPGOBY
Amblyeleotris latifasciata Polunin & Lubbock, 1979
Inhabits sand-rubble; shares burrow with alpheid shrimp; distinguished by broad brown bars, blue spotting on head, and large pale-rimmed dark spots on first dorsal fin; Indonesia; to 10 cm.

9 YELLOW SHRIMPGOBY - (LIGHT & DARK)
Cryptocentrus cinctus (Herre, 1936)
Inhabits sand-rubble fringe of coral reefs; shares burrow with alpheid shrimp; has two colour phases - one primarily yellow, and another that is whitish with brown bars on side; Great Barrier Reef, offshore reefs of N.W. Australia, and S.E. Asia; Indo-W. Pacific; to 8 cm.

10 FLAGTAIL SHRIMPGOBY
Amblyeleotris yanoi Aonuma & Yoshino, 1996
Inhabits sand-rubble fringe of coral reefs; shares burrow with alpheid shrimp; distinguished by brightly marked tail fin; Indonesia to Malanesian Archipelago; W. Pacific; to 8 cm.

11 SADDLED SHRIMPGOBY
Cryptocentrus leucostictus (Günther, 1871)
Inhabits sand-rubble; shares burrow with alpheid shrimp; distinguished by overall dark colour, white snout, and white saddles on back; Great Barrier Reef and S.E. Asia; W. Pacific; to 7 cm.

12 TARGET SHRIMPGOBY
Cryptocentrus strigilliceps (Jordan & Seale, 1906)
Inhabits sand-rubble; shares burrow with alpheid shrimp; distinguished by large black spots along middle of side; Great Barrier Reef, offshore reefs of N.W. Australia, and throughout S.E. Asia; Indo-C. Pacific; to 7 cm.

13 WHITECAP GOBY
Lotilia graciliosa Klausewitz, 1960
Inhabits sand-rubble fringe of coral reefs; shares burrow with shrimp, distinguished by black body with broad white band on top of head; Great Barrier Reef and throughout S.E. Asia; Indo-W. Pacific; to 4.5 cm.

14 MASTED SHRIMPGOBY
Ctenogobiops tangaroae Lubbock & Polunin, 1977
Inhabits sandy areas; pairs share burrow with alpheid shrimp; distinguished by very tall first dorsal fin; Great Barrier Reef, offshore reefs of N.W. Australia, and throughout S.E. Asia; W. Pacific; to 5.5 cm.

15 GOLD-STREAKED SHRIMPGOBY
Ctenogobius aurocingulus (Herre, 1935)
Inhabits sand-rubble; shares burrow with shrimp; distinguished by yellow-centred dark markings on head and vertical orange bands on lower side; Philippines, E. Indonesia, and New Guinea; W. Pacific; to 6 cm.

16 TWINSPOT GOBY
Signigobius biocellatus Hoese & Allen, 1977
Inhabits sandy fringe of coral reefs; Great Barrier Reef, offshore reefs of N.W. Australia, and throughout S.E. Asia; Indo-Australian Archipelago; to 6.5 cm.

17 OLD GLORY GOBY
Amblygobius rainfordi (Whitley, 1940)
Inhabits caves and holes with sand-rubble bottom at base of coral reefs; similar to 18 below, but is paler, and lacks dark spot on first dorsal fin; Great Barrier Reef, New Guinea, offshore reefs of N.W. Australia, E. Indonesia, and Philippines; mainly W. Pacific; to 6.5 cm.

18 HECTOR'S GOBY
Amblygobius hectori (Smith, 1956)
Inhabits coral reefs; similar to 17 above, but body is darker, and has pale-edged dark spot on first dorsal fin; Great Barrier Reef and throughout S.E. Asia; Indo-W. Pacific; to 5.5 cm.

19 NOCTURNA GOBY
Amblygobius nocturna (Herre, 1945)
Inhabits sheltered inshore reefs; is pale greyish with orange to pinkish-red stripes on upper half of body and greyish spots along base of dorsal fin; W. Pacific; to 8 cm.

20 ORANGE-STRIPED GOBY
Amblygobius decussatus (Bleeker, 1855)
Inhabits silty-sandy edge of sheltered reefs; pale-edged orange spot on tail base; Great Barrier Reef, offshore reefs of N.W. Australia, and throughout S.E. Asia; W. Pacific; to 8 cm.

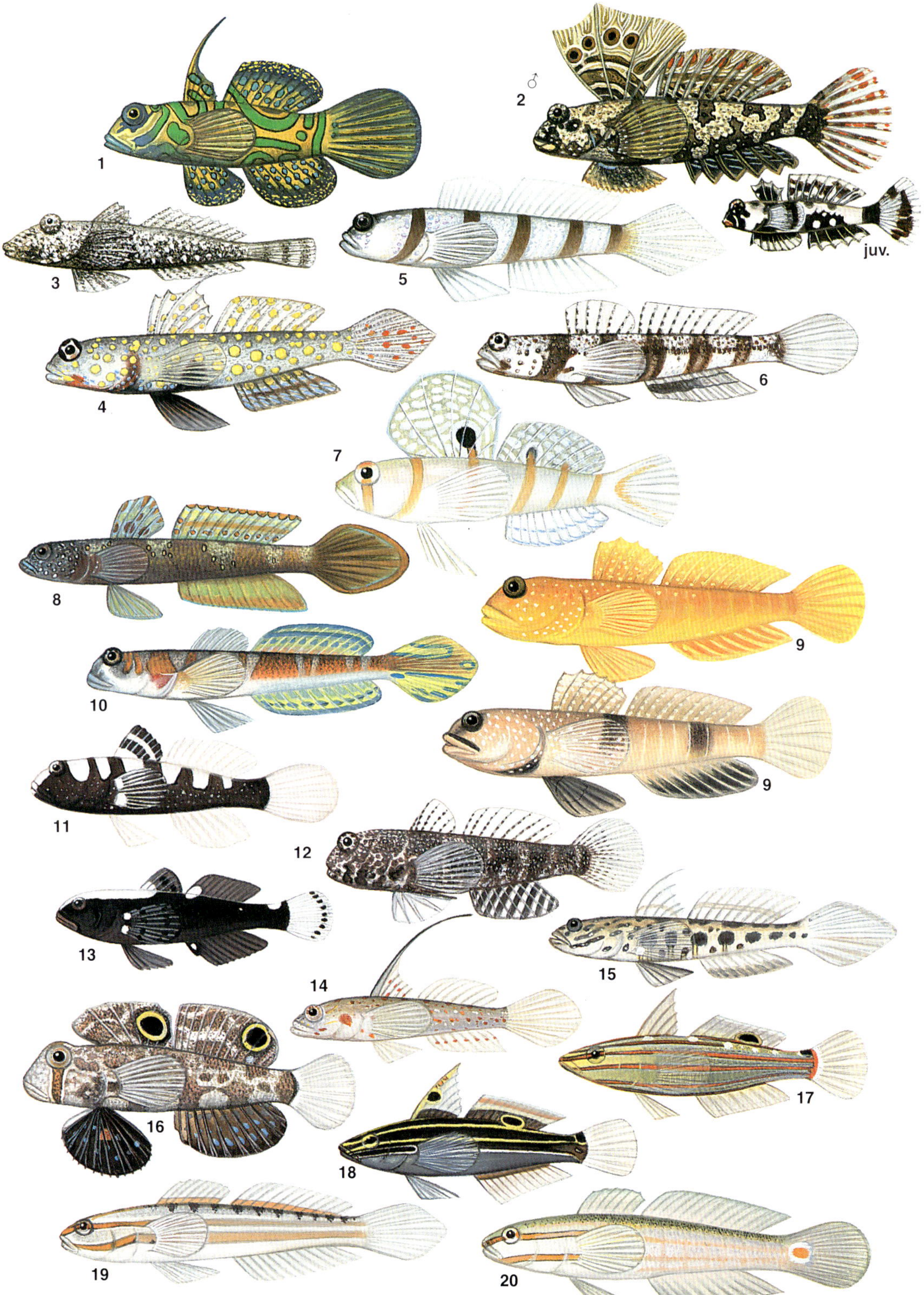

1
♂
2
juv.
3
5
4
6
7
8
9
10
9
11
12
13
15
14
16
17
18
19
20

PLATE 89: GOBIES (FAMILY GOBIIDAE)

1 MANGROVE GOBY
Acentrogobius gracilis (Bleeker, 1875)
Inhabits mangrove estuaries; distinguished by overall pale colour with midlateral stripe composed of brown blotches and bright blue spots, also brown spots and blotches on back and fins; found throughout the region; Indo-Australian Archipelago; to 8 cm.

2 WHEELER'S SHRIMPGOBY
Amblyeleotris wheeleri (Polunin & Lubbock, 1977)
Inhabits sand bottoms near coral reefs, shares its burrow with an alpheid shrimp; distinguished by combination of wine-red bars and small bluish spots; Great Barrier Reef, N.W. Australia, and throughout S.E. Asia; Indo-W. Pacific; to 6 cm.

3 STARRY GOBY
Asterropteryx semipunctatus Rüppell, 1830
Inhabits rubble bottoms; distinglushed by mottled green-brown colour with numerous small blue spots on head, body and fins, also has microscopic spines on edge of cheek; found throughout the region; Indo-C. Pacific to 5 cm.

4 BANDED GOBY
Amblygobius phalaena (Valenciennes, 1837)
Inhabits sand-weed areas; distinguished by stripes that are interrupted by narrow dark bars on side, a large pale-edged dark spot on first dorsal fin, and smaller dark spot on upper part of tail; also known as White spotted goby and Barred goby; found throughout the region; Indo-W. Pacific; to 15 cm.

5 COCOS GOBY
Bathygobius cocosensis (Bleeker, 1854)
Inhabits shallow beach-rock reefs and tide pools; distinguished by diffuse dark saddles and row of elongated dark spots slightly below middle of side, also scales on top of head extend only to rear margin of cheek (preopercle); Great Barrier Reef, N.W. Australia, and throughout S.E. Asia; Indo-W. Pacific; to 8 cm.

6 COMMON GOBY
Bathygobius fuscus (Rüppell, 1831)
Inhabits shallow beach-rock reefs and tide pools distinguished by blunt snout, bulbous cheeks, and often with large irregular shaped blotches and small pale spots on head and side, scales on top of head extend almost to rear of eyes; found throughout the region; Indo-C. Pacific; to 12 cm.

7 WHISKERED GOBY
Callogobius sp.
Inhabits rubble areas and coral reef crevices; distinguished by rows of raised papillae ("whiskers") on head, easily shed scales, and blotchy colour pattern; possibly Australia only; to 6 cm.

8 DOUBLE-BARRED GOBY
Callogobius sclateri (Steindachner, 1880)
Inhabits coral reefs; distinguished by pair of dark bars, one below each dorsal fin, and rows of raised papillae ("whiskers") on head; Great Barrier Reef, N.W. Australia, and throughout S.E. Asia; Indo-C. Pacific; to 5 cm.

9 GREEN SHRIMPGOBY
Cryptocentrus octafasciatus Regan, 1908
Inhabits sand-rubble areas near coral reefs, shares its burrow with an alpheid shrimp; distinguished by broad green-brown bars with narrower pale areas between them, also blackish spot in middle of each dark bar; *C. caeruleomaculatus* is a synonym; N.W. Australia and throughout S.E. Asia; Indo-W. Pacific; to 5 cm.

10 PINK SHRIMPGOBY
Cryptocentrus leptocephalus Bleeker, 1876
Inhabits sandy areas near coral reefs, shares its burrow with an alpheid shrimp; distinguished by pink or red spots and stripes on head and dorsal fins, and oblique dark bars on body, also small white spots on head and side; *C. obliquus* and *C. singapurensis* are synonyms; N.W. Australia and throughout S.E. Asia; Indo-Australian Archipelago; to 10 cm.

11 SPOTFIN SHRIMPGOBY
Ctenogobiops pomasticus Lubbock & Polunin, 1977
Inhabits sandy areas near coral reefs, shares its burrow with an alpheid shrimp; distinguished by brown spots on head and body, and white blotch on lower edge of pectoral fins; Great Barrier Reef, N.W. Australia, and throughout S.E. Asia; E. Indian Ocean and W. Pacific; to 6 cm.

12 MUD GOBY
Drombus sp.
Inhabits coastal waters, including estuaries; distinguished by abruptly pale back and small white spots on head and side; N. Australia; to 5 cm.

13 RED CORAL GOBY
Eviota sp.
Inhabits coral reefs; distinguished by tiny size, semi-transparent appearance, red spots on head, black spot above gill cover, diffuse dark bars above anal fin, and dark spot on middle of tail base; possibly N.W. Australia only; to 2 cm.

14 SMITH'S CORAL GOBY
Eviota infulata (Smith, 1956)
Inhabits coral reefs; distinguished by tiny size, filamentous first and second dorsal spines (males) and black spot or blotch above pectoral fin base; Great Barrier Reef and W. Austalia; Indo-C. Pacific; to 2 cm.

15 TWOSPOT GOBY
Fusigobius duospilus Hoese & Reader, 1985
Inhabits sand bottoms adjacent to coral reefs; distinguished by semi-transparent appearance, 2 dark blotches on first dorsal fin, small brown spots on head and body, and black spot at middle of tail base; Great Barrier Reef, N.W. Australia, and throughout S.E. Asia; Indo-C. Pacific; to 6 cm.

16 ESTUARY GOBY
Glossogobius biocellatus (Valenciennes, 1837)
Inhabits brackish estuaries and tidal creeks; distinguished by dark blotches or bars on lower lobe of tail, also anal and pelvic fins and lower part of head frequently dark; found throughout the region; mainly W. Pacific: to 10 cm

17 CIRCUMSPECT GOBY
Glossogobius circumspectus (Macleay, 1883)
Inhabits brackish estuaries and tidal creeks; similar to **16** above, but has blunter snout, is generally lighter in colour, and lacks darkened areas on pelvic, anal, and tail fins (except narrow dark bands across tail); found throughout the region; Indo-Australian Archipelago; to 12 cm.

18 YELLOWSPOT GOBY
Gnatholepis scapulostigma Herre, 1953
Inhabits sand bottoms adjacent to coral reefs; distinguished by narrow dark stripes on side, thin bar below eye, and small orange spot above pectoral fin base; found throughout the region; Indo-Australian Archipelago; to 7 cm.

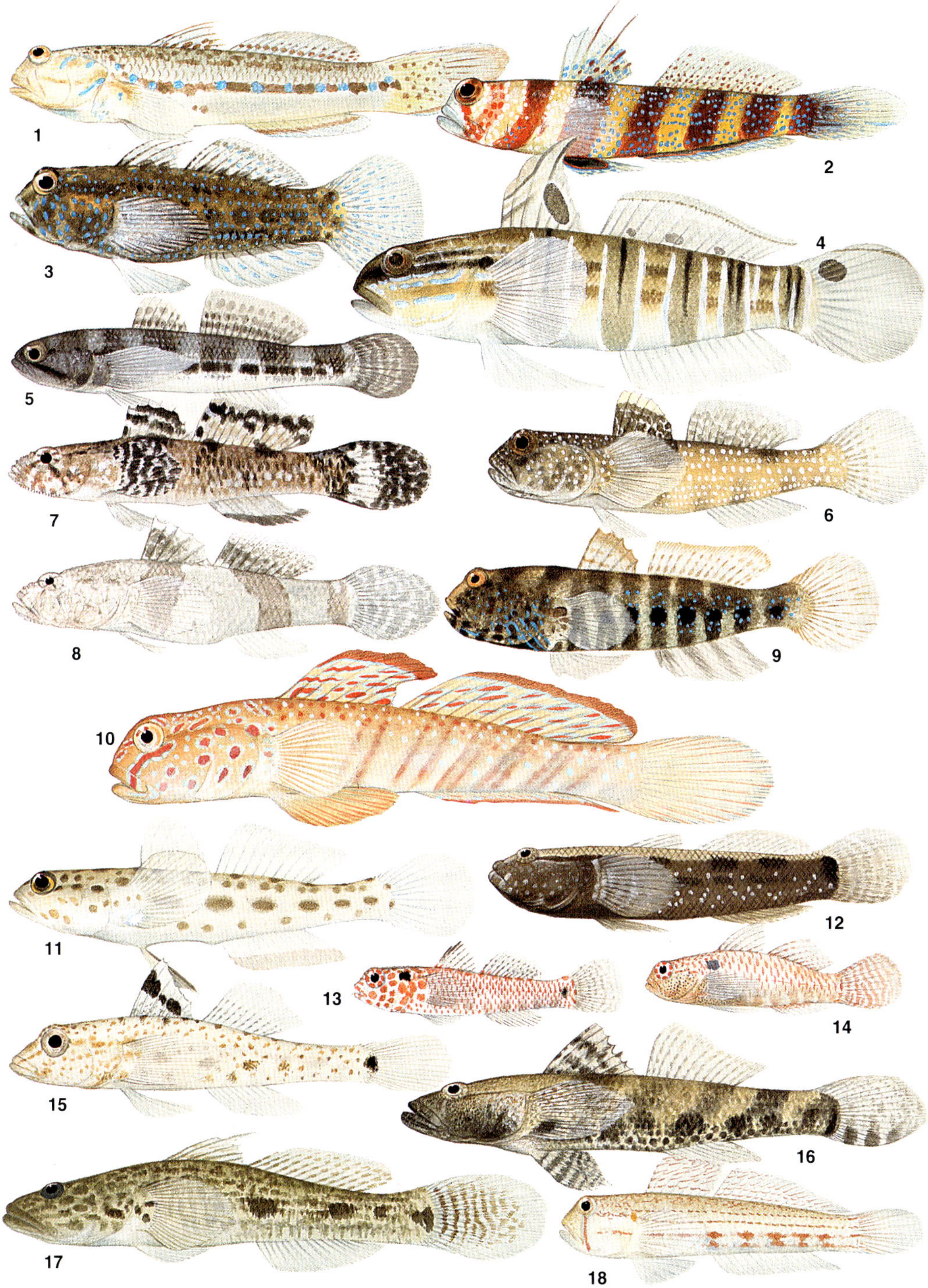
1
2
3
4
5
7
6
8
9
10
11
12
13
14
15
16
17
18

PLATE 90: GOBIES (FAMILY GOBIIDAE)

1 BROAD-BARRED MAORI GOBY
Gobiodon histrio (Valenciennes, 1837)
Wedges amongst branching corals; distinguished by large rounded head, and red-brown bars and stripes; found throughout the region; W. and C. Pacific; to 6 cm.

2 FIVE-BAR CORALGOBY
Gobiodon quinquestrigatus (Valenciennes, 1837)
Wedges amongst branching corals; similar shape to **1** above, but overall dark with narrow blue lines on head; found throughout the region; Indo-C. Pacific; to 4 cm.

3 DECORATED GOBY
Istigobius decoratus (Herre, 1927)
Inhabits sand bottoms in the vicinity of coral reefs; similar to **4** below, but lacks black spot on rear part of first dorsal fin and less than 5 dark spots on top of head between eyes and first dorsal fin; Great Barrier Reef, W. Australia, and S. E. Asia; Indo-W. Pacific; to 12 cm.

4 BLACK-SPOTTED GOBY
Istigobius nigroocellatus (Günther, 1873)
Inhabits sand or silt bottoms in turbid water near inshore reefs; similar to **3** above, but has black spot at rear of first dorsal fin, and about 10-15 small black spots on top of head between eyes and first dorsal fin; Great Barrier Reef, W. Australia, and Philippines; mainly W. Pacific; to 7 cm.

5 SPECTACLED GOBY
Istigobius perspicillatus (Herre, 1945)
Inhabits sand-rubble areas near inshore reefs; distinguished by reddish colour, and narrow dark line connecting eyes over top of head; found throughout the region; Great Barrier Reef and N.W. Australia; E. Indian Ocean to N. Australia; to 11 cm.

6 ORNATE GOBY
Istigobius ornatus (Rüppell, 1830)
Inhabits mangrove areas and shallow rubble reefs; distinguished by 3-4 free (filamentous) upper pectoral rays and enlarged canine tooth on each side of lower jaw, colour variable; found throughout the region; Indo-W. Pacific; to 10 cm.

7 SCHOOLING GOBY
Parioglossus formosus (Smith, 1931)
Inhabits mangrove areas, tidal creeks, and inshore reefs, occurring in schools; distinguished by prominent black stripe along lower side and extending on to tail; found throughout the region; mainly W. Pacific; to 3.5 cm.

8 GIRDLED GOBY
Priolepis cinctus (Regan, 1908)
Inhabits reef crevices; distinguished by series of dark-edged brown bars with narrower pale bars between; found throughout the region; Great Barrier Reef, N.W. Australia, and throughout S. E. Asia; Indo-C. Pacific; to 6 cm.

9 HEAD-BARRED GOBY
Priolepis semidoliatus (Valenciennes, 1837)
Inhabits coral reef crevices and caves; distinguished by narrow blue bars on head and elongate filament on first dorsal fin; N.W. Australia, New Guinea, S. Indonesia, and Philippines; Indo-W. Pacific; to 3 cm.

10 REDHEAD GOBY
Paragobiodon echinocephalus Rüppell, 1830
Inhabits coral reefs, found amongst branches of *Stylophora* coral; distinguished by red or pinkish head and black body, also has tiny bristles covering head; found throughout the region; Indo-W. Pacific; to 3.5 cm.

11 BLUE GOBY
Ptereleotris hanae (Jordan & Snyder, 1901)
Inhabits sand bottoms near coral reefs, distinguished by slender shape, pale blue colour, and elongate filaments on first dorsal fin and tail; Great Barrier Reef, N.W. Australia, and Philippines; W. and C. Pacific; to 10 cm.

12 RED AND WHITE GOBY
Trimma sp.
Inhabits coral reef crevices and caves; distinguished by tiny size, irregular red and white blotches on head and body, and red spots on fins; Great Barrier Reef, N.W. Australia, and Indonesia; entire distribution unknown; to 2.5 cm.

13 ORANGE-SPOTTED GOBY
Trimma okinawae (Aoyagi, 1949)
Inhabits coral reef crevices and caves, distinguished by dense network of red-orange spots on body and fins; Great Barrier Reef, N.W. Australia, and throughout S. E. Asia; E. Indian Ocean and W. Pacific; to 4 cm.

14 HOESE'S GOBY
Silhouettea hoesei Larson & Miller, 1986
Inhabits sand bottoms close to shore; distinguished by eyes on dorsal profile of head and overall pale colour with brown speckling; N.W. Australia only; to 3.5 cm.

15 PATCHWORK GOBY
Gobiopsis bravoi (Herre, 1940)
Inhabits inshore reefs; distinguished by flattened head, rows of raised papillae on head, short skin tentacles around mouth, and irregular dark bars and pale patches on body; N. W. Australia, Philippines, E. Indonesia, and New Guinea; mainly W. Pacific; to 4 cm.

16 STRIPED GOBY
Valenciennea muralis (Valenciennes, 1837)
Inhabits sand-rubble areas near coral reefs; distinguished by broad white band and series of narrower brown stripes along side and black spot at rear of first dorsal fin; found throughout the region; mainly W. Pacific; to 13 cm.

17 LONG-FINNED GOBY
Valenciennea longipinnis (Lay & Bennett, 1839)
Inhabits sand-rubble areas near coral reefs; distinguished by blue-edged saddles or bars on side with dark spot at lower part of each bar; Great Barrier Reef, N.W. Australia, and throughout S. E. Asia; mainly W. Pacific; to 15 cm.

18 ORANGE-DASHED GOBY
Valenciennea puellaris (Tomiyama, 1936)
Inhabits sand-rubble areas near coral reefs; distinguished by orange longitudinal band from mouth to tail and orange spots on back; Great Barrier Reef, N.W. Australia, and throughout S.E. Asia; Indo-W. Pacific; to 14 cm.

19 OCELLATED GOBY
Vanderhorstia ornatissimus Smith, 1959
Inhabits sand bottoms; lives in burrow with alpheid shrimp; has long filament on first dorsal fin and blue-edged yellow spots on body; Great Barrier Reef, N.W. Australia, and S.E. Asia; Indo-C. Pacific; to 8 cm.

20 SHADOW GOBY
Yongeichthys nebulosus (Forsskål, 1775)
Inhabits silt bottoms; large brown spots on side; found throughout region; Indo-W. Pacific; to 18 cm.

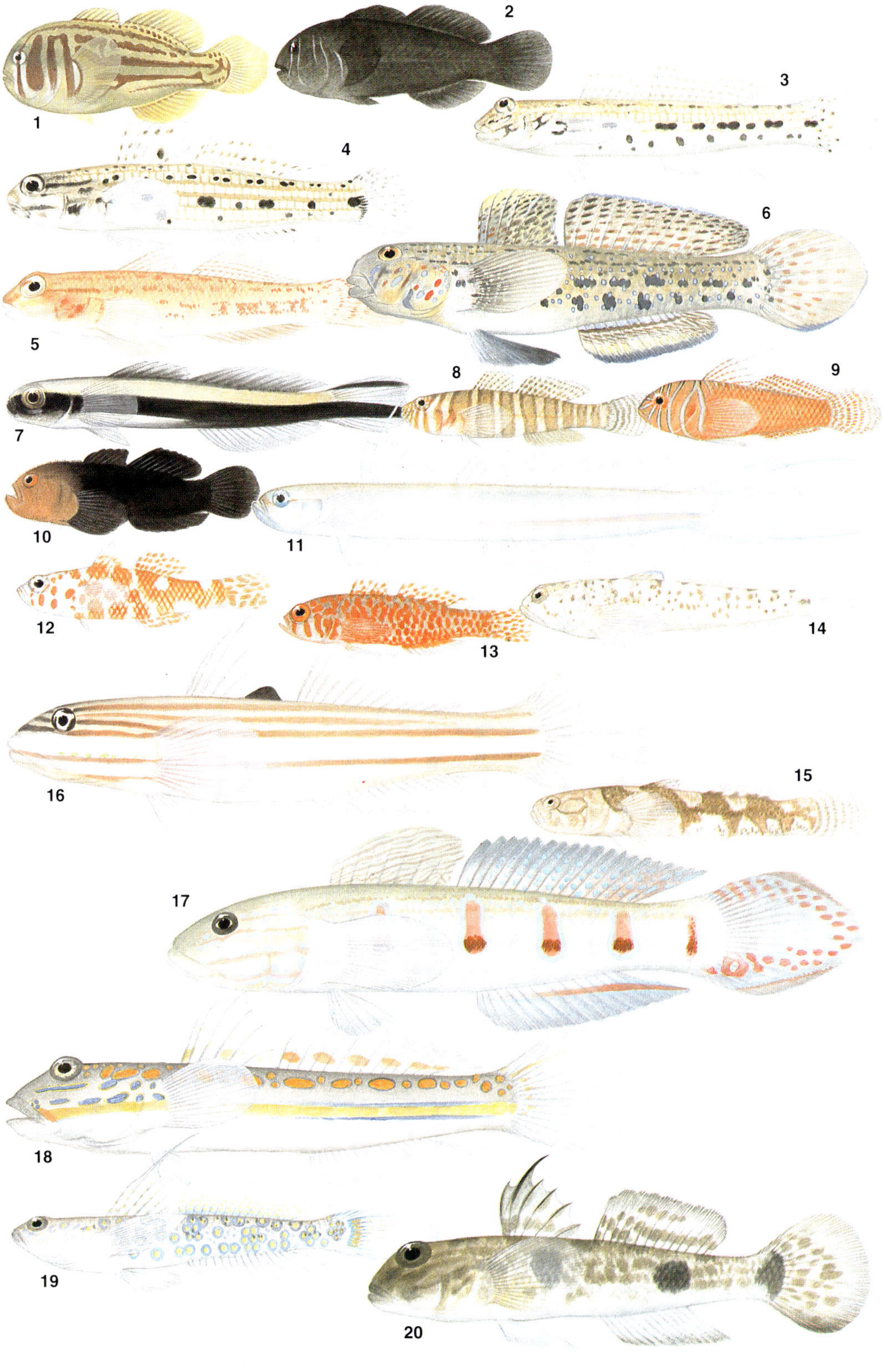
1
2
3
4
5
6
7
8
9
10
11
12
13
14
15
16
17
18
19
20

PLATE 91: GOBIES, DARTFISHES, SAND-DIVERS & CONVICT BLENNIES

1 BEAUTIFUL GOBY
Exyrias bellissimus (Smith, 1959)
Inhabits sandy or silty fringe of coral reefs; distinguished by elevated dorsal and anal fins, filamentous tips on first dorsal fin, about 10 diffuse brown bars on side, and small white spots on head; *E. puntang* (not shown) is very similar, but has 10 or more predorsal scales (8-10 in *bellissimus*); Great Barrier Reef, offshore reefs of N.W. Australia, and throughout S.E. Asia; Indo-W. Pacific; to 13 cm. (GOBIIDAE)

2 ALLEN'S GOBY
Valenciennea alleni Hoese & Larson, 1994
Inhabits sand-rubble fringe of coral reefs; distinguished by general pale colour with white stripe on side of head and body, also a pair of dark-edged orange stripes on side of head; N.W. Australia and Great Barrier Reef; to 9 cm. (GOBIIDAE)

3 TWOSTRIPE GOBY
Valenciennea heldingenii (Bleeker, 1858)
Inhabits sand-rubble fringe of coral reefs; usually seen in pairs that share sandy burrow; distinguished by pair of dark stripes on head and body; Great Barrier Reef and S.E. Asia; Indo-W. Pacific; to 16 cm. (GOBIIDAE)

4 SIXSPOT GOBY
Valenciennea sexguttata (Valenciennes, 1837)
Inhabits sand-rubble fringe of coral reefs; generally pale, but has black tip on first dorsal fin; Great Barrier Reef, offshore reefs of W. Australia, and throughout S.E. Asia; Indo-W. Pacific; to 14 cm. (GOBIIDAE)

5 BLUEBAND GOBY
Valenciennea strigata (Broussonet, 1782)
Inhabits sand-rubble fringe of coral reefs; often seen in pairs that share burrow; distinguished by yellow colour on head and blue stripe below eye; Great Barrier Reef, offshore reefs of W. Australia, and throughout S.E. Asia; Indo-W. Pacific; to 18 cm. (GOBIIDAE)

6 PARVA GOBY
Valenciennea parva Hoese & Larson, 1994
Inhabits sand-rubble fringe of coral reefs; distinguished by pair of thin orange stripes on side with dark "lattice" pattern on upper back; yellow colour on head and blue stripe below eye; Great Barrier Reef, offshore reefs of W. Australia, New Guinea, and Philippines; Indo-W. Pacific; to 7 cm. (GOBIIDAE)

7 DECORA GOBY
Valenciennea decora Hoese & Larson, 1994
Inhabits sand-rubble fringe of coral reefs; distinguished by thin dark "moustache" and orange "lattice" pattern (sometimes faint, but orange bar just behind pectoral base usually evident); Great Barrier Reef; SW Pacific to Fiji; to 12 cm. (GOBIIDAE)

8 IMMACULATE GOBY
Valenciennea immaculata (Ni, 1981)
Inhabits sand-rubble fringe of coral reefs; distinguished by pair of blue-edged orange stripes on head and body, and pointed, dark-edged tail; W. Australia, S. Great Barrier Reef, and Indonesia (Komodo); W. Pacific; to 10 cm. (GOBIIDAE)

9 ELEGANT DARTFISH
Nemateleotris decora Randall & Allen, 1973
Inhabits outer reef slopes to 68 m depth; lives in sandy burrow; distinguished by short dorsal fin "spike", violet snout and magenta hue on posterior part of body; Great Barrier Reef and throughout S.E. Asia; Indo-W. Pacific; to 8.5 cm. (MICRODESMIDAE)

10 FIRE DARTFISH
Nemateleotris magnifica Fowler, 1928
Inhabits outer reef slopes to 60 m depth; pairs live in sandy burrow; distinguished by dorsal fin "spike", yellow snout and reddish hue on posterior part of body and adjoining fins; Great Barrier Reef, offshore reefs of N.W. Australia, and throughout S.E. Asia; Indo-W. Pacific; to 8 cm. (MICRODESMIDAE)

11 ZEBRA DARTFISH
Ptereleotris zebra Fowler, 1928
Inhabits exposed coral reefs subject to surge; sometimes in schools that retreat to holes in bottom; distinguished by dark-edged light bars on side; Great Barrier Reef, offshore reefs of N.W. Australia, and throughout S.E. Asia; Indo-C. Pacific; to 12 cm. (MICRODESMIDAE)

12 TWOTONE DARTFISH
Ptereleotris evides (Jordan & Hubbs, 1925)
Inhabits sandy fringe of coral reefs; pairs often share burrow; distinguished by bicolour pattern and moderately elevated dorsal and anal fins; Great Barrier Reef, offshore reefs of N.W. Australia, and throughout S.E. Asia; Indo-C. Pacific; to 14 cm. (MICRODESMIDAE)

13 PALE DARTFISH
Ptereleotris microlepis (Bleeker, 1856)
Inhabits sandy fringe of coral reefs; occurs in pairs or groups; distinguished by plain pale pattern with dark elongate dark across pectoral fin base; Great Barrier Reef, N.W. Australia, and throughout S.E. Asia; Indo-C. Pacific; to 13 cm. (MICRODESMIDAE)

14 SPOT-TAIL DARTFISH
Ptereleotris heteroptera (Bleeker, 1855)
Inhabits sandy fringe of coral reefs; occurs in pairs or groups; distinguished by plain pale pattern with dark blotch on middle of tail; Great Barrier Reef, offshore reefs of N.W. Australia, and throughout S.E. Asia; Indo-C. Pacific; to 12 cm. (MICRODESMIDAE)

15 ONESPOT WORMFISH
Gunnelichthys monostigma Smith, 1958
Inhabits sand-rubble fringe of coral reefs; buries into sand when disturbed; distinguished by long slender body and small dark "ear" spot; Great Barrier Reef and throughout S.E. Asia; Indo-W. Pacific; to 11 cm. (MICRODESMIDAE)

16 CURIOUS WORMFISH
Gunnelichthys curiosus Dawson, 1968
Inhabits sand-rubble fringe of coral reefs; buries into sand when disturbed; distinguished by long slender body and broad orange stripe along middle of side; Coral Sea and Indonesia; Indo-C. Pacific; to 11 cm. (MICRODESMIDAE)

17 BLUE-BARRED RIBBON GOBY
Text on page 222.

18 CONVICT BLENNY
Text on page 222.

19 SPOTTED SAND-DIVER
Text on page 222.

20 ELEGANT SAND-DIVER
Text on page 222.

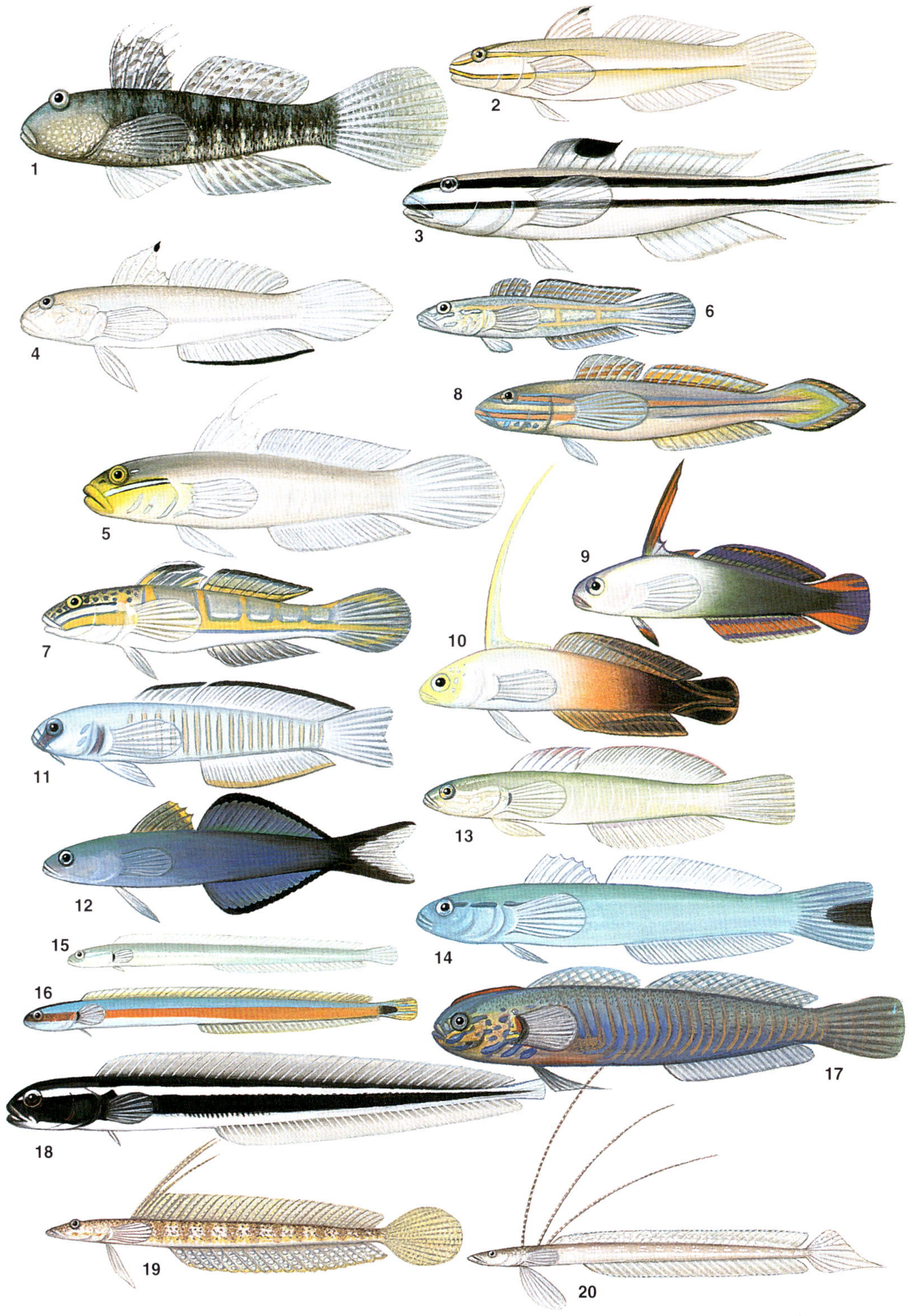
1
2
3
4
5
6
7
8
9
10
11
12
13
14
15
16
17
18
19
20

PLATE 92: GOBIES, GUDGEONS, AND SPINEFEET

1 MADURA GOBY
Apocryptodon madurensis (Bleeker, 1849)
Inhabits brackish mangrove estuaries; distinguished by overall whitish colour, scattered dark spots on head and body, and dark-edged pointed tail; found throughout the region; Indo-W. Pacific; to 8 cm. (GOBIIDAE)

2 MUDSKIPPER
Periophthalmus argentilineatus (Valenciennes, 1837)
Inhabits brackish mangrove estuaries, often seen resting on muddy banks; several similar species in region, but this is one of the most common; distinguished by protruding eyes and sail-like dorsal fin; found throughout the region; Indo-W. Pacific; to 27 cm. (GOBIIDAE)

3 BEARDED GOBY
Scartelaos histiophorus (Valenciennes, 1837)
Inhabits brackish mangrove estuaries; distinguished by a row of short barbels ("whiskers") along lower surface of head and antenna-like dorsal fin; found throughout the region; mainly W. Pacific; to 15 cm. (GOBIIDAE)

4 GOGGLE-EYED GOBY
Boleophthalmus caeruleomaculatus McCulloch & Waite, 1918
Inhabits brackish mangrove estuaries; distinguished by huge first dorsal fin with filamentous edge; N. Australia only, between Onslow, W. Australia and Gulf of Carpentaria; mainly W. Pacific; to 12 cm. (GOBIIDAE)

5 SMALL-EYED SLEEPER
Prionobutis microps (Weber, 1908)
Inhabits brackish mangrove estuaries; distinguished by diagonal bands on head, striped fins, and robust shape; N. Austrlalia and New Guinea; to 23 cm. (GOBIIDAE)

6 OLIVE FLATHEAD GUDGEON
Butis amboinensis (Bleeker, 1853)
Inhabits brackish mangrove estuaries; distinguished by flattened shovel-like snout; found throughout the region; Indo-Australian and Melanesian archipelagos; to 14 cm. (GOBIIDAE)

7 CHINESE GUDGEON
Bostrichthys sinensis (Lacepède, 1801)
Inhabits coastal mudflats and estuaries; distinguished by pale-rimmed spot at upper tail base; found throughout the region; mainly W. Pacific and Andaman Sea; to 12 cm. (GOBIIDAE)

8 BLIND GOBY
Brachyamblyopus coecus (Weber, 1913)
Inhabits soft mud, usually buried below surface; distinguished by reddish-pink colour and lack of eyes; found throughout the region; N. Australia, New Guinea, and Indonesia; Indo-Australian Archipelago; to 5 cm. (GOBIOIDIDAE)

9 MOORISH IDOL Càmo 10/98
Zanclus cornutus Linnaeus, 1758
Inhabits coral reefs; distinguished by elongate snout, dorsal fin filament and conspicuous pattern; found throughout the region; Indo C. Pacific; to 24 cm. (ZANCLIDAE)

10 SPOTTED SPINEFOOT
Siganus punctatus (Forster, 1801)
Inhabits coral and rocky reefs; distinguished by numerous spots on head, body, and fins; found throughout the region; Great Barrier Reef, W. Australia, and throughout S.E. Asia; mainly W. Pacific; to 40 cm. (SIGANIDAE) ★★★

11 BLACK SPINEFOOT
Siganus fuscescens (Houttyn, 1782)
Inhabits rock and weed areas, sometimes found in estuaries, occurs in schools; similar to **13** below, but body colour more uniform (generally lacks spotting) and edge of gill cover darkly outlined; found throughout the region; mainly W. Pacific; to 41 cm. (SIGANIDAE) ★★★

12 GOLDEN-LINED SPINEFOOT
Siganus lineatus (Linnaeus, 1835)
Inhabits coral reefs and mangrove estuaries; distinguished by yellow-orange lines and large spot below rear part of dorsal fin; N. Australia, New Guinea, and Philippines, mainly W. Pacific; to 30 cm. (SIGANIDAE) ★★

13 SMUDGESPOT SPINEFOOT
Siganus canaliculatus (Park, 1797)
Inhabits sand-weed areas; similar to **11** above, but has more prominent spots on body and fins, frequently with dark blotch behind upper edge of gill cover; N.W. Australia and throughout S.E. Asia; Indo-W. Pacific; to 20 cm. (SIGANIDAE) ★★★

14 THREESPOT SPINEFOOT
Siganus trispilos Woodland & Allen, 1977
Inhabits coral reefs, often found amongst branching *Acropora* corals; distinguished by 3 black blotches on upper side; N.W. Australia only; to 23 cm. (SIGANIDAE)

15 DOUBLEBAR SPINEFOOT
Siganus doliatus (Cuvier, 1830)
Inhabits coral reefs; distinguished by pair of diagonal dark bars on head and front of body; found throughout the region; *S. virgatus* is a similar species from Malay Peninsula, W. Indonesia and Philippines; N. Australia, E. Indonesia, and New Guinea; mainly W. Pacific; to 30 cm. (SIGANIDAE) ★★

PLATE 91

17 BLUE-BARRED RIBBON GOBY
Oxymetopon cyanoctenosum Klausewitz & Conde, 1981
Inhabits silty bottoms; in pairs that live in burrow; distinguished by laterally compressed body, and neon blue bars and bands on head and body; Indonesia and Philippines; to 20 cm.

18 CONVICT BLENNY
Pholidichthys leucotaenia Bleeker, 1856
Inhabits coral reefs, often seen along dropoffs; swims in mid-water close to reef, sometimes in large aggregations; similar to young of Striped catfish (Pl. 11-5), but lacks "whiskers" around mouth; Indonesia and Philippines to Solomon Islands; to 12 cm. (PHOLIDICHTHYIDAE)

19 SPOTTED SAND-DIVER
Trichonotus setigerus (Bloch & Schneider, 1801)
Inhabits sandy fringe of coral reefs; occurs in schools, which hover near the bottom; sometimes in sand when threatened; distinguished from **20** below by about 12 faint brown bars and longitudinal rows of small pale spots; Great Barrier Reef and SE Asia; Indo-W. Pacific; to 15 cm (TRICHONOTIDAE)

20 ELEGANT SAND-DIVER
Trichonotus elegans Shimada & Yoshino, 1984
Inhabits sandy fringe of coral reefs; behaviour same as **19** above; distinguished from **19** by black spot on first few dorsal-fin rays, male has extremely elongate rays at beginning of dorsal fin; Great Barrier Reef, offshore reefs of N.W. Australia, and SE Asia; mainly W. Pacific; to 18 cm (TRICHONOTIDAE)

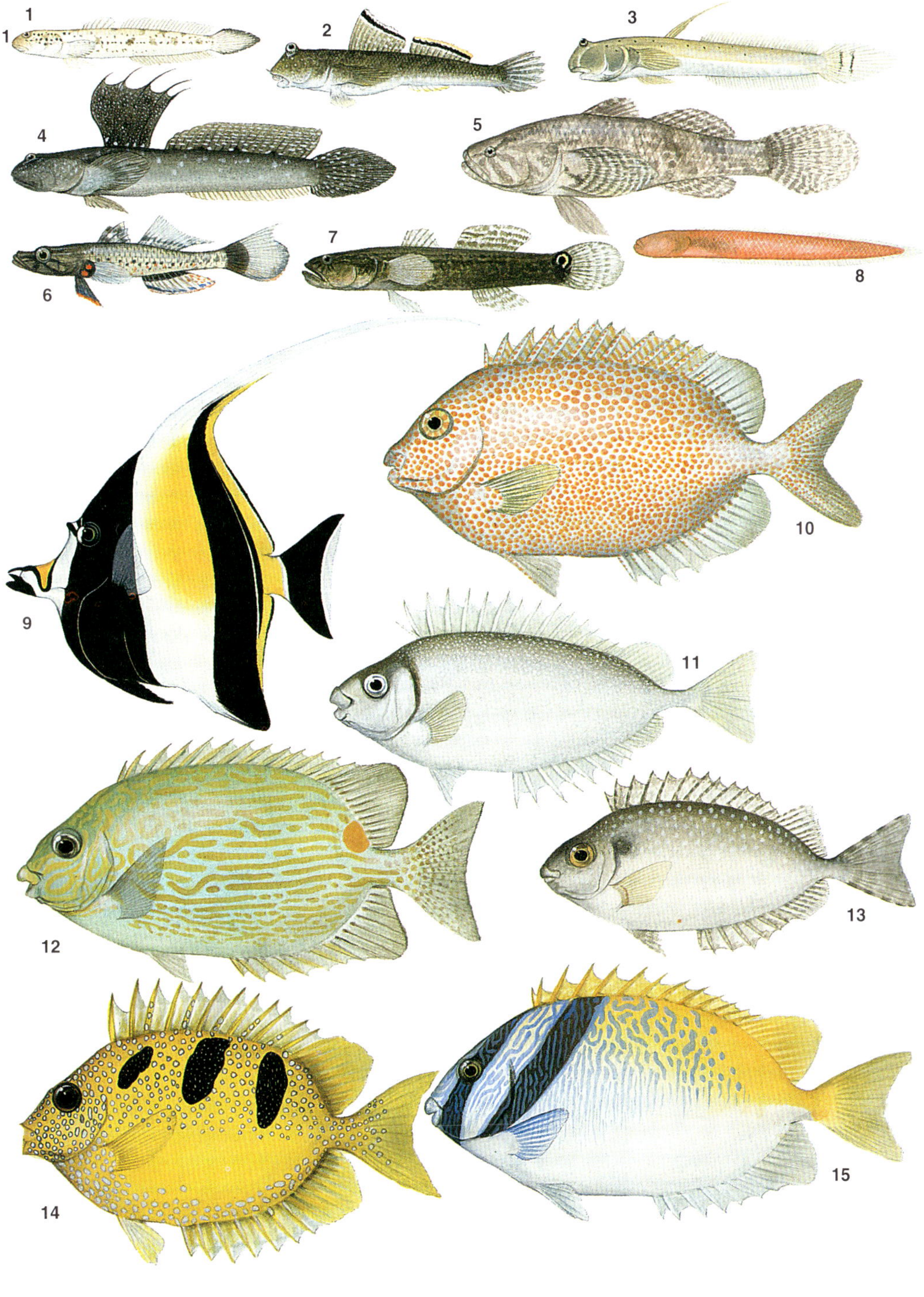
1
1
2
3
4
5
6
7
8
9
10
11
12
13
14
15

PLATE 93: SURGEONFISHES AND SPINEFEET

1 TOMINI BRISTLETOOTH
Ctenochaetus tominiensis Randall, 1955
Inhabits coral reefs; distinguished by pale tail and yellow outer portion of dorsal and anal fins; Indonesia, Philippines, New Guinea, and Solomon Islands; to 22 cm. (ACANTHURIDAE) ★★

2 GOLDRING BRISTLETOOTH
Ctenochaetus strigosus (Bennett, 1828)
Inhabits coral reefs; distinguished by yellow ring around eye and small pale spots covering head; juvenile entirely bright yellow; Great Barrier Reef, offshore reefs of W. Australia, and scattered localities throughout S.E. Asia; Indo-C. Pacific. (ACANTHURIDAE) ★★

3 RINGTAILED UNICORNFISH
Naso annulatus (Quoy & Gaimard, 1825)
Inhabits steep outer reef slopes, adults usually below 20 m depth; distinguished long slender forehead "spike", and "webbed" markings on tail; juvenile has white ring around tail base; Great Barrier Reef, offshore reefs of W. Australia, and throughout S.E. Asia; Indo-C. Pacific; to 100 cm. (ACANTHURIDAE) ★★

4 HUMPBACK UNICORNFISH
Naso brachycentron (Valenciennes, 1835)
Inhabits coral reefs; distinguished by angular profile of back; adult male has long tapering "spike" in front of eye, female (shown here) with only a bump; Great Barrier Reef, offshore reefs of W. Australia, and throughout S.E. Asia; Indo-W. Pacific; to 60 cm. (ACANTHURIDAE) ★★

5 SLEEK UNICORNFISH
Naso hexacanthus (Bleeker, 1855)
Inhabits steep outer reef slopes, usually seen in schools; brown to bluish grey, but can quickly change to pale blue; has black markings on cheek; Great Barrier Reef, offshore reefs of W. Australia, and throughout S.E. Asia; Indo-C. Pacific; to 75 cm. (ACANTHURIDAE) ★★

6 SPINY SPINEFOOT
Siganus spinus (Linnaeus, 1758)
Inhabits coral reefs, usually seen in aggregations on shallow outer reefs; distinguished by "maze" colour pattern; Great Barrier Reef and throughout S.E. Asia; Andaman Sea to C. Pacific. (SIGANIDAE) ★★

7 ELONGATE UNICORNFISH
Naso lopezi Herre, 1927
Inhabits coral reefs, usually seen on edge of dropoffs below 20-30 m depth; distinguished by elongate shape and bluish-grey coloration with numerous dark grey spots; Great Barrier Reef and throughout S.E. Asia; Indo-C. Pacific; to 50 cm. (ACANTHURIDAE) ★★

8 SINGLE-SPINED UNICORNFISH
Naso thynnoides (Valenciennes, 1835)
Inhabits coral reefs, usually along the edge of lagoon and outer slopes to depth of 30 m; distinguished by elongate shape and numerous fine bars on side of body; throughout S.E. Asia; Indo-W. Pacific. (ACANTHURIDAE) ★★

9 VLAMING'S UNICORNFISH
Naso vlamingii (Valenciennes, 1835)
Inhabits steep outer reef slopes; feeds on zooplankton; distinguished by vertical blue lines on side of body, blue band through eye, and prolonged filament on upper and lower corner of tail; Great Barrier Reef, offshore reefs of W. Australia, and throughout S.E. Asia; Indo-C. Pacific; to 55 cm. (ACANTHURIDAE) ★★

10 JAVA SPINEFOOT
Siganus javus (Linnaeus 1766)
Inhabits silty coastal reefs and brackish areas; distinguished by white spots and lines on side and large black spot on tail; Queensland and throughout S.E. Asia; Indo-W. Pacific; to 53 cm. (SIGANIDAE) ★★

11 VERMICULATE SPINEFOOT
Siganus vermiculatus (Valenciennes, 1835)
Inhabits shallow coastal reefs and brackish areas, usually seen in small aggregations; distinguished by vermiculate colour pattern; Great Barrier Reef and throughout S.E. Asia; E. Indian Ocean and W. Pacific; to 45 cm. (SIGANIDAE) ★★

12 OCELLATED SPINEFOOT
Siganus corallinus (Valenciennes, 1835)
Inhabits coral reefs, usually seen in pairs; distinguished by dense covering of blue spots on yellow background colour; also known as Coral Spinefoot; Great Barrier Reef, offshore reefs of W. Australia, and throughout S.E. Asia; Indo-W. Pacific; to 28 cm. (SIGANIDAE) ★★

13 BLUE-LINED SPINEFOOT
Siganus puellus Schlegel, 1852
Inhabits coral reefs, usually seen in pairs; distinguished by diagonal dark bar through eye and narrow blue lines on side; Great Barrier Reef, offshore reefs of W. Australia, and throughout S.E. Asia; mainly W. Pacific; to 38 cm. (SIGANIDAE) ★★

14 SILVER SPINEFOOT
Siganus argenteus (Quoy & Gaimard, 1825)
Inhabits coral reefs, frequently seen in aggregations; distinguished by numerous yellow spots on side, yellow "wash" on top of head, and deeply forked tail; also known as Forktail Spinefoot; Great Barrier Reef, offshore reefs of W. Australia, and throughout S.E. Asia; Indo-W. Pacific; to 37 cm. (SIGANIDAE) ★★★

15 GOLDEN SPINEFOOT
Siganus guttatus (Bloch, 1787)
Inhabits coastal reefs and lagoons, frequently among mangroves and in brackish water; distinguished by spotted pattern and large golden spot below posterior part of dorsal fin; *S. lineatus* (Queensland and S.E. Asia) is very similar, but has horizontal lines on the body instead of spots; throughout S.E. Asia; E. Indian Ocean and W. Pacific; to 42 cm. (SIGANIDAE) ★★

16 FOXFACE
Siganus vulpinus (Schlegel & Muller, 1845)
Inhabits coral reefs, adults seen in pairs; distinguished by elongate snout, diagonal dark bar through eye, black breast, and bright yellow body; Great Barrier Reef, Kimberley coast and offshore reefs of W. Australia, and throughout S.E. Asia; W. Pacific; to 30 cm. (SIGANIDAE) ★★

RABBITFISHES
Rabbitfishes (named because of their snout shape) of the family Siganidae (Plates 92-93) are plant feeders, sometimes forming large schools as they roam over the reef. They are also sometimes called "spinefeet" in reference to the unusual arrangement of two pelvic fin spines separated by three soft rays. Another peculiarity is the high (seven) number of anal fin spines. All dorsal, anal, and pelvic fin spines are grooved and contain venom glands. If handled carelessly they are capable of inflicting very painful wounds.

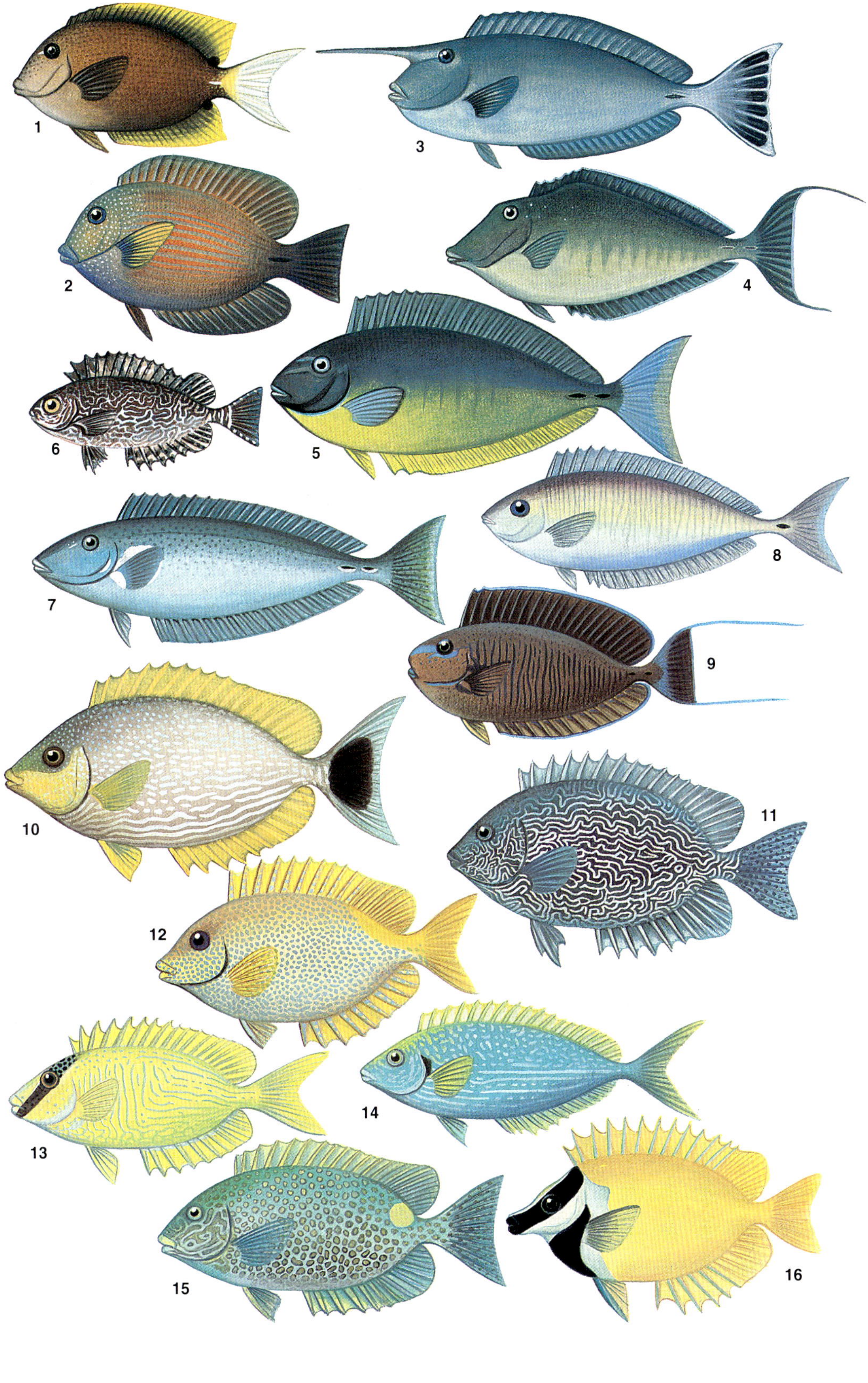
1
3
2
4
6
5
7
8
9
10
11
12
13
14
15
16

PLATE 94: SURGEONFISHES (FAMILY ACANTHURIDAE)

1 YELLOWMASK SURGEONFISH
Acanthurus mata (Cuvier, 1829)
Inhabits coral reef areas; distinguished by narrow stripes on side and yellow bands in front of eye; sharp spine on each side of tail base; formerly known as *A. bleekeri*; Great Barrier Reef, N.W. Australia, and throughout S.E. Asia; Indo-C. Pacific; to 50 cm. ★★

2 WHITE-CHEEKED SURGEONFISH
Acanthurus nigricans (Linnaeus, 1758)
Inhabits coral reefs; distinguished by white patch below eye; sharp spine on each side of tail base; formerly known as *A. glaucoparieus*; Great Barrier Reef, N.W. Australia, and throughout S.E. Asia; mainly tropical Pacific Ocean to Central America and the Galapagos; to 21 cm. ★★

3 ORNATE SURGEONFISH
Acanthurus dussumieri Valenciennes, 1835
Inhabits coral reefs; distinguished by yellowish fins and black outline around spine on base of tail; Great Barrier Reef, N.W. Australia, and throughout S.E. Asia; Indo-W. Pacific; to 54 cm. ★★

4 BLUE-LINED SURGEONFISH
Acanthurus lineatus (Linnaeus, 1758)
Inhabits shallow coral reefs usually where there is some wave action; distinguished by blue stripes on side; sharp spine on each side of tail base; Great Barrier Reef, N.W. Australia, and throughout S.E. Asia; Indo-C. Pacific; to 38 cm. ★★

5 RING-TAILED SURGEONFISH
Acanthurus grammoptilus Richardson, 1843
Inhabits coral and rocky reefs, often in silty inshore areas; distinguished by uniform dark colour and pale bar at base of tail; sharp spine on each side of tail base; N. Australia, Indonesia, and Philippines; Indo-Australian Archipelago; to 30 cm. ★★

6 ORANGE-SPOT SURGEONFISH
Acanthurus olivaceous Bloch & Schneider, 1801
Inhabits sand and rubble areas adjacent to coral reefs; distinguished by elliptical orange band behind eye, juveniles entirely yellow; sharp spine on each side of tail base; Great Barrier Reef, N.W. Australia, and throughout S.E. Asia; Indo-C. Pacific; to 32 cm. ★★

7 DUSKY SURGEONFISH
Acanthurus nigrofuscus (Forsskål, 1775)
Inhabits coral and rocky reefs; distinguished by black spot at rear base of dorsal and anal fins; sharp spine on each side of tail base; found throughout the region; Great Barrier Reef, N.W. Australia, and throughout S.E. Asia; Indo-C. Pacific; to 20 cm. ★★

8 CONVICT SURGEONFISH
Acanthurus triostegus (Linnaeus, 1758)
Inhabits coral and rocky reefs; distinguished by bold black bars on side; sharp spine on each side of tail base; Great Barrier Reef, N.W. Australia, and throughout S.E. Asia; Indo-E. Pacific; to 25 cm. ★★

9 LINED BRISTLETOOTH
Ctenochaetus striatus (Quoy & Gaimard, 1825)
Inhabits coral reefs; distinguished by overall dark colour and flexible bristle-like teeth (versus fixed teeth); sharp spine on each side of tail base; Great Barrier Reef, N.W. Australia, and throughout S.E. Asia; Indo-C. Pacific; to 18 cm. ★★

10 STRIPE-FACE UNICORNFISH
Naso lituratus (Bloch & Schneider, 1801)
Inhabits coral reefs; distinguished by dark stripe on snout and yellow patches at tail base; pair of sharp spines on each side of tail base; Great Barrier Reef, N.W. Australia, and throughout S.E. Asia; Indo-C. Pacific; to 15 cm. ★★

11 LONGNOSED UNICORNFISH
Naso brevirostris (Valenciennes, 1835)
Inhabits coral reefs; distinguished by long spike on snout and short white tail; a pair of sharp spines on each side of tail base; Great Barrier Reef, N.W. Australia, and throughout S.E. Asia; Indo-C. Pacific; to 50 cm. ★★

12 HUMPHEAD UNICORNFISH
Naso tuberosus Lacepède, 1801
Inhabits coral reefs; distinguished by bulbous snout and plain greyish colour; pair of sharp spines on each side of tail base; Great Barrier Reef, N.W. Australia, and throughout S.E. Asia; found throughout the region; Indo-W. Pacific; to 60 cm.

13 BROWN UNICORNFISH
Naso unicornis Forsskål, 1775
Inhabits coral reefs; distinguished by relatively short forehead spike, blue spots around pair of spines on each side of tail base, and elongate tail filaments; Great Barrier Reef, N.W. Australia, and throughout S.E. Asia; Indo-C. Pacific; to 70 cm. ★★

14 SAILFIN TANG
Zebrasoma veliferum Bloch, 1797
Inhabits coral reefs; distinguished by sail-like fins and bars on side; sharp spine on each side of tail base; Great Barrier Reef, N.W. Australia, and throughout S.E. Asia; Indo-C. Pacific; to 40 cm. ★★

15 BLUE-LINED TANG
Zebrasoma scopas Cuvier, 1829
Inhabits coral reefs; distinguished by overall dark colour, protruding snout and white spine on each side of tail base; Great Barrier Reef, N.W. Australia, and throughout S.E. Asia; Indo-C. Pacific; to 20 cm. ★★

THE MEAN SURGEON

Most surgeonfishes have a peaceful disposition, travelling in schools, as they graze on algae from the reef's surface. However, the beautiful and pugnacious Blue-lined Surgeon (4) is an exception. It is strongly territorial, aggressively chasing away intruders that enter its domain. The territory of each adult occupies about 6-8 square metres of bottom. Attacks are mainly directed against algal-feeding fishes, particularly fellow Blue-lined surgeons and other members of the family, as well as parrotfishes, and triggerfishes. This behaviour ensures an adequate supply of seaweed-covered turf, which is its sole food source. The fish often occurs in colonies, with adjacent territories in close proximity. Adults occupy the central area and juvenile are scattered around the periphery. Like other surgeonfishes, this species possesses a collapsible scalpel-like spine on each side of the tail base. However, the spines of the Blue-lined Surgeon differ by being venomous and are capable of inflicting painful wounds if handled carelessly.

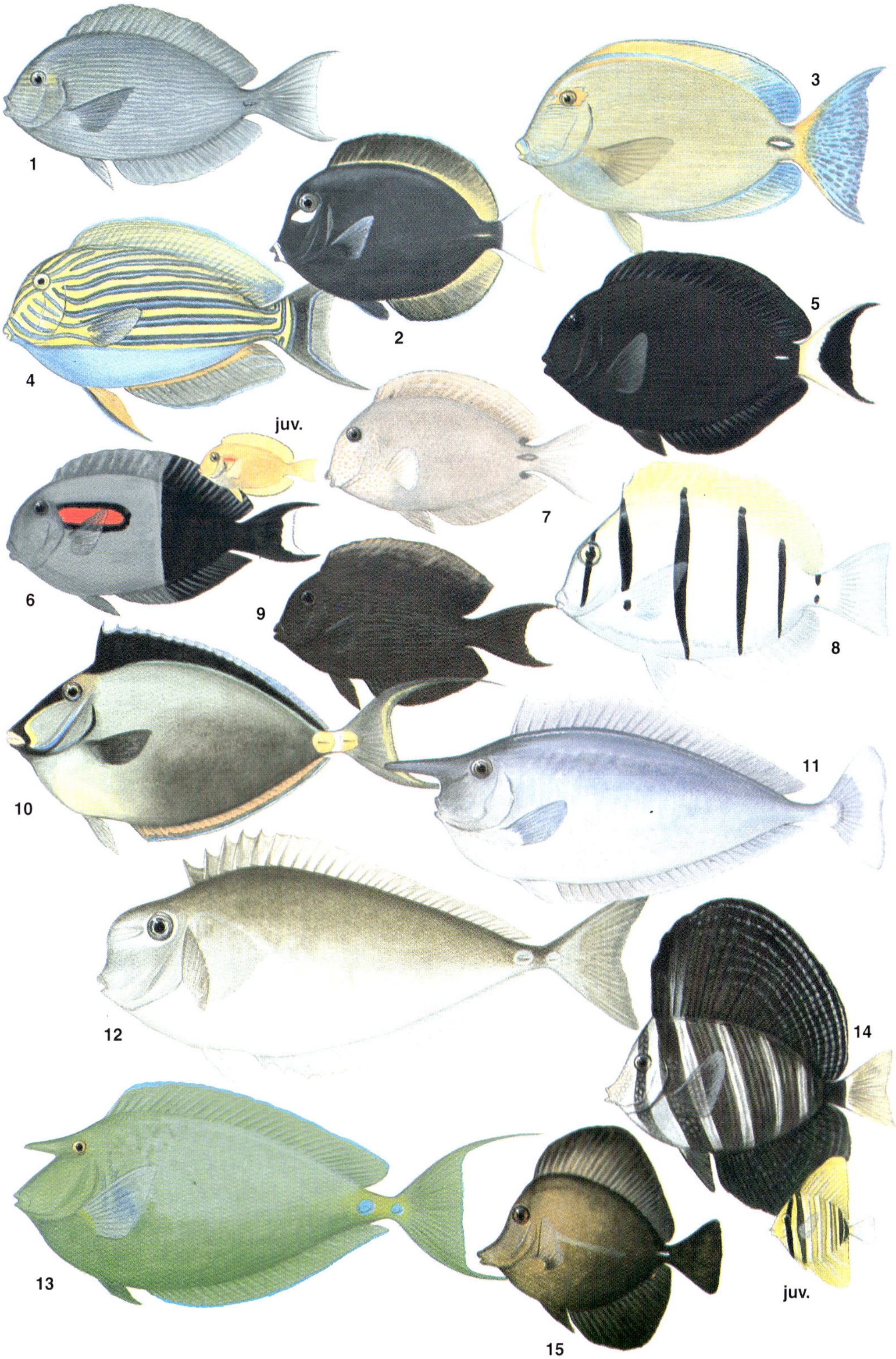

1
3
2
4
5
juv.
7
6
9
8
10
11
12
14
13
15
juv.

PLATE 95: SURGEONFISHES (ACANTHURIDAE)

1 POWDERBLUE SURGEONFISH
Acanthurus leucosternon Bennett, 1832
Inhabits coral reefs, usually on upper seaward slope or on nearby reef flats; distinguished by overall blue colour, white area on chin, and yellow dorsal fin; S.W. Indonesia; widely distributed in Indian Ocean; to 23 cm. ★★

2 MIMIC SURGEONFISH
Acanthurus pyroferus Kittlitz, 1834
Inhabits coral reefs; distinguished by dark brown area on lower and rear part of head and yellow edge on tail; juvenile mimics the pygmy angelfish, *Centropyge vrolikii*; also known as Orange-Gilled Surgeonfish; Great Barrier Reef, offshore reefs of W. Australia, and throughout S.E. Asia; W. and C. Pacific; to 25 cm. ★★

3 INDO-PACIFIC BLUETANG
Paracanthurus hepatus (Linnaeus, 1766)
Inhabits coral reefs, young hide amongst branching corals; feeds on zooplankton; distinguished by deep blue colouration, and black marking on side; also known as Palette Surgeonfish; Great Barrier Reef and throughout S.E. Asia; Indo-W. Pacific; to 31 cm. ★

4 WHITE-SPOTTED SURGEONFISH
Acanthurus guttatus Forster, 1801
Inhabits coral reefs, usually found in surge zone, often in schools; distinguished by pair of white bars, small white spots on side, and yellow pelvic fins; Great Barrier Reef and scattered localities in other parts of S.E. Asia; Indo-C. Pacific; to 26 cm. ★★

5 RINGTAIL SURGEONFISH
Acanthurus blochii Valenciennes, 1835
Inhabits coral reefs; distinguished by yellow patch behind eye, white bar at base of tail, and narrow blue and orange stripes on dorsal and anal fin; Great Barrier Reef, offshore reefs of W. Australia, and throughout S.E. Asia; Indo-C. Pacific; to 42 cm. ★★

6 WHITEFIN SURGEONFISH
Acanthurus albipectoralis Allen & Ayling, 1987
Inhabits coral reefs on steep outer slopes between 5-20 m depth; swims well above the bottom while feeding on zooplankton; outer half of pectorals white; Great Barrier Reef and Coral Sea eastward to Tonga; to 33 cm. ★★

7 THOMPSON'S SURGEONFISH
Acanthurus thompsoni (Fowler, 1923)
Inhabits steep outer reef slopes; feeds on zooplankton; distinguished by dark brown coloration with white tail; Great Barrier Reef, offshore reefs of W. Australia, and throughout S.E. Asia; Indo-C. Pacific; to 27 cm. ★★

8 EARBAR SURGEONFISH
Acanthurus maculiceps (Ahl, 1923)
Inhabits coral reefs; distinguished by blue bar behind upper rear corner of gill cover, pale spots on head and pale bar across tail base; throughout S.E. Asia; E. Indian Ocean and W. and C. Pacific; to 35 cm. ★★

9 BLACKSTREAK SURGEONFISH
Acanthurus nigricauda Duncker & Mohr, 1929
Inhabits coral reefs; distinguished by horizontal black band behind eye, white bar at tail base (sometimes absent), and black streak on middle of tail base; Great Barrier Reef, offshore reefs of W. Australia, and throughout S.E. Asia; Indo-C. Pacific; to 40 cm. ★★

10 ROUNDSPOT SURGEONFISH
Acanthurus bariene Lesson, 1830
Inhabits coral reefs, usually seen on outer slope below 30 m depth; distinguished by round marking behind eye, yellow bar behind gill cover, and yellow dorsal fin; Great Barrier Reef and throughout S.E. Asia; C. Indian Ocean to W. Pacific; to 50 cm. ★★

11 PALE-LIPPED SURGEONFISH
Acanthurus leucocheilus Herre, 1927
Inhabits coral reefs in the vicinity of dropoffs between 5-35 m depth; distinguished by whitish lips, white bar on chin, and pale ring around tail base; throughout S.E. Asia; Indo-W. Pacific; to about 40 cm. ★★

12 DARK SURGEONFISH
Acanthurus nubilus (Fowler and Bean, 1929)
Inhabits coral reefs in the vicinity of dropoffs in 5-40 m depth; distinguished by pattern of fine blue horizontal lines on side and spot on head; overall colouration changeable from relatively pale to very dark, nearly blackish; shape is relatively rounded; throughout S.E. Asia; Indo-C. Pacific Pacific; to 26 cm. ★★

13 YELLOWFIN SURGEONFISH
Acanthurus xanthopterus Valenciennes, 1835
Inhabits sandy areas near coral reefs; distinguished by yellow pectoral fins and yellow area in front and behind eye; also known as Ringtail Surgeonfish; Great Barrier Reef, offshore reefs of W. Australia, and throughout S.E. Asia; Indo-E. Pacific; to 56 cm. ★★

14 FOWLER'S SURGEONFISH
Acanthurus fowleri de Beaufort, 1951
Inhabits coral reefs, usually seen on seaward slopes and adjacent to dropoffs in 10-50 m depth; distinguished by horseshoe-shaped marking behind head; E. Indonesia and Philippines to New Britain; to 27 cm. ★★

15 ORANGE-SOCKET SURGEONFISH
Acanthurus auranticavus Randall, 1956
Inhabits coral reefs, usually seen in small schools at shallow depths in lagoons and on outer reefs; distinguished by eliptical mark behind head and orange border around caudal spine; Great Barrier Reef and throughout S.E. Asia; Maldive Islands to Indonesia and Philippines; to 35 cm. ★★

16 TWOSPOT BRISTLETOOTH
Ctenochaetus binotatus Randall, 1955
Inhabits coral reefs; distinguished by blue iris, narrow pale lines on side, and small black spot at rear base of dorsal and anal fin; juvenile has yellowish tail; Great Barrier Reef, offshore reefs of W. Australia, and throughout S.E. Asia; Indo-C. Pacific; to 22 cm. ★★

SURGEONFISHES

The estimated 75 species of surgeonfishes in the family Acanthuridae are widely distributed in tropical and sub-tropical seas. Their common name is derived from the scalpel-like spines on each side of the tail base, a handy defensive weapon. Surgeons are solitary in habit or form schools, The largest genus, *Acanthurus*, contains mainly algal feeders that graze widely over their home reefs. However, most members of the genus *Naso* and a few *Acanthurus* feed high above the bottom on zooplankton.

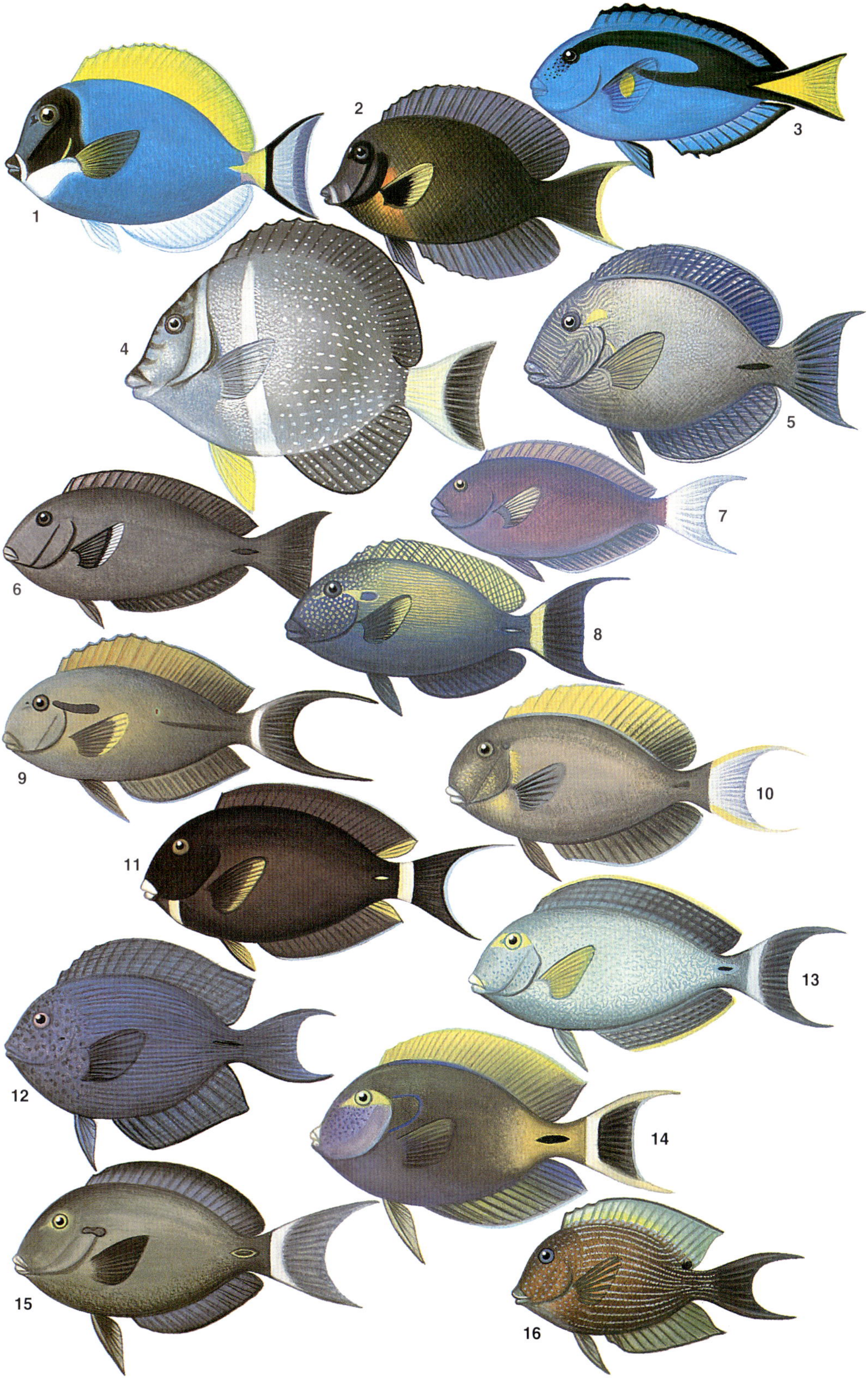
1
2
3
4
5
6
7
8
9
10
11
12
13
14
15
16

PLATE 96: BILLFISHES AND TUNAS

1 BLACK MARLIN
Makaira indica (Cuvier, 1832)
Inhabits oceanic waters, generally well offshore; distinguished by lack of cross bars and the rigid pectoral fin which cannot be folded against side of body; found throughout the region; Indo-C. Pacific; to 500 cm. All tackle world record 707.61 kg, Australian record 199.580 kg. (ISTIOPHORIDAE) ★★★

2 INDO-PACIFIC BLUE MARLIN
Makaira mazara (Jordan & Snyder, 1901)
Inhabits oceanic waters, generally well offshore; distinguished from **1** above by narrow bars on sides and non-rigid pectoral fin, and from **4** below by its lower dorsal fin; found throughout the region; Indo-E. Pacific; to 500 cm. All tackle world record 498.95 kg Australian record 275.0 kg. (ISTIOPHORIDAE) ★★★

3 SHORT BILL SPEARFISH
Tetrapturus angustirostris Tanaka, 1915
Inhabits oceanic waters, generally well offshore; distinguished by very short bill and slender shape; found throughout the region; Indo-E. Pacific; to 200 cm and 52 kg. (ISTIOPHORIDAE) ★★

4 STRIPED MARLIN
Tetrapturus audux (Philippi, 1887)
Inhabits oceanic waters, generally well offshore; similar to 2 above, but has taller dorsal fin; found throughout the region; Indo-E. Pacific; to 420 cm. All tackle world record 189.37 kg, Australia record 148 kg. (ISTIOPHORIDAE) ★★★

5 INDO-PACIFIC SAILFISH
Istiophorus platypterus (Shaw & Nodder, 1792)
Inhabits oceanic waters, generally well offshore; distinguished by sail-like dorsal fin; found throughout the region; Indo-E. Pacific to 360 cm. All tackle world record 100.24 kg, Australian record 55. 79kg. (ISTIOPHORIDAE) ★★

6 SWORDFISH
Xiphias gladius Linnaeus, 1758
Inhabits oceanic waters, generally well offshore; distinguished by the long, flattened sword; found throughout the region; worldwide in temperate and tropical seas; to 450 cm. All tackle world record 536.15 kg, Australian record 95.25 kg. (XIPHIIDAE) ★★★

7 ALBACORE
Thunnus alalunga (Bonnaterre, 1788)
Inhabits oceanic waters, generally well offshore; occurs in schools; distinguished by the very long pectoral fins, these are shorter in juveniles and sub-adults which resemble **8** and **9** below, but differ from them by having a white rear border on the tail; found throughout the region; worldwide in temperate and tropical seas; to 150 cm. All tackle world record 40 kg, Australian record 15 kg. (SCOMBRIDAE) ★★★

8 YELLOWFIN TUNA
Thunnus albacares (Bonnaterre, 1788)
Inhabits oceanic waters, generally well offshore; occurs in schools; distinguished by yellow dorsal and anal fins that become elongated with increased age; found throughout the region; worldwide tropical and temperate seas, to 210 cm. All tackle world record 176.4 kg, Australian record 97 kg. (SCOMBRIDAE) ★★★

9 BIGEYE TUNA
Thunnus obesus (Lowe, 1839)
Inhabits oceanic waters, generally well offshore; occurs in schools; similar to **8** above, but has much larger eye and lacks elongate dorsal and anal rays; found throughout the region; worldwide in tropical and temperate seas; to 240 cm. All tackle world angling record 197.3 kg, Australian record 37.13 kg. (SCOMBRIDAE) ★★

10 SOUTHERN BLUEFIN TUNA
Thunnus maccoyii (Castelnau, 1872)
Inhabits oceanic waters, generally well offshore; occurs in schools; distinguished by short pectoral fins and robust body shape; S. Queensland, W. Australia, and S. Indonesia; Southern Hemisphere in mainly temperate seas, but ranges to tropics in Australia and S.Indonesia; to 240 cm. All tackle world angling record 158 kg, Australian record 71.2 kg. (SCOMBRIDAE) ★★

11 NORTHERN BLUEFIN TUNA
Thunnus tonggol (Bleeker, 1851)
Inhabits oceanic waters, generally well offshore; occurs in schools; distinguished by short pectoral fins and slender body shape; found throughout the region; mainly W. Pacific and N. Indian Ocean; to 150 cm. All tackle world record 35.9 kg, Australian record 31.26 kg. (SCOMBRIDAE) ★★

12 SKIPJACK TUNA
Katsuwonis pelamis (Linnaeus, 1758)
Inhabits oceanic waters, generally well offshore; occurs in schools; distinguished by dark stripes on sides; found throughout the region; worldwide in tropical and temperate seas. All tackle world record 18.93 kg, Australian record 9.3 kg. (SCOMBRIDAE) ★★

KINGS OF THE SEA

Marlins are swift predatory inhabitants of the high seas, well known among anglers for their fighting ability. The Indo-Pacific Blue Marlin is the largest species, reported to reach unofficial weights in excess of 906 kilograms (2000 pounds). Numerous Black Marlin weighing in excess of 500 kilograms have been caught in waters off Cairns, Queensland. These fishes range widely in tropical and temperate seas of the Indo-Pacific region. There is still much to be learned about their biology, but tagging programs by sport anglers have been helpful in establishing their extensive migration routes. At certain times of the year they migrate to spawning grounds on the edge of continental shelves or off oceanic islands. One of the best known Black Marlin grounds in our region is the outer Great Barrier Reef off Cairns, Queensland, where dense concentrations appear between August and November.

Fully grown marlin probably have few enemies, but they are occasionally attacked and killed by large pelagic sharks and killer whales while struggling after being hooked. Humans are definitely their biggest threat - although sport anglers often tag and release their catches, commercial fishing operations take a heavy toll, utilising long-lines. Large quantities of marlin flesh are marketed in Japan, Taiwan, and other Asian countries.

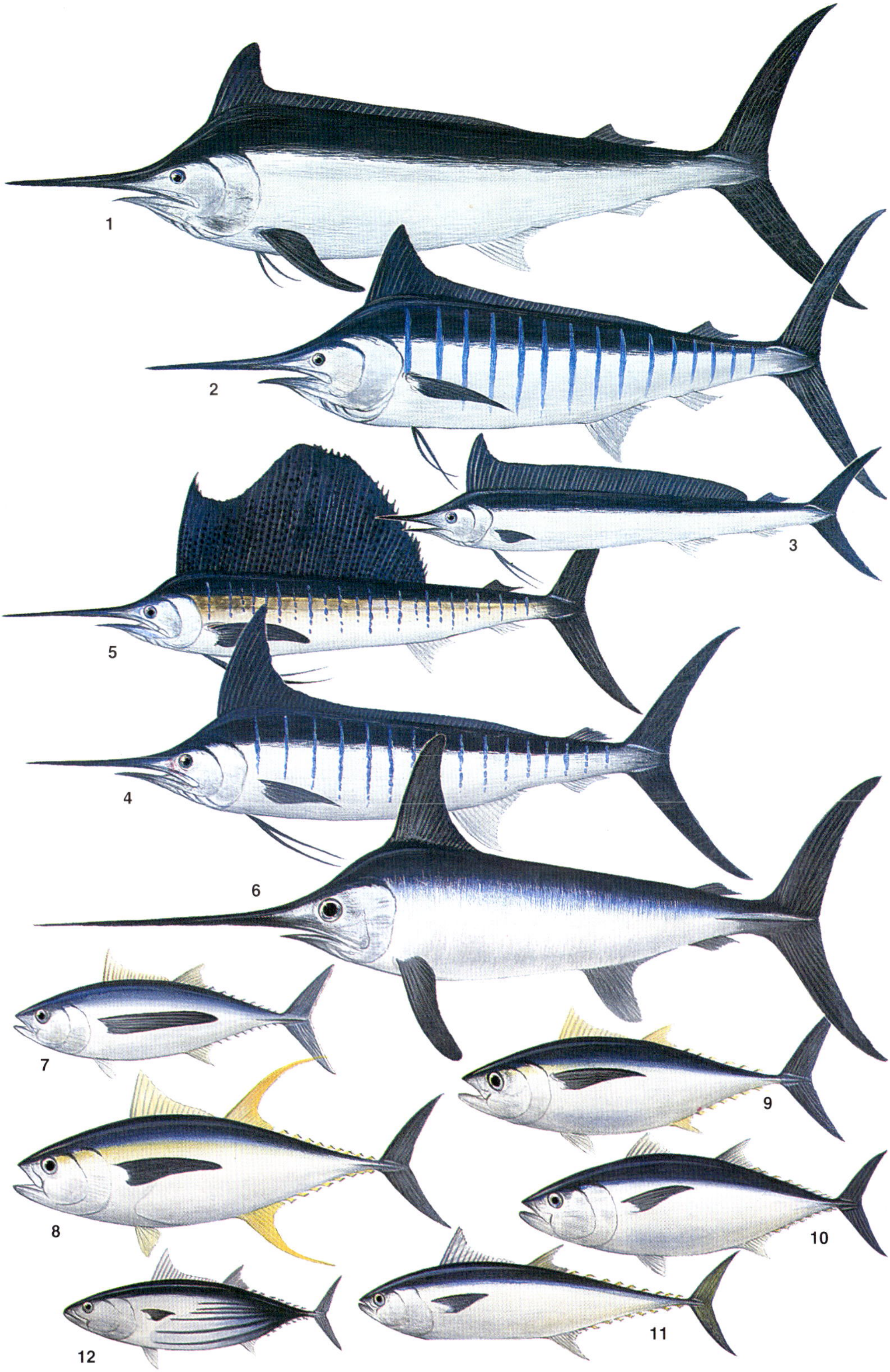
1
2
3
5
4
6
7
9
8
10
12
11

PLATE 97: MACKERELS AND TUNAS (FAMILY SCOMBRIDAE)

1 WAHOO
Acanthocybium solandri (Cuvier, 1831)
Inhabits oceanic waters, generally well offshore; occurs solitarily or in loose aggregations; distinguished from mackerels by its more elongate shape, more numerous spines in the first dorsal fin and by its banded pattern (except also in **2** below); found throughout the region; all tropical seas; to 210 cm. All tackle world record 67.6 kg, Australian record 46.15 kg. ★★★

2 NARROW-BARRED SPANISH MACKEREL
Scomberomorus commerson (Lacepède, 1800)
Inhabits coastal seas, frequently near reefs; similar to **1** above, but has fewer dorsal spines (15-18 versus 23-27) and shorter first dorsal fin; found throughout the region; Indo-W. Pacific; to 235 cm. All tackle world record 44.9 kg Australian record 42.2kg Cairns 10/98 ★★★★

3 BROAD-BARRED SPANISH MACKEREL
Scomberomorus semifasciatus (Macleay, 1884)
Inhabits coastal seas in the vicinity of reefs; distinguished by dark bars extending part way down sides and black area at front of dorsal fin; N. Australia and S. New Guinea only; to 120 cm and 8.5 kg. ★★★

4 AUSTRALIAN SPOTTED MACKEREL
Scomberomorus munroi Collette & Russo, 1980
Inhabits coastal seas; distinguished by broad band of small spots along middle of sides; tropical and subtropical Australia and S. New Guinea only; to 104 cm and 10 kg. ★★★

5 QUEENSLAND SCHOOL MACKEREL
Scomberomorus queenslandicus Munro, 1943
Inhabits inshore coastal waters; distinguished by large dark spots on sides and black area at front of dorsal fin; N. Australia and S. New Guinea only; to 100 cm and ; 12 kg. ★★★

6 ORIENTAL BONITO
Sarda orientalis (Temminck & Schlegel, 1844)
Inhabits coastal seas, sometimes in large schools; distinguished by narrow stripes on upper half of body; found throughout the region; Indo-E. Pacific; to 102 cm and 3.5 kg. ★★

7 MACKEREL TUNA
Euthynnus affinis (Cantor, 1849)
Inhabits oceanic waters, sometimes well offshore; colour pattern is similar to **10** below, but distinguished by the lack of space between the dorsal fins; found throughout the region; Indo-C. Pacific; to 100 cm. All tackle world and Australian record 12.0 kg. ★★

8 LEAPING BONITO
Cybiosarda elegans (Whitley, 1935)
Inhabits coastal waters, sometimes entering estuaries; distinguished by spots on back and stripes on lower side, also has tall, dark-coloured first dorsal fin; Australia (except south coast) and S. New Guinea; to 54 cm and 1.15 kg. ★★

9 CORSELETTED FRIGATE MACKEREL
Auxis rochei (Risso, 1810)
Inhabits coastal and oceanic waters, forming large schools; distinguished by widely separated dorsal fins, elongate patch of short bars or blotches on back at rear half of body, and by its slender shape; found throughout the region; worldwide tropical and subtropical seas; to 50 cm. ★★

10 FRIGATE MACKEREL
Auxis thazard (Lacepède, 1803)
Inhabits coastal and oceanic waters, forming large schools; similar to **7** above, but has wide gap between dorsal fins and more slender shape; found throughout the region; worldwide tropical and subtropical seas; to 58 cm; 4.536 kg. ★★

11 LONG-JAWED MACKEREL
Rastrelliger kanagurta (Cuvier, 1816)
Inhabits coastal waters, near reefs; forms large schools; distinguished by widely separate dorsal fins; narrow lines or rows of spots on upper part of body, and black spot near lower margin of pectoral fin; found throughout the region; Indo-W. Pacific; to 35 cm. ★★★

12 DOGTOOTH TUNA
Gymnosarda unicolor (Rüppell, 1836)
Inhabits offshore waters, usually in the vicinity of coral reefs; distinguished by large conical teeth, relatively large eye, and undulating lateral line; found throughout the region; Indo-C. Pacific, to 150 cm. All tackle world record 131 kg.

13 SHARK MACKEREL
Grammatorcynus bicarinatus (Quoy & Gaimard, 1824)
Inhabits offshore waters, usually in the vicinity of coral reefs; has double lateral line, similar to **14** below, but eye smaller and frequently has dark spots along belly; Australia only (except south coast); to 130 cm and at least 12 kg. ★★

14 DOUBLE-LINED MACKEREL
Grammatorcynus bilineatus (Quoy & Gaimard, 1824)
Inhabits offshore waters, usually in the vicinity of coral reefs; has double lateral line, similar to **13** above, but eye larger and lacks spotting along belly, found throughout the region; Indo-W. Pacific; to 70 cm. ★★

TONNES OF TUNA

The fishes featured on Plates 96-97 occur worldwide, primarily in tropical and temperate seas. They occur both inshore and far out to sea. Tunas, mackerels, and bonitos (family Scombridae; 49 species worldwide) are well known for their fine eating qualities. Over the past 10 years world catches have generally fluctuated between about 5 and 6 million tonnes annually. These fishes are also highly prized by recreational anglers and throughout much of the world they support important subsistence fisheries. All species are powerful swimmers and some undergo extensive annual migrations. The largest species is about 300 cm, but most grow to between 100-200 cm TL.

Billfishes exhibit similar habits and are close relatives of tunas being distinguished by their elongate, spear-like snout. The group is comprised of the swordfish (Xiphidae; 1 species worldwide) and the marlins and sailfishes (Istiophoridae; 11 species worldwide). They are favourite angling fishes and some may reach the massive size of over 450 cm TL and 700 kg in weight. Although the billfishes and larger tunas are distributed widely throughout the N. Australian-S.E. Asian region, their occurance is dependent on deep, clear blue, oceanic seas.

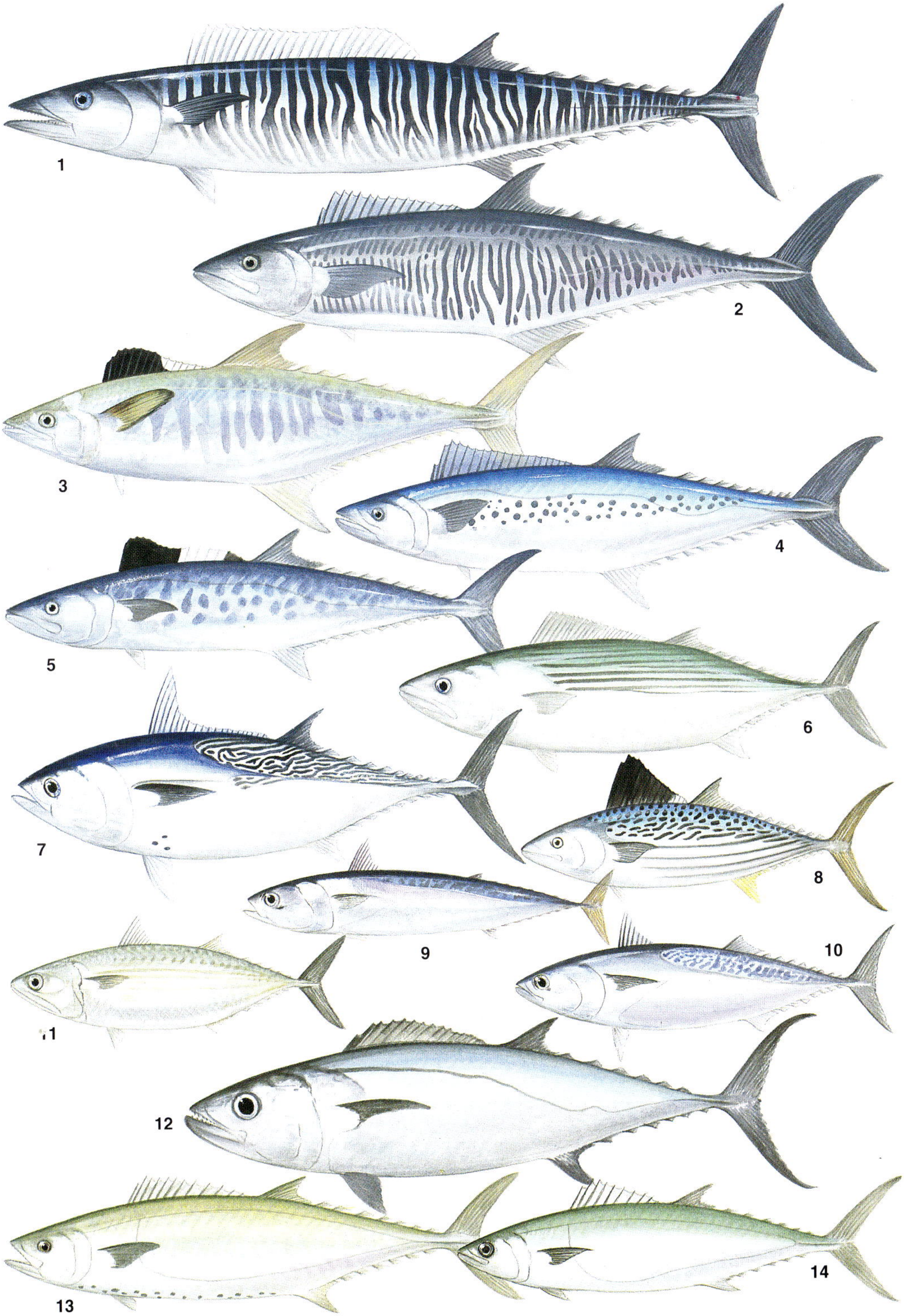
1
2
3
4
5
6
7
8
9
10
11
12
13
14

PLATE 98: FLOUNDERS AND SOLES

1 QUEENSLAND HALIBUT
Psettodes erumei (Bloch & Schneider, 1801)
Inhabits sand or mud bottoms; distinguished by large mouth and large sharp teeth; found throughout the region; Indo-W. Pacific; to 64 cm. (PSETTODIDAE) ★★★

2 LARGE-TOOTHED FLOUNDER
Pseudorhombus arsius (Hamilton, 1822)
Inhabits sand or mud bottoms, sometimes in estuaries; distinguished by several enlarged teeth at front of mouth, large brown blotch behind pectoral fin, smaller blotch about half way between first spot and tail base, and scattered dark-edged pale spots; found throughout the region; Indo-W. Pacific; to 31 cm. (BOTHIDAE) ★★★

3 TWINSPOT FLOUNDER
Pseudorhombus diplospilus Norman, 1926
Inhabits sand bottoms; distinguished by 4 pairs of "eye-spots"; other similar species (not shown) include *P. argus* (same "eye-spot" pattern except 1 or 2 additional sets of spots on middle of body near tail base), *P. dupliocellatus* (same "eye-spot" pattern as *P. diplospilus* but lacks enlarged canine teeth [versus strong canines]), *P. quinquocellatus* (similar pattern to *P. argus* but only a single eye-spot instead of double spots in each marking) and *P. spinosus* (3 well separated, single spots on back); N. Australia and Indo-Malay Archipelago; to 25 cm. (BOTHIDAE) ★★

4 DEEP-BODIED FLOUNDER
Pseudorhombus elevatus Ogilby, 1912
Inhabits sand or mud bottoms; similar to **2** above, but lacks enlarged teeth, is deeper-bodied (rounder), and generally has more pale spots; N. Australia and Indo-Malay Archipelago; N. Indian Ocean and Indo-Australian Archipelago; to 15 cm. (BOTHIDAE)

5 SMALL-TOOTHED FLOUNDER
Pseudorhombus jenynsii Bleeker, 1855
Inhabits sand or mud bottoms; distinguished by 6 pale-edged spots as shown in illustration; Australia only; to 34 cm. (BOTHIDAE) ★★★

6 INTERMEDIATE FLOUNDER
Asterhombus intermedius (Bleeker, 1866)
Inhabits sand bottoms; distinguished by dense scattering of dark speckles and blotches on fins and body; also known as Blotched flounder; N.W. Australia and Indonesia; Indian Ocean and Indo-Australian Archipelago; to 13 cm. (BOTHIDAE)

7 SPINY-HEADED FLOUNDER
Engyprosopon grandisquama (Temminck & Schlegel, 1846)
Inhabits sand bottoms; distinguished by well separated eyes and pair of black spots on margin of tail; also known as Mottled wide-eyed flounder; found throughout the region; Indo-W. Pacific; to 13 cm. (BOTHIDAE)

8 THREE-SPOT FLOUNDER
Grammatobothus polyophthalmus (Bleeker, 1866)
Inhabits sand bottoms; distinguished by mottled pattern and 3 pale-edged dark spots; N. Australia and Indo-Malay Archipelago; N.E. Indian Ocean and Indo-Australian Archipelago; to 14 cm. (BOTHIDAE)

9 PANTHER FLOUNDER
Bothus pantherinus (Rüppell, 1830)
Inhabits sand bottoms near coral reefs; distinguished by mottled appearance with numerous brown and white spots, and large brown patch in middle of rear part of body, males have filamentous pectoral rays reaching tail base; also known as Leopard flounder; found throughout the region; Indo-W. Pacific; to 24 cm. (BOTHIDAE) ★★

10 DARK THICK-RAYED SOLE
Aesopia cornuta Kaup, 1858
Inhabits sand or mud bottoms; similar to 14 and 15 below, but has darker stripes, *A. heterorhinos* (not shown) also similar but has irregular-shaped broken bars and white black-tipped tail; N. Australia only; to 18 cm. (SOLEIDAE)

11 TUFTED SOLE
Dexillichthys muelleri (Steindachner, 1879)
Inhabits sand bottoms; distinguished by light brown colour with patches of dark skin filaments; N. Australia and S. New Guinea only; to 20 cm. (SOLEIDAE)

12 DARK-SPOTTED SOLE
Aseraggodes melanospilus (Bleeker, 1854)
Inhabits sand bottoms; distinguished by numerous small dark spots and pair of large dark blotches superimposed on mottled pattern; found throughout the region; mainly W. Pacific; to 15 cm. (SOLEIDAE)

13 PEACOCK SOLE
Pardachirus pavoninus (Lacepède, 1802)
Inhabits sand bottoms; distinguished by numerous black spots with broad pale margins; experiments have shown that the mucus of this fish has shark-repellant qualities; found throughout the region; Indo-C. Pacific; to 25 cm. (SOLEIDAE)

14 WICKER-WORK SOLE
Zebrias craticula (McCulloch, 1916)
Inhabits sand bottoms; similar to **10** and **15**, but dark bars more numerous and narrower; N. Australia only; to 15 cm. (SOLEIDAE)

15 HARROWED SOLE
Strabozebrias cancellatus (McCulloch, 1916)
Inhabits sand bottoms; bars lighter and body more elongate than **10** above; bars wider and less numerous than **14** above; N. Australia only; to 27 cm. (SOLEIDAE)

16 SPOTTED SOLE
Phyllichthys punctatus McCulloch, 1916
Inhabits sand bottoms; distinguished by dense covering of diffuse white blotches; N. Australia only; to 24 cm. (SOLEIDAE)

17 COCKATOO FLOUNDER
Samaris cristatus Gray, 1831
Inhabits sand bottoms; distinguished by filamentous dorsal fin-rays on top of head; also known as Cockatoo righteye flounder; N. Australia and Indo-Malay Archipelago; N. Indian Ocean and W. Pacific; to 17 cm. (PLEURONECTIDAE)

18 OCELLATED FLOUNDER
Psammodiscus ocellatus Günther, 1862
Inhabits sand bottoms; distinguished by dense mottled or speckled pattern, also 3 or 4 paled edged dark spots usually present; found throughout the region; Indo-Australian Archipelago; to 15 cm. (PLEURONECTIDAE)

19 PATTERNED TONGUE SOLE
Paraplagusia bilineata (Bloch, 1787)
Inhabits sand bottoms; distinguished by elongate shape and fringe of branched tentacles on lips; other elongate soles (not shown) from this region without fringe of tentacles belong to the genus *Cynoglossus*; found throughout the region; Indo-W. Pacific; to 40 cm. (CYNOGLOSSIDAE) ★★★

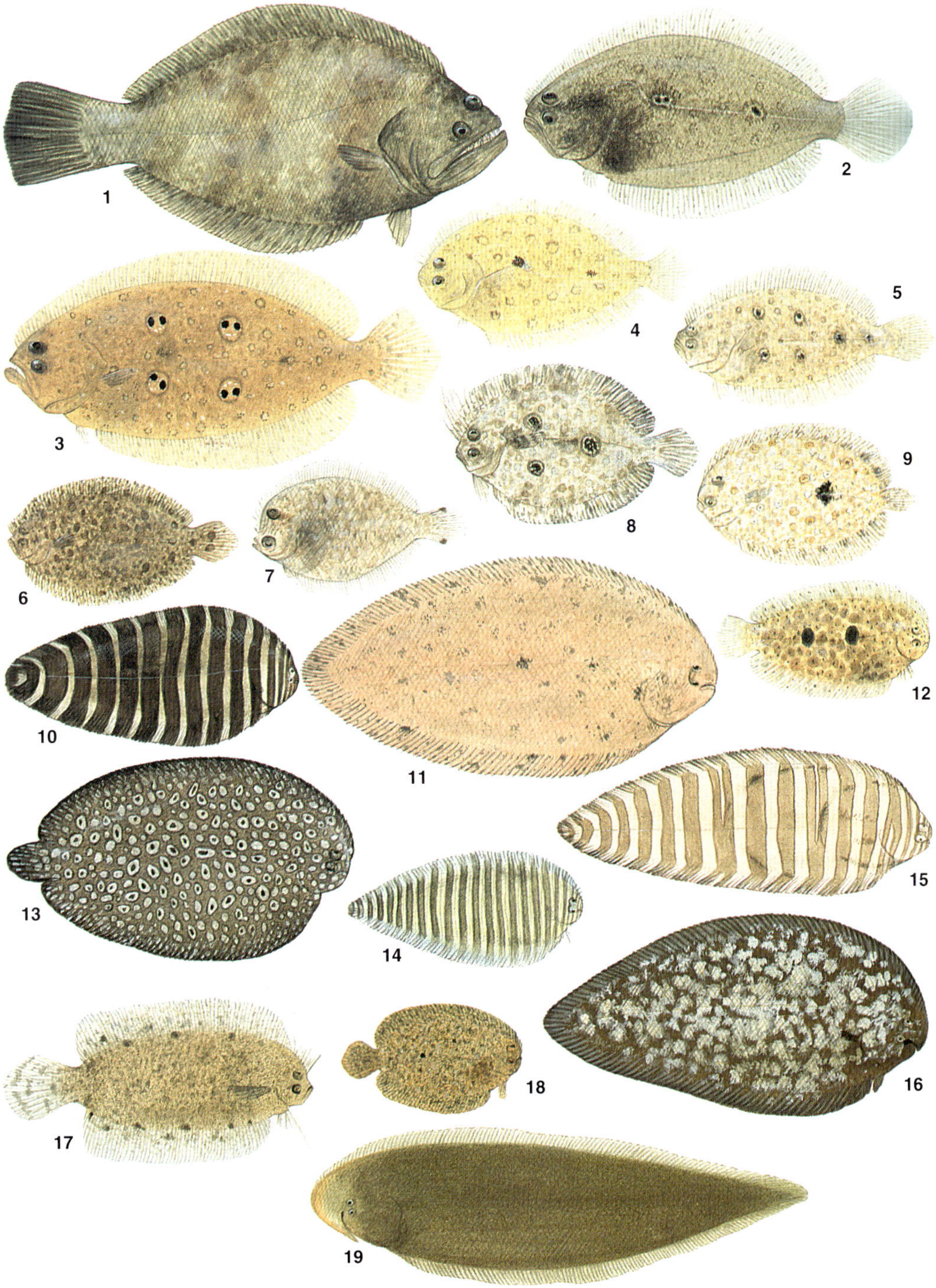
1
2
3
4
5
6
7
8
9
10
11
12
13
14
15
16
17
18
19

PLATE 99: DUCKBILLS, DRIFTFISHES, AND TRIPODFISHES

1 SPOTTED DUCKBILL
Bembrops aethalea McKay, 1971
Inhabits offshore trawling grounds; distinguished by 2 spines on gill cover, a loose flap of skin on rear part of upper jaw, and small black spots on back; N.W. Australia only; to 20 cm. (PERCOPHIDAE)

2 SHARPNOSED DUCKBILL
Bembrops filodorsalia Okada & Suzuki, 1952
Inhabits offshore trawling grounds; similar to **1** above, but lacks distinct spotting and tail has dark margin with dark spot on upper part of base; found throughout the region; mainly W. Pacific; to 20 cm. (PERCOPHIDAE)

3 BLOTCHED DUCKBILL
Chironema chlorotaenia McKay, 1971
Inhabits offshore trawling grounds; distinguished by 2 spines on gill cover, no flap of skin at rear of upper jaw, elongate orange spots, and diffuse brown blotches on sides; N.W. Australia only; to 21 cm. (PERCOPHIDAE)

4 INDIAN DRIFTFISH
Ariomma indica (Day, 1870)
Inhabits oceanic waters usually well offshore; distinguished by ovate silvery body and rounded snout with small mouth; similar to **5** below, but has 2 dorsal fins; found throughout the region; Indo-W. Pacific; to 25 cm.

5 NORTH-WEST RUFFE
Psenopsis humerosa Munro, 1958
Inhabits oceanic waters usually well offshore; similar to **4** above, but has only one dorsal fin; N.W. Australia only; to 20 cm.

6 WHITELEGGE'S EYEBROWFISH
Cubiceps whiteleggii (Waite, 1894)
Inhabits oceanic waters usually well offshore; distinguished by elongate shape, 2 dorsal fins, and overall darkish colour; N. Australia only; to 13 cm.

7 THREE-TOOTH PUFFER
Triodon macropterus Lesson, 1829
Inhabits coastal waters and offshore trawling grounds; distinguished by large skin flap on belly and black spot on middle of side; found throughout the region; Indo-W. Pacific; to 60 cm. **P**

8 BLACKTAIL TRIPODFISH
Triacanthus biaculeatus (Bloch, 1786)
Inhabits trawling grounds; similar to **9** below, but spiny dorsal fin entirely black (versus only front half of fin black); found throughout the region; Indo-W. Pacific; to 25 cm.

9 SILVER TRIPODFISH
Triacanthus nieuhofi Bleeker, 1852
Inhabits trawling grounds; similar to **8** above, but rear half of spiny dorsal fin pale (versus black); N. W. Australia and Indo-Malay Archipelago; to 28 cm.

10 BLACKTIP TRIPODFISH
Trixiphichthys weberi (Chadhuri, 1910)
Inhabits trawling grounds; distinguished by black tip on spiny dorsal fin and diffuse dark blotches on side similar to **11** below, but has longer, narrower snout; N.W. Australia, Indo-Malay Archipelago, and Philippines; Andaman Sea and Indo-Australian Archipelago; to 20 cm.

11 BLOTCHED TRIPODFISH
Pseudotriacanthus strigilifer (Cantor, 1850)
Inhabits trawling grounds; similar to **10** above, but has more pronounced pattern of blotches on side and shorter snout; N.W. Australia and througout S.E. Asia; Indian Ocean and Indo-Australian Archipelago; to 24 cm.

12 LONGSNOUT SPIKEFISH
Halimochirurgus centriscoides Alcock, 1899
Inhabits trawling grounds; similar to **14** below, but snout is directed upwards and lacks enlarged mouth opening at tip; N.W. Australia and Indo-Malay Archipelago; Andaman Sea and Indo-Australian Archipelago; to 15 cm.

13 SHORTSNOUT SPIKEFISH
Triacanthodes ethiops Alcock, 1894
Inhabits trawling grounds; similar in general shape to **8-11** above, but has 6 spines (versus 5) in first dorsal fin and much shorter and thicker tail base, usually has 3 diverging stripes on side; N.W. Australia and throughout S.E. Asia; Indo-W. Pacific; to 9 cm.

14 TRUMPETSNOUT SPIKEFISH
Macrorhamphosodes platycheilus Fowler, 1934
Inhabits trawling grounds; similar to **12** above, but has snout directed straight forward and enlarged mouth opening at tip; N.W. Australia, Indonesia, and Philippines; Andaman Sea and Indo-Australian Archipelago; to 13 cm.

FLATFISHES
(See Plate 98)

Flatfishes are amazing bottom-dwelling creatures with highly compressed bodies and both eyes oddly situated on the same side of the head. The larvae actually begin life looking very much like normal, symmetrical fish larvae - but profound changes occur during their first few weeks. One of the eyes slowly migrates across the top of the head, and at the same time the mouth and other body parts become asymmetrically distorted. The end result is a creature that looks more like a camouflaged pancake than a fish. Camouflage is the operative word as flatfishes possess chameleon-like powers enabling them to quickly match their surroundings. In stark contrast, the underside of these fishes is usually devoid of pigment. Their spectacular colour changes are useful for hiding from larger predators or for deceiving its own prey, usually small fishes and small invertebrates. The flatfish group includes seven families with about 540 species worldwide. Most are marine fishes, inhabiting continental shelves and slopes, but fresh and brackish waters are also inhabited.

TRASH FISHES

Most of the species featured on this plate particularly the duckbills, tripodfishes and spikefishes, are classed as "trash" by commercial fishermen. These species, along with many other small bottom fishes are often captured in large numbers while trawling. Because of their small size or poor edibility they are dumped overboard after sorting out the marketable species. Unfortunately most of the trash catch dies during this process, resulting in tremendous waste of this resource. However, in some parts of the world the trash catch is effectively harvested for use as fish meal and fertilizer.

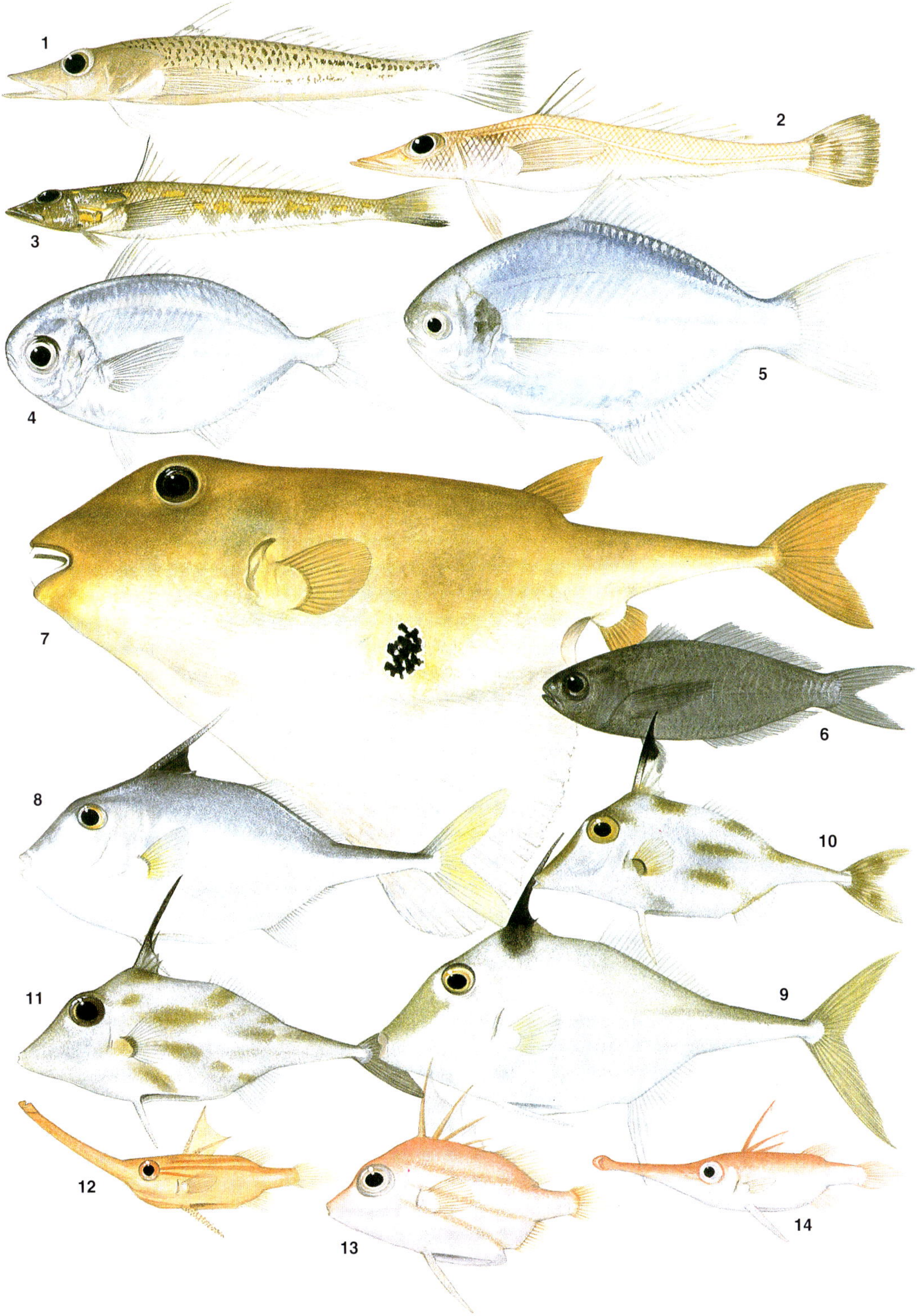
1
2
3
4
5
7
6
8
10
11
9
12
13
14

PLATE 100: TRIGGERFISHES (FAMILY BALISTIDAE)

1 RED-LINED TRIGGERFISH
Balistapus undulatus (Park, 1797)
Inhabits coral reefs; distinguished by red diagonal stripes; Great Barrier Reef, N.W. Australia, and throughout S.E. Asia; Indo-C. Pacific; to 35 cm.

2 STARRY TRIGGERFISH
Abalistes stellatus (Bloch & Schneider, 1801)
Inhabits coral and rocky reefs; distinguished by relatively elongate shape, blue to orange spotting on body, and frequently with 3 white patches on back at base of dorsal fin; large adults have elongate filaments on tail; found throughout the region; Indo-W. Pacific; to 60 cm and 2.4 kg.

3 BLUE-FINNED TRIGGERFISH
Balistoides viridescens (Bloch & Schneider, 1801)
Inhabits coral and rocky reefs; nest guarding females may attack divers inflicting painful bites; distinguished by relatively large size, dark area with pale "bridle" around upper part of mouth, yellowish cheeks, diffuse dark bar through eye and dark fin margins; Great Barrier Reef, N.W. Australia, and throughout S.E. Asia; Indo-C. Pacific; to 60 cm and 2.3 kg.

4 EBONY TRIGGERFISH
Melichthys niger (Bloch, 1786)
Inhabits coral reefs; distinguished by black colour of body and fins; *Melichthys vidua* (Pl.106) is a similar species, but has pale dorsal, anal, and tail fins; Great Barrier Reef, N.W. Australia, and throughout S.E. Asia; Indo-E. Pacific; to 35 cm.

5 WHITE-BARRED TRIGGERFISH
Rhinecanthus aculeatus (Linnaeus, 1758)
Inhabits coral reefs; distinguished by white diagonal bars above anal fin base; Great Barrier Reef, N.W. Australia, and throughout S.E. Asia; Indo-C. Pacific; to 30 cm.

6 WEDGE-TAILED TRIGGERFISH
Rhinecanthus rectangulus (Bloch & Schneider, 1801)
Inhabits coral reefs on limestone platforms exposed to wave action; distinguished by black wedge-shaped mark at tail base; Great Barrier Reef, N.W. Australia, and throughout S.E. Asia; Indo-C. Pacific; to 24 cm.

7 YELLOW-SPOTTED TRIGGERFISH
Pseudobalistes fuscus (Bloch & Schneider, 1801)
Inhabits coral reefs; distinguished by dense network of yellow-orange spots on body and pointed lobes of tail; Great Barrier Reef, N.W. Australia, and throughout S.E. Asia; Indo-C. Pacific; to 50 cm.

8 BLACK TRIGGERFISH
Sufflamen chrysopterus (Bloch & Schneider, 1801)
Inhabits coral reefs; distinguished by dark colour of body, pale line below eye and white-edged tail; Great Barrier Reef, N.W. Australia, and throughout S.E. Asia; Indo-C. Pacific; to 30 cm.

9 PALLID TRIGGERFISH
Sufflamen bursa (Bloch & Schneider, 1801)
Inhabits coral reefs, often in rubble areas; distinguished by general grey colouration with narrow brown to orange bar through eye and another from dorsal fin to pectoral base; Great Barrier Reef, N.W. Australia, and throughout S.E. Asia; Indo-C. Pacific; to 30 cm.

10 BROWN TRIGGERFISH
Sufflamen fraenatus Latreille, 1804
Inhabits coral reefs and hard flat-bottom areas with occasional outcrops; distinguished by brown colour and bridle-like marking behind mouth; Great Barrier Reef, N.W. Australia, and throughout S.E. Asia; Indo-C. Pacific; to 30 cm.

11 LINED TRIGGERFISH
Xanthichthys lineopunctatus (Hollard, 1854)
Inhabits offshore reefs, often seen well above the bottom; distinguished by narrow lines on head and body and orange outline on tail; N.W. Australia and possibly S. Indonesia; Indo-W. Pacific; to 30 cm.

TRIGGERFISHES

Triggerfishes (family Balistidae) are characterised by a rugby-ball shape, leathery skin, and small mouth with powerful crushing jaws. Their common name is derived from the peculiar mechanism by which the first dorsal spine can be locked into an erect position by the second dorsal spine - if pressure is exerted on the trigger-like second spine, the first spine can be "unlocked" and depressed. This device is used to good advantage at night when the fish wedges into a coral crevice and "locks" itself in. Most of the 40 triggerfish species inhabit the topical Indo-Pacific region, but a few are found in other warm seas. They are usually solitary in habit, although the White-spotted Trigger (Plate 101-4) forms large schools. Swimming is slow and deliberate, usually accomplished by gentle undulations of the second dorsal and anal fins, but when threatened they retreat quickly to a hole in the reef, by using their tail. In many species the same hole is always used and also serves as a nocturnal resting place. Some triggers, for example 5 and 6, are capable of producing grunt-like sounds when disturbed. Although the mouth is relatively small, triggers possess powerful jaws and strong teeth. These are used for crushing a variety of hard-shelled prey including crabs, molluscs, and echinoderms. Sponges, gorgonians, hydroids, corals, and algae are also consumed by several species. A few triggers including the Redtooth (Plate 101-3), White-spotted (103-4), and the species of *Xanthichthys* (100-11 and 101-5) feed heavily on zooplankton. Triggerfishes lay eggs on the bottom, which are aggressively guarded by the female. Divers should exercise caution when swimming through a nest area. It is best to give the Blue-finned or "Titan" Trigger (Plate 100-3) a wide berth on these occasions as it is particularly belligerent and can inflict nasty bites.

The Clown Trigger (Plate 101-2) is easily the most spectacular member of the family. It has long been a favourite of marine aquarists. Several decades ago, when it first appeared in pet shops, the price tag was over US$300 per fish. Nowadays it is more common, but still fetches at least $50.

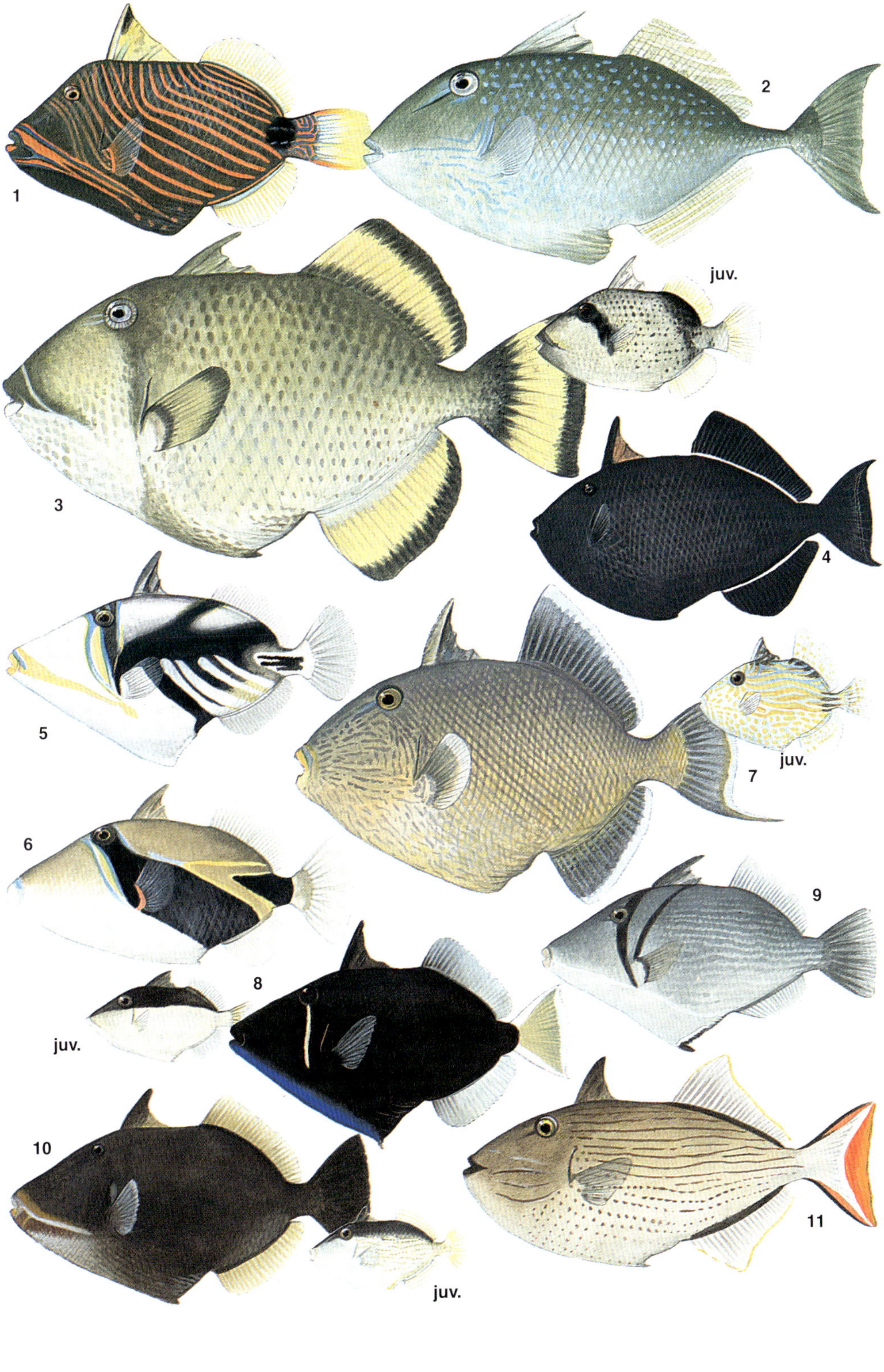
1
2
juv.
3
4
5
7
juv.
6
9
8
juv.
10
11
juv.

PLATE 101: TRIGGERFISHES AND LEATHERJACKETS (FILEFISHES)

1 YELLOWMARGIN TRIGGERFISH
Pseudobalistes flavimarginatus (Rüppell, 1829)
Inhabits coral reefs; distinguished by raised spines on scales of tail base, yellow margins on fin, and cross-hatch pattern on side; Great Barrier Reef, offshore reefs of W. Australia, and throughout S.E. Asia; Indo-C. Pacific; to 60 cm. (BALISTIDAE)

2 CLOWN TRIGGERFISH
Balistoides conspicillum (Bloch & Schneider, 1801)
Inhabits coral reefs, usually seen on outer slopes; distinguished by yellow lips and large white spots on lower half of head and body; Great Barrier Reef, offshore reefs of W. Australia, and throughout S.E. Asia; Indo-C. Pacific; to 50 cm. (BALISTIDAE)

3 REDTOOTH TRIGGERFISH
Odonus niger (Rüppell, 1837)
Inhabits coral reefs, sometimes seen in groups on outer reefs; feeds on zooplankton; distinguished by dark blue to purplish blue colour, lunate tail (with prolonged lobes in adult), blue fin margins, and red teeth; Great Barrier Reef, offshore reefs of W. Australia, and throughout S.E. Asia; Indo-C. Pacific; to 35 cm. (BALISTIDAE)

4 WHITE-SPOTTED TRIGGERFISH
Canthidermis maculatus (Bleeker, 1865)
Inhabits steep outer reef slopes, sometimes seen far out to sea around logs or other floating debris; distinguished by relatively elongate body, triangular dorsal and anal fins, rounded caudal fin with upper and lower lobes slightly elongate, colour changeable from grey to blackish, often with small white spots covering head and body; islands adjacent to deep water throughout S.E.Asia; tropical circumglobal; to 35 cm. (BALISTIDAE)

5 GILDED TRIGGERFISH
Xanthichthys auromarginatus (Bennett, 1831)
Inhabits steep outer reef slopes in 15-70 m depth; feeds on zooplankton; male distinguished by blue patch on cheek and yellow fin margins; female (not shown) is bluish grey with pale scale centres; Great Barrier Reef and throughout S.E. Asia; Indo-C. Pacific; to 22 cm. (BALISTIDAE)

6 BLACKPATCH TRIGGERFISH
Rhinecanthus verrucosus (Linnaeus, 1758)
Inhabits sheltered coral reefs, including silty inshore areas; distinguished by large black patch on lower side and narrow dark bar through eye to pectoral fin base; Great Barrier Reef and throughout S.E. Asia; Indo-W. Pacific; to 23 cm. (BALISTIDAE)

7 HALFMOON TRIGGERFISH
Rhinecanthus lunula Randall & Steene, 1983
Inhabits coral reefs, usually seen in depths below 10 m; distinguished by black first dorsal fin, black spot at pectoral-fin base, another in front of anal fin, and black ring around tail base; Great Barrier Reef and islands of the South Pacific eastward to Pitcairn Group; to 28 cm. (BALISTIDAE)

8 LARGE-SCALED LEATHERJACKET
Cantheschenia grandisquamis Hutchins, 1977
Inhabits coral and rocky reefs; distinguished by orange tail with broad blue margins and dark blotch in front of anal fin; S.Great Barrier Reef and New South Wales; to 26 cm. (MONACANTHIDAE)

9 BRUSH-SIDED LEATHERJACKET
Amanses scopas Cuvier, 1829
Inhabits coral reefs; distinguished by overall dark brown colouration with several incomplete blackish bars on middle of side; males with patch of numerous long spines and females with toothbrush-like patch of bristles in front of tail base; Great Barrier Reef and throughout S.E. Asia; Indo-W. Pacific; to 20 cm. (MONACANTHIDAE)

10 YELLOWEYE LEATHERJACKET
Cantherhinus dumerilii (Hollard, 1854)
Inhabits coral reefs; distinguished by yellowish iris and tail, also by two pairs of forward-curved spines on tail base; Great Barrier Reef, offshore reefs of W. Australia, and throughout S.E. Asia; Indo-E. Pacific; to 35 cm. (MONACANTHIDAE)

11 SMALL-SPOTTED LEATHERJACKET
Pseudomonacanthus macrurus (Bleeker, 1857)
Inhabits shallow weedy areas and trawling grounds; a nondescript species which can change its colours to match the surroundings; found throughout S.E. Asia; to 24 cm. (MONACANTHIDAE)

12 RHINOCEROS LEATHERJACKET
Pseudaluteres nasicornis (Schlegel, 1846)
Inhabits trawling grounds to 75 m depth, also found in shallow weedy areas; distinguished by pair of dark stripes on back and unusual placement (in front of eye) of dorsal spine; Queensland and throughout S.E. Asia; W. Pacific; to 18 cm. (MONACANTHIDAE)

13 ORANGE-TAILED LEATHERJACKET
Pervagor aspricaudus (Hollard, 1854)
Inhabits coral reefs; distinguished by orange tail and numerous fine dark spots on head and body; similar to *P. janthinosoma*, but lacks prominent black blotch above gill opening; Great Barrier Reef; anti-equatorial distribution, mainly around oceanic islands from Mauritius to New Caledonia and Taiwan to Hawaii; to 12 cm. (MONACANTHIDAE)

14 DARK-HEADED LEATHERJACKET
Pervagor melanocephalus (Bleeker, 1853)
Inhabits coral reefs; distinguished by bright orange body and brown head; also known as Lace-Finned Leatherjacket; Great Barrier Reef and throughout S.E. Asia; W. Pacific; to 10 cm. (MONACANTHIDAE)

15 BLACK-LINED LEATHERJACKET
Pervagor nigrolineatus (Herre, 1927)
Inhabits sheltered coral reefs; distinguished by pale stripe extending from cheek to middle of body and by prolonged rays at front of second dorsal fin; offshore reefs of W. Australia, and throughout S.E. Asia; W. Pacific; to 7 cm. (MONACANTHIDAE)

16 BRISTLE-TAILED LEATHERJACKET
Acreichthys tomentosus (Linnaeus, 1758)
Inhabits weed-rubble bottoms on shallow reefs; distinguished by chameleon-like colour pattern and numerous skin-flap appendages on body; found throughout S.E. Asia; W. edge of Pacific, excluding Australia; to 12 cm. (MONACANTHIDAE)

17 RADIAL LEATHERJACKET
Acreichthys radiatus (Popta, 1901)
Inhabits coral reefs, frequently seen among soft corals; distinguished by narrow white bars (may be faint) on side and filamentous appendages on dorsal spine; Great Barrier Reef and throughout S.E. Asia; W. Pacific; to 7 cm.

18 MIMIC LEATHERJACKET
Paraluteres prionurus (Bleeker, 1851). Text on page 242.

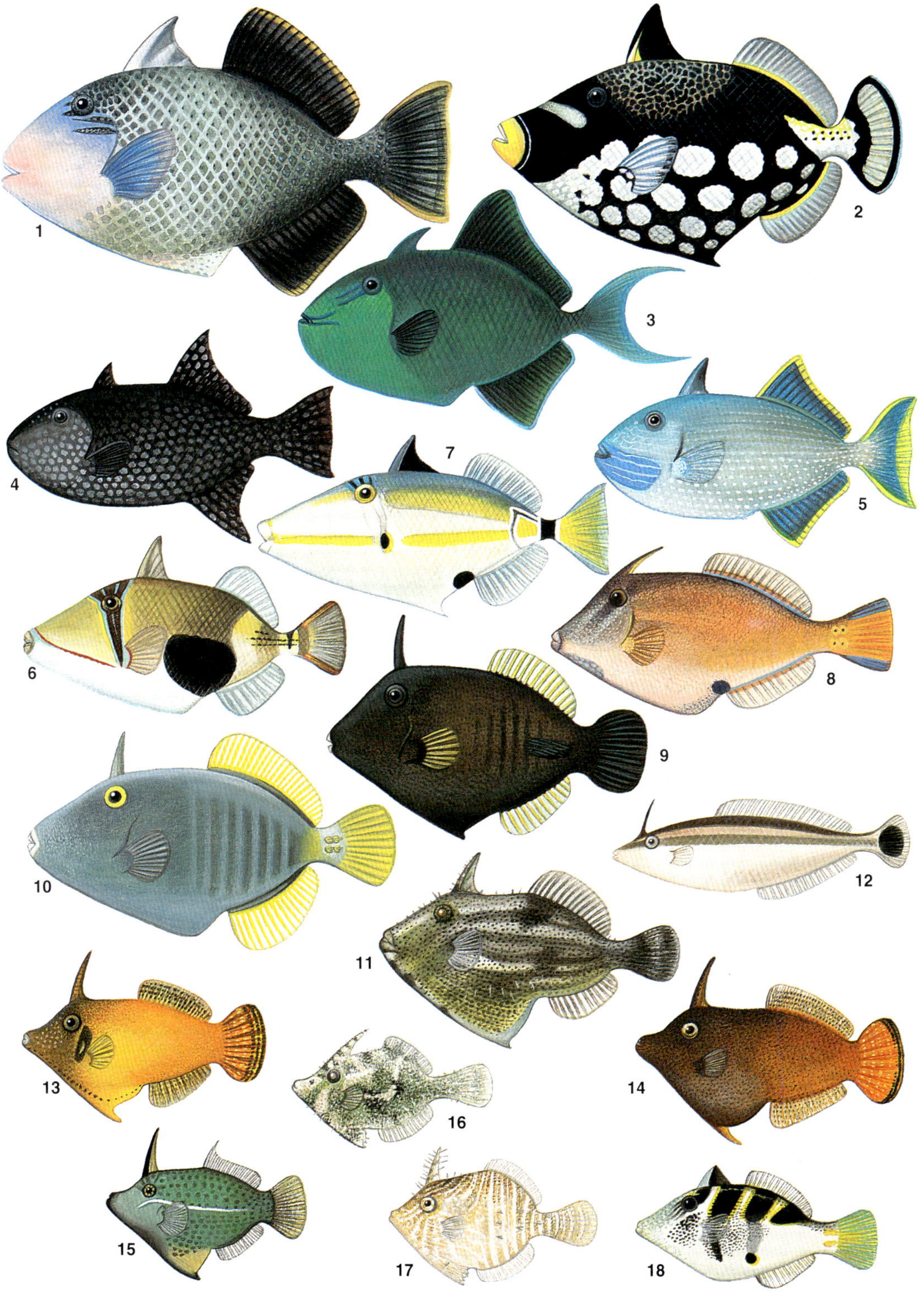
1
2
3
4
7
5
6
8
9
10
12
11
13
14
16
15
17
18

PLATE 102: LEATHERJACKETS (FAMILY MONACANTHIDAE)

1 UNICORN LEATHERJACKET
Aluterus monoceros (Linnaeus, 1758)
Inhabits coral reefs and flat bottom trawling grounds; distinguished by large ovate shape and slender dorsal spine, similar to **2** below, but has rounder head, shorter tail and lacks ornate pattern; found throughout the region; worldwide tropical and subtropical seas; to 76 cm. ★★★

2 SCRIBBLED LEATHERJACKET
Aluterus scriptus (Osbeck, 1765)
Inhabits coral reefs; distinguished by relatively elongate shape, slender dorsal spine, spotted pattern, and elongate tail; found throughout the region; world-wide tropical and subtropical seas; to 100 cm. ★★★

3 HONEYCOMB LEATHERJACKET
Cantherhines pardalis (Rüppell, 1866)
Inhabits coral reefs; distinguished by network of dark spots on sides; Great Barrier Reef, N.W. Australia, and throughout S.E. Asia; Indo-C. Pacific; to 20 cm. ★★★

4 SPECTACLED LEATHERJACKET
Cantherhines fronticinctus (Günther, 1866)
Inhabits coral reefs; distinguished by white bar on tail base, sometimes with irregular dark blotches on sides; N.W. Australia, S. New Guinea, Indonesia, and Philippines; Indo-W. Pacific, to 23cm. ★★★

5 BEARDED LEATHERJACKET
Anacanthus barbatus Gray, 1831
Inhabits sand-weed bottoms; distinguished by unusually elongate shape and barbel on chin; found throughout the region; E. Indian Ocean and Indo-Australian Archipelago; to 35 cm.

6 BLUE-SPOTTED LEATHERJACKET
Eubalichthys caeruleoguttatus Hutchins, 1977
Inhabits trawling grounds; male has distinctive pattern of stripes on head and body and elevated dorsal and anal fin lobes, female distinguished by blue to whitish spots on side; W. Australia only; to 38 cm. ★★★

7 FAN-BELLIED LEATHERJACKET
Monacanthus chinensis (Osbeck, 1765)
Inhabits reef and weed bottoms, also trawling grounds; distinguished by triangular back profile, huge skin flap on belly, and filamentous extension on tail; found throughout the region; mainly W. Pacific; to 38 cm kg. ★★★

8 PRICKLY LEATHERJACKET
Chaetoderma penicilligera (Cuvier, 1817)
Inhabits sea grass and trawling grounds; distinguished by round shape and tentacles on head and body; found throughout the region; mainly W. Pacific; to 31 cm. ★★★

9 BEAKED LEATHERJACKET
Oxymonacanthus longirostris Bloch & Schneider, 1801
Inhabits coral reefs, usualy associated with branching or plate corals; distinguished by tubular snout and orange spots; Great Barrier Reef, N.W. Australia, and throughout S.E. Asia; Indo-W. Pacific; to 9 cm

10 THREADFIN LEATHERJACKET
Paramonacanthus filicauda (Günther, 1880)
Inhabits trawling grounds; distinguished by relatively deep body, dark blotch below front of dorsal fin, and filament on upper lobe of tail; tropical and subtropical Australia only; to 22 cm. ★★★

11 HAIR-FINNED LEATHERJACKET
Paramonacanthus choirocephalus (Bleeker, 1852)
Inhabits trawling grounds; distinguished by slightly diagonal dark stripes on side with intense blotch in middle stripe just behind pectoral fin; female shown here, male is much more slender; N. Australia and Indo-Malay Archipelago; N.E. Indian Ocean and Indo-Australian Archipelgo; to 14 cm. ★★★

12 RED-TAILED LEATHERJACKET
Pervagor janthinosoma (Bleeker, 1854)
Inhabits coral reefs; distinguished by brownish body and fan-shaped red tail; Great Barrier Reef, W. Australia, and throughout S.E. Asia; Indo-W. Pacific; to 14 cm.

13 MODEST LEATHERJACKET
Thamnaconus modestoides (Barnard, 1927)
Inhabits trawling grounds; distinguished by relatively elongate shape and non-descript pattern; N.W. Australia and Indonesia; Indian Ocean and W. Pacific; to 30 cm.

14 BROWN-BLOTCHED LEATHERJACKET
Stephanolepis sp.
Inhabits trawling grounds; an undescribed species distinguished by a relatively deep body and dense pattern of dark blotches; W. Australia only; to 34 cm. ★★★

15 POT-BELLIED LEATHERJACKET
Pseudomonacanthus peroni (Hollard, 1854)
Inhabits trawling grounds; distinguished by moderately large skin flap on belly and small spots on body and tail; N. Australia only; to 40 cm. ★★★

16 PAXMAN'S LEATHERJACKET
Colurodontis paxmani Hutchins, 1977
Inhabits sea grass; distinguished by movable pelvic spine and flattened teeth at front of jaws; W. Australia only, between Shark Bay and Dampier Archipelago; to 15 cm.

PLATE 101

18 MIMIC LEATHERJACKET
Paraluteres prionurus (Bleeker, 1851)
Inhabits coral reefs; a mimic of *Canthigaster valentini* (Pl. 104-13), a poisonous puffer; distinguished from the nearly identical puffer by the presence of the spinous first dorsal fin and the longer-based dorsal and anal fins (these have 22-28 rays, compared to 9 rays in the puffer); Great Barrier Reef and throughout S.E. Asia; Indo-W. Pacific; to 10 cm. (MONACANTHIDAE)

LEATHER JACKETS

Leatherjackets (family Monacanthidae) are closely related to triggerfishes (Plates 100-101) and share many of their anatomical peculiarities. However they are generally more laterally compressed and usually have two dorsal spines instead of three. Members of both groups are able to swim slowly by undulating movements of the soft dorsal and anal fins. Rapid bursts are achieved mainly by vigorous tail movement. Australia has more leatherjackets than any other region with nearly 60 of the estimated total of 85 species being represented. However, the majority are confined to temperate and subtropical seas. The flesh of many species is good eating and in places such as Australia and Japan they are important commercial fishes. Food items are similar to those of triggerfishes, mainly consisting of benthic invertebrates.

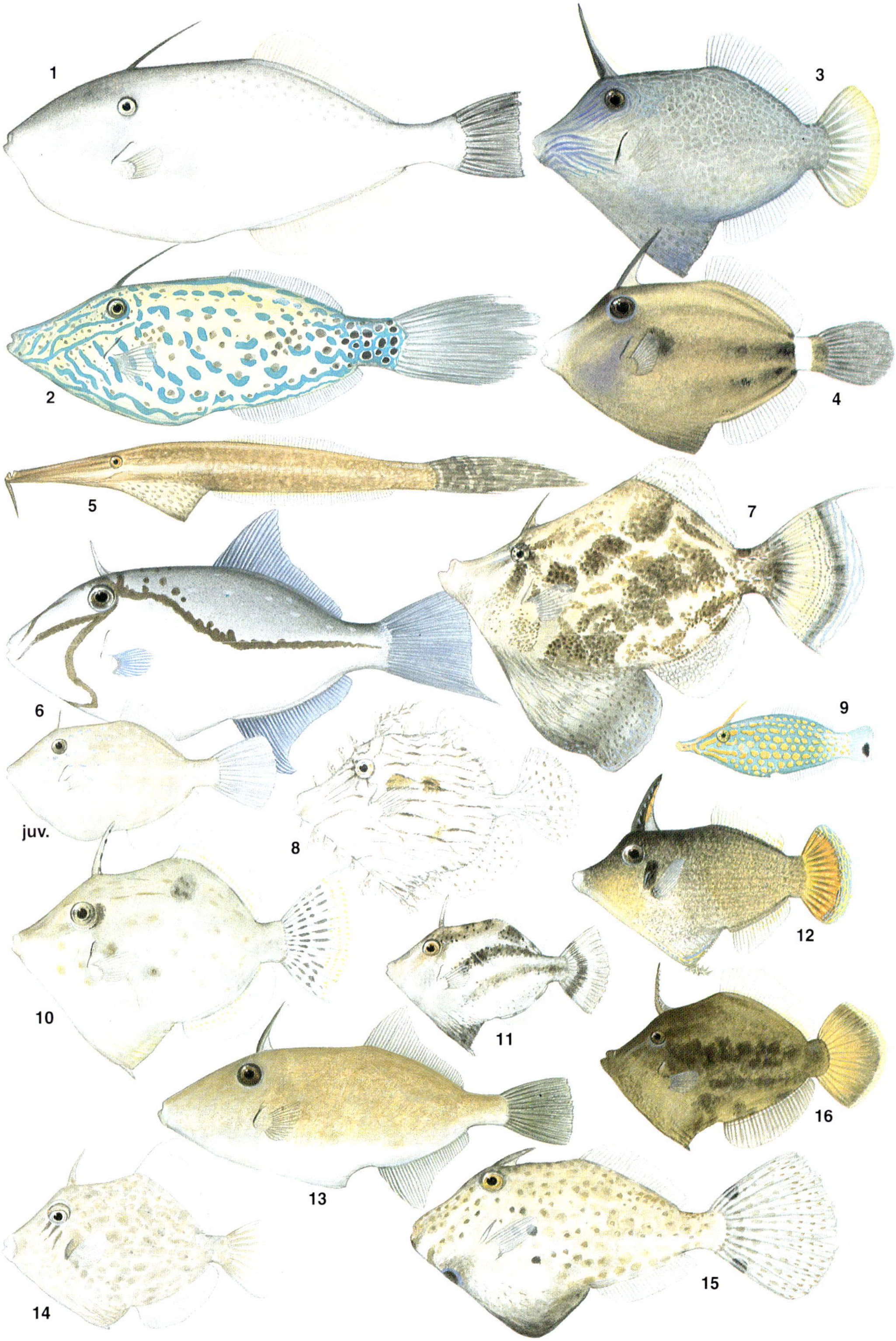

1
3
2
4
5
7
6
9
juv.
8
12
10
11
16
13
15
14

PLATE 103: BOXFISHES AND PUFFERS

1 LONG-HORNED COWFISH
Lactoria cornuta (Linnaeus, 1758)
Inhabits weed-sand areas near rock or coral reefs; distinguished from **2** below by longer horns and flatter belly that is not semi-transparent; found throughout the region; Indo-C. Pacific; to 46 cm. (OSTRACIIDAE) **P**

2 ROUNDBELLY COWFISH
Lactoria diaphana (Bloch & Schneider, 1801)
Inhabits coastal waters, the young sometimes in estuaries; distinguished from **1** above by shorter horns; "thorn" on back, and semi-transparent, rounded belly; found throughout the region; Indo-W. Pacific; to 25 cm. (OSTRACIIDAE) **P**

3 YELLOW BOXFISH
Ostracion cubicus Linnaeus, 1758
Inhabits coral and rocky reefs; distinguished by yellow to brown colour with small black spots arranged in clusters on side, juvenile bright yellow with black spots; found throughout the region; Indo-W. Pacific; to 45 cm. (OSTRACIIDAE) **P**

4 SPOTTED BOXFISH
Ostracion meleagris Shaw, 1796
Inhabits coral reefs; male distinguished by blue sides with yellow-orange spots, female by black colour and numerous white spots, found throughout the region; Great Barrier Reef, N.W. Australia, and throughout S.E. Asia; Indo-C. Pacific; to 16 cm. (OSTRACIIDAE) **P**

5 HORN-NOSED BOXFISH
Rhynchostracion rhinorhynchus (Bleeker, 1852)
Inhabits reefs and flat bottom areas; distinguished by bump on snout; found throughout the region; mainly Indo-Australian Archipelago; to 35 cm. (OSTRACIIDAE) **P**

6 SMALL-NOSED BOXFISH
Rhynchostracion nasus (Bloch, 1785)
Inhabits reefs and flat bottom areas; distinguished by lack of horns, a ridged back, and dull spots on side; found throughout the region; Indo-W. Pacific; to 30 cm. (OSTRACIIDAE) **P**

7 BLACK-BLOTCHED TURRETFISH
Tetrasomus gibbosus (Linnaeus, 1758)
Inhabits coastal waters, sometimes trawled; distinguished by prominent peak on back; found throughout the region; Indo-W. Pacific; to 30 cm. (OSTRACIIDAE) **P**

8 TURRETFISH
Tetrasomus concatenatus (Bloch, 1786)
Inhabits coastal waters; distinguished by ridge on back with 2 stout spines, and blue lines on body; young have peak on back similar to **7** above, but with 2 spines at apex; *Trioris reipublicae* is a synonym; found throughout the region; Indo-W. Pacific; to 22 cm. (OSTRACIIDAE) **P**

9 FINE-SPINED PUFFERFISH
Tylerius spinosissimus (Regan, 1908)
Inhabits trawling grounds; distinguished by short bristles covering body; N.W. Australia and S. Indonesia; Indo-W. Pacific; to 12 cm. (TETRAODONTIDAE) **P**

10 MANY-STRIPED PUFFERFISH
Anchisomus multistriatus Richardson, 1854
Inhabits trawling grounds and offshore reefs; distinguished by curved bars on top half of head and body with spots below; N.W. Australia; mainly S.W. Pacific; to 15 cm. (TETRAODONTIDAE) **P**

11 NARROW-LINED TOADFISH
Arothron manillensis de Procé, 1822
Inhabits relatively turbid inshore waters over silty sand or mud bottoms, sometimes amongst weeds; distinguished by numerous thin stripes on side; found throughout the region; mainly W. Pacific; to 31 cm. (TETRAODONTIDAE) **P**

12 RETICULATED PUFFERFISH
Arothron reticularis (Bloch & Schneider, 1801)
Inhabits shallow coastal waters, usually over sand or mud bottoms; similar to **13** below, but with dark lines on snout and cheek; N.W. Australia and S.E. Asia; W. Pacific and E. Indian Ocean; to 30 cm. (TETRAODONTIDAE) **P**

13 STARS AND STRIPES TOADFISH
Arothron hispidus (Linnaeus, 1758)
Inhabits shallow waters near rock or coral reefs; similar to **12** above, but without lines on snout and cheek; Queensland, N.W. Australia, and throughout S.E. Asia; Indo-C. Pacific; to 51 cm. (TETRAODONTIDAE) **P**

14 IMMACULATE PUFFERFISH
Arothron immaculatus (Bloch & Schneider, 1801)
Inhabits sand or mud bottoms, often amongst weed; distinguished by lack of markings and dark rim around tail; Indo-Malay Archipelago; mainly Indian Ocean; to 30 cm. (TETRAODONTIDAE)

15 STARRY PUFFERFISH
Arothron stellatus (Schneider, 1801)
Inhabits sand and mud bottoms, often amongst weeds; distinguished by large size and dense spotting on head and body, juvenile with curved black bars on belly; Great Barrier Reef, N.W. Australia, and throughout S.E. Asia; Indo-C. Pacific; to 90 cm. (TETRAODONTIDAE) **P**

16 BLACK-SPOTTED TOAD FISH
Arothron nigropunctatus (Bloch & Schneider, 1801). Text on page 246.

BOXFISHES AND BLOWIES

Boxfishes of the family Ostraciidae (Plates 103-104) are strange creatures found mainly on tropical and subtropical reefs. Their bodies are encased in a bony carapace and the fins are relatively small. They are slow swimmers, but capable of short, rapid bursts. When feeding, boxfishes sometimes squirt a jet of water into the sand to uncover small plants and invertebrates that are then sucked into the mouth. Some species produce a toxic mucus that can kill other fishes or even themselves when confined in a small aquarium.

The puffers or blowies (family Tetraodontidae) and the related porcupinefishes (family Diodontidae) can inflate their body by swallowing water (or air if out of water). Presumably this adaptation serves as a deterrent to potential predators. The estimated 140 species of puffers occur worldwide in tropical and warm temperate seas and estuaries. The flesh, especially the viscera, contains a potent toxin that has caused many human fatalities. However, they are considered a great delicacy in Japan, where they are prepared by specially trained and licensed chefs. The largest species grow to about 1 m TL, but most are considerably less.

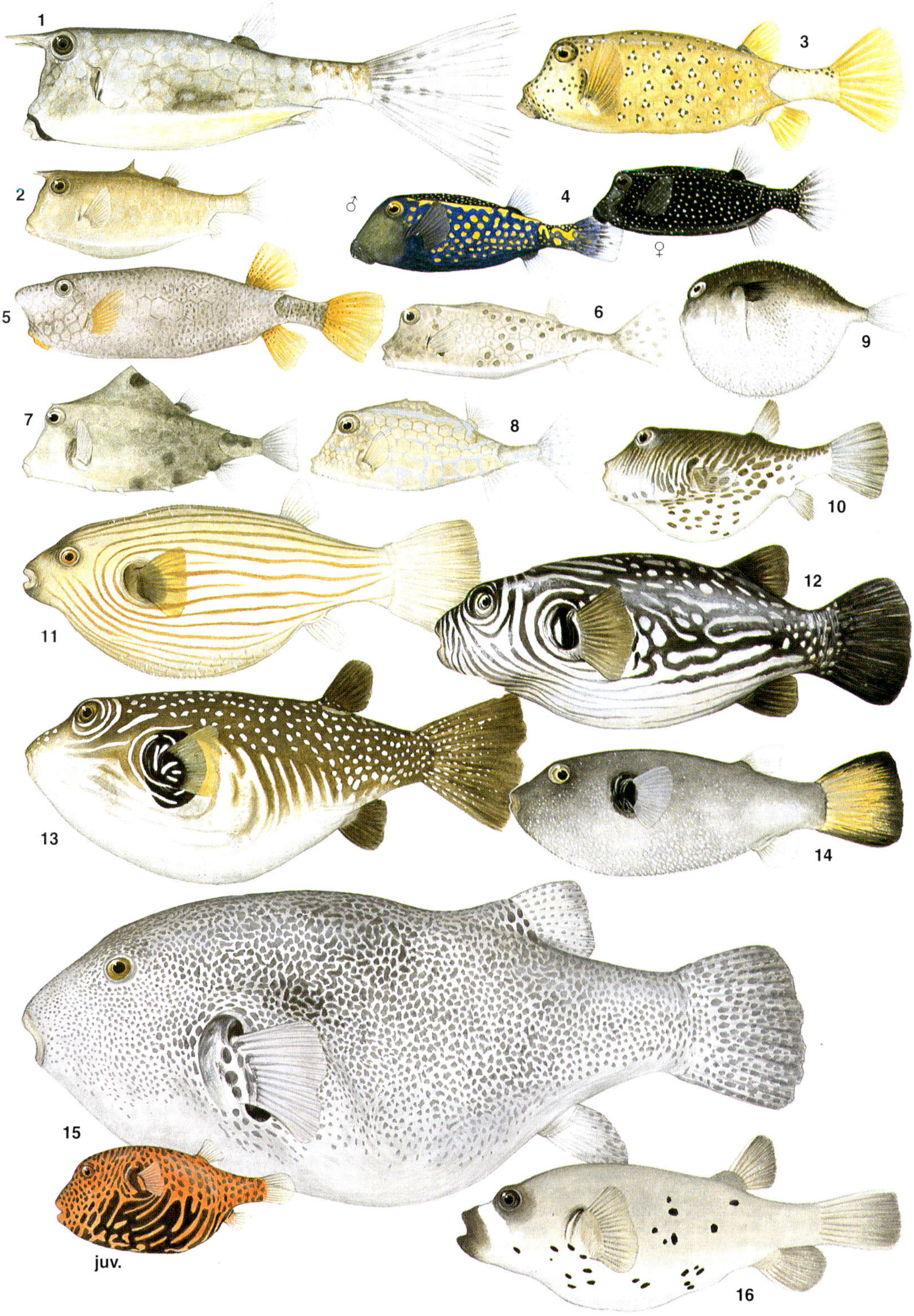
1
3
2
♂
4
♀
5
6
9
7
8
10
11
12
13
14
15
juv.
16

PLATE 104: BOXFISHES AND PUFFERFISHES

1 THORNBACK COWFISH
Lactoria fornasini (Bianconi, 1846)
Inhabits weed bottoms near coral reefs; distinguished by spine in front of each eye, "thorn" on back, and blue dots and dashes covering head and body; Great Barrier Reef and throughout S.E. Asia; Indo-C. Pacific; to 15 cm. (OSTRACIIDAE) **P**

2 STRIPED BOXFISH
Ostracion solorensis Bleeker, 1853
Inhabits coral reefs; female distinguished by reticulum of gold lines on side, male by bluish head and anterior body region, and by dark-edged pale stripes and pale spots on side; also known as Striped Boxfish; Great Barrier Reef, offshore reefs of N.W. Australia, and throughout S.E. Asia; Indo-Australian Archipelago; to 11 cm. (OSTRACIIDAE) **P**

3 BLUE-SPOTTED PUFFER
Arothron caeruleopunctatus Matsuura, 1994
Inhabits coral reefs, distinguished by bluish-white spots on body and dark rings around eye; Indonesia, New Guinea, and Coral Sea; Indo-W. Pacific; to 50 cm. (TETRAODONTIDAE) **P**

4 MAP PUFFER
Arothron mappa (Lesson, 1830)
Inhabits coral reefs; distinguished by spoke-like black lines radiating from eye, black blotch around gill opening and pectoral fin base, and highly irregular reticulum of black lines with pale spots; Great Barrier Reef and throughout S.E. Asia; Indo-W. Pacific; to 60 cm. (TETRAODONTIDAE) **P**

5 AMBON TOBY
Canthigaster amboinensis (Bleeker, 1865)
Inhabits coral reefs in areas exposed to surge; distinguished by ocellated dark spot below dorsal fin and white to blue-white spots and lines on head and body, also prominent spotting on tail;Great Barrier Reef and throughout S.E. Asia; Indo-C. Pacific; to 11 cm. (TETRAODONTIDAE) **P**

6 BENNETT'S TOBY
Canthigaster bennetti (Bleeker, 1854)
Inhabits vicinity of coral reefs, usually seen on rubble or sand bottoms or in sea grass and algal beds; distinguished by brown upper half and white lower half, also has large pale-edged dark spot at base of dorsal fin; Great Barrier Reef, offshore reefs of N.W. Australia, and throughout S.E. Asia; Indo-C. Pacific; to 10 cm. (TETRAODONTIDAE) **P**

7 LANTERN TOBY
Canthigaster epilampra (Jenkins, 1903)
Inhabits outer reef slopes, usually below 20 m depth; distinguished by yellow area around eye with blue "spokes", blue spotting on side, and brown back with ocellus-like mark below dorsal fin; Great Barrier Reef, New Guinea, and Solomon Islands; E. Indian Ocean (Christmas Island) and W. and C. Pacific; to 11 cm. (TETRAODONTIDAE) **P**

8 TYLER'S TOBY
Canthigaster tyleri Allen & Randall, 1977
Inhabits outer reef slopes, usually below 20 m depth; distinguished by white ground colour and numerous brown spots on head and body; Christmas Island (Indian Ocean) and Indonesia (Moluccas); mainly Indian Ocean; to 7.5 cm. (TETRAODONTIDAE) **P**

9 LEOPARD TOBY
Canthigaster leoparda Lubbock & Allen, 1979
Inhabits outer reef slopes, usually below 25 m depth; distinguished by leopard-like brown spotting on side; Phillipines and Indonesia (Molucca Islands); E. Indian Ocean (Christmas Island) and W. Pacific; to 7 cm. (TETRAODONTIDAE) **P**

10 SHY TOBY
Canthigaster ocellicincta Allen & Randall, 1977
Inhabits coral reefs, very secretive and seldom seen; distinguished by pair of dark bars with narrow white bar between them on middle of side, and ocellated spot below dorsal fin; Great Barrier Reef, Indonesia, Philippines, and Melanesian Archipelago; to 6.5 cm. (TETRAODONTIDAE) **P**

11 SOLANDER'S TOBY
Canthigaster solandri (Richardson, 1844)
Inhabits coral reefs; distinguished by ocellated dark spot below dorsal fin and white to blue-white spots and lines on head and body, also prominent spotting on tail; fish from Oceania are slightly diffent in colour, having smaller, more numerous spots; Indo-C. Pacific; to 11 cm. (TETRAODONTIDAE) **P**

12 COMPRESSED TOBY
Canthigaster compressa (Marion de Procé, 1822)
Inhabits coral reefs, frequently seen in silty bays or harbours, often around wharf pilings; distinguished by wavy lines on sides and ocellated spot below dorsal fin, also spots on tail are frequently joined to form vertical bands; Indo-Malay Archipelago; to 10 cm. (TETRAODONTIDAE) **P**

13 BLACK-SADDLED TOBY
Canthigaster valentini (Bleeker, 1853)
Inhabits coral reefs; mimicked by the leatherjacket *Paraluteres prionurus* (Pl. 101-18); distinguished by dark saddle-like markings; Great Barrier Reef, offshore reefs of N.W. Australia, and throughout S.E. Asia; Indo-C. Pacific; to 11 cm. (TETRAODONTIDAE) **P**

14 PORCUPINEFISH
Diodon hystrix Linnaeus, 1758
Inhabits coral reefs; distinguished by pepper-like spotting on head and body, and absence of dark saddles or bars; Great Barrier Reef, N.W. Australia, and throughout S.E. Asia; worldwide in tropical seas; to 71 cm. (DIODONTIDAE)

PLATE 103

16 BLACK-SPOTTED TOADFISH
Arothron nigropunctatus (Bloch & Schneider, 1801)
Inhabits clear waters of offshore coral reefs; distinguished by scattered black spots and dark lips, background colour ranges from grey to yellow; found throughout the region; Indo-C. Pacific; to 30 cm. (TETRAODONTIDAE) **P**

"P" IS FOR POISON

Under no circumstance should any fish be eaten that is indicated by a "**P**" in the text accompanying Plates 103-105. These fishes produce toxins and may cause serious illness or death. If caught while fishing it is advisable to release them immediately.

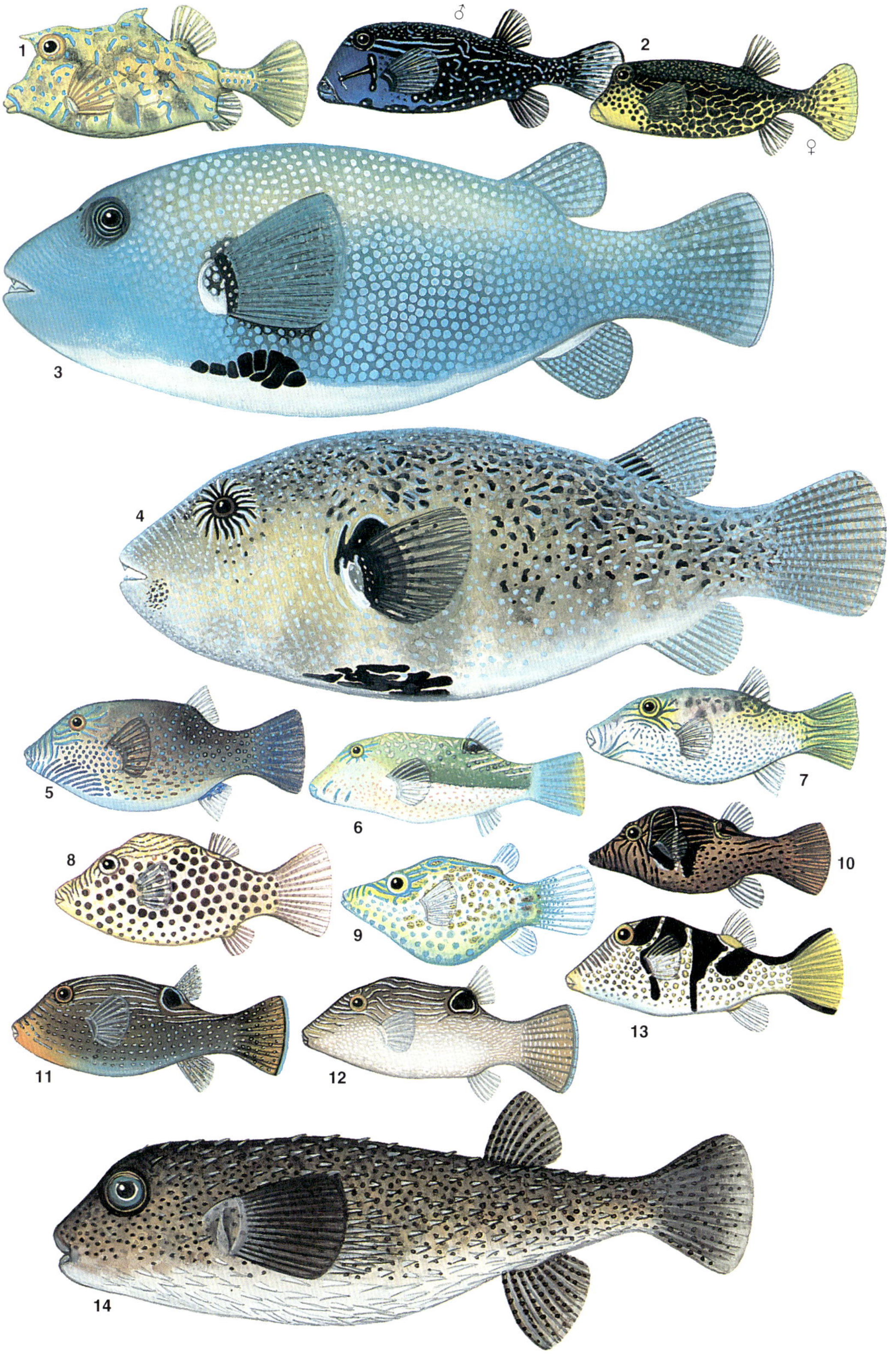

1
♂
2
♀
3
4
5
6
7
8
9
10
11
12
13
14

PLATE 105: PUFFERS AND PORCUPINEFISHES

1 THREE-BARRED PUFFER
Canthigaster coronata (Vaillant & Sauvage, 1875)
Inhabits flat bottom areas with coral, sponge, and rocky outcrops; distinguished by 3 dark saddles on upper half of body; Great Barrier Reef, N.W. Australia, and Philippines; Indo-C. Pacific; to 13 cm. (TETRAODONTIDAE) **P**

2 SPOTTED PUFFER
Canthigaster janthinoptera (Bleeker, 1855)
Inhabits caves and crevices of coral reefs; distinguished by dense network of white lines and spots on head and body; Great Barrier Reef, N.W. Australia, and throughout S.E. Asia; Indo-C. Pacific; to 8.5 cm. (TETRAODONTIDAE) **P**

3 BROWN-LINED PUFFER
Canthigaster rivulata (Schlegel, 1850)
Inhabits offshore reefs and trawling grounds; distinguished by wavy lines on back and brown stripe on side that curves around front of pectoral fin base; N.W. Australia and S. China Sea; Indo-W. Pacific; to 20 cm. (TETRAODONTIDAE) **P**

4 MILK-SPOTTED TOADFISH
Chelonodon patoca (Hamilton-Buchanan, 1822)
Inhabits bays and brackish mangrove estuaries; distinguished by network of large dark-centred white spots on head and body; found throughout the region; Indo-W. Pacific; to 20 cm. (TETRAODONTIDAE) **P**

5 ROUGH GOLDEN TOADFISH
Lagocephalus lunaris (Bloch & Schneider, 1801)
Inhabits coastal waters; similar to **6** and **7** below, but has bristles on dorsal surface between snout and dorsal fin; found throughout the region; Indo-W. Pacific; to 30 cm. (TETRAODONTIDAE) **P**

6 SMOOTH GOLDEN TOADFISH
Lagocephalus inermis (Temminck & Schlegel, 1850)
Inhabits coastal waters; similar to **5** and **7**, but lacks bristles on dorsal surface (and elsewhere); N.W. Australia and S. Indonesia; Indo-W. Pacific; to 20 cm. (TETRAODONTIDAE) **P**

7 BROWN-BACKED TOADFISH
Lagocephalus spadiceus (Richardson, 1844)
Inhabits coastal waters; similar to **5** and **6** above, but has patch of bristles on dorsal surface from snout to about half way to dorsal fin; found throughout the region; Indo-W. Pacific; to 30 cm. (TETRAODONTIDAE) **P**

8 SILVER TOADFISH
Lagocephalus scleratus (Gmelin, 1788)
Inhabits coastal waters, usually in schools; an aggressive species that can inflict painful bites, in a feeding frenzy it is very dangerous, attacking everything in sight; distinguished by spots and faint blotches on back and silvery stripe on sides, is also more elongate than **5-7** above; found throughout the region; Indo-W. Pacific; to 85 cm. (TETRAODONTIDAE) **P**

9 DARWIN'S TOADFISH
Marilyna darwinii (Castelnau, 1873)
Inhabits mud bottom areas, frequently in mangrove estuaries or the lower reaches of freshwater streams; N. Australia and S. New Guinea; to 17 cm. (TETRAODONTIDAE) **P**

10 HICK'S TOADFISH
Torquigener hicksi Hardy, 1983
Inhabits coastal waters; similar to **11** and **12** below, but bristles sparsely distributed on head and body; N. Australia only; to 13 cm. (TETRAODONTIDAE) **P**

11 YELLOW-EYED TOADFISH
Torquigener parcuspinus Hardy, 1983
Inhabits coastal waters; similar to **10** above, but covered with well developed bristles, lacks distinctive spotting of **12** below, which has shorter bristles, also distinguished by small yellow areas above each eye; N. Australia only; to 10 cm. (TETRAODONTIDAE) **P**

12 ORANGE-SPOTTED TOADFISH
Torquigener pallimaculatus Hardy, 1983
Inhabits coastal waters; distinguished by pattern of spots and bristle development that is intermediate to **10** and **11** above, also by orange-brown spots on lower side; N. Australia only; to 15 cm. (TETRAODONTIDAE) **P**

13 FRECKLED PORCUPINEFISH
Diodon holacanthus (Linnaeus, 1758)
Inhabits vicinity of coral reefs; distinguished by long movable spines on head and body, and combination of scattered small dark spots and larger dark blotches (often much darker than shown); *Diodon hystrix* (Pl. 104-14) is similar but lacks the larger dark patches, and the small spots are more numerous; Great Barrier Reef, N.W. Australia, and throughout S.E. Asia; worldwide in tropical seas; to 35 cm. (DIODONTIDAE) **P**

14 BLOTCHED PORCUPINEFISH
Diodon liturosus Shaw, 1804
Inhabits vicinity of coral reefs; distinguished by long movable spines on head and body, and large dark patches on side and dorsal surface, a solid broad bar across the top of the head just behind eyes is also distinctive; Great Barrier Reef, N.W. Australia, and throughout S.E. Asia; mainly W. and C. Pacific; to 40 cm. (DIODONTIDAE) **P**

15 SPOTBASE BURRFISH
Chilomycterus spilostylus Leis & Randall, 1982
Inhabits coastal waters in the vicinity of reefs; distinguished by short non-movable spines on head and body, and small black spots at the base of most spines; N.W. Australia, Indonesia, and Philippines; N. Indian Ocean to S. China Sea; to 35 cm. (DIODONTIDAE) **P**

16 SPOTFIN PORCUPINEFISH
Chilomycterus reticulatus (Linnaeus, 1758)
Inhabits coastal waters in the vicinity of reefs; distinguished by short non-movable spines on head and body, faint dark bars on side, and spotted tail; *C. affinis* is a synonym; found throughout the region, but rare; worldwide in tropical seas; to 55 cm. (DIODONTIDAE) **P**

17 LONG-SPINED PORCUPINEFISH
Cyclichthys jaculiferus (Cuvier, 1818)
Inhabits coastal waters; distinguished by long non-movable spines, 3-4 dark patches on back, and a few dark spots on side; N. Australia and Arafura Sea, to 30 cm. (DIODONTIDAE) **P**

18 SHORT-SPINED PORCUPINEFISH
Cyclichthys orbicularis (Bloch, 1785)
Inhabits coastal waters; distinguished by short non-movable spines on head and body, and relatively large black spots on back and side; Indo-W. Pacific; Great Barrier Reef, N.W. Australia, and throughout S.E. Asia; to 30 cm. (DIODONTIDAE) **P**

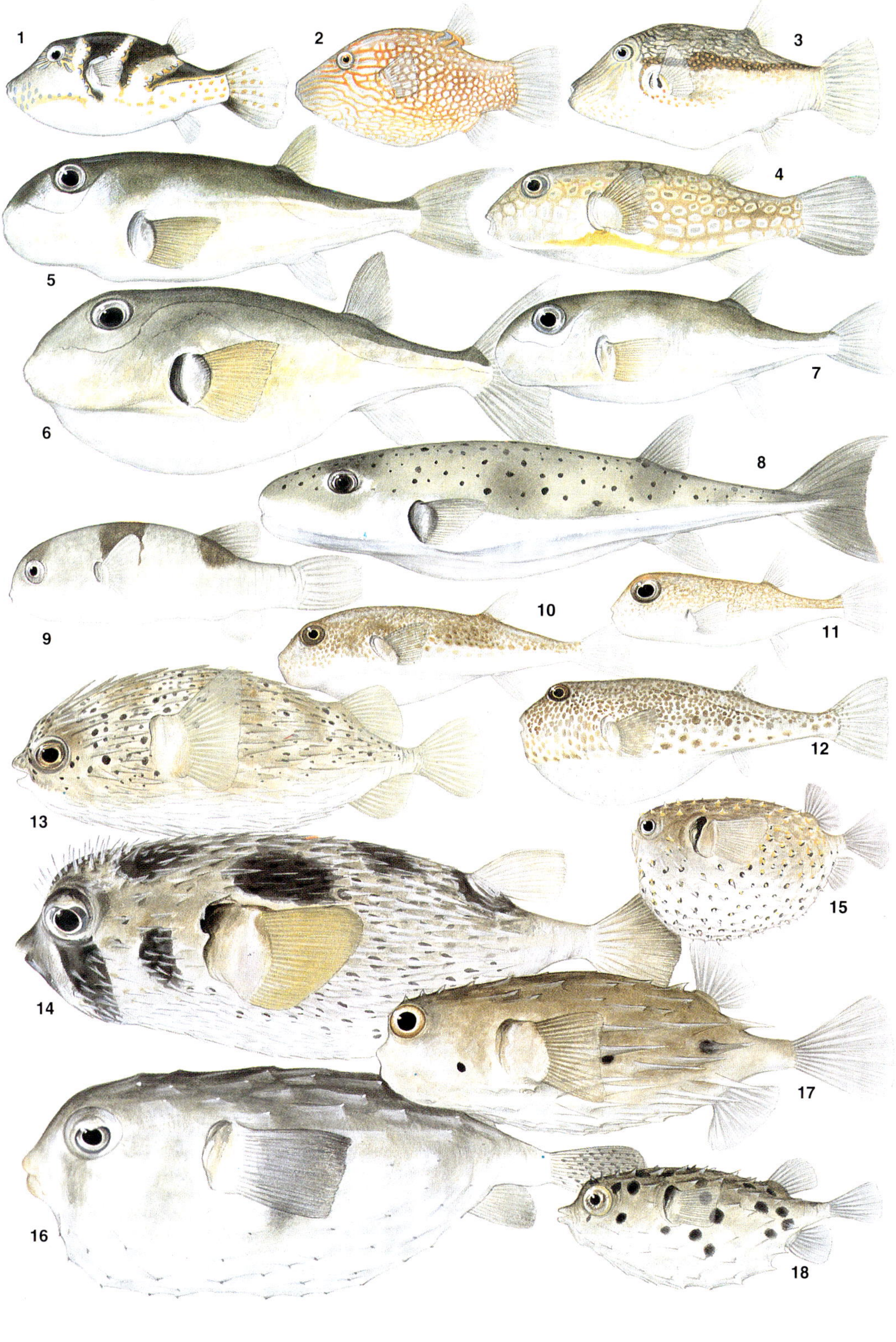
1
2
3
4
5
6
7
8
9
10
11
12
13
14
15
16
17
18

PLATE 106: SUPPLEMENT

NOTE – After the paintings and text for the book were completed several important omissions were discovered. These are included on this plate.

1 GOLD-SPOTTED SWEETLIPS
Plectorhinchus flavomaculatus (Cuvier, 1830)
Inhabits coral reefs and subtropical rocky reefs; similar to *P. multivittatum* (Pl.45-3), but lacks yellow on fins and has small dark spots on dorsal fin and tail; found throughout the region; Indo-W. Pacific; to 50 cm and 3 kg. (HAEMULIDAE) ★★

2 YELLOW-TAILED EMPEROR
Lethrinus atkinsoni Seale, 1909
Inhabits coral reefs; distinguished by steep forehead and yellow area along middle of sides extending on to tail; formerly confused with *L. mahsena* from the W. Indian Ocean; found throughout the region; E. Indian Ocean and W. and C. Pacific; to 43 cm and 2 kg. (LETHRINIDAE) ★★★★

3 BLACKSPOT PIGFISH
Bodianus vulpinus (Richardson, 1850)
Inhabits mainly rocky reefs; male shown here distinguished by black blotch at base of middle dorsal spines and reddish margin on upper and lower edges of tail; female shown on Plate 70; W. Australia only from Cape Naturaliste northwards to Shark Bay; to 60 cm. (LABRIDAE) ★★★★

4 BLUNT UNICORNFISH
Naso fageni Morrow, 1954
Inhabits coral reefs, often in schools; distinguished by lack of forehead spike (although large males may develop protruding snout) and pearl-white ring around tail base (rapidly disappears after death); found throughout the region; N.W. Australia, Indonesia, and Philippines; Indo-W. Pacific; to 80 cm. (ACANTHURIDAE) ★★

5 PINKTAIL TRIGGERFISH
Melichthys vidua (Solander, 1844)
Inhabits coral reefs; similar to *M. niger* (Pl. 100-4), but distinguished by pale dorsal, anal, and tail fins; found throughout the region; Indo-C. Pacific; to 38 cm. (BALISTIDAE)

▲ Sipadan Island off Sabah, Malaysia is perched on top of a steep underwater mountain. It's famous for its labyrinth of submerged caves and steep drop-offs. (G. Allen)

▲ Extreme low tide showing exposed corals at Matabuan Reef at the Bodgaya Islands, northeast Borneo (Sabah). Unexplored areas such as this contain a wealth of fishes, including species still unknown to science. (G. Allen)

▲ Western Australia's unique Kimberley coast features huge tidal fluctuations and deep fjord-like inlets in desert surroundings. The region contains a wealth of fishes and although waters are turbid, it is a paradise for anglers. (G. Allen)

▲ Scenic vistas, superb diving, and great fishing abound throughout Indonesia. This is Kungkungan Bay near Manado, Sulawesi. (G. Allen)

▲ Agung Volcano provides a spectacular backdrop to the idyllic bays and quiet villages scattered along Bali's northern coast. (G. Allen)

▲ Rich coral reefs and palm-fringed beaches at Madang, Papua New Guinea. (R. Steene).

INDEX

	Species No.		Species No.		Species No.
A		*Aethaloperca rogaa*	23-1	*steinitzi*	88-5
Abalistes stellatus	100-2	*Aetobatus narinari*	5-10	*wheeleri*	89-2
abbreviata, Acanthocepola	68-11	*Aetomyleus nichofii*	5-9	*yanoi*	88-10
Ablabys taenianotus	19-11	affinis, Euthynnus	97-7	*Amblygaster leiogaster*	6-6
Ablennes hians	13-9	Hyporhamphus	13-6	*sirm*	6-7
abocellata, Fowleria	34-12	Scolopsis	51-7	*Amblygobius decussatus*	88-20
Abroholos Jawfish	83-9	akindynos, Amphiprion	73-9	*hectori*	88-18
Abudefduf bengalensis	61-4	alalunga, Thunnus	96-7	*nocturna*	88-19
lorenzi	63-2	albacares, Thunnus	96-8	*phalaena*	89-4
notatus	63-4	Albacore	96-7	*rainfordi*	88-17
septemfasciatus	61-1	albilabrus, Paraplotosus	11-6	*Amblyglyphidodon batunai*	63-5
sexfasciatus	61-5	albimaculosus, Apogon	33-14	*curacao*	61-6
sordidus	61-2	albimarginatus, Carcharhinus	2-1	*Amblypomacentrus breviceps*	67-14
vaigiensis	61-3	albipectoralis, Acanthurus	95-6	amblyrhynchos, Carcharhinus	2-3
whitleyi	73-3	albofasciata, Wetmorella	79-11	amblyuropterus, Pseudamia	34-8
Acanthocepela abbreviata	68-11	alboguttata, Parapercis	83-1	Ambon Blenny	86-21
Acanthochromis polyacanthus	63-1	albomarginata, Gracilia	26-1	Chromis	65-3
Acanthocybium solandri	97-1	*Albula vulpes*	6-2	Damsel	66-6
Acanthopagrus latus	47-15	*Alectes ciliaris*	37-1	Emperor	48-13
palmaris	47-14	*indicus*	37-2	Toby	98-2
Acanthurus albipectoralis	95-6	*Alepes* sp.	37-4	amboinensis, Butis	92-6
auranticavus	95-15	alexanderae, Pomacentrus	66-5	Canthigaster	104-5
bariene	95-10	Alexander's Damsel	66-5	Carcharhinus	2-4
blochii	95-5	*Allanetta mugiloides*	13-14	Chromis	65-3
dussumieri	94-3	alleni, Cheilodipterus	34-4	Lethrinus	48-13
fowleri	95-14	Cirripectes	86-7	Paralticus	86-21
grammoptilus	94-5	Ecsenius	86-10	Pomacentrus	66-6
guttatus	95-4	Labropsis	73-9	Pterogogus	77-7
leucocheilus	95-11	Valenciennea	91-2	Amethyst Anthias	27-9
leucosternon	95-1	Allen's Blenny	86-10	*Amniataba caudovittatus*	30-9
lineatus	94-4	Cardinalfish	34-4	*Amphiprion akindynos*	63-9
maculiceps	95-8	Goby	91-2	*chrysopterus*	63-12
mata	94-1	Tubelip	73-9	*clarkii*	64-1
nigricans	94-2	Almaco Jack	39-14	*ephippium*	63-10
nigricauda	95-9	*Alopias pelagicus*	1-5	*frenatus*	63-11
nigrofuscus	94-7	Alpha Chromis	65-1	*melanopus*	63-7
nubilus	95-12	alpha, Chromis	65-1	*ocellaris*	64-2
olivaceous	94-6	altimus, Carcharhinus	2-2	*perideraion*	64-3
pyroferus	95-2	altipinnis, Scarus	82-1	*polymnus*	63-8
thompsoni	95-7	altivelis, Calloplesiops	28-8	*rubrocinctus*	64-5
triostegus	94-8	Coradion	58-8	*sandaracinos*	64-4
xanthopterus	95-13	Cromileptes	22-10	amplex, Corythoichthys	18-8
Acentrogobius gracilis	89-1	altivelus, Synchiropus	87-9	anabatoides, Neopomacentrus	62-15
Acentronurus larsonae	17-7	*Aluterus monoceros*	102-1	*Anacanthus barbatus*	102-5
Acreichthys radiatus	101-17	*scriptus*	102-2	analis, Chromis	65-2
tomentosus	101-16	*Amanses scopas*	101-9	Pempheris	53-9
aculeatus, Rhinecanthus	100-5	*Ambassis interruptus*	30-4	Selenanthias	22-12
acuminatus, Heniochus	56-13	*nalua*	30-5	Sillago	30-14
acutidens, Negaprion	3-3	*vachelli*	30-6	*Anampses caeruleopunctatus*	70-1
acutus, Rhizoprionodon	3-6	Amberjack	39-16	*femininus*	73-1
adelus, Pomacentrus	67-5	amblycephalum,, Thalassoma	78-2	*geographicus*	70-2
adetii, Lutjanus	43-2	amblycephalus, Epinephelus	23-3	*lennardi*	70-3
adiergastos, Chaetodon	55-3	Johnius	53-4	*melanurus*	73-2
adustus, Myripristis	14-1	*Amblycirrhitus bimacula*	68-2	*meleagrides*	70-4
Adventor elongatus	21-1	*Amblyeleotris guttata*	88-4	*neoguinaicus*	73-3
Aeoliscus strigatus	18-1	*latifasciata*	88-8	*twistii*	73-4
Aesopia cornuta	98-10	*periophthalma*	88-6	*Anaora tentaculata*	88-3
aethalea, Bembrops	99-1	*randalli*	88-7	*Anchisomus multistriatus*	103-10

255

	Species No.		Species No.		Species No.
Anchor Tuskfish	71-12	*annularis, Pomacanthus*	60-13	*frassendai*	36-2
anchorago, Choerodon	71-12	*annulatus, Hologymnosus*	77-1	*fuscomaculatus*	36-5
Anchovy, Bareback	6-18	Naso	93-3	*fuscus*	32-11
Hamilton's	6-21	*Anodontostoma chacunda*	6-11	*gilberti*	36-10
Indian	6-19	*Antennarius coccineus*	12-11	*griffini*	35-15
Longfin	6-20	*hispidus*	12-12	*hartzfeldii*	36-9
ancylostoma, Rhina	3-14	*nummifer*	12-10	*hoeveni*	35-13
aneitensis, Xyrichtys	79-14	*pictus*	12-13	*hyalosoma*	36-13
Anemonefish, Barrier Reef	63-9	*striatus*	12-9	*kallopterus*	33-15
Black	63-7	*antennata, Pterois*	19-3	*kiensis*	36-15
Clark's	64-1	Anthias, Amethyst	27-9	*lateralis*	36-14
False Clown	64-2	Bicolor	27-11	*leptacanthus*	35-8
Orange	64-4	Hawk	29-1	*margaritophorus*	36-12
Orangefin	63-12	Longfin	27-13	*melanopus*	33-13
Pink	64-3	Lori's	27-2	*melas*	36-6
Red	64-5	Princess	27-1	*moluccensis*	33-1
Red Saddleback	63-10	Purple	27-10	*multilineatus*	35-6
Saddleback	63-8	Randall's	27-8	*nanus*	35-19
Spine-cheek	63-13	Redbar	27-16	*neotes*	35-18
Tomato	63-11	Redfin	27-6	*nigripinnis*	33-4
Angelfish, Bicolor	69-2	Scalefin	27-4	*nigrofasciatus*	35-9
Black Velvet	60-7	Squarespot	27-12	*notatus*	36-16
Blackspot	60-10	Stocky	27-14	*novemfasciatus*	35-10
Blue	59-10	Striped	27-15	*ocellicaudus*	36-1
Blue-girdled	59-8	Threadfin	27-3	*pallidofasciatus*	32-7
Blue-ringed	60-13	Twospot	27-16	*parvulus*	36-3
Eibl's	59-3	Yellow-lined	27-5	*perlitus*	36-11
Emperor	59-9	*anthioides, Bodianus*	71-1	*poeciloperus*	33-10
Flame	58-14	*Antigonia rhomboidea*	16-6	*properupta*	35-2
Herald's	60-2	*Anyperodon leucogrammicus*	22-1	*quadrifasciatus*	32-8
Keyhole	59-4	*aphanes, Rhinopias*	18-13	*rhodopterus*	33-8
Lamarck's	60-9	*Aphareus furca*	41-1	*ruppelli*	32-17
Lemonpeel	60-1	*rutilans*	43-1	*sangiensis*	35-11
Midnight	60-4	*Apistops coloundra*	20-1	*savayensis*	32-11
Mulit-barred	60-3	*Apsitus carinatus*	20-2	*sealei*	36-7
Pearl-scaled	60-5	*Aploactis aspersa*	21-2	*selas*	36-17
Queensland Yellowtail	60-6	*Apocryptodon madurensis*	92-1	*semilineatus*	32-6
Regal	59-11	*Apogon albimaculosus*	33-14	*semiornatus*	33-3
Scribbled	59-5	*angustata*	32-1	*septemstriatus*	32-12
Six-banded	59-7	*apogonides*	35-1	sp.	33-2
Three-spot	59-1	*aureus*	33-12	*taeniophorus*	32-2
Two-spined	58-13	*bandanensis*	32-10	*thermalis*	36-18
Vermiculated	60-8	*brevicaudatus*	33-7	*timorensis*	32-18
Watanabe's	60-11	*capricornis*	35-17	*trimaculatus*	35-12
White-tail	58-15	*carinatus*	33-11	*truncatus*	33-9
Yellowmask	60-12	*chrysopomus*	36-8	*unicolor*	33-6
Yellowtail	59-6	*chrysotaenia*	32-16	*unitaeniata*	35-5
Anglerfish, Freckled	12-11	*coccineus*	33-5	*ventrifasciatus*	35-14
Humpback	12-8	*compressus*	35-3	*Apolemichthys trimaculatus*	59-1
Marbled	12-15	*cooki*	32-5	*aprins, Cirrhitichthys*	68-3
Painted	12-13	*cyanosoma*	32-14	*Aprion virescens*	41-2
Shaggy	12-12	*dispar*	36-4	*Apterichtus klazingai*	10-1
Spotted-tail	12-14	*doederleini*	32-3	*apterygia, Conniella*	72-15
Striped	12-9	*ellioti*	33-9	*Aptychotrema* sp.	4-2
White-finger	12-10	*endekataenia*	33-4	*arcatus, Paracirrhites*	68-7
angustata, Apogon	32-1	*evermanni*	32-9	*Archamia biguttata*	34-19
angustatus, Hippocampus	17-5	*exostigma*	35-4	*fucata*	33-17
angustirostris, Tetrapturus	96-3	*flavus*	35-16	*macroptera*	34-18
aniara, Kanekonia	21-6	*fraenatus*	32-13	*melasma*	33-18
annotatus, Dasyatis	4-9	*fragilis*	35-7	*zosterophora*	34-17

	Species No.
Archerfish, Banded	54-2
Spotted	54-1
arenarius, Platycephalus	21-16
areolatus, Epinephelus	23-3
Plectropomus	24-8
argenteus, Monodactylus	53-6
Neoniphon	15-6
Siganus	93-14
argentea, Liza	69-1
argentimaculatus, Lutjanus	41-10
argentilineatus, Periophthalmus	92-2
Argus Wrasse	76-12
argus, Cephalopholis	22-4
Halichoeres	76-12
Lepidotrigla	21-7
Scatophagus	55-1
Argyrops spinifer	47-16
Ariomma indica	99-4
Arius graefei	11-2
thalassinus	11-1
armatus, Neocirrhites	67-17
Arothron caeruleopunctatus	104-3
hispidus	103-13
immaculatus	103-14
manillensis	103-11
mappa	104-4
nigropunctatus	103-16
reticularis	103-12
stellatus	103-15
Armoured-gurnard, Slender	21-10
Spotted	21-11
Arramphus sclerolepis	13-2
Arrowhead Soapfish	29-2
arsius, Pseudorhombus	98-2
artus, Cheilodipterus	31-9
aruanus, Dascyllus	64-15
Aseraggodes melanospilus	98-12
asper, Rogadius	21-20
aspersa, Aploactis	21-2
Aspidontus dussumieri	84-1
taeniatus	84-2
aspircadus, Pervagor	101-13
assarius, Chaetodon	55-4
Assessor flavissimus	29-5
macneilii	29-4
Assiculus punctatus	28-13
Asterhombus intermedius	98-6
Asterropteryx semipunctatus	89-3
Asymmetrical Goatfish	52-15
asymmetricus, Upeneus	52-15
ataenia, Psuedocheilonops	72-14
Atelomycterus macleayi	1-12
Atherinomorus endrachtensis	13-17
ogilbyi	13-18
atkinsoni, Lethrinus	106-2
atripectoralis, Chromis	64-7
atripes, Chromis	65-4
Atrobrucca brevis	53-3
atrocinctus Mimoblennius	84-5

	Species No.
atrodorsalis, Meiacanthus	86-1
Atrosalarias fuscus	84-3
Atule mate	37-5
audax, Tetrapturus	96-4
audleyi, Gymnocranius	48-2
Aulostomus chinensis	16-10
auranticavus, Acanthurus	95-15
aurantifasciata, Pseudocoris	79-5
auratus, Scolopsis	51-8
aureofasciatus, Chaetodon	55-6
aureus, Apogon	33-12
auricilla, Pristipomoides	44-5
auriga, Chaetodon	55-5
Aurita Cardinalfish	31-15
aurita, Fowleria	31-15
aurochs, Ulua	40-1
aurocingulus, Ctenogobius	88-15
aurolineatus, Gnathodentex	48-1
Neoniphon	15-7
auromarginatus, Xanthichthys	101-5
Australian Blenny	86-11
Spotted Mackerel	97-4
Stinkfish	87-6
australianus, Ecsenius	86-11
australis, Gymnura	5-8
Labropsis	73-11
Auxis rochei	97-9
thazard	97-10
axelrodi, Ecsenius	86-13
Axelrod's Blenny	86-13
axillaris, Bodianus	70-5
aygula, Coris	74-14
Azure Demoiselle	62-10
azysron, Neopomacentrus	61-11
B	
bailloni, Trachinotus	39-10
Baldchin Groper	74-5
Bali Threadfin-Bream	49-1
balinensis, Nemipterus	49-1
Balistapus undulatus	100-1
Balistoides conspicillum	101-2
viridescens	100-3
balteatus, Dipterygonatus	46-12
Banda Blenny	86-14
Banana Fusilier	46-11
bandanensis, Apogon	32-10
bandanus, Ecsenius	86-14
Banded Archerfish	54-2
Blenny	85-15
Damsel	62-14
Frogfish	12-4
Goatfish	52-8
Goby	89-4
Grubfish	83-6
Grunter	30-13
Ilisha	6-12
Longfin	28-7
Moray	8-15
Pipefish	17-11

	Species No.
Scad	38-6
Sergeant	61-1
Silver-biddy	50-16
Stargazer	83-14
Wobbegong	1-14
bandanensis, Apogon	32-10
Stethojulius	77-10
Bandfish	68-11
Band-spot Cardinalfish	36-17
Banggai Cardinalfish	34-7
Banjofish	45-14
Banjos banjos	45-14
banjos, Banjos	45-14
Bannerfish, Humphead	56-14
Longfin	56-12
Masked	58-12
Pennant	56-10
Schooling	56-13
Singular	56-11
barbatus, Anacanthus	102-5
barberinoides, Parupeneus	52-2
barberinus, Parupeneus	51-16
Barbless Eagle Ray	5-9
Bar-cheeked Coral Trout	24-9
Bareback Anchovy	6-18
bariene, Acanthurus	95-10
Barhead Damsel	67-2
barnesi, Gorgasia	10-11
Barnes's Garden Eel	10-11
baronessa, Chaetodon	58-1
Barracuda	69-12
barracuda, Sphyraena	69-12
Barramundi	30-1
Cod	22-10
Barred Cardinalfish	34-13
Garfish	13-4
Longtom	13-9
Queenfish	39-4
Barred-face Spinecheek	50-7
Barrier Reef Anemonefish	63-9
Chromis	65-14
Bartail Moray	7-13
Bar-tailed Flathead	21-17
Goatfish	52-14
Bass, Red	41-12
Sand	30-2
Spiky	30-3
Bassett-Hulls Trevally	40-3
batavianus, Platax	54-4
Batfish, Hump-headed	54-4
Long-finned	54-5
Orbicular	54-7
Short-finned	54-8
Silver	53-6
Teira	54-6
bathi, Ecsenius	86-12
Bath's Blenny	86-12
bathybius, Nemipterus	49-2
Bathygobius cocosensis	89-5
fuscus	89-6

	Species No.		Species No.		Species No.
Batrachomoeus dahli	12-1	Long-finned	31-8	Black-blotched Moray	7-7
occidentalis	12-2	Lunar-tailed	31-4	Stingray	5-3
trispinosus	12-3	Red	31-3	Turretfish	103-7
batuensis, Coris	72-13	Robust	31-6	Black-crested Trevally	40-1
batunai, Amblyglyphidodon	63-5	Threadfin	31-5	Black-dotted Groper	25-2
Batuna's Damsel	63-5	White-barred	31-1	Black-edged Conger	9-1
bayeri, Enchelycore	8-4	Bigeyed Lizardfish	11-11	Cuskeel	12-18
Beach Salmon	53-7	Bignose Shark	2-2	Blackfin Pigfish	71-5
Beaked Coralfish	58-5	Big-lip Damsel	64-6	Soldierfish	14-1
Leatherjacket	102-9	Big Red Cardinalfish	33-6	Black-finned Gurnard	21-8
Bearded Cuskeel	12-19	Bigscale Soldierfish	15-1	Snake-Eel	9-11
Goby	92-3	*bigibbus, Kyphosus*	53-12	Threadfin	69-9
Leatherjacket	102-5	*biguttata, Archamia*	34-19	Black-lined Leatherjacket	101-15
Velvetfish	21-5	*biguttatus, Lutjanus*	43-4	Blacknape Large-eye Bream	48-4
Beautiful Goby	91-1	*bilineata, Paraplagusia*	98-19	Black-nosed Cardinalfish	33-15
Behn's Damsel	61-14	*bilineatus, Grammatorcynus*	97-14	Blackpatch Triggerfish	101-6
bellissimus, Exyrias	91-1	*Scolopsis*	50-8	Black-saddled Toby	104-13
Belly-barred Cardinalfish	35-1	*bilobatus, Epinephelus*	324-1	Black-shouldered Lizardfish	11-12
Belonepterygium fasciolatum	28-7	*bilunulatus, Bodianus*	70-6	Blackspot Angelfish	60-10
Beloneperca chabanaudi	29-2	*bimacula, Amblycirrhitus*	68-2	Cardinalfish	33-18
Bembrops aethalea	99-1	*bimaculatus, Bodianus*	71-2	Damsel	67-11
filodorsalia	99-2	*Oxycheilinus*	71-6	Goatfish	52-5
Bengal Seaperch	43-3	*bindus, Leiognathus*	40-11	Pigfish	106-3
Sergeant	61-4	*binotatus, Ctenochaetus*	95-16		& 70-8
bengalensis, Abudefduf	61-4	*biocellata, Chrysiptera*	62-6	Razorfish	78-11
Lutjanus	43-3	*biocellatus, Dendrochirus*	18-12	Seaperch	42-8
bennetti, Canthigaster	104-6	*Glossogobius*	89-16	Sergeant-Major	62-1
Chaetodon	57-7	*Halichoeres*	75-4	Shark	2-15
Bennett's Butterflyfish	57-7	*Signigobius*	88-16	Squirrelfish	15-15
Toby	104-6	*bipinnulata, Elegatis*	39-2	Tuskfish	74-8
berndti, Myripristis	15-1	*birostis, Manta*	5-6	Waspfish	19-12
biaculeatus, Premnas	63-13	*bispinosus, Centropyge*	58-13	Black-spotted Dart	39-10
Syngnathoides	17-21	*bitaeniata, Pseudochromis*	29-16	Goby	90-4
Triacanthus	99-8	*bitaeniatus, Lutjanus*	41-11	Moray	8-1
bicarinatus, Grammatorcynus	97-13	Black Anemonefish	63-7	Stingray	5-5
bicinctus, Uranoscopus	83-16	Cardinalfish	36-7	Toadfish	103-16
bicoarctatus, Trachyrhamphus	17-22	Damsel	61-13	Blackstreak Surgeonfish	95-9
Bicolor Angelfish	59-2	Jawfish	83-12	Blackstripe Cardinalfish	35-9
Anthias	27-11	Jew	53-1	Black-striped Cardinalfish	34-2
Blenny	85-3	Kingfish	40-6	Dottyback	29-21
Chromis	64-11	Leopard Wrasse	77-9	Snake Eel	10-4
Cleanerfish	77-2	Lizardfish	11-10	Wrasse	74-15
Fangblenny	86-5	Marlin	96-1	Blacktail Sergeant	63-2
bicolor, Centropyge	59-2	Pomfret	38-18	Wrasse	73-2
Cetoscarus	80-2	Spinefoot	92-11	Tripodfish	99-8
Ecsenius	85-3	Stingray	5-1	Black-tailed Dascyllus	63-6
Labroides	77-2	Trevally	38-2	Black-throated Threefin	87-1
Pseudanthias	27-11	Triggerfish	100-8	Blacktip Reef Shark	2-11
bifasciatum, Diploprion	28-1	Velvet Angelfish	60-7	Shark	2-10
bifasciatus, Parupeneus	52-3	Whaler	2-13	Tripodfish	99-10
Pentapodus	51-1	Black and White Snapper	43-15	Black-tipped Cardinalfish	32-6
Big Longnose Butterflyfish	58-11	Blackass Blenny	86-15	Cod	23-6
Bigeye Humpnose Bream	47-1	Black-axil Chromis	64-7	Fusiler	45-13
Seaperch	42-13	Black-back Butterflyfish	57-10	Blackvent Damsel	62-16
Trevally	38-8	Black-banded Blenny	84-10	Blanquillo, Blue	46-14
Tuna	96-9	Damsel	67-14	Flagtail	40-4
Bigeye, Deepsea	31-7	Kingfish	39-13	*bleekeri, Chlorurus*	82-2
Duskyfin	31-2	Seaperch	42-8	*Epinephelus*	25-5
Glass	31-2	Black-bar Chromis	65-13	Bleeker's Parrotfish	82-2
Large-spined	31-3	Black-blotch Emperor	48-16	*Blenniella periophthalmus*	85-14

258

	Species No.		Species No.		Species No.
Anchor Tuskfish	71-12	*annularis, Pomacanthus*	60-13	*frassendai*	36-2
anchorago, Choerodon	71-12	*annulatus, Hologymnosus*	77-1	*fuscomaculatus*	36-5
Anchovy, Bareback	6-18	Naso	93-3	*fuscus*	32-11
Hamilton's	6-21	*Anodontostoma chacunda*	6-11	*gilberti*	36-10
Indian	6-19	*Antennarius coccineus*	12-11	*griffini*	35-15
Longfin	6-20	*hispidus*	12-12	*hartzfeldii*	36-9
ancylostoma, Rhina	3-14	*nummifer*	12-10	*hoeveni*	35-13
aneitensis, Xyrichtys	79-14	*pictus*	12-13	*hyalosoma*	36-13
Anemonefish, Barrier Reef	63-9	*striatus*	12-9	*kallopterus*	33-15
Black	63-7	*antennata, Pterois*	19-3	*kiensis*	36-15
Clark's	64-1	Anthias, Amethyst	27-9	*lateralis*	36-14
False Clown	64-2	Bicolor	27-11	*leptacanthus*	35-8
Orange	64-4	Hawk	29-1	*margaritophorus*	36-12
Orangefin	63-12	Longfin	27-13	*melanopus*	33-13
Pink	64-3	Lori's	27-2	*melas*	36-6
Red	64-5	Princess	27-1	*moluccensis*	33-1
Red Saddleback	63-10	Purple	27-10	*multilineatus*	35-6
Saddleback	63-8	Randall's	27-8	*nanus*	35-19
Spine-cheek	63-13	Redbar	27-16	*neotes*	35-18
Tomato	63-11	Redfin	27-6	*nigripinnis*	33-4
Angelfish, Bicolor	69-2	Scalefin	27-4	*nigrofasciatus*	35-9
Black Velvet	60-7	Squarespot	27-12	*notatus*	36-16
Blackspot	60-10	Stocky	27-14	*novemfasciatus*	35-10
Blue	59-10	Striped	27-15	*ocellicaudus*	36-1
Blue-girdled	59-8	Threadfin	27-3	*pallidofasciatus*	32-7
Blue-ringed	60-13	Twospot	27-16	*parvulus*	36-3
Eibl's	59-3	Yellow-lined	27-5	*perlitus*	36-11
Emperor	59-9	*anthioides, Bodianus*	71-1	*poecilopterus*	33-10
Flame	58-14	*Antigonia rhomboidea*	16-6	*properupta*	35-2
Herald's	60-2	*Anyperodon leucogrammicus*	22-1	*quadrifasciatus*	32-8
Keyhole	59-4	*aphanes, Rhinopias*	18-13	*rhodopterus*	33-8
Lamarck's	60-9	*Aphareus furca*	41-1	*ruppelli*	32-17
Lemonpeel	60-1	*rutilans*	43-1	*sangiensis*	35-11
Midnight	60-4	*Apistops couloundra*	20-1	*savayensis*	32-11
Mulit-barred	60-3	*Apsitus carinatus*	20-2	*sealei*	36-7
Pearl-scaled	60-5	*Aploactis aspersa*	21-2	*selas*	36-17
Queensland Yellowtail	60-6	*Apocryptodon madurensis*	92-1	*semilineatus*	32-6
Regal	59-11	*Apogon albimaculosus*	33-14	*semiornatus*	33-3
Scribbled	59-5	*angustata*	32-1	*septemstriatus*	32-12
Six-banded	59-7	*apogonides*	35-1	sp.	33-2
Three-spot	59-1	*aureus*	33-12	*taeniophorus*	32-2
Two-spined	58-13	*bandanensis*	32-10	*thermalis*	36-18
Vermiculated	60-8	*brevicaudatus*	33-7	*timorensis*	32-18
Watanabe's	60-11	*capricornis*	35-17	*trimaculatus*	35-12
White-tail	58-15	*carinatus*	33-11	*truncatus*	33-9
Yellowmask	60-12	*chrysopomus*	36-8	*unicolor*	33-6
Yellowtail	59-6	*chrysotaenia*	32-16	*unitaeniata*	35-5
Anglerfish, Freckled	12-11	*coccineus*	33-5	*ventrifasciatus*	35-14
Humpback	12-8	*compressus*	35-3	*Apolemichthys trimaculatus*	59-1
Marbled	12-15	*cooki*	32-5	*aprins, Cirrhitichthys*	68-3
Painted	12-13	*cyanosoma*	32-14	*Aprion virescens*	41-2
Shaggy	12-12	*dispar*	36-4	*Apterichtus klazingai*	10-1
Spotted-tail	12-14	*doederleini*	32-3	*apterygia, Conniella*	72-15
Striped	12-9	*ellioti*	33-9	*Aptychotrema* sp.	4-2
White-finger	12-10	*endekataenia*	33-4	*arcatus, Paracirrhites*	68-7
angustata, Apogon	32-1	*evermanni*	32-9	*Archamia biguttata*	34-19
angustatus, Hippocampus	17-5	*exostigma*	35-4	*fucata*	33-17
angustirostris, Tetrapturus	96-3	*flavus*	35-16	*macroptera*	34-18
aniara, Kanekonia	21-6	*fraenatus*	32-13	*melasma*	33-18
annotatus, Dasyatis	4-9	*fragilis*	35-7	*zosterophora*	34-17

259

	Species No.		Species No.		Species No.
Blue-spotted Large-eye	48-5	Bull Shark	2-7	Vagabond	57-11
Collared Large-eye	48-2	Bullhead Shark	1-6	Western	55-4
Humpnose Big-eye	47-1	Bullseye, Bronze	53-9	Yellow-dotted	57-8
Japanese Large-eye	48-3	Oualan	53-10	Yellow-tailed	57-5
Monocle	50-7	Slender	53-8		
North West Black	47-14	Striped	53-11	C	
Striped Large-eye	48-1	Bump-nosed Trevally	37-12	*caerulaurea, Caesio*	46-3
Western Buffalo	53-13	*burgeri, Glaucosoma*	30-7	*caeruleoguttatus, Eubalichthys*	102-6
Western Yellowfin	47-15	*burgessi, Chaetodon*	58-6	*caeruleomaculatus, Boleophthalmus*	
Yellowsnout Large-eye	48-6	Burgess's Butterflyfish	58-6		92-4
Breastspot Cleanerfish	73-7	*buroensis, Gymnothorax*	7-4	*caeruleopinnatus, Carangoides*	37-7
Wrasse	76-8	Burrfish, Spotbase	105-15	*caeruleopunctatus, Anampses*	70-1
brevicaudatus, Apogon	33-7	*burroughi, Pomacentrus*	67-12	Arothron	104-3
breviceps, Amblypomacentrus	67-14	Burrough's Damsel	67-12	Epinephelus	23-4
Petroscirtes	84-8	Burrowing Snake-eel	9-13	*Caesio caerulaurea*	46-3
brevipinna, Carcharhinus	2-6	*bursa, Sufflamen*	100-9	cuning	45-11
brevirostris, Malacanthus	40-4 &	*Butis amboinensis*	92-6	lunaris	45-12
	46-15	Butterfish, Blue	51-5	teres	46-1
Naso	94-11	Japanese	50-4	xanthonota	46-2
brevis, Atrobucca	53-3	Striped	55-2	*calamus, Cirrhimuraena*	9-5
Exallias	84-7	Western	50-6	*calcarifer, Lates*	30-1
Bridled Monocle Bream	50-8	Butterfly Cod	19-6	Cale Cale Trevally	40-2
Spinecheek	50-8	Ray	5-8	*Callechelys catastomus*	10-4
Bristle-tailed Leatherjacket	101-16	Scorpionfish	19-2	marmoratus	9-2
Bristletooth, Goldring	93-2	Whiptail	51-4	*Callionymus goodladi*	87-4
Lined	94-9	Butterflyfish, Bennett's	57-7	grossi	87-8
Tomini	93-1	Big Longnose	58-11	japonicus	87-7
Twospot	95-16	Black-back	57-10	margaretae australis	87-6
Broad-banded Cardinalfish	32-8	Bluespot	55-12	margaretae margaretae	87-6
Broad-barred Maori Goby	90-1	Burgess's	58-6	moretonensis	87-5
Spanish Mackerel	97-3	Chevroned	56-5	*Callogobius sclateri*	89-8
brocki, Halicampus	17-15	Collare	57-12	sp.	89-7
Bronze Bullseye	53-9	Dot-and-Dash	57-6	*Calloplesiops altivelis*	28-8
Whaler	2-5	Dotted	57-4	*Calotomus spinidens*	80-3
Brotula multibarbata	12-19	Doublesaddle	56-4	*canadus, Rachycentron*	40-6
Brown Coral Blenny	84-3	Dusky	58-2	*canaliculatus, Siganus*	92-13
Demoiselle	61-10	Goldenstriped	55-6	*cancellatus, Strabozebrias*	98-15
Dottyback	28-11	Gunther's	58-3	*cancrivorous, Pisodonophis*	9-13
Reticulated Stingray	4-10	Klein's	55-11	Candystripe Cardinalfish	32-4
Stingaree	4-3	Latticed	57-3	*canina, Enchelynassa*	8-2
Stingray	4-9	Lined	55-9	*caninus, Pentapodus*	51-2
Sweetlips	45-5	Long-nosed	56-9	*Cantherhines fronticinctus*	102-4
Triggerfish	100-10	Merten's	57-1	pardalis	102-3
Unicornfish	94-13	Meyer's	55-10	*Cantherhinus dumerilii*	101-10
Brown-backed Toadfish	105-7	Ornate	55-14	*Cantheschenia grandisquamis*	101-8
Brown-banded Cat Shark	1-11	Ovalspot	56-3	*Canthidermis maculatus*	101-4
Pipefish	18-8	Philippine	55-3	*Canthigaster amboinensis*	104-5
Rockcod	22-3	Pyramid	58-10	bennetti	104-6
Brown-blotched Leatherjacket	102-14	Racoon	55-13	compressa	104-12
Brown-flecked Reef-eel	7-6	Rainford's	58-4	coronata	105-1
Brown-lined Puffer	105-3	Redfin	56-6	epilampra	104-7
Brown-spotted Cardinalfish	36-5	Reticulated	57-13	janthinoptera	105-2
brummeri, Pseudechidna	8-17	Saddled	55-8	leoparda	104-9
Brush-sided Leatherjacket	101-9	Speckled	55-7	ocellicincta	104-10
bucculentus, Caranx	38-7	Spot-banded	56-2	rivulata	105-3
buchanani, Valamugil	69-6	Spotnape	57-9	solandri	104-11
Buffalo Bream, Western	53-13	Spotted	57-2	tyleri	104-8
buffonis, Zenarchopterus	13-3	Teardrop	56-1	valentini	104-13
Buffon's Garfish	13-3	Threadfin	55-5	Capricorn Cardinalfish	35-17
Bulbonaricus brauni	17-8	Triangular	58-1	*capricornis, Apogon*	35-17

	Species No.		Species No.		Species No.
Carangoides caeruleopinnatus	37-7	Doederlein's	32-3	Spotted-gill	36-8
chrysophrys	37-6	Dusky-tailed	34-18	Striped	32-1
equula	37-10	Eight-lined	31-10	Tailspot	36-1
ferdau	37-8	Eight-spined	34-15	Thermal	36-18
fuluoguttatus	37-9	Estuary	34-11	Threadfin	35-8
gymnostethus	37-15	Evermann's	32-9	Three-saddle	32-10
hedlandensis	37-12	False Three-spot	33-8	Three-spot	33-8
humerosus	37-11	Five-lined	31-11	Timor	32-18
malabaricus	37-16	Flagfin	33-9	Tiny	35-19
orthogrammus	37-14	Flores	36-2	Twinspot	34-19
talamparoides	37-13	Fragile	35-7	Two-eyed	33-4
uii	37-17	Frostfin	35-13	Variegated	31-14
Caranx bucculentus	38-7	Gelantinous	34-9	Weed	31-13
ignobilis	38-1	Gilbert's	36-10	Wolf	31-9
lugubris	38-2	Girdled	34-17	Yellow	35-16
melampygus	38-3	Glassy	34-16	Yellowbelly	34-6
papuensis	38-4	Goldbelly	35-1	Yellow-lined	32-16
para	38-6	Hartzfeldt's	36-9	*carinatus, Apogon*	33-11
sexfasciatus	38-8	Hookfin	35-15	*Apistus*	20-2
tille	38-5	Hump-backed	36-13	*carnolabrum, Lipocheilus*	41-5
carbunculus, Etelis	41-3	Intermediate	34-1	Carpet Eel-Blenny	68-14
Carcharhinus albimarginatus	2-1	Iridescent	32-15	Wrasse	78-12
altimus	2-2	Larval	35-18	*carponotatus, Lutjanus*	42-2
amblyrhynchos	2-3	Little Red	33-5	*castaneus, Cirripectes*	85-2
amboinensis	2-4	Mangrove	36-14	*catastomus, Callechelys*	10-4
brachyurus	2-5	Many-banded	33-7	Catfish, Giant Salmon	11-1
brevipinna	2-6	Many-lined	35-6	Long-tailed	11-4
dussumieri	2-9	Mimic	34-5	Naked-headed	11-3
falciforimes	2-8	Moluccan	33-1	Sailfin	11-7
leucas	2-7	Narrow-lined	33-17	Smaller Salmon	11-2
limbatus	2-10	Narrowstripe	35-4	Striped	11-5
longimanus	2-12	Nine-banded	35-10	White-lipped	11-6
melanopterus	2-11	Oblique-banded	33-3	Catshark, Brown-banded	1-11
obscurus	2-13	Ocellated	33-11	Marbled	1-12
plumbeus	2-14	Orbicular	36-20	Speckled	1-9
sealei	2-15	Paddlefin	34-10	*Cattapistus cottoides*	20-4
sorrah	2-16	Pajama	36-19	*caudalis, Chromis*	65-16
Cardinalfish, Allen's	34-4	Pale-striped	32-7	*caudimacula, Coris*	74-16
Aurita	31-15	Pearly	36-11	*caudimaculatum, Sargocentron*	15-10
Band-spot	36-17	Pearly-finned	33-10	*caudovittatus, Amniataba*	30-9
Banggai	34-7	Pearl-jawed	34-8	*cauteroma, Choerodon*	74-9
Barred	34-13	Redspot	36-3	Cavite Cardinalfish	33-2
Belly-barred	35-14	Red-striped	36-12	*cavitiensis, Apogon*	33-2
Big Red	33-6	Reef-flat	32-2	Celebes Sweetlips	45-4
Black	36-6	Rifle	36-15	Wrasse	71-8
Black-nosed	33-15	Ring-tailed	33-12	*celebicus, Nemipterus*	49-3
Blackspot	33-18	Sailfin	31-12	Oxycheilinus	71-8
Blackstripe	35-9	Samoan	32-11	Plectorhinchus	45-4
Black-striped	34-2	Sangi	35-11	*centriscoides, Halimochirurgus*	99-12
Black-tipped	32-6	Seale's	36-7	*Centriscus scutatus*	17-1
Blue-striped	32-14	Seven-banded	32-12	*Centrogenys vaigiensis*	22-2
Broad-banded	32-8	Singapore	34-3	*Centropyge bicolor*	59-2
Brown-spotted	36-5	Single-striped	35-5	*bispinosus*	58-13
Candystripe	32-4	Slender	33-16	*eibli*	59-3
Capricorn	35-17	Spindle-egg	33-13	*flavicauda*	58-15
Cavite	33-2	Spiny-eyed	32-13	*flavissimus*	60-1
Cook's	32-5	Split-banded	35-3	*heraldi*	60-2
Coral	35-2	Spotless	34-12	*loriculus*	58-14
Cream-spotted	33-14	Spotnape	36-16	*multifasciatus*	60-3
Dispar	36-4	Spotted	34-14	*nox*	60-4

	Species No.		Species No.		Species No.
tibicen	59-4	*Chaetodomtoplus duboulayi*	59-5	*cauteroma*	74-9
vroliki	60-5	*melanosoma*	60-7	*cephalotes*	71-13
Cephalopholis argus	22-4	*meredithi*	60-6	*cyanodus*	74-3
boenack	22-3	*mesoleucus*	60-8	*fasciatus*	71-14
cyanostigma	26-6	*personifer*	59-6	*graphicus*	71-15
formosa	22-6	Chain-lined Wrasse	76-5	*jordani*	74-6
igarashiensis	26-3	*chameleon, Scarus*	80-9	*monostigma*	74-11
leopardis	22-7	Chameleon Parrotfish	80-9	*rubescens*	74-5
microprion	26-2	*Chanos chanos*	6-3	*schoenlenii*	74-8
miniata	22-5	*chanos, Chanos*	6-3	*sugillatum*	74-10
pachycentron	22-3	Charcoal Damsel	67-9	*venustus*	71-16
sexmaculata	26-5	*chatareus, Toxotes*	54-1	*vitta*	74-7
sonnerati	23-12	Checkered Lizardfish	11-17	*zamboangae*	74-12
spiloparaea	26-4	Seaperch	42-3	*zosterophorus*	71-17
urodeta	22-9	*Cheilinus chlorurus*	70-10	*choerorhynchus, Lethrinus*	47-7
Cephaloscyllium fasciatum	1-13	*fasciatus*	70-9	*choirocephalus, Paramonacanthus*	
cephalotes, Choerodon	71-13	*oxycephalus*	71-10		102-11
cephalozona, Ophichthus	9-9	*trilobatus*	70-12	Chromis, Alpha	65-1
cephalus, Mugil	69-5	*undulatus*	71-11	Ambon	65-3
Cetoscarus bicolor	80-2	*unifasciatus*	70-11	Barrier Reef	65-14
chabanaudi, Belonoperca	29-2	*Cheilio inermis*	77-1	Bicolor	64-11
chacunda, Anodontostoma	6-11	*Cheilodipterus alleni*	34-4	Black-axil	64-7
Chaetoderma penicilligera	102-8	*artus*	31-9	Black-bar	65-13
Chaetodon adiergastos	55-3	*intermedius*	34-1	Blue-green	64-8
assarius	55-4	*lineatus*	31-10	Darkfin	65-4
aureofasciatus	55-6	*nigrotaeniatus*	34-2	Dusky	65-16
auriga	55-5	*parazonatus*	34-5	Green	64-9
baronessa	58-1	*quinquelineatus*	31-11	Half and half	65-7
bennetti	57-7	*singapurensis*	34-3	Lined	65-11
burgessi	58-6	*zonatus*	34-6	Pale-tail	65-15
citrinellus	55-7	*Cheiloprion labiatus*	64-6	Philippines	65-9
collare	57-12	*Chelmon marginalis*	56-8	Scaly	65-8
ephippium	55-8	*muelleri*	*58-7*	Smokey	64-10
flavirostris	58-2	*rostratus*	58-5	Spiny	63-1
guentheri	58-3	*Chelonodon patoca*	105-4	Stout-body	65-6
guttatissimus	57-2	Chevroned Butterflyfish	56-5	Ternate	65-10
kleinii	55-11	*Chilomycterus reticulatus*	105-16	Twinspot	65-5
lineolatus	55-9	*spilostylus*	105-15	Weber's	64-12
lunula	55-13	*Chiloscyllium punctatum*	1-11	Yellow	65-2
melannotus	57-10	Chinaman Fish	41-9	Yellow-axil	65-12
mertensii	57-1	Chinaman Rockcod	23-14	*Chromis alpha*	65-1
meyeri	55-10	Chinese Footballer	25-12	*amboinensis*	65-3
ornatissimus	55-14	Gudgeon	92-7	*analis*	65-2
oxycephalus	57-9	*chinensis, Aulostomus*	16-10	*atripectoralis*	64-7
pelewensis	57-6	*Monacanthus*	102-7	*atripes*	65-4
plebeius	55-12	Chingilt	9-14	*caudalis*	65-16
punctatofasciatus	56-2	*Chirocentrus dorab*	6-5	*chrysura*	65-6
rafflesi	57-3	*Chironema chlorotaenia*	99-3	*cinerascens*	64-9
rainfordi	58-4	Chiseltooth Wrasse	79-7	*elerae*	65-5
reticulatus	57-13	*Choeroichthys brachysoma*	17-10	*fumea*	64-10
selene	57-8	*latispinosus*	17-9	*iomelas*	65-7
semion	57-4	*chloropterus, Halichoeres*	76-1	*lepidolepis*	65-8
speculum	56-3	*chlorotaenia, Chironema*	99-3	*lineata*	65-11
trifascialis	56-5	*Chlorurus bleekeri*	82-2	*margaritifer*	64-11
trifasciatus	56-6	*frontalis*	82-5	*nitida*	65-14
ulietensis	56-4	*chlorurus, Cheilinus*	70-10	*retrofasciata*	65-13
unimaculatus	56-1	*chlorotaenia, Chiromema*	99-3	*scotochilopterus*	65-9
vagabundus	57-11	*choati, Macropharyngodon*	73-13	*ternatensis*	65-10
xanthurus	57-5	Choat's Wrasse	73-13	*viridis*	64-8
chaetodontoides, Plectorhinchus	45-2	*Choerodon anchorago*	71-12	*weberi*	64-12

	Species No.		Species No.		Species No.
xanthochira	65-12	Clark's Anemonefish	64-1	*commerson, Scomberomorus*	97-2
xanthura	65-15	*clathrata, Parapercis*	83-3	*commersonnianus, Scomberoides*	
chryseres, Myripristis	15-2	Cleanerfish	77-3		39-5
Chrysiptera biocellata	62-6	Bicolor	77-2	*commersonii, Fistularia*	16-8
cyanea	62-1	Breastspot	73-7	Common Dart	39-12
flavipinnis	62-8	False	84-2	Dolphinfish	39-1
glauca	62-2	Clearfin Lionfish	18-11	Goby	89-6
leucopoma	62-9	*Cleidopus gloriamaris*	16-1	Grinner	11-18
parasema	62-10	Clingfish, Shark Bay	12-7	Ponyfish	40-12
rex	62-3	Urchin	12-6	Silver-Biddy	50-15
rollandi	62-5	Clouded Reef-eel	7-1	*compressa, Canthigaster*	104-12
springeri	62-11	Clown Triggerfish	101-2	*compressus, Apogon*	35-3
talboti	62-4	Club-nosed Trevally	37-6	Compressed Toby	104-12
unimaculata	62-7	Wrasse	75-2	*concatenatus, Tetrasomus*	103-8
chrysophrys, Carangoides	37-6	Coachwhip Stingray	5-5	*Conger cinereus*	9-1
chrysopleuron, Parapeneus	52-4	Cobia	40-6	Conger, Black-edged	9-1
chrysopoecilus, Dischistodus	62-13	Cobra Garden Eel	10-6	*Congrogadus spinifer*	68-13
chrysopomus, Apogon	36-8	*cobra, Heteroconger*	10-6	*subducens*	69-14
chrysopterus, Amphiprion	63-12	*coccineus, Antennarius*	12-11	*Conniella apterygia*	72-15
Sufflamen	100-8	*Apogon*	33-5	Connie's Wrasse	72-15
chrysospilos, Istiblennius	85-10	*Coccotropus* sp.	21-3	*conspicillum, Balistoides*	101-2
chrysostomus, Heinochus	56-10	Cockatoo Flounder	98-17	Convict Blenny	91-18
Lethrinus	47-6	Waspfish	19-11	Surgeonfish	94-8
chrysotaenia, Apogon	32-16	Cocos Goby	89-5	*Cookeolus boops*	31-8
chrysozona, Pterocaesio	46-10	*cocosensis, Bathygobius*	89-5	*cooki, Apogon*	32-5
chrysozonus, Coradion	56-7	Cod Scorpionfish	20-3	Cook's Cardinalfish	32-5
chrysura, Chromis	65-6	Cod, Barramundi	22-10	Cooktown Salmon	69-10
chrysus, Halichoeres	76-2	Black-tipped	23-6	*cooperi, Pseudanthias*	22-11
ciliaris, Alectes	37-1	Blue Maori	26-8	Cooper's Fairy Basslet	22-11
ciliatus, Scolopsis	51-9	Butterfly	19-6	*Coradion altivelis*	58-8
cinctus, Cryptocentrus	88-9	Coral	22-5	*chrysozonus*	56-7
Priolepis	90-8	Estuary	24-4	*melanopus*	58-9
cinerascens, Chromis	64-9	Flowery	23-8	Coral Blenny	85-6
cinereus, Conger	9-1	Frostback	24-1	Cardinalfish	35-2
Circle-cheek Wrasse	76-10	Honeycomb	23-10	Cod	22-5
Circumspect Goby	89-17	Leopard	24-7	Monocle Bream	50-12
circumspectus, Glossogobius	89-17	Polkadot	24-8	Pigfish	70-5
Cirrhilabrus cyanopleura	72-1	Potato	24-5	Rockcod	23-5
exquisitus	72-3	Reef	26-11	Trout	24-7
filamentosus	72-9	Scorpionfish	20-3	Coral Trout, Bar-cheeked	24-9
laboutei	72-2	Small-toothed	23-11	Coralfish, Beaked	58-5
lineatus	72-5	Trout	26-7	Highfin	58-8
lubbocki	72-4	Vermicular	24-10	Margined	56-8
punctatus	72-6	*coecus, Brachyamblyopus*	92-8	Mueller's	56-7
randalli	72-8	*coelestis, Pomacentrus*	66-7	Ocellate	56-15
scottorum	72-7	*coeruleolineatus, Plesiops*	28-9	Orange-banded	56-7
temmincki	74-2	*cognatus, Uranoscopus*	83-17	Two-eyed	58-9
Cirrhimuraena calamus	9-5	*coioides, Epinephelus*	24-4	*corallicola, Epinephalus*	23-5
Cirrhitichthys aprinus	68-3	Coitor Croaker	53-3	*corallinus, Siganus*	93-12
falco	67-16	*coitor, Johnius*	53-3	*cordyla, Megalaspis*	38-15
oxycephalus	68-4	Collare Butterflyfish	57-12	*Coris aygula*	74-14
Cirrhitus pinnulatus	67-15	*collare, Chaetodon*	57-12	*batuensis*	72-13
Cirripectes alleni	86-7	Collared Knifefish	79-16	*caudimacula*	74-16
castaneus	85-2	Large-eye Bream	48-2	*dorsomacula*	72-12
filamentosus	85-1	*coloundra, Apistops*	20-1	*gaimardi*	74-13
stigmaticus	86-6	*colubrinus, Myrichthys*	9-7	*pictoides*	74-15
cirrocheilos, Brachysomophis	9-3	*Colurodontis paxmani*	102-16	*cornelii, Kyphosus*	53-13
citrinellus, Chaetodon	55-7	*come, Nematalosa*	6-14	*cornuta, Aesopia*	98-10
Citron Perchlet	22-13	Comet	28-8	*Lactoria*	103-1
clarkii, Amphiprion	64-1	Groper	25-8	*cornutum, Sargocentron*	15-11

	Species No.		Species No.		Species No.
cornutus, Zanclus	92-9	Cuskeel, Bearded	12-19	Charcoal	67-9
coronata, Canthigaster	105-1	Black-edged	12-18	Cross's	67-3
Coronation Trout	24-11	Golden	12-20	Darwin	61-7
Corselletted Frigate Mackerel	97-9	Red	12-16	Dick's	66-1
Coryphaena hippurus	39-1	*cuspidatus, Pristis*	3-12	Dusky	67-5
Corythoichthys amplex	18-8	Cutribbon Wrasse	79-3	Goldback	67-4
haematopterus	18-5	*cuvier, Galeocerdo*	3-1	Grey	62-2
intestinalis	18-6	*cyaena, Chrysiptera*	62-1	Gulf	66-16
schultzi	18-7	*cyanoctenosum, Oxymetopon*	91-17	Honeyhead	61-8
Cottapistus cottoides	20-4	*cyanodus, Choerodon*	74-3	Java	67-1
praepositus	19-12	*cyanomos, Neopomacentrus*	61-9	Jewel	66-3
cottoides Cottapistus	20-4	*cyanopleura, Cirrhilabrus*	72-1	Johnston	66-2
Cowfish, Long-horned	103-1	*Leptojulis*	77-5	Lagoon	61-12
Round-belly	103-2	*cyanopodus, Epinephelus*	26-8	Lemon	66-9
Thornback	104-1	*cyanopterus, Solenostomus*	17-3	Miller's	66-8
Cowtail Stingray	5-2	*cyanosoma, Apogon*	32-14	Neon	66-7
Craterocephalus pauciradiatus	13-15	*cyanostigma, Cephalopholis*	26-6	Philippine	67-8
craticula, Zebrias	98-14	*Sebastapistes*	18-18	Princess	66-10
Cream-spotted Cardinalfish	33-14	*cyanotaenia, Pseudochromis*	28-15	Richardson's Reef	62-12
Crescent Perch	30-11	*Cybiosarda elegans*	97-8	Sandy	66-11
criboris Gymnothorax	7-6	*Cyclichthys jaculiferus*	105-17	Scaly	67-6
Crimson Seaperch	42-1	*orbicularis*	105-18	Staghorn	61-6
Soldierfish	14-5	*cyclophthalmus, Labracinus*	29-6	Threespot	67-13
cristatus, Samaris	98-17	*cyclostomos, Parupeneus*	52-1	White	62-17
Croaker, Coitor	53-3	*Cymbacephalus nematophthalmus*		White-banded	66-4
Green-backed	53-4		21-12	Whitepatch	62-13
Orange	53-2	*Cymolutes praetextatus*	79-13	Dark Surgeonfish	95-12
Spotted	584	*torquatus*	79-16	Dark-banded Fusilier	46-6
Crocodile Snake Eel	10-3	*Cypho purpurescens*	29-17	Dark-finned Velvetfish	21-6
Crocodilian Longtom	13-12	*cypho, Cyttopsis*	16-3	Dark-headed Leatherjacket	101-14
crocodilinus, Brachysomophis	10-3	*Cyprinocirrhites polyactis*	68-5	Darkspot Tuskfish	74-11
crocodilus, Tylosurus	13-12	*cyprinoides, Megalops*	6-4	Dark-spotted Sole	98-12
Cromileptes altivelis	22-10	*Cypselurus* sp.	13-1	Dark-tailed Seaperch	42-14
crossi, Neoglyphidodon	67-3	*cypselurus, Rhabdamia*	33-15	Dark Thick-rayed Sole	98-10
Crossosalarias macrospilus	86-8	*Cyttopsis cypho*	16-3	Dart, Black-spotted	39-10
Cross's Damsel	67-3			Common	39-12
Crowned Squirrelfish	14-9	**D**		Snub-nosed	39-11
crumenthalmops, Selar	39-8	*dactyliophorus, Doryrhamphus*		Dartfish, Elegant	91-9
Cryptic Wrasse	79-4		17-11	Fire	91-10
Cryptocentrus cinctus	88-9	*Dactyloptaenia orientalis*	16-12	Pale	91-13
leucostictus	88-11	*Dactylopus dactylopus*	87-14	Spot-tail	91-14
obliquus	89-10	*dactylopus, Dactylopus*	87-14	Twotone	91-12
octafasciatus	89-9	*dahli, Batrachomoeus*	12-1	Zebra	91-11
strigilliceps	88-12	Dahl's Frogfish	12-1	*daruma, Dampierosa*	19-10
cryptus, Pterogogus	79-4	Dampier Stonefish	19-10	Darwin Damsel	61-7
Ctenochaetus binotatus	95-16	*Dampierosa daruma*	19-10	Jawfish	83-10
striatus	94-3	Damsel, Alexander's	66-5	Darwin's Toadfish	105-9
strigosus	93-2	Ambon	66-6	*darwiniensis, Dischistodus*	61-7
tominiensis	93-1	Banded	62-14	*Opistognathus*	83-10
Ctenogobiops pomasticus	89-11	Barhead	67-2	*darwinii, Marilyna*	105-9
tangaroae	88-14	Batuna's	63-5	*Dascyllus aruanus*	64-15
Ctenogobius aurocingulus	88-15	Behn's	61-14	*melanurus*	63-6
Cubiceps whiteleggii	99-6	Biglip	64-6	*reticulatus*	64-14
cubicus, Ostracion	103-3	Black	61-13	*trimaculatus*	64-13
Culverin	9-6	Black-banded	67-14	Dascyllus, Black-tailed	63-6
cuniculus, Hoplatilus	46-16	Blackspot	67-11	Humbug	64-15
cuning, Caesio	45-11	Blackvent	62-16	Reticulated	64-14
curacao, Amblyglyphidodon	61-6	Blue	67-7	Three-spot	64-13
curiosus, Gunnelichthys	91-16	Bluespot	67-10	Dash-dot Goatfish	51-16
Curious Wormfish	91-16	Burrough's	67-12	*Dasyatis annotatus*	4-9

		Species No.		Species No.		Species No.
	kuhlii	4-12	*Sargocentron*	14-9	*janssi*	17-12
	leylandi	4-10	*Diademichthys lineatus*	12-6	*multiannulatus*	18-2
	sephen	5-2	Diagonal-banded	44-13	*pessuliferus*	18-3
	sp.	4-11	Sweetlips		Dot-and-Dash Butterflyfish	57-6
	thetidis	5-1	Diagonal-lined Wrasse	79-11	Dothead Rockcod	26-2
dasypogon, Eucrossorhinus		1-16	*Diagramma labiosom*	45-6	Dotted Butterflyfish	57-4
dea, Xyrichtys		78-11	*pictum*	44-9	Sweetlips	44-12
Decapterus kurroides		38-10	*diagramma, Pterocaesio*	45-13	Wrasse	72-6
	macarellus	38-11	*diagrammus, Oxycheilinus*	71-7	Dottyback, Black-striped	29-21
	macrosoma	38-13	Diamond Fish	53-6	Brown	28-11
			Trevally	37-2	Double-striped	29-16
Decora Goby		91-7	Wrasse	75-8	Firetail	29-6
decora, Nemateleotris		91-9	Diamond-scale Mullet	69-3	Howson's	29-13
	Valenciennea	91-7	*diana, Bodianus*	71-4	Imperial	29-15
Decapterus russelli		38-12	Diana's Wrasse	71-4	Lined	28-10
	tabl	38-9	*diaphana, Lactoria*	103-2	Longfin	28-13
Decorated Goby		90-3	*dickii, Plectroglyphidodon*	66-1	Long-finned	29-12
decoratus, Istigobius		90-3	Dick's Damsel	66-1	Magenta	29-8
decurrens, Helcogramma		87-1	*didactylus, Inimicus*	19-15	Marshall	28-12
decussatus, Amblygobius		88-20	*diemensis, Halophryne*	12-4	Moore's	29-14
	Entomacrodus	85-8	*dimidiatus, Labroides*	77-3	Multi-coloured	29-18
	Lutjanus	42-3	*Scarus*	80-7	Oblique-lined	29-17
Deep-bodied Flounder		98-4	*Diodon holocanthus*	105-13	Purpletop	29-9
Deepsea Bigeye		31-7	*hystrix*	104-14	Queensland	29-19
	Fairy Basslet	28-2	*liturosus*	105-14	Royal	29-7
	Jewfish	30-7	*diphreutes, Heniochus*	56-12	Sailfin	29-20
	Leatherskin	39-4	*Diplogrammus xenicus*	87-13	Splendid	29-10
	Scorpionfish	20-14	*Diploprion bifasciatum*	28-1	Spotted	28-14
	Snapper	47-17	*diplospilus, Parapercis*	83-2	Steene's	29-11
Deepwater Firefish		19-4	*Pseudorhombus*	98-3	Yellowfin	28-16
	Squirrelfish	14-7	*Diproctacanthus xanthurus*	73-5	Yellowhead	28-15
Demoiselle, Azure		62-10	*Dipterygonatus balteatus*	46-12	Double-banded Stargazer	83-15
	Brown	61-10	Disappearing Wrasse	72-17	Doublebar Goatfish	52-3
	King	62-3	*Dischistodus chrysopoecilus*	62-13	Spinefoot	92-15
	Onespot	62-7	*darwiniensis*	61-7	Double-barred Goby	89-8
	Regal	61-9	*fasciatus*	62-14	Double-ended Pipefish	17-21
	Rolland's	62-5	*melanotus*	62-16	Double-headed Maori Wrasse	71-11
	Silver	62-15	*perspicillatus*	62-17	Parrotfish	80-1
	Springer's	62-11	*prosopotaenia*	61-8	Double-lined Mackerel	97-14
	Surge	62-9	Dispar Cardinalfish	36-4	Monocle Bream	50-8
	Talbot's	62-4	*dispar, Apogon*	36-4	Doublepore Fangblenny	86-2
	Twospot	62-6	*Pseudanthias*	27-6	Doublesaddle Butterflyfish	56-4
	Violet	62-18	*ditchela, Pellona*	6-13	Doublespot Grubfish	83-2
	Yellowtail	61-11	Ditchelee	6-13	Lionfish	18-12
Demon Stinger		19-15	*ditrema, Meiacanthus*	86-2	Double-spotted Queenfish	39-6
Dendrochirus biocellatus		18-12	*djiddensis, Rhynchobatus*	3-15	Double-striped Dottyback	29-16
	brachypterus	19-1	*doederleini, Apogon*	32-3	Doubletooth Soldierfish	14-2
	zebra	19-2	Doederlein's Cardinalfish	32-3	Doublewhip Threadfin-Bream	49-10
dentex, Pseudocaranx		38-19	Dogtooth Tuna	97-12	*draconis, Eurypegasus*	16-14
Dentex tumifrons		47-17	*doliatus, Hologymnosus*	73-6	Dragon Moray	8-5
dermopterus, Triso		24-6	*dolosus, Peristrominous*	20-3	Dragonet, Fingered	87-14
Devil, Blue		62-1	Dolphinfish, Common	39-1	Highfinned	87-11
Devilfish, Blue		29-4	*dorab, Chirocentrus*	6-5	Morrison's	87-10
	Rose	28-17	Dorado	39-1	Northern	87-13
	Yellow	29-5	*dorsomacula, Coris*	72-12	Ocellated	88-2
Dexillichthys muelleri		98-11	Dory, John	16-5	Rosy	87-9
diabolis, Scorpaenopsis		20-11	Little	16-3	Weedy	88-3
diacanthus, Protonibea		53-1	Mirror	16-4	Western	87-12
	Pygoplites	59-1	*Doryrhamphus dactyliophorus*	17-11	*Drepane punctata*	54-3
diadema, Pseudochromis		29-9	*excisus excisus*	18-4	Driftfish, Indian	99-4

	Species No.		Species No.		Species No.
Drombus sp.	89-12	Girdled Reef	7-2	Thumbprint	48-15
Drummer, Low-finned	53-14	Leopard	7-8	Variegated	47-13
Southern	53-12	Ribbon	8-16	Yellow	28-1
duboulayi, Chaetodontoplus	59-5	Starry	7-1	Yellowlip	48-10
Duckbill, Blotched	99-3	Vulture	9-15	Yellow-spotted	48-11
Sharpnosed	99-2	White Ribbon	8-17	Yellow-tailed	106-2
Spotted	99-1	Zebra	7-3	*Enchelycore bayeri*	8-4
ductor, Naucrates	38-16	Eel Pipefish	17-8	*canina*	8-2
dumerili, Seriola	39-16	Eel-Blenny, Carpet	68-14	*pardalis*	8-5
dumerilli, Cantherhinus	101-10	Ocellated	68-12	*schismatorhynchus*	8-3
duospilus, Fusigobius	89-15	Shark Bay	68-16	*endekataenia, Apogon*	32-4
Dusky Blenny	85-1	Spiny	68-13	Endracht Hardyhead	13-17
Butterflyfish	58-2	Spotted	68-15	*endrachtensis, Atherinomorus*	13-17
Chromis	65-16	*ehrenbergi, Lutjanus*	43-6	Platycephalus	21-17
Damsel	67-5	Ehrenberg's Seaperch	43-6	*Engyprosopon grandisquama*	98-7
Gregory	66-15	*eibli, Centropyge*	59-3	*enigmaticus, Gymnothorax*	9-12
Parrotfish	81-4	Eibl's Angelfish	59-3	*Entomacrodus decussatus*	85-8
Shark	2-13	Eight-lined Cardinalfish	31-10	*striatus*	85-9
Surgeonfish	94-7	Wrasse	72-16	*thalassinus*	85-7
Velvetfish	21-2	Eight-spined Cardinalfish	34-15	Epaulet Trevally	37-11
Duskyfin Bigeye	31-2	*Elates ransonnetti*	21-13	Epaulette Shark	1-10
Duskytail Groper	25-5	*elegans, Cybiosarda*	97-8	*Epibulus insidiator*	75-1
Dusky-tailed Cardinalfish	34-18	*Trichonotus*	91-20	*ephippium, Amphiprion*	63-10
dussumieri, Acanthurus	94-3	Elegant Dartfish	91-9	Chaetodon	55-8
Aspidontus	84-1	Sand-diver	91-20	*epilampra, Canthigaster*	104-7
Carcharhinus	2-9	*Elegatis bipinnulata*	39-2	*Epinephelus amblycephalus*	23-2
Dussumieria elopsoides	6-8	*elerae, Chromis*	65-5	*areolatus*	23-3
Dwarf Flathead	21-13	*Eleutheronema tetradactylum*	69-10	*bilobatus*	24-1
Hawkfish	67-16	*elevatus, Pseudorhombus*	98-4	*bleekeri*	25-5
Lionfish	19-1	*ellioti, Apogon*	33-9	*caeruleopunctatus*	23-4
Moray	8-13	Elongate Unicornfish	93-7	*coioides*	24-4
		elongatus, Adventor	21-1	*corallicola*	23-5
E		Gymnocranius	47-2	*cyanopodus*	26-8
Eagle Ray, Barbless	5-9	Leiognathus	40-13	*fasciatus*	23-6
Spotted	5-10	Lethrinus	47-5	*fuscoguttatus*	23-8
Earbar Surgeonfish	95-8	Myxus	69-4	*heniochus*	23-9
Ebony Triggerfish	100-4	Pseudojuloides	78-5	*hexagonatus*	26-9
Echeneis naucrates	40-7	*Elops hawaiiensis*	6-1	*lanceolatus*	24-12
Echidna nebulosa	7-1	*elopsoides, Dussumieria*	6-8	*latifasciatus*	23-7
polyzona	7-2	Ember Parrotfish	81-6	*macrospilos*	26-10
zebra	7-3	*emeryii, Pentapodus*	50-3	*maculatus*	26-7
echinocephalus, Paragobiodon	90-10	Emperor Angelfish	59-9	*magniscuttis*	25-7
Ecsenius alleni	86-10	Emperor, Ambon	48-13	*malabaricus*	25-1
australianus	86-11	Black-blotch	48-16	*merra*	23-10
axelrodi	86-13	Blue-lined	47-9	*microdon*	23-11
bandanus	86-14	Grass	47-7	*miliaris*	25-6
bathi	86-12	Longfin	48-12	*morrhua*	25-8
bicolor	85-3	Long-nosed	47-5	*multinotatus*	24-2
lineatus	85-4	Longspine	47-10	*ongus*	25-10
lividanalis	86-15	Orange-striped	48-7	*polylepis*	25-9
melarchus	86-16	Ornate	48-9	*quoyanus*	23-13
midas	86-9	Pink-eared	47-11	*radiatus*	23-15
namiyei	86-17	Purple-headed	47-11	*rivulatus*	23-14
oculus	85-5	Red	42-10	*sexfasciatus*	24-3
pictus	86-18	Redsnout	48-14	*stictus*	25-2
yaeyamensis	85-6	Smalltooth	48-8	*tauvina*	26-11
edentululus, Istiblennius	85-11	Spangled	47-8	*timorensis*	25-3
Eel, Brown-flecked Reef	7-6	Spotcheek	47-12	*trophis*	25-4
Clouded Reef	7-1	Sweetlip	47-6	*tukula*	24-5
Giraffe	7-7	Threadfin	47-10	*undulostriatus*	25-11

	Species No.		Species No.		Species No.
equula, Carangoides	37-10	Smith's	86-3	Seaperch	42-7
equulus, Leiognathus	40-12	Yellow	86-4	Threadfin-Bream	49-3
eriomma, Parascolopsis	50-1	Yellowtail	86-1	Fivefinger Razorfish	79-15
Erosa erosa	19-9	*far, Hemirhamphus*	13-4	Fivestripe Wrasse	79-9
erosa, Erosa	19-9	*fasciatum, Cephaloscyllum*	1-13	Flagfin Cardinalfish	33-9
erumei, Psettodes	98-1	*Stegostoma*	1-2	Wrasse	77-7
erythropterus, Lethrinus	48-12	*fasciatus, Cheilinus*	70-9	Flagtail Blanquillo	40-4
Lutjanus	42-1	*Choerodon*	71-14	Shrimpgoby	88-10
Estuarine Stonefish	19-8	*Dischistodus*	62-14	Flag-tailed Rockcod	22-9
Estuary Cardinalfish	34-11	*Epinephelus*	23-6	Flame Angelfish	58-14
Cod	24-4	*Hemigymnus*	75-11	Hawkfish	67-17
Goby	89-16	*Ichthyscopus*	83-14	Flappy Snake-eel	9-8
Snake-eel	9-12	*Limnichthys*	68-10	Flathead, Bar-tailed	21-17
Whaler	2-7	*Pseudanthias*	27-15	Dwarf	21-13
Etelis carbunculus	41-3	*Salarias*	85-15	Fringe-eyed	21-12
radiosis	41-4	*fasciolatum, Belonepterygium*	28-7	Ghost	16-11
ethiops, Triacanthodes	99-13	*fasciolatus, Stegastes*	66-12	Harris's	21-14
Eubalichthys caeruleoguttatus	102-6	*favagineus, Gymnothorax*	7-7	Heart-headed	21-19
Eucrossorhinus dasypogon	1-16	Feminine Wrasse	73-1	Northern Sand	21-16
Eugomphodus taurus	1-4	*femininus, Anampses*	73-1	Olive-tailed	21-20
Euleptorhamphus viridis	13-8	*ferdau, Carangoides*	37-8	Rusty	21-18
Euristhmus lepturus	11-4	*ferox, Omobranchus*	84-12	Flathead, Spiny	21-15
nudiceps	11-3	*ferrugineus, Nebrius*	1-3	Flat-tail Mullet	69-1
eurostus, Gymnothorax	7-5	*Festucalex scalaris*	17-13	Flat-tailed Longtom	13-10
Eurypegasus draconis	16-14	Few-rayed Hardyhead	13-15	*flavicauda, Centropyge*	58-15
Euphyra blochii	3-8	*filamentosus, Cirrhilabrus*	72-9	*flavimarginatus, Gymnothorax*	7-8
Euthynnus affinis	97-7	*Cirripectes*	85-1	*Pseudobalistes*	101-1
evanidus, Pseudocheilinus	72-17	*Gerres*	50-14	*flavipectoralis, Scarus*	82-3
evermanni, Apogon	32-9	*Neopomacentrus*	61-10	*flavipinnis, Chrysiptera*	62-8
Evermann's Cardinalfish	32-9	*Paracheilinus*	72-10	*Pristipomoides*	44-6
evides, Pterleotris	91-12	*Pristipomoides*	41-6	*flavirostris, Chaetodon*	58-2
Eviota infulata	89-14	Filamentous Wrasse	72-10	*flavissimus, Assessor*	29-5
sp.	89-13	*Filicampus tigris*	17-14	*Centropyge*	60-1
Exallias brevis	84-7	*filicauda, Paramonacanthus*	102-10	*Forcipiger*	56-9
Exquisite Wrasse	72-3	*filodorsalia, Bembrops*	99-2	*flavolineatus, Mulloidichthys*	51-14
exquisitus, Cirrhilabrus	72-3	Fimbriated Moray	7-9	*flavomaculatus, Plectorhinchus*	106-1
Exyrias bellissimus	91-1	*fimbriatus, Gymnothorax*	7-9	*flavus, Apogon*	35-16
Eyebrowfish, Whitlegge's	99-6	Fine-spined Pufferfish	103-9	Flores Cardinalfish	36-2
Eyed Skate	4-7	Fingered Dragonet	87-14	Flounder, Cockatoo	98-17
		Fingermark Seaperch	43-9	Deep-bodied	98-4
F		Finny Scad	38-15	Intermediate	98-6
faber, Zeus	16-5	Fire Dartfish	91-10	Large-toothed	98-2
fageni, Naso	106-4	Firefish, Deepwater	19-4	Ocellated	98-18
Fairy Basslet, Cooper's	22-11	Ragged-finned	19-3	Panther	98-9
Deepsea	28-2	Red	19-6	Small toothed	98-5
Little	28-3	Roundface	19-7	Spiny-headed	98-7
Pearl-spotted	22-12	Spotless	19-5	Three-spot	98-8
falciformes, Carcharhinus	2-8	Firetail Dottyback	29-6	Twinspot	98-3
falco, Cirrhitichthys	67-16	Fish, Chinaman	41-9	Flowery Cod	23-8
False Bluefin Trevally	37-14	Diamond	53-6	Flutemouth, Painted	16-10
Cleanerfish	84-2	Knight	16-1	Rough	16-9
Clown Anemonefish	64-2	Pilot	38-16	Smooth	16-8
Gramma	28-6	Sargassum	12-15	*fluvus, Lutjanus*	43-7
Scorpionfish	22-2	Fishnet Lizardfish	11-13	Flyingfish	13-1
Stonefish	20-11	*Fistularia commersonii*	16-8	*Foa brachygramma*	31-13
Three-spot Cardinalfish	33-8	*petimba*	16-9	Footballer, Chinese	35-12
Whiptail	50-5	*fitchi, Priacanthus*	31-7	*Forcipiger flavissimus*	56-9
Fan-bellied Leatherjacket	102-7	Five-banded Wrasse	75-11	*longirostris*	58-11
Fangblenny, Bicolor	86-5	Five-bar Coral Goby	90-2	*formosa, Cephalopholis*	22-6
Doublepore	86-2	Five-lined Cardinalfish	31-11	*formosus, Parioglossus*	90-7

	Species No.		Species No.		Species No.
fornasini, Lactoria	104-1	Dark-banded	46-6	Tetrasomus	103-7
forsteni, Scarus	82-4	Goldband	46-10	*gibbus, Lutjanus*	43-8
forsteri, Paracirrhites	68-6	Marr's	46-5	Scarus	81-1
Fourspot Wrasse	75-3	Mottled	46-12	*gilberti, Apogon*	36-10
fourmanori, Hoplolatilus	46-15	One-striped	46-9	Gilbert's Cardinalfish	36-10
fowleri, Acanthurus	95-14	Randall's	46-8	Gilded Triggerfish	101-5
Fowler's Surgeonfish	95-14	Red-bellied	45-11	Giraffe Eel	7-7
Fowleria aurita	31-15	Slender	46-13	Girdled Goby	90-8
marmorata	34-13	Three-striped	46-4	Cardinalfish	34-17
punctulata	34-14	Wideband	46-7	Reef Eel	7-2
vaiulae	34-12	Yellow and Blueback	46-1	Gizzard Shad	6-11
variegatus	31-14	Yellowback	46-2	*gladius, Xiphias*	96-6
Foxface	93-16	*fusovatus, Apogon*	33-13	Glassy Bigeye	31-2
fraenatus, Apogon	32-13			Bombay Duck	11-19
Fragile Cardinalfish	35-7	**G**		Cardinalfish	34-16
fragilis, Apogon	35-7	*gaimardi, Coris*	74-13	*glauca, Chrysiptera*	62-2
frassendai, Apogon	36-2	*Galeocerdo cuvier*	3-1	Prionace	3-4
Freckled Anglerfish	12-11	Garden Eel, Barnes's	10-11	*glaucoparieus, Acanthurus*	94-2
Hawkfish	68-6	Cobra	10-6	*Glaucosoma burgeri*	30-7
Moray	7-15	Many-toothed	10-7	*hebracium*	30-7
Porcupinefish	105-13	Speckled	10-12	*magnificum*	30-8
frenatus, Amphiprion	63-11	Splendid	10-8	Glittering Pipefish	18-9
Gymnocranius	48-6	Spotted	10-9	*globiceps, Scarus*	81-2
Lepadichthys	12-7	Taylor's	10-10	*gloriamaris, Cleidopus*	16-1
Lethrinus	47-9	Garfish, Barred	13-4	*Glossogobius biocellatus*	89-16
Scarus	80-6	Buffon's	13-3	*circumspectus*	89-17
Sufflamen	100-10	Long-finned	13-8	*Gnathanodon speciosus*	38-14
Frigate Mackerel	97-10	Quoy's	13-7	*Gnathodentex aurolineatus*	48-1
Fringe-eyed Flathead	21-12	Robust	13-5	*Gnatholepis scapulostigma*	89-18
Fringe-finned Trevally	37-3,	Snub-nosed	13-2	Goatfish, Asymmetrical	52-15
	38-17	Three by Two	13-5	Banded	52-8
Fringe-lipped Snake-eel	9-5	Tropical	13-6	Bar-tailed	52-14
Frogfish, Banded	12-4	Garish Rockcod	26-3	Blackspot	52-5
Dahl's	12-1	*gavialoides, Tylosurus*	13-13	Dash-dot	51-16
Ocellated	12-5	*Gazza minuta*	40-10	Doublebar	52-3
Three-spined	12-3	*gelatinosa, Pseudamia*	34-9	Goldband	52-10
Western	12-2	Gelatinous Cardinalfish	34-9	Gold-saddled	52-1
frontalis, Chlorurus	82-5	*Genicanthus lamarck*	60-9	Indian	52-6
fronticinctus, Cantherhines	102-4	*melanospilos*	60-10	Ochre-banded	52-12
Frostback Cod	24-1	*watanabei*	60-11	Sidespot	52-9
Frostfin Cardinalfish	35-13	*genivittatus, Lethrinus*	47-10	Spotted Golden	52-7
fucata, Archamia	33-17	*germaini, Omobranchus*	84-11	Striped	52-13
fulviflamma, Lutjanus	42-8	Germain's Blenny	84-11	Stripe-spot	51-17
fulvoguttatus, Carangoides	37-9	*geographicus, Anampses*	70-2	Sunrise	52-11
fumea, Chromis	64-10	*Gerres filamentosus*	50-14	Swarthy-headed	52-2
furca, Aphareus	41-1	*oyena*	50-15	Yellowfin	51-15
furcosus, Nemipterus	49-4	*subfasciatus*	50-16	Yellowstripe	51-14
fuscescens, Siganus	92-11	*ghobban, Scarus*	80-8	Yellow striped	52-4
fuscoguttatus, Epinephelus	23-8	Ghost Flathead	16-11	Gobbleguts	32-17
fuscomaculatus, Apogon	36-5	Shark	5-11	*Gobiodon histrio*	90-1
fuscus, Apogon	32-11	Giant Herring	6-1	*quinquestrigatus*	90-2
Atrosalarias	84-3	Moray	7-10	Goby, Allen's	91-2
Bathygobius	89-6	Salmon Catfish	11-1	Banded	89-4
Pseudobalistes	100-7	Seapike	69-11	Bearded	92-3
Pseudochromis	28-11	Shovelnose Ray	3-17	Beautiful	91-1
Fusigobius duospilus	89-15	Sweetlips	44-11	Black-spotted	90-4
Fusiler, Banana	46-11	Threadfin	69-10	Blind	92-8
Black-tipped	45-13	Trevally	38-1	Blue	90-11
Blue	45-12	*gibbosa, Sardinella*	6-17	Blueband	91-5
Blue and Gold	46-3	*gibbosus, Plectorhinchus*	45-5	Blue-barred Ribbon	91-17

Broad-barred Maori	90-1	Whiting	30-14	Grey Nurse Shark	1-4
Circumspect	89-17	Goldflag Jobfish	44-5	Reef Shark	2-3
Cocos	89-5	Goldring Bristletooth	93-2	Tilefish	46-16
Common	89-6	Gold-saddled Goatfish	52-1	*griffini, Apogon*	35-15
Decora	91-7	Goldspot Pigfish	70-7	Grinner, Common	11-18
Decorated	90-3	Gold-spotted Sweetlips	106-1	Large-scaled	11-17
Double-barred	89-8	Trevally	37-9	Painted	11-15
Estuary	89-16	Gold-streaked Shrimp Goby	88-15	Slender	11-16
Five-bar Coral	90-2	Goldstripe Sardine	6-17	*griseus, Gymnocranius*	47-3
Girdled	90-8	Wrasse	76-4	Hexanchus	1-8
Goggle-eyed	92-4	Golden-striped Butterflyfish	55-6	Grooved Razorfish	17-1
Green Shrimp	89-9	*Gomphosus varius*	75-2	Groper, Baldchin	74-5
Head-barred	90-9	*goodladi, Callionymus*	87-4	Black-dotted	25-2
Hector's	88-18	Goodlad's Stinkfish	87-4	Comet	25-8
Hoese's	90-14	*Gorgasia barnesi*	10-11	Duskytail	25-5
Immaculate	91-8	sp.	10-12	Malabar	25-1
Long-finned	90-7	*Gracilia albomarginata*	26-1	Maori	25-11
Madura	92-1	*glacilicaudus, Gymnothorax*	8-12	Netfin	25-6
Mangrove	89-1	*graciliosa, Lotilia*	88-13	Plump	25-4
Mud	89-12	*gracilis, Acentrogobius*	89-1	Queensland	24-12
Nocturna	88-19	Rhabdamia	33-16	Small-scaled	25-9
Ocellated	90-19	Saurida	11-16	Speckled	25-7
Old Glory	88-17	Spratelloides	6-15	*grossi, Callionymus*	87-8
Orange-dashed	90-18	*graefei, Arius*	11-2	Gross's Stinkfish	87-8
Orange-spotted	90-13	Gramma, False	28-6	Grubfish, Banded	83-6
Orange-striped	88-20	*Grammatobothus polyophthalmus*		Blue-nosed	83-1
Ornate	90-6		98-8	Doublespot	83-2
Parva	91-6	*Grammatorcynus bicarinatus*	97-13	Narrow-barred	83-4
Patchwork	90-15	*bilineatus*	97-14	Red-banded	83-7
Pink Shrimp	89-10	*Grammistes sexlineatus*	28-5	Red-barred	83-8
Red and White	90-12	*grammistes, Meiacanthus*	84-10	Rosy	83-5
Red Coral	89-13	*grammorhynchus, Pomacentrus*		Spothead	83-3
Redhead	90-10		67-10	Grunter, Banded	30-13
Schooling	90-7	*grammoptilus, Acanthurus*	94-5	Spiny-cheeked	30-12
Shadow	90-20	*grandisquama, Engyprosopon*	98-7	Three-lined	30-12
Sixspot	91-4	*grandisquamis, Cantheschenia*	101-8	Yellowtail	30-9
Smith's Coral	89-14	*grandoculus, Gymnocranius*	47-4	Guam Scorpionfish	20-6
Spectacled	90-5	Monotaxis	47-1	*guamensis, Scorpaenodes*	20-6
Spotted Shrimp	89-11	Graphic Tuskfish	71-15	Gudgeon, Chinese	92-7
Starry	89-3	*graphicus, Choerodon*	71-15	Olive Flathead	92-6
Striped	90-16	Grass Emperor	47-7	*guentheri, Chaetodon*	58-3
Twinspot	88-16	Tuskfish	71-13	Setarches	20-14
Twostripe	91-3	Great Hammerhead	3-9	Guineafowl Moray	8-6
Twospot	89-15	Green Chromis	64-9	Gulf Damsel	66-16
Wheeler's Shrimp	89-2	Jobfish	41-2	*Gunnelichthys curiosus*	91-16
Whiskered	89-7	Moon Wrasse	78-5	*monostigma*	91-15
Whitecap	88-13	Sawfish	3-11	Gunther's Butterflyfish	58-3
Yellowspot	89-18	Shrimp Goby	89-9	Threadfin	69-8
Goggle-eyed Goby	92-4	Wrasse	76-16	Gurnard, Black-finned	21-8
Goldback Damsel	67-4	Greenback Mullet	69-2	Half-spotted	21-9
Goldband Fusilier	46-10	Greenbacked Croaker	53-4	Long-finned	21-7
Goatfish	52-10	Greensnout Parrotfish	82-9	*gushikeni, Parapercis*	83-5
Gold-banded Jobfish	41-8	Green-finned Parrotfish	81-8	*guttata, Amblyeleotris*	88-4
Goldbelly Cardinalfish	35-1	Green-spotted Wrasse	76-1	*guttatissimus Chaetodon*	57-2
Golden Cuskeel	12-20	*gregoryi, Notograptus*	68-16	*guttatus, Acanthurus*	95-4
Spinefoot	93-15	Gregory, Blunt-snout	66-13	Notograptus	68-15
Trevally	38-14	Dusky	66-15	Siganus	93-15
Wrasse	76-2	Pacific	66-12	*Gymnocaesio gymnopterus*	46-13
Goldeneye Jobfish	44-6	Western	66-14	*Gymnocranius audleyi*	48-2
Golden-lined Spinefoot	92-12	Grey Damsel	62-2	*elongatus*	47-2

	Species No.		Species No.		Species No.
euanus	48-3	*nigrescens*	75-8	*hedlandensis, Carangoides*	37-12
frenatus	48-6	*ornatissimus*	76-13	*Helcogramma decurrens*	87-1
grandoculus	47-4	*podostigma*	76-8	*striata*	87-2
griseus	47-3	*prosopeion*	76-14	*heldingenii, Valenciennea*	91-3
microdon	48-5	*purpurescens*	76-7	*Hemiglyphidodon plagiometopon*	
sp.	48-4	*rubricephalus*	76-3		61-12
gymnopterus, Gymnocaesio	46-13	*scapularis*	76-15	*Hemigymnus fasciatus*	75-11
gymnotus, Muraenichthys	9-4	*schwartzi*	76-11	*melapterus*	75-12
Gymnosarda unicolor	97-12	*solorensis*	76-16	*Hemirhamphus far*	13-4
Gymnothorax buroensis	7-4	*trimaculatus*	75-6	*robustus*	13-5
criboris	7-6	*Haliichthys taeniophorus*	17-17	*Hemiscyllium ocellatum*	1-10
enigmaticus	8-9	*Halimochirurgus centriscoides*	99-12	*trispeculare*	1-9
eurostus	7-5	*Halophyrne diemensis*	12-4	*hemisticta, Pterygotrigla*	21-9
favagineus	7-7	*ocellatus*	12-5	*hemistictus, Paracirrhites*	68-8
fimbriatus	7-9	*hamiltonii, Thyrssa*	6-21	*Hemitaurichthys polylepis*	58-10
flavimarginatus	7-8	Hamilton's Anchovy	6-21	*Heniochus acuminatus*	56-13
gracilicaudus	8-12	Hammerhead, Great	3-9	*chrysostomus*	56-10
javanicus	7-10	Scalloped	3-7	*diphreutes*	56-12
margaritophorus	7-11	Smooth	3-10	*monoceros*	58-12
melanospilus	8-1	*hamrur, Priacanthus*	31-4	*singularius*	66-11
melatremus	8-13	*hanae, Pterelotris*	90-11	*varius*	56-14
meleagris	8-6	*Hapalogenys kishinouyei*	45-10	*heniochus, Epinephelus*	23-9
moluccensis	8-7	*harak, Lethrinus*	48-15	*hepatus, Paracanthurus*	95-3
nudivomer	8-8	*hardwickei, Solegnathus*	78-6	*heptacanthus, Parupeneus*	52-7
pseudothrysoideus	8-10	*Thalassoma*	17-19	*heraldi, Centropyge*	60-2
richardsoni	8-11	Hardyhead, Endracht	13-17	Herald's Angelfish	60-2
rueppelliae	8-15	Few-rayed	13-15	*Herklotsichthys koningsbergeri*	6-10
undulatus	7-12	Ogilby's	13-18	*quadrimaculatus*	6-9
zonipectus	7-13	Samoan	13-16	Herring, Bluestripe	6-9
Gymnura australis	5-8	Spotted	13-14	Giant	6-1
		Harlequin Ghost Pipefish	17-2	Hairback	6-14
H		Snake-eel	9-7	Koningsberger's	6-10
haematopterus, Corythoichthys	18-5	Tuskfish	71-14	Oxeye	6-4
Lethrinus	47-7	*Harpodon transluceus*	11-19	Wolf	6-5
Hairback Herring	6-14	*harrisii, Inegocia*	21-14	*Heteroconger cobra*	10-6
Hair-finned Leatherjacket	102-11	Harris's Flathead	21-14	*hassi*	10-9
Hairtail Blenny	84-17	Harrowed Sole	98-15	*perissodon*	10-7
Hairy Scorpionfish	18-16	*hartzfeldi, Halichoeres*	76-4	*preclara*	10-8
Half and Half Chromis	65-7	*hartzfeldii, Apogon*	36-9	*taylori*	10-10
Halfmoon Triggerfish	101-7	Hartzfeldt's Cardinalfish	36-9	*Heteropriacanthus cruentatus*	31-2
Half-spotted Gurnard	21-9	*hassi, Heteroconger*	10-9	*heteroptera, Pseudocoris*	79-2
Halibut, Queensland	98-1	*hasta, Pomadasys*	45-8	*Pterelotris*	91-14
Halicampus brocki	17-15	*hawaiiensis, Elops*	6-1	*Heterodontus zebra*	1-6
nitidus	18-9	Hawk Anthias	29-1	*hexacanthus, Naso*	93-5
spinirostris	17-16	Hawkfish, Blotched	68-3	Hexagon Rockcod	26-9
Halichoeres argus	76-12	Dwarf	67-16	*hexagonatus, Epinephelus*	26-9
biocellatus	75-4	Flame	67-17	*Myripristis*	14-2
chloropterus	76-1	Freckled	68-6	*Hexanchus griseus*	1-8
chrysus	76-2	Longnose	67-18	*hexataenia, Pseudocheilinus*	77-6
hartzfeldi	76-4	Lyretail	68-5	*hexodon, Nemipterus*	49-5
hortulanus	75-3	Marbled	67-15	*hians, Ablennes*	13-9
leucurus	76-5	Ornate	68-8	Hick's Toadfish	105-10
margaritaceous	75-9	Ring-eyed	68-7	*hicksi, Torquigener*	105-10
marginatus	75-5	Sharp-headed	68-4	Highfin Coralfish	58-8
melanochir	75-7	Twinspot	68-2	Moray	8-10
melanurus	76-6	Head-barred Goby	90-9	Parrotfish	82-6
melasmapomus	76-9	Heart-headed Flathead	21-19	High-finned Blenny	84-9
miniatus	76-10	*hebraicum, Glaucosoma*	30-7	Dragonet	87-11
nebulosus	75-10	*hectori, Amblygobius*	88-18	Veilfin	16-7
		Hector's Goby	88-18	*Himantura toshi*	5-5

	Species No.		Species No.		Species No.
undulata	5-6	**I**		*Istigobius decoratus*	90-3
Hippichthys penicillus	17-18	*Ichthyapus vulturis*	9-15	*nigroocellatus*	90-4
Hippocampus angustatus	17-5	*Ichthyscopus fasciatus*	83-14	*ornatus*	90-6
hystrix	17-6	*insperatus*	83-15	*perspicillatus*	90-5
kuda	17-4	Idol, Moorish	92-9	*Istiophorus platypterus*	96-5
Hipposcarus longiceps	80-5	*igarashiensis, Cephalopholis*	26-3	*Isurus oxyrinchus*	1-7
hippurus, Coryphaena	39-1	*ignobilis, Caranx*	38-1	*ittodai, Sargocentron*	15-12
hirsutus, Scorpaenodes	18-16	Ilisha, Banded	6-12		
hispidus, Antennarius	12-12	*Ilisha striatula*	6-12	**J**	
Arothron	103-13	Imperial Dottyback	29-15	Jack, Almaco	39-14
histiophorus, Scartelaos	92-3	*immaculata, Valenciennea*	91-8	Mangrove	41-10
Histiopterus typus	68-1	Immaculate Goby	91-8	*jaculatrix, Toxotes*	54-2
histrio, Gobiodon	90-1	Pufferfish	103-14	*jaculiferus, Cyclichthys*	105-17
Histrio	12-15	*immaculatus, Arothron*	103-14	*jaculum, Synodus*	11-9
Histrio histrio	12-15	*imperator, Pomacanthus*	59-9	*janseni, Thalassoma*	78-7
hoesei, Silhouettea	90-14	Indian Anchovy	6-19	Jansen's Wrasse	78-7
Hoese's Goby	90-14	Driftfish	99-4	*jansii, Doryrhamphus*	17-12
hoeveni, Apogon	35-13	Goatfish	52-6	Janss's Pipefish	17-12
holocanthus, Diodon	105-13	Lizardfish	11-8	*janthinoptera, Canthigaster*	105-2
Hologymnosus annulatus	77-1	Scad	38-12	*janthinosoma, Pervagor*	102-12
doliatus	73-6	Seaperch	43-11	Japanese Butterfish	50-4
Honeycomb Cod	23-10	*indica, Ariomma*	99-4	Large-eye Bream	48-3
Leatherjacket	102-3	*Makaira*	96-1	Perchlet	22-15
Honeyhead Damsel	61-8	*indicus, Alectes*	37-2	Pineapplefish	16-2
Hookfin Cardinalfish	35-15	*Parupeneus*	52-6	Stinkfish	87-7
Hookjaw Moray	8-4	*Stolephorus*	6-19	Threadfin Bream	49-7
Hoplichthys regani	16-11	*Synodus*	11-8	Trevally	37-17
Hoplolatilus cuniculus	46-16	Indonesian Seaperch	41-11	Wrasse	79-1
fourmanori	46-15	Indo-Pacific Blue Marlin	96-2	*japonica, Monocentris*	16-2
starcki	46-17	Bluetang	95-3	*Inegocia*	21-18
Horn-nosed Boxfish	103-5	Sailfish	96-5	*japonicus, Callionymus*	87-7
horrida, Synanceja	19-8	*Inegocia harrisii*	21-14	*Nemipterus*	49-7
hortulanus, Halichoeres	75-3	*japonica*	21-18	*Plectranthias*	22-15
hoshinonis, Synodus	11-12	*inermis, Cheilio*	74-1	*jaruba, Terapon*	30-11
howsoni, Pseudochromis	29-13	*Lagocephalus*	105-6	Java Damsel	67-1
Howson's Dottyback	29-13	*infulata, Eviota*	89-14	Spinefoot	93-10
huchtii, Pseudanthias	27-3	*Inimicus didactylus*	19-15	*javanicus, Gymnothorax*	7-10
Humbug Dascyllus	64-15	*sinensis*	19-16	Javelinfish, Blotched	45-9
humerosa, Psenopsis	99-5	Inky-tail Shark	1-6	Lined	45-10
humerosus, Carangoides	37-11	*inornatus, Opistognathus*	83-12	Spotted	45-8
Humpback Anglerfish	12-8	*insidiator, Epibulus*	75-1	*javus, Siganus*	93-10
Unicornfish	93-4	*Secutor*	40-16	Jawfish, Abrolhos	83-9
Hump-backed Cardinalfish	36-13	*insperatus, Ichthyscopus*	83-15	Black	83-12
Humphead Bannerfish	56-14	Intermediate Cardinalfish	34-1	Blotched	83-11
Unicornfish	94-12	Flounder	98-6	Darwin	83-10
Hump-headed Batfish	54-4	*intermedius, Asterhombus*	98-6	Leopard	83-13
Wrasse	74-14	*Cheilodipterus*	34-1	*jello, Sphyraena*	69-13
Humpnose Bigeye Bream	47-1	*interrupta, Stethojulis*	79-3	*jenynsii, Pseudorhombus*	98-5
Hussar	43-2	*interruptus Ambassis*	30-4	*jerdoni, Pristotis*	66-16
hyalosoma, Apogon	36-13	*intestinalis, Corythoichthys*	18-6	Jew, Black	53-1
Hydrolagus sp.	5-11	*intonsa, Paraploactis*	21-5	Jewel Damsel	66-3
Hypnos monopterygium	4-6	*iomelas, Chromis*	65-7	Jewfish, Deepsea	30-7
Hypoatherina temminckii	13-16	Iridescent Cardinalfish	32-15	Little	53-5
Hypopterus macropterus	30-3	*Irolito* sp.	4-8	Spotted	53-1
Hyporhamphus affinis	13-6	*isacanthus, Nemipterus*	49-6	Jobfish, Goldbanded	41-8
quoyi	13-7	*Istiblennius chrysospilos*	85-10	Goldeneye	44-6
hypselopterus, Velifer	16-7	*edentulus*	85-11	Goldflag	44-5
hypselosoma, Pseudanthias	27-14	*lineatus*	85-13	Green	41-2
hystrix, Diodon	104-14	*meleagris*	85-12	Rusty	43-1
Hippocampus	17-6			Sharptooth	41-7

	Species No.		Species No.		Species No.
Small-toothed	41-1	*lineatus*	28-10	Brown-blotched	102-14
John Dory	16-5	*Labrichthys unilineatus*	77-4	Brush-sided	101-9
johnii, Lutjanus	43-9	*Labroides bicolor*	77-2	Dark-headed	101-14
Johnius amblycephalus	53-4	*dimidiatus*	77-3	Fan-bellied	102-7
coitor	53-3	*pectoralis*	73-7	Hair-finned	102-11
vogleri	53-5	*Labropsis alleni*	73-9	Honeycomb	102-3
Johnston Damsel	66-2	*australis*	73-11	Large-scaled	101-8
johnstonianus, Plectroglyphidodon		*manabei*	73-8	Mimic	101-18
	66-2	*xanthonota*	73-10	Modest	102-13
jordani, Choerodon	74-6	*lacrymatus, Plectroglyphidodon*		Orange-tailed	101-13
Jordan's Wrasse	74-6		66-3	Paxman's	102-16
Jumping Mullet	69-1	*Lactoria cornuta*	103-1	Pot-bellied	102-15
Jurgen	50-12	*diaphana*	103-2	Prickly	102-8
		fornasini	104-1	Radial	101-17
K		Ladder Pipefish	17-13	Red-tailed	102-12
kaakan, Pomadasys	45-8	Wrasse	79-10	Rhinoceros	101-12
Kai Stargazer	83-18	*laevis, Plectropomus*	25-12	Scribbled	102-2
kaianus, Ostichthys	14-7	*Lagocephalus inermis*	105-6	Small-spotted	101-11
Synodus	11-10	*lunaris*	105-5	Spectacled	102-4
Uranoscopus	83-18	*scleratus*	105-8	Threadfin	102-10
kallopterus, Apogon	32-15	*spadiceus*	105-7	Unicorn	102-1
kanagurta, Rastrelliger	97-11	Lagoon Damsel	61-12	Yelloweye	101-10
Kanekonia aniara	21-6	*Laiphognathus multimaculatus*	84-4	Leatherskin	39-5
kasmira, Lutjanus	42-6	*lalandi, Seriola*	39-15	Leatherskin, Deep	39-4
Katsuwonis pelamis	96-12	*lamarck, Genicanthus*	60-9	Slender	39-3
kauderni, Pterapogon	34-7	Lamarck's Angelfish	60-9	*leiogaster, Amblygaster*	6-6
Keyhole Angelfish	59-4	*lanceolatus, Epinephelus*	24-12	*Leiognathus bindus*	40-11
kiensis, Apogon	36-15	Lancer	47-10	*elongatus*	40-13
Kimberley Blenny	86-7	Lano	69-4	*Leiognathus equulus*	40-12
King Demoiselle	62-3	*landanus, Plagiotremus*	86-5	*leuciscus*	40-14
Kingfish, Black	40-6	Lantern Toby	104-7	*smithhursti*	40-15
Black-banded	39-13	Large-scaled Grinner	11-17	*leiura, Strongylura*	13-11
Rockhampton	69-10	Leatherjacket	98-2	*Leiuranus semicinctus*	9-6
Yellowtail	39-15	Large-spined Bigeye	31-3	*lemniscatus, Lutjanus*	42-14
kishinouyei, Hapalogenys	45-10	Large-spotted Rockcod	26-10	Lemon Damsel	66-9
klazingai, Apterichtus	10-1	Large-toothed Flounder	99-2	Shark	3-3
kleinii, Chaetodon	55-11	*larsonae, Acentronurus*	17-7	Lemonpeel Angelfish	60-1
Klein's Butterflyfish	55-11	Larson's Pipehorse	17-7	*lennardi, Anampses*	70-3
Knifefish	79-13	Larval Cardinalfish	35-18	*lentjan, Lethrinus*	47-11
Knifefish, Collared	79-16	*lateralis, Apogon*	36-14	Leopard Blenny	84-7
Knightfish	16-1	*Lates calcarifer*	30-1	Cod	24-7
koningsbergeri, Herklotsichthys		*laticaudis, Lethrinus*	47-7	Eel	7-8
	6-10	*latifasciata, Amblyeleotris*	88-8	Jawfish	83-13
Koningsberger's Herring	6-10	*latifasciatus, Epinephelus*	23-7	Shark	1-2
kuda, Hippocampus	17-4	*latispinosus, Choeroichthys*	17-9	Toby	104-9
kuhlii Dasyatis	4-12	*latitabundus, Opistognathus*	83-11	*leoparda, Canthigaster*	104-9
kuiteri, Macropharyngodon	73-14	*lativittata, Pterocaesio*	46-7	*leopardus, Cephalopholis*	22-7
Kuiter's Wrasse	73-14	*latovittatus, Malacanthus*	46-14	*Plectropomus*	24-7
kuntee, Myriptistis	14-3	Latticed Butterflyfish	57-3	*Lepadichthys frenatus*	12-7
Kuntee Soldierfish	14-3	Soldierfish	15-4	*lepidogenys, Pomacentrus*	67-6
kurroides, Decapterus	38-10	Lattice-tail Moray	7-4	*lepidolepis, Chromis*	65-8
Kyphosus bigibbus	53-12	*latus, Acanthopagrus*	47-15	*Lepidotrigula argus*	21-7
cornelii	53-13	*Serranocirrhitus*	29-1	*lepros, Sargocentron*	15-14
vaigiensis	53-14	Leaf Scorpionfish	20-13	*leptacanthus, Apogon*	35-8
		Leaping Bonito	97-8	*Pterygotrigla*	21-8
L		Leatherjacket, Beaked	102-9	*Leptobramma mulleri*	53-7
labiatus, Cheiloprion	64-6	Bearded	102-5	*Leptojulis cyanopleura*	77-5
labiosom, Diagramma	45-6	Black-lined	101-15	*leptolepis, Selaroides*	39-9
laboutei, Cirrhilabrus	72-2	Blue-spotted	102-6	*Leptoscarus vaigiensis*	80-4
Labracinus cyclophthalmus	29-6	Bristle-tailed	101-16	*lepturus, Euristhmus*	11-4

	Species No.			Species No.			Species No.
lessoni, Plectorhinchus	44-8	*lineolatus, Chaetodon*	55-9	*longirostris, Forcipiger*	58-11
Lethrinus amboinensis	48-13	*Lutjanus*	42-13	*Oxymonacanthus*	102-9
atkinsoni	106-2	*Omobranchus*	84-13	*Trachyrhamphus*	17-23
chrysostomus	47-6	*lineopunctatus, Xanthichthys*	100-11	Long-jawed Mackerel	97-11
erythracanthus	48-11	Lionfish, Clearfin	18-11	Longnose Hawkfish	67-18
erythropterus	48-12	Doublespot	18-12	Long-nosed Butterflyfish	56-9
genivittatus	47-10	Dwarf	19-1	Emperor	47-5
haematopterus	47-7	Spotfin	19-3	Grey Shark	2-6
harak	48-15	Volitans	19-6	Parrotfish	80-5
laticaudis	47-7	Zebra	19-2	Unicornfish	94-11
lentjan	47-11	*liorhynchus, Peristedion*	21-10	Longsnout Spikefish	99-12
mahsena	106-2	*Lipocheilus carnolabrum*	41-5	Longspine Emperor	47-10
microdon	48-8	Little Dory	16-3	Long-spined Snapper	47-16
miniatus	47-6	Fairy Basslet	28-3	Porcupinefish	105-17
nebulosus	47-8	Jewfish	53-5	Long-tailed Catfish	11-4
nematacanthus	47-10	Red Cardinalfish	33-5	Longtom, Barred	13-9
obsoletus	48-7	Scorpioncod	20-8	Crocodilian	13-12
olivaceus	47-5	*lituratus, Naso*	94-10	Flat-tailed	13-10
ornatus	48-9	*liturosus, Diodon*	105-14	Slender	13-11
reticulatus	48-14	*lividanalis, Ecsenius*	86-15	Stout	13-13
rubrioperculatus	47-12	*lividus, Stegastes*	66-13	*lopezi, Naso*	93-7
semicinctus	48-16	*Liza argentea*	69-1	*Lophiocharon trisignatus*	12-14
sp.	47-9	*subviridis*	69-2	*lorenzi, Abudefduf*	63-2
variegatus	47-13	*vaigiensis*	69-3	*lori, Pseudanthias*	27-2
xanthocheilus	48-10	Lizardfish, Bigeyed	11-11	*loriculus, Centropyge*	59-14
leucas, Carcharhinus	2-7	Black	11-10	Lori's Anthias	27-2
leuciscus, Leiognathus	40-14	Black-shouldered	11-12	*Lotilia graciliosa*	88-13
leucocheilus, Acanthurus	95-11	Checkered	11-17	*louti, Variola*	24-11
leucogaster, Richardsonichthys	20-12	Fishnet	11-13	Low-finned Drummer	53-14
leucogrammicus, Anyperodon	22-1	Indian	11-8	Lowly Trevally	38-1
leucopoma, Chrysiptera	62-9	Netted	11-13	*Loxodon macrorhinus*	3-2
leucosternon, Acanthurus	95-1	Tailspot	11-9	*loxozonus, Bodianus*	71-5
leucostictus, Cryptocentrus	88-11	Variagated	11-14	*lubbocki, Cirrhilabrus*	72-4
leucotaenia, Pholidichthys	91-18	*Lobotes surinamensis*	40-17	Lubbock's Wrasse	72-4
leucozona, Plectroglyphidodon	66-4	Long-bodied Scad	38-13	*lugubris, Caranx*	38-2
leucurus, Halichoeres	76-5	Longfang Moray	8-2	*lumbricoides, Yirrkala*	9-14
lewini, Sphyrna	3-7	Longfin Anchovy	6-20	*lumniscatus, Lutjanus*	42-14
leylandi, Dasyatis	4-10	Anthias	27-13	*lunare, Thalassoma*	78-4
lima, Plectrypops	15-9	Bannerfish	56-12	*lunaris, Caesio*	45-12
limbatus, Carcharhinus	2-10	Dottyback	28-13	*Lagocephalus*	105-5
Limnichthys fasciatus	68-10	Emperor	48-12	Lunartail Seaperch	43-10
lineata, Chromis	65-11	Longfin, Banded	28-7	Lunar-tailed Bigeye	31-4
lineatus, Acanthurus	94-4	Red-tipped	28-9	*lunula, Chaetodon*	55-13
Cirrhilabrus	72-5	Long-finned Batfish	54-5	*Rhinecanthus*	101-7
Diademichthys	12-6	Bigeye	31-8	*lutescens, Thalassoma*	78-5
Ecsensis	85-4	Dottyback	29-12	*luteus, Meiacanthus*	86-4
Istiblennius	85-13	Garfish	13-8	*Lutjanus adetii*	43-2
Labracinus	28-10	Goby	90-17	*argentimaculatus*	41-10
Plectorhinchus	44-13	Gurnard	21-7	*bengalensis*	43-3
Plotosus	11-5	Rockcod	23-13	*biguttatus*	43-4
Scolopsis	51-10	Silver-biddy	50-13	*bitaeniatus*	41-11
Siganus	92-12	Waspfish	20-2	*bohar*	41-12
Lined Blenny	85-4	Long Green Wrasse	78-8	*boutton*	43-5
Bristletooth	94-9	Long-jawed Mackerel	97-11	*carponotatus*	42-2
Butterflyfish	55-9	Long-horned Cowfish	103-1	*decussatus*	42-3
Chromis	65-11	*longiceps, Hipposcarus*	80-5	*ehrenbergi*	43-6
Dottyback	28-10	*longimanus, Carcharhinus*	2-12	*erythropterus*	42-1
Javelinfish	45-10	*Pentaprion*	50-13	*fulviflamma*	42-8
Monocle Bream	51-10	*longipinnis, Scarus*	82-6	*fulvus*	43-7
Triggerfish	100-11	*Valenciennea*	90-17	*gibbus*	43-8

	Species No.
johnii	43-9
kasmira	42-6
lemniscatus	42-14
lineolatus	42-13
lunulatus	43-10
lutjanus	42-13
lutjanus, Lutjanus	42-13
Lutjanus *madras*	43-11
malabaricus	42-4
monostigma	43-12
quinquelineatus	42-7
rivulatus	42-9
russelli	42-12
rufolineatus	42-5
sebae	42-10
semicinctus	43-13
timorensis	43-14
vitta	42-11
Luzonichthys waitei	27-7
luzonensis, Pseudanthias	27-5
lymma, Taeniura	5-4
Lyretail Hawkfish	68-6
Pigfish	71-1
lysan, Scomberoides	39-6

M

macarellus, Decapterus	38-11
maccoskeri, Paracheilinus	72-11
maccoyii, Thunnus	96-10
Mackerel, Australian Spotted	97-4
Broad-barred Spanish	97-3
Corselletted Frigate	97-9
Double-lined	97-14
Frigate	97-10
Long-jawed	97-11
Narrow-barred Spanish	97-2
Queensland School	97-5
Scad	38-11
Shark	97-13
Tuna	97-7
macleayi, Atelomycterus	1-12
macneilli, Assessor	29-4
Macolor *macularis*	43-16
niger	43-15
macracanthus, Priacanthus	31-3
macrodon, Cheilodipterus	31-10
macronema, Parupeneus	51-17
Macropharyngodon *choati*	73-13
kuiteri	73-14
meleagris	73-15
negrosensis	77-9
ornatus	77-8
macrophthalma, Parapercis	83-4
macrops, Synodus	11-11
macroptera, Archamia	34-18
macropterus, Hypopterus	30-3
Triodon	99-7
Macrorhamphosodes *platycheilus*	99-14

	Species No.
macrorhinus, Loxodon	3-2
macrosoma, Decapterus	38-13
macrospilos, Epinephelus	26-10
macrospilus, Crossosalarias	86-8
macrurus, Pseudomonacanthus	101-11
macularis, Macolor	43-16
maculata, Mene	40-9
maculata burrus, Sillago	30-16
maculatum, Pomadasys	45-9
maculatus, Cantherhines	101-4
Epinephelus	26-7
Plectropomus	24-9
maculiceps, Acanthurus	95-8
maculosus, Myrichthys	10-5
madras, Lutjanus	43-11
Madura Goby	92-1
madurensis, Apocryptodon	92-1
Magenta Dottyback	29-8
Magenta-streaked Wrasse	72-2
magnifica, Nemateleotris	91-10
magnificum, Glaucosoma	20-8
magniscuttis, Epinephelus	25-7
Mahimahi	39-1
mahsena, Lethrinus	106-2
majimai, Siphamia	31-16
Major, Sergeant	61-3
Makaira indica	96-1
mazara	96-2
Mako, Shortfin	1-7
Malabar Groper	25-1
Trevally	37-16
malabaricus Carangoides	37-16
Epinephelus	25-1
Lutjanus	42-4
Malacanthus *brevirostris*	40-4
latovittatus	46-14
manabei, Labropsis	73-8
Mandarinfish	88-1
Mangrove Cardinalfish	36-14
Goby	89-1
Jack	41-10
Pipefish	17-18
manillensis, Arothron	103-11
Manta *birostis*	5-6
Manta Ray	5-6
Many-banded Cardinalfish	33-7
Many-lined Cardinalfish	35-6
Sweetlips	45-3
Many-spotted Blenny	84-4
Sweetlips	45-2
Many-striped Pufferfish	103-10
Many-toothed Garden Eel	10-7
Maori Groper	25-11
Seaperch	42-9
Map Puffer	104-4
mappa, Arothron	104-4
Marbled Anglerfish	12-15
Cat Shark	1-12
Hawkfish	67-15
Moray	8-14

	Species No.
Snake-eel	9-2
Stargazer	83-16
Stingfish	20-4
marcolepidotus, Novaculichthys	79-17
margaretae australis, Callionymus	87-6
margaretae, Callionymus	87-6
margaritaceous, Halichoeres	75-9
Xenojulis	77-13
margaritifer, Chromis	64-11
Scolopsis	51-11
margaritiferae, Onuxodon	12-17
margaritophorus, Apogon	36-12
Gymnothorax	7-11
marginalis, Chelmon	56-8
marginatus, Halichoeres	75-5
Margined Coralfish	56-8
Marilyna darwinii	105-9
Marlin, Black	96-1
Indo-Pacific Blue	96-2
Striped	96-4
marmorata, Fowleria	34-13
marmoratus, Callechelys	9-2
Uropterygius	8-14
marri, Pterocaesio	46-5
Marr's Fusilier	46-5
Marshall Dottyback	28-12
marshallensis, Pseudochromis	28-12
Masked Bannerfish	58-12
Masted Shrimp Goby	88-14
mata, Acanthurus	94-1
mate, Atule	37-5
mazara, Makaira	96-2
mcadamsi, Parascorpaena	18-14
Mcadam's Scorpionfish	18-14
McCosker's Wrasse	72-11
Megalaspis cordyla	38-15
megalophthalmus, Plectranthias	22-13
Megalops cyprinoides	6-4
Meiacanthus atrodorsalis	86-1
ditrema	86-2
grammistes	84-10
luteus	86-4
smithi	86-3
melampygus, Caranx	38-3
melannotus, Chaetodon	57-10
melanocephalus, Pervagor	101-14
melanochir, Halichoeres	75-7
Ophichthus	9-11
melanopterus, Carcharhinus	2-11
melanopus, Amphiprion	63-7
Apogon	33-13
Coradion	58-9
melanosoma, Chaetodontoplus	60-7
melanospila, Taeniura	5-3
melanospilos, Genicanthus	60-10
Sargocentron	15-15
melanospilus, Aseraggodes	98-12

274

	Species No.		Species No.		Species No.
Gymnothorax	8-1	*Mimoblennius atrocinctus*	84-5	Black-spotted	8-1
melanostictus, Myripristis	14-4	miniata, Cephalopholis	22-5	Dragon	8-5
melanotus, Dischistodus	62-16	miniatus, Halichoeres	76-10	Dwarf	8-13
melanurus, Anampses	73-2	Lethrinus	47-6	Fimbriated	7-9
Dascyllus	63-6	Minifin Parrotfish	82-1	Freckled	7-15
Halichoeres	76-6	*Minous versicolor*	19-13	Giant	7-10
melapterus, Hemigymnus	75-12	Minstrel Sweetlips	45-1	Guineafowl	8-6
melarchus, Ecsenius	86-16	minuta, Gazza	40-10	Highfin	8-10
melas, Apogon	36-6	mirifica, Pterapogon	31-12	Hookjaw	8-4
Paraglyphidodon	61-13	Mirror Dory	16-4	Lattice-tail	7-4
melasma, Archamia	33-18	mitosis, Urolophus	4-4	Longfang	8-2
melasmapomus, Halichoeres	76-9	mitratus, Petroscirtes	84-9	Marbled	8-14
melatremus, Gymnothorax	8-13	Modest Leatherjacket	102-13	Moluccan	8-7
meleagrides, Anampses	70-4	modestoides, Thamnaconus	102-13	Mottled	7-12
meleagris, Gymnothorax	8-6	mokarran, Sphyrna	3-9	Painted	7-14
Istiblennius	85-12	Moluccan Cardinalfish	33-1	Pearly	7-11
Macropharyngodon	73-15	Moray	8-7	Peppered	7-14
Ostracion	103-4	Seaperch	43-5	Richardson's	8-11
Melichthys niger	100-4	moluccanus, Pseudodax	79-7	Sieve-patterned	7-6
vidua	106-5	moluccensis, Apogon	33-1	Slender	8-12
Mene maculata	40-9	*Gymnothorax*	8-7	Spotted	7-5
mentalis, Ulua	40-2	*Pomacentrus*	66-9	Tesselated	7-7
meredithi, Chaetodontoplus	60-6	*Upeneus*	52-10	Tiger	8-9
merra, Epinephalus	23-10	mombasae, Pterios	19-4	White-margined	8-3
mertensii, Chaetodon	57-1	*Monacanthus chinensis*	102-7	Yellow-edged	7-8
Merten's Butterflyfish	57-1	Monkeyfish	19-9	Yellowmouth	8-8
mesoleucus, Chaetodontoplus	60-8	*Monocentrus japonica*	16-2	moretonensis, Callionymus	87-5
mesoprion, Nemipterus	49-8	monoceros, Aluterus	102-1	morrhua, Epinephelus	25-8
mesothorax, Bodianus	71-3	*Heniochus*	58-12	morrisoni, Synchiropus	87-10
meyeri, Chaetodon	55-10	Monocle Bream	50-7	Morrison's Dragonet	87-10
Meyer's Butterflyfish	55-10	Monocle-Bream, Bridled	50-8	Moses Perch	42-12
Microcanthus strigatus	53-15	Coral	50-12	Mottled Fusilier	46-12
microdon, Epinephelus	23-11	Double-lined	50-8	Moray	7-12
Gymnocranius	48-5	Lined	51-10	Mud Goby	89-12
Lethrinus	48-8	Pearl-streaked	50-11	Mudskipper	92-2
Pristis	3-13	Pearly	51-11	muelleri, Chelmon	58-7
Micrognathus micronotopterus	17-20	Peter's	51-7	*Dexillichthys*	98-11
microlepis, Pterereotris	91-13	Rainbow	51-12	Mueller's Coralfish	58-7
micronotopterus, Micrognathus	17-20	Red-spot	50-10	*Mugil cephalus*	69-5
micropinna, Papuengraulis	6-18	Rosy Dwarf	50-1	mugiloides, Allanetta	13-14
microprion, Cephalopholis	26-2	Saddled Dwarf	50-2	mulleri, Leptobrama	53-7
microps, Prionobutis	92-5	Three-lined	51-13	Mullet, Blue-tail	69-6
Midas Blenny	86-9	Whitecheek	50-9	Diamond-scale	69-3
midas, Ecsenius	86-9	Whitestreak	51-9	Flat-tail	69-1
Midget Wrasse	72-14	Yellowstripe	51-8	Greenback	69-2
Midnight Angelfish	60-4	*Monodactylus argenteus*	53-6	Sand	69-4
Snapper	43-16	monogramma, Scolopsis	50-7	Sea	69-5
miliaris, Epinephelus	25-6	monopterygium, Hypnos	4-6	*Mulloidichthys flavolineatus*	51-14
milii, Scaevius	50-12	monostigma, Choerodon	74-11	*vanicolensis*	51-15
Military Seapike	69-13	*Gunnelichthys*	91-15	multiannulatus, Doryrhamphus	18-2
Milkfish	6-3	*Lutjanus*	43-12	multibarbata, Brotula	12-19
Milk Shark	3-6	*Monotaxis grandoculus*	47-1	Multi-banded Pipefish	18-2
Milk-spotted Toadfish	105-4	Moonfish	40-9	Multi-barred Angelfish	60-3
milleri, Pomacentrus	66-8	Moon Wrasse	78-4	Multi-coloured Dottyback	29-18
Miller's Damsel	66-8	moorei, Pseudochromis	29-14	multidens, Pristipomoides	41-8
mimaseana, Parapercis	83-6	Moore's Dottyback	29-14	multifasciata, Selenotoca	55-2
Mimic Blenny	84-5	Moorish Idol	92-9	multifasciatus, Centropyge	60-3
Cardinalfish	34-5	Moray, Banded	8-15	*Parupeneus*	52-8
Leatherjacket	101-18	Bartail	7-13	multilineatus, Apogon	35-6
Surgeonfish	95-2	Black-blotched	7-7	multimaculatus, Laiphognathus	84-4

	Species No.		Species No.		Species No.
multinotatus, Epinephelus	24-2	*nasus, Rhynchostracion*	103-6	*neotes, Apogon*	35-18
multiplicata, Parapercis	83-7	*natans, Parapegasus*	16-13	Netfin Groper	25-6
multiradiatus, Polydactylus	69-8	*Naucrates ductor*	38-16	Netted Lizardfish	11-13
multistriatus, Anchisomus	103-10	*naucrates, Echeneis*	40-7	New Guinea Wrasse	73-3
multivittatum, Plectorhinchus	45-3	*navarchus, Pomacanthus*	59-8	*nichofii, Aetomyleus*	5-9
munroi, Scomberomorus	97-4	*Neamia octospina*	34-15	*nieuhofi, Triacanthus*	99-9
Muraenichthys gymnotus	9-4	*Nebrius ferrugineus*	1-3	*niger, Macolor*	43-15
muraenolepis, Ophidion	12-18	*nebulosa, Echidna*	7-1	*Melichthys*	100-4
muralis, Valenciennea	90-16	*Parapercis*	83-8	*Odonus*	101-3
murdjan, Myripristis	14-5	*nebulosus, Halichoeres*	75-10	*Parastromateus*	38-18
muricatum, Bolbometopom	80-1	*Lethrinus*	47-8	*Scarus*	82-7
Muiron Pipefish	17-9	*Yongeichthys*	90-20	*nigra, Pseudamia*	34-11
Muzzled Blenny	84-14	*Zenopsis*	16-4	*nigrescens, Halichoeres*	75-8
myops, Trachinocephalus	11-15	Nebulous Wrasse	75-10	*nigricans, Acanthurus*	94-2
Myrichthys colubrinus	9-7	Needleskin Queenfish	39-3	*Stegastes*	66-15
Myripristis adustus	14-1	*Negaprion acutidens*	3-3	*nigricauda, Acanthurus*	95-9
berndti	15-1	*negrosensis, Macropharyngodon*		*nigripinnis, Apogon*	33-4
chryseres	15-2		77-9	*Polydactylus*	69-9
colubrinus	9-7	*nematacanthus, Lethrinus*	47-10	*nigrofasciata, Seriolina*	39-13
hexagonatus	14-2	*Nematalosa come*	6-14	*nigrofasciatus, Apogon*	35-9
kuntee	14-3	*Nemateleotris decora*	91-9	*nigrofuscus, Acanthurus*	94-7
maculosus	10-5	*magnifica*	91-10	*nigrolineatus, Pervagor*	101-15
melanostictus	14-4	*nematophorus, Nemipterus*	49-10	*Siokunichthys*	18-10
murdjan	14-5	*Symphorus*	41-9	*nigromanus, Pomacentrus*	67-4
pralinia	15-3	*nematophthalmus, Papilloculiceps*		*nigroocellatus, Istigobius*	90-4
violacea	15-4		21-12	*nigropinnata, Wetmorella*	79-12
vittata	15-5	*nematophorus, Nemipterus*	49-10	*nigropunctatus, Arothron*	103-16
Myxus elongatus	69-4	*nematoptera, Sphaeramia*	36-19	*nigroris, Paraglyphidodon*	61-14
		nematopus, Nemipterus	49-14	*nigrosensis, Macropharyngodon*	
		Nemipterus baliensis	49-1		77-9
N		*bathybius*	49-2	*nigrotaeniatus, Cheilodipterus*	34-2
nagasakiensis, Pentapodus	50-4	*celebicus*	49-3	Nine-banded Cardinalfish	35-10
Pomacentrus	66-11	*furcosus*	49-4	*niphonia, Pristigenys*	31-1
Naked-headed Catfish	11-3	*hexodon*	49-5	*nitida, Chromis*	65-14
Seabream	47-3	*isacanthus*	49-6	*nitidus, Halicampus*	18-9
nalua, Ambassis	30-5	*japonicus*	49-7	*nocturna, Amblygobius*	88-19
namiyei, Ecsenius	86-17	*mesoprion*	49-8	Nocturna Goby	88-19
Namiye's Blenny	86-17	*nematophorus*	49-10	*Norfolkia brachylepis*	87-3
nanus, Apogon	35-19	*nematopus*	49-14	Northern Bluefin Tuna	96-11
Narcine westaustraliensis	4-5	*peronii*	49-11	Dragonet	87-13
narinari, Aetobatus	5-10	sp.	49-12	Pilchard	6-7
Narrow Sawfish	3-12	*virgatus*	49-13	Sand Flathead	21-16
Narrow-banded Sergeant-Major		*zysron*	49-9	Scorpionfish	20-5
	61-4	*Neoaploactis tridorsalis*	21-4	Threadfin	69-7
Narrow-barred Grubfish	83-4	*Neocirrhites armatus*	67-17	Whaler	2-14
Spanish Mackerel	97-2	*Neoglyphidodon crossi*	67-3	Whiting	30-17
Narrow-lined Cardinalfish	33-17	*oxyodon*	67-1	Wobbegong	1-15
Toadfish	103-11	*thoracotaeniatus*	67-2	North West Black Bream	47-14
Narrowstripe Cardinalfish	35-4	*neoguinaicus, Anampses*	73-3	Ruffe	99-5
nasicornis, Pseudaluteres	101-12	Neon Damsel	66-7	Snapper	47-8
Naso annulatus	93-3	Threefin	87-2	*notatus, Abudefduf*	63-4
brachycentron	93-4	*Neoniphon argenteus*	15-6	*Apogon*	36-16
brevirostris	94-11	*aurolineatus*	15-17	Notched Threadfin-Bream	49-11
fageni	106-4	*opercularis*	15-8	*Notograptus gregoryi*	68-16
hexacanthus	93-5	*sammara*	14-6	*guttatus*	68-15
lituratus	94-10	*Neopomacentrus anabatoides*	62-15	*Novaculichthys macrolepidotus*	79-17
lopezi	93-7	*azysron*	61-11	*taeniurus*	78-12
thynnoides	93-8	*cyanomos*	61-9	*novaehollandiae, Ogilbyina*	29-18
tuberosus	94-12	*filamentosus*	61-10	*novemfasciatus, Apogon*	35-10
unicornis	94-13	*violascens*	62-18	*novemaculeatus, Zabidius*	54-8
vlamingii	93-9				

	Species No.
nox, Centropyge	60-4
nubilus, Acanthurus	95-12
nudiceps, Euristhmus	11-3
nudivomer, Gymnothorax	8-8
Numbfish	4-6
Numbfish, Banded/Ornate	4-5
nummifer, Antennarius	12-10

O

obesus, Thunnus	96-9
Triaenodon	3-5
Oblique-banded Cardinalfish	33-3
Snapper	44-7
Oblique-lined Dottyback	29-17
obliquus, Cryptocentrus	89-10
obreptus, Stegastes	66-14
obscurum, Plectorhinchus	44-11
obscurus, Carcharhinus	2-13
obsoletus, Lethrinus	48-7
obtusata, Sphyraena	69-14
Squamicreedia	68-9
Obtuse Sandfish	68-9
occidentalis, Batrachomoeus	12-2
Oceanic Whitetip Shark	2-12
Ocellate Coralfish	56-15
ocellaris, Amphiprion	64-2
Ocellated Cardinalfish	33-11
Dragonet	88-2
Eel-Blenny	68-12
Flounder	98-18
Frogfish	12-5
Goby	90-19
Rockcod	23-4
Spinefoot	93-12
Wrasse	76-9
ocellatum, Hemiscyllium	1-10
Tetrabrachium	12-8
ocellatus, Halophryne	12-5
Parachaetodon	56-15
Psammodiscus	98-18
Synchiropus	88-2
ocellicaudus, Apogon	36-1
ocellicincta, Canthigaster	104-10
Ochre-banded Goatfish	52-12
octafasciatus, Cryptocentrus	89-9
octospina, Neamia	34-15
octotaenia, Pseudocheilinus	72-16
Ocular Blenny	85-5
oculus, Ecsenius	85-5
Odonus niger	101-3
Ogilbia sp.	12-16
ogilbyi, Atherinomorus	13-18
Ogilbyina novaehollandiae	29-18
queenslandiae	29-19
velifera	29-20
Ogilby's Hardyhead	13-18
okinawae, Trimma	90-13
Old Glory Goby	88-17
oligocanthus, Plectropomus	24-10
oligolepis, Pristilepis	14-8

	Species No.
olivacerous, Acanthurus	94-6
olivaceus, Lethrinus	47-5
Olive Flathead-Gudgeon	92-6
Olive Snake-eel	9-10
Olive-tailed Flathead	21-20
Omobranchus ferox	84-12
germaini	84-11
lineolatus	84-13
punctatus	84-14
Onespot Demoiselle	62-7
Seaperch	43-12
Wormfish	91-15
One-banded Snake-eel	9-9
One-lined Wrasse	77-4
One-striped Fusilier	46-9
ongus, Epinephelus	25-10
Onigocia spinosa	21-15
Onion Trevally	37-7
Onuxodon margaritiferae	12-17
opercularis, Neoniphon	15-8
Rainfordia	28-4
Ophichthus bonaparti	10-2
cephalozona	9-9
melanochir	9-11
rutiodermatoides	9-10
Ophidion muraenolepis	12-18
Opistognathus darwiniensis	83-10
inornatus	83-12
latitabundus	83-11
reticulatus	83-13
sp.	83-9
Orange Anemonefish	64-4
Croaker	53-2
Orangebanded Coralfish	56-7
Orange-banded Pipefish	18-3
Orange-dashed Goby	90-18
Orangefin Anemonefish	63-12
Ponyfish	40-11
Orange-socket Surgeonfish	95-15
Orange-spot Surgeonfish	94-6
Orange-spotted Goby	90-13
Toadfish	105-12
Orange-striped Goby	88-20
Emperor	48-7
Orange-tailed Leatherjacket	101-13
Orbicular Batfish	54-7
Cardinalfish	36-20
orbicularis, Cyclichthys	105-18
Platax	54-7
Sphaeramia	36-20
Orectolobus ornatus	1-14
wardi	1-15
Oriental Bonito	97-6
Searobin	16-12
Sweetlips	44-10
Wrasse	71-9
orientalis, Dactyloptaenia	16-12
Oxycheilinus	71-9
Plectorhinchus	44-10
Sarda	97-6

	Species No.
Ornamental Wrasse	76-13
Ornate Butterflyfish	55-14
Emperor	48-9
Goby	90-6
Hawkfish	68-8
Numbfish	4-5
Scorpionfish	20-7
Surgeonfish	94-3
Threadfin-Bream	49-5
Wrasse	77-8
ornatissimus, Chaetodon	55-14
Halichoeres	76-13
Vanderhorstia	90-19
ornatus, Istigobius	90-6
Lethrinus	48-9
Macropharyngodon	77-8
Orectolobus	1-14
orthogrammus, Carangoides	37-14
Ostichthys kaianus	14-7
Ostracion cubicus	103-3
meleagris	103-4
solorensis	104-2
Oualan Bullseye	53-10
oualensis, Pempheris	53-10
Ovalspot Butterflyfish	56-3
Oxeye Herring	6-4
Scad	39-7
oxycephala, Scorpaenopsis	18-17
oxycephalus, Chaetodon	57-9
Cheilinus	71-10
Cirrhitichthys	68-4
Oxycheilinus bimaculatus	71-6
celebicus	71-8
diagrammus	71-7
orientalis	71-9
Oxycirrhitus typus	67-18
Oxymetopon cyanoctenosum	91-17
Oxymonacanthus longirostris	102-9
oxyodon, Neoglyphididon	67-1
oxyrinchus, Isurus	1-7
oyena, Gerres	50-15

P

paccagnellae, Pseudochromis	29-7
pachycentron, Cephalopholis	22-3
Pacific Gregory	66-12
Paddlefin Cardinalfish	34-10
Paddletail	43-8
Painted Anglerfish	12-13
Flutemouth	16-10
Grinner	11-15
Moray	7-14
Sweetlips	45-6
Pajama Cardinalfish	36-19
Pale Dartfish	91-13
Snapper	41-4
Soldierfish	14-4
Pale-lipped Surgeonfish	95-11
Palenose Parrotfish	81-5
Pale-striped Cardinalfish	32-7

	Species No.
Pale-tail Chromis	65-15
Pallid Pipefish	17-19
Triggerfish	100-9
pallidofasciatus, Apogon	32-7
pallimaculatus, Torquigener	105-12
palmaris, Acanthopagrus	47-14
Panther Flounder	98-9
pantherinus, Bothus	98-9
Pantolabus radiatus	37-3, 38-17
Papuan Trevally	38-4
Papuengraulis micropinna	6-18
papuensis, Caranx	38-4
para, Caranx	38-6
Paracaesio sordidus	44-3
xanthurus	44-4
Paracanthus hepatus	95-3
Paracentropogon vespa	19-14
Parachaetodon ocellatus	56-15
Paracheilinus filamentosus	72-10
maccoskeri	72-11
Paracirrhites arcatus	68-7
forsteri	66-6
hemistictus	68-8
Paradise Whiptail	51-3
paradiseus, Pentapodus	51-3
paradoxus, Solenostomus	17-2
Paraglyphidodom melas	61-13
nigroris	61-14
Paragobiodon echinocephalus	90-10
Paralticus amboinensis	86-21
Paraluteres prionurus	101-18
Paramonacanthus choirocephalus	102-11
filicauda	102-10
Parapegasus natans	16-13
Parapercis alboguttata	83-1
clathrata	83-3
diplospilus	83-2
gushikeni	83-5
macrophthalma	83-4
mimaseana	83-6
multiplicata	83-7
nebulosa	83-8
Paraplagusia bilineata	98-19
Paraploactis intonsa	21-5
pulvinus	21-5
Paraplotosus albilabris	11-6
sp.	11-7
Parapriacanthus ransonneti	53-8
Parascolopsis eriomma	50-1
rufomaculatus	50-2
tanyactis	50-2
Parascorpaena mcadamsi	18-14
picta	20-5
parasema, Chrysiptera	62-10
Parastromateus niger	38-18
parazonatus, Cheilodipterus	34-5
parcuspinus, Torquigener	105-11

	Species No.
Pardachirus pavoninus	98-13
pardalis, Cantherhines	102-3
Enchelycore	8-5
Parioglossus formosus	90-7
Parrotfish, Bleeker's	82-2
Blue	81-3
Blue-barred	80-8
Blue-spotted	80-4
Chameleon	80-9
Double-headed	80-1
Dusky	81-4
Ember	81-6
Green-finned	81-8
Greensnout	82-9
Highfin	82-6
Long-nosed	80-5
Minifin	82-1
Palenose	81-5
Quoy's	82-10
Red-speckled	80-2
Redtail	82-8
Reefcrest	82-5
Saddled	80-7
Schlegel's	81-7
Six-banded	80-6
Spinytooth	80-3
Steephead	81-1
Surf	81-9
Swarthy	82-7
Tricolor	82-11
Violet-lined	81-2
Whitespot	82-4
Yellowfin	82-3
Parupeneus barberinoides	52-2
barberinus	51-16
bifasciatus	52-3
chrysopleuron	52-4
cyclostomus	52-1
heptacanthus	52-7
indicus	52-6
macronema	51-17
multifasciatus	52-8
pleurostigma	52-9
signatus	52-5
Parva Goby	91-6
parva, Sacura	28-3
Valenciennea	91-6
parvipinnis, Scorpaenodes	18-15
parvulus, Apogon	36-3
pascalus, Pseudanthias	27-9
Pastel Ringwrasse	73-6
Patchwork Goby	90-15
Stingaree	5-7
patoca, Chelonodon	105-4
Patterned-Tongue Sole	98-19
patzneri, Salarias	86-20
Patzner's Blenny	86-20
pauciradiatus, Craterocephalus	13-15
Pavo Razorfish	78-10

	Species No.
pavo, Pomacentrus	67 7
Xyrichtys	78-10
pavoninus, Pardachirus	98-13
paxmani, Colurodontis	102-16
Paxman's Leatherjacket	102-16
Peacock Rockcod	22-4
Sole	98-13
Wrasse	74-2
Pearlfish	12-17
Pearl-jawed Cardinalfish	34-8
Pearl-Perch, Threadfin	30-8
Pearl-scaled Angelfish	60-5
Pearl-spotted Fairy Basslet	22-12
Pearl-streaked Monocle Bream	50-11
Pearly Cardinalfish	36-11
Monocle Bream	51-11
Moray	7-11
Rainbowfish	77-13
Pearly-finned Cardinalfish	33-10
pectoralis, Labroides	73-7
Pegasus volitans	16-13
pelagicus, Alopias	1-5
pelamis, Katsuwomis	96-12
Pelates quadrilineatus	30-10
pelewensis, Chaetodon	57-6
Pellona ditchela	6-13
Pempheris analis	53-9
oualensis	53-10
schwenkii	53-11
penicillus, Hippichthys	17-18
penicilligera, Chaetoderma	102-8
Pennant Bannerfish	56-10
Pennantfish	37-1
pentadactylus, Xyrichtys	79-15
pentanemus, Rhinoprenes	54-9
Pentapodus bifasciatus	51-1
caninus	51-2
emeryii	50-3
nagasakiensis	50-4
paradiseus	51-3
porosus	50-5
setosus	51-4
sp.	51-5
trivittatus	51-6
vitta	50-6
Pentaprion longimanus	50-13
Peppered Moray	7-14
Perch, Crescent	30-14
Moses	42-12
Rainford's	28-4
Six-lined	28-5
Two-banded	28-1
Yellow-tailed	30-9
Perchlet, Citron	22-13
Japanese	22-15
Sailfin	30-4
Scalloped	30-5
Spotted	22-14
Telkara	30-6

	Species No.	

perdito, Bodianus	70-7
perideraion, Amphiprion	64-3
Periophthalma Shrimp Goby	88-6
periophthalma, Amblyeleotris	88-6
Periophthalmus argentilineatus	92-2
periophthalmus, Blenniella	85-14
perissodon, Heteroconger	10-7
Peristedion liorhynchus	21-10
Peristrominous dolosus	20-3
perlitus, Apogon	36-11
peroni, Pseudomonacanthus	102-15
peronii, Nemipterus	49-11
personifer, Chaetodontoplus	59-6
perspicillatus, Dischistodus	62-17
Istigobius	90-5
Pseudochromis	29-21
Pervagor aspircaudus	101-13
janthinosoma	102-12
melanocephalus	101-14
nigrolineatus	101-15
pessuliferus, Doryrhamphus	18-3
Peter's Monocle Bream	51-7
petimba, Fistularia	16-9
Petroscirtes breviceps	84-8
mitratus	84-9
phalaena, Amblygobius	89-4
Phillipine Butterflyfish	55-3
Damsel	67-8
Philippines Chromis	65-9
Wrasse	79-6
philippina, Pseudocoris	79-6
philippinus, Pomacentrus	67-8
Pholidichthys leucotaenia	91-18
Phyllichthys punctatus	98-16
Phyllophichthus xenodontus	9-8
picta, Parascorpaena	20-5
Siderea	7-14
pictoides, Coris	74-15
pictum, Diagramma	44-9
picturatus occidentalis, Synchiropus	87-12
picturatus picturatus, Synchiropus	87-12
pictus, Antennarius	12-13
Ecsenius	86-18
picus, Plectorhinchus	44-12
Pigeye Shark	2-4
Pigfish, Blackfin	71-5
Blackspot	70-8 & 106-3
Coral	70-5
Goldspot	70-7
Lyretail	71-1
Saddleback	70-6
Splitlevel	71-3
Twospot	71-2
Pilchard, Northern	6-7
Pilot Fish	38-16
Pineapplefish, Japanese	16-2
Pinjalo	44-1

Pinjalo pinjalo	44-1
pinjalo, Pinjalo	44-1
Pink Anemonefish	64-3
Boarfish	16-6
Shrimp Goby	89-10
Squirrelfish	15-13
Pink-breasted Siphonfish	31-17
Pink-eared Emperor	47-11
Pinkfin Threadfin-Bream	49-12
Pinktail Triggerfish	106-5
pinnatus, Platax	54-5
pinnulatus, Cirrhitus	67-15
Pipefish, Banded	17-11
Blue-striped	18-4
Brown-banded	18-8
Double-ended	17-21
Eel	17-8
Glittering	18-9
Harlequin Ghost	17-2
Janss's	17-12
Ladder	17-13
Mangrove	17-18
Multi-banded	18-2
Muiron	17-9
Orange-banded	18-3
Pallid	17-19
Ribboned	17-17
Ringed	18-6
Rubble	18-5
Schultz's	18-7
Short-bodied	17-10
Short-nosed	17-16
Short-tailed	17-22
Slender	17-23
Tasselled	17-15
Tiger	17-14
White	18-10
White-saddled	17-20
Pipehorse, Larson's	17-7
Pipidonia bravoi	90-15
pisang, Pterocaesio	46-11
Pisodonophis boro	9-12
cancrivorous	9-13
Pitted Scorpionfish	19-9
plagiometopon, Hemiglyphidodon	61-12
Plagiotremus landanus	86-5
rhinorhynchus	84-16
tapeinosoma	84-15
Platax batavianus	54-4
orbicularis	54-7
pinnatus	54-5
tiera	54-6
Platybelone platyura	13-10
Platycephalus arenarius	21-16
endrachtensis	21-17
platycheilus, Macrorhamphosodes	99-14
platypterus, Istiophorus	96-5
platyura, Platybelone	13-10

plebius, Chaetodon	55-12
Polydactylus	69-7
Plectorhinchus chrysotaenia	45-4
chaetodontoides	45-2
flavomaculatus	106-1
gibbosus	45-5
lessoni	44-8
lineatus	44-13
multivittatum	45-3
nigrus	45-5
obscurum	44-11
orientalis	44-10
picus	44-12
polytaenia	45-7
schotaf	45-1
Plectranthias japonicus	22-15
megalophthalmus	22-13
wheeleri	22-14
Plectroglyphidodon dicki	66-1
johnstonianus	66-2
lacrymatus	66-3
leucozona	66-4
Plectropomus areolatus	24-8
laevis	25-12
leopardus	24-7
maculatus	24-9
oligocanthus	24-10
Plectrypops lima	15-9
Plesiops verecundus	28-9
pleurostigma, Parupeneus	52-9
pleurotaenia, Pseudanthias	27-12
Plotosus lineatus	11-5
plumbeus, Carcharhinus	2-14
Plumb-striped Stingfish	19-13
Plump Groper	25-4
podostigma, Halichoeres	76-8
poecilopterus, Apogon	33-10
Pogonoperca punctata	29-3
Pointed-head Maori Wrasse	71-10
Polkadot Cod	24-8
polyacantha, Pseudogramma	28-6
polyacanthus, Acanthochromis	63-1
polyactis, Cyprinocirrhites	68-5
Polydactylus multiradiatus	69-8
nigripinnis	69-9
plebius	69-7
polylepis, Epinephelus	25-9
Hemitaurichthys	58-10
polymnus, Amphiprion	63-8
polynemus, Pseudochromis	29-12
polyophthalmus, Grammatobothus	98-8
polytaenia, Plectorhinchus	45-7
polyzona, Echidna	7-2
Pomacanthus annularis	60-13
imperator	59-9
navarchus	59-8
semicircularis	59-10
sexstriatus	59-7
xanthometopon	60-12

	Species No.		Species No.		Species No.
Pomacentrus adelus	67-5	*prionurus, Paraluteres*	101-18	*perspicillatus*	29-21
alexanderae	66-5	*Pristigenys niphonia*	31-1	*polynemus*	29-12
amboinensis	66-6	*Pristilepis oligolepis*	14-8	*porphyreus*	29-8
brachialis	67-9	*Pristipomoides auricilla*	44-5	*quinquedentatus*	28-14
burroughi	67-12	*filamentosus*	41-6	*splendens*	29-10
coelestis	66-7	*flavipinnis*	44-6	*steenei*	29-11
grammorhynchus	67-10	*multidens*	41-8	sp.	29-15
lepidogenys	67-6	*typus*	41-7	*wilsoni*	28-16
milleri	66-8	*zonatus*	44-7	*Pseudocoris aurantifasciata*	79-5
moluccensis	66-9	*Pristis cuspidatus*	3-12	*heteroptera*	79-2
nagasakiensis	66-11	*microdon*	3-13	*philippina*	79-6
nigromanus	67-4	*zijsron*	3-11	*yamashiroi*	79-1
pavo	67-7	*Pristotis jerdoni*	66-16	*Pseudodax moluccanus*	79-7
philippinus	67-8	*Promicrops*	24-12	*Pseudogramma polyacantha*	28-6
stigma	67-11	*properupta, Apogon*	35-2	*Pseudojuloides elongatus*	78-8
tripunctatus	67-13	*prosopeion, Halichoeres*	76-14	*Pseudomonacanthus macrurus*	101-11
vaiuli	66-10	*prosopotaenia, Dischistodus*	61-8	*peroni*	102-15
Pomachromis richardsoni	62-12	*Protonibea diacanthus*	51-3	*Pseudoplesiops rosae*	28-17
Pomadasys hasta	45-8	*Psammodiscus ocellatus*	98-18	*Pseudorhombus arsius*	98-2
kaakan	45-8	*Psammoperca waigiensis*	30-2	*diplospilus*	98-3
maculatum	45-9	*Psenopsis humerosa*	99-5	*elevatus*	98-4
pomastictus, Ctenogobiops	89-11	*Psettodes erumei*	99-1	*jenynsii*	98-5
Pomatomus saltator	40-5	*Pseudaluteres nasicornis*	101-12	*pseudothrysoideus, Gymnothorax*	
Pomfret, Black	38-18	*Pseudamia amblyuropterus*	34-8		8-10
Ponyfish, Common	40-12	*gelatinosa*	34-9	*Pseudotriacanthus strigilifer*	99-11
Orangefin	40-11	*nigra*	34-11	*Pseudechidna brummeri*	8-17
Pugnose	40-16	*zonata*	34-10	*psittacus, Scarus*	81-5
Slender	40-13	*Pseudanthias bicolor*	27-11	*Pterapogon kauderni*	34-7
Smithhurst's	40-15	*bimaculatus*	27-16	*mirifica*	31-12
Whipfin	40-14	*cooperi*	22-11	*Ptereleotris evides*	91-12
Porcupinefish	104-14	*dispar*	27-6	*hanae*	90-11
Porcupinefish, Blotched	105-14	*fasciatus*	27-15	*heteroptera*	91-14
Freckled	105-13	*huchtii*	27-3	*microlepis*	91-13
Long-spined	105-17	*hypselosoma*	27-14	*zebra*	91-11
Short-spined	105-18	*lori*	27-2	*Pterocaesio chrysozona*	46-10
Spotfin	105-16	*luzonensis*	27-5	*diagramma*	45-13
porosus, Pentapodus	50-5	*pascalus*	27-9	*lativittata*	46-7
porphyreus, Pseudochromis	29-8	*pleurotaenia*	27-12	*marri*	46-5
Port Hedland Trevally	37-12	*randalli*	27-8	*pisang*	46-11
Potato Cod	24-5	*rubrizonatus*	28-2	*randalli*	46-8
Pot-bellied Leatherjacket	102-15	*smithvanizi*	27-1	*tessellata*	46-9
Powderblue Surgeonfish	95-1	*squamipinnis*	27-4	*tile*	46-6
praepositum, Cottapistus	19-12	*tuka*	27-10	*trilineata*	46-4
praetextatus, Cymolutes	79-13	*ventalis*	27-13	*Pterogogus amboinensis*	77-7
pralinia, Myripristis	15-3	*Pseudobalistes flavimarginatus*		*cryptus*	79-4
prasiognathus, Scarus	81-4		101-1	*Pterois antennata*	19-3
preclara, Heteroconger	10-8	*fuscus*	100-7	*mombasae*	19-4
Premnas biaculeatus	63-13	*Pseudocaranx dentex*	38-19	*radiata*	18-11
Priacanthus fitchi	31-7	*Pseudocheilinus evanidus*	72-17	*russelli*	19-5
hamrur	31-4	*hexataenia*	77-6	*volitans*	19-6
macracanthus	31-3	*octotaenia*	72-16	*Pterygotrigla hemisticta*	21-9
sagittarius	31-6	*Pseudocheilonops ataenia*	72-14	*leptacanthus*	21-8
tayenus	31-5	*Pseudochromis bitaeniata*	29-16	*puellaris, Valenciennea*	90-18
Prickly Leatherjacket	102-8	*cyanotaenia*	28-15	*puellus, Siganus*	93-13
Princess Anthias	27-1	*diadema*	29-9	Puffer, Blue-spotted	104-3
Damsel	66-10	*fuscus*	28-11	Brown-lined	105-3
Priolepis cinctus	90-8	*howsoni*	29-13	Map	104-4
semidoliatus	90-9	*marshallensis*	28-12	Spotted	105-2
Prionace glauca	3-4	*moorei*	29-14	Three-barred	105-1
Prionobutis microps	92-5	*paccagnellae*	29-7	Three-tooth	99-7

	Species No.
Pufferfish, Fine-spined	103-9
Immaculate	103-14
Many-striped	103-10
Reticulated	103-12
Starry	103-15
Pugnose Ponyfish	40-16
pulvinus, Paraploactis	21-5
punctata, Drepane	54-3
Pogonoperca	29-3
punctatissimum, Sargocentron	14-10
punctatofasciatus, Chaetodon	56-2
punctatum, Chiloscyllium	1-11
punctatus, Assiculus	28-13
Cirrhilabrus	72-6
Omobranchus	84-14
Phyllichthys	98-16
Siganus	92-10
punctulata, Fowleria	34-14
Purple Anthias	27-10
Threadfin-Bream	50-3
Tuskfish	74-4
Wrasse	75-7
Purple-headed Emperor	47-11
Purple-lined Wrasse	72-5
Purpletop Dottyback	29-9
purpurescens, Cypho	29-17
Halichoeres	76-7
purpureum, Thalassoma	78-1
Purse-eyed Scad	39-8
puta, Terapon	30-12
putnamiae, Sphyraena	69-11
Pygmy Scorpionfish	20-9
Pygoplites diacanthus	59-11
Pyramid Butterflyfish	58-10
pyroferus, Acanthurus	95-2
pyrrhurus, Scarus	82-8

Q

qenia, Sphyraena	69-13
quadrifasciatus, Apogon	32-8
quadrilineatus, Pelates	30-10
quadrimaculatus, Herklotsichthys	6-9
quaesita, Rhinomuraena	8-16
Queenfish, Barred	39-4
Double-spotted	39-6
Needleskin	39-3
Talang	39-5
Queensland Dottyback	29-19
Groper	24-12
Halibut	98-1
School Mackerel	97-5
Stinkfish	87-5
Yellowtail Angelfish	60-6
queenslandiae, Ogilbyina	29-19
queenslandicus, Scomberomorus	97-5
quinquedentatus, Pseudochromis	28-14
quinquelineatus, Cheilodipterus	31-11

	Species No.
Lutjanus	42-7
quinquestrigatus, Gobiodon	90-2
quinquevittatum, Thalassoma	79-9
quoyanus, Epinephelus	23-13
quoyi, Hyporhamphus	13-7
Scarus	82-10
Quoy's Garfish	13-7
Parrotfish	82-10

R

Rachycentron canadus	40-6
Racoon Butterflyfish	55-13
Radial Leatherjacket	101-17
Radiant Rockcod	23-15
radiata, Pterois	18-11
radiatus, Acreichthys	101-17
Epinephelus	23-15
Pantolobus	37-3, 38-17
radiosus, Etelis	41-4
rafflesi, Chaetodon	57-3
Ragged-finned Firefish	19-3
Raggy Scorpionfish	20-10
Rainbow Monocle Bream	51-12
Runner	39-2
Rainbowfish, Pearly	77-13
Red-finned	74-13
Saddled	75-9
Speckled	75-5
rainfordi, Amblygobius	88-17
Chaetodon	58-4
Rainfordia opercularis	28-4
Rainford's Butterflyfish	58-4
Perch	28-4
Raja sp.	4-7
rameus, Synchiropus	87-11
ramosus, Salarias	85-16
randalli, Amblyeleotris	88-7
Cirrhilabrus	72-8
Pseudanthias	27-8
Pterocaesio	46-8
Randall's Anthias	27-8
Fusilier	46-8
Shrimp Goby	88-7
Wrasse	72-8
Rankin's Rockcod	24-2
ransonneti, Parapriacanthus	53-8
ransonnetti, Elates	21-13
Rastrelliger kanagurta	97-11
Rat-tailed Ray	5-8
Ray, Barbless Eagle	5-9
Giant Shovelnose	3-17
Manta	5-6
Rat-tailed	5-8
Shark	3-14
Spotted Eagle	5-10
Spotted Shovelnose	3-16
White-spotted Shovelnose	3-15
Yellow Shovelnose	4-2

	Species No.
Razorfish	18-1
Razorfish, Blackspot	78-11
Fivefinger	79-15
Grooved	17-1
Pavo	78-10
Whitepatch	79-14
rectangulus, Rhinecanthus	100-6
Red Anemonefish	64-5
Bass	41-12
Bigeye	31-3
Coral Goby	89-13
Cuskeel	12-16
Emperor	42-10
Firefish	19-6
Soldierfish	15-5
Squirrelfish	14-11
Red and Green Wrasse	78-1
Red and White Goby	90-12
Red-banded Grubfish	83-7
Redbar Anthias	27-16
Red-barred Grubfish	83-8
Red-bellied Fusiler	45-11
Redfin Anthias	27-6
Butterflyfish	56-6
Red-finned Rainbowfish	74-13
Red-flushed Rockcod	23-1
Redhead Goby	90-10
Wrasse	76-3
Red-lined Triggerfish	100-1
Wrasse	75-4
Red Saddleback Anemonefish	63-10
Redsnout Emperor	48-14
Red-speckled Parrotfish	80-2
Redspot Cardinalfish	36-3
Monocle Bream	50-10
Wrasse	77-10
Red-spotted Blenny	85-10
Rockcod	22-7
Redstripe Tuskfish	74-7
Red-striped Cardinalfish	36-12
Redtail Parrotfish	82-8
Scad	38-10
Red-tailed Leatherjacket	102-12
Red-tipped Longfin	28-9
Redtooth Triggerfish	101-3
Reef Cod	26-11
Stonefish	18-19
Reefcrest Parrotfish	82-5
Reef-flat Cardinalfish	32-2
Regal Angelfish	59-11
Demoiselle	61-9
regani, Hoplichthys	16-11
Remora	40-8
Remora remora	40-8
remora, Remora	40-8
reticularis, Arothron	103-12
Reticulated Blenny	86-6
Butterflyfish	57-13
Dascyllus	64-14
Pufferfish	103-12

Species	No.	Species	No.	Species	No.
Swellshark	1-13	*robustus, Hemirhamphus*	13-5	Royal Dottyback	29-7
Wrasse	73-15	*Spratelloides*	6-16	Rubble Pipefish	18-5
reticulatus, Chilomycterus	105-16	*rochei, Auxis*	97-9	*rubescens, Choerodon*	74-5
Dascyllus	64-14	Rockcod, Blue-lined	22-6	*rubricephalus, Halichoeres*	76-3
Lethrinus	48-14	Blue-spotted	26-6	*rubrioperculatus, Lethrinus*	47-12
Opistognathus	83-13	Blunt-headed	23-2	*rubrizonatus, Pseudanthias*	28-2
retrofasciata, Chromis	65-13	Brown-banded	22-3	*rubrocinctus, Amphiprion*	64-5
rex, Chrysiptera	62-3	Chinaman	23-14	*rubroviolaceus, Scarus*	81-6
Rhabdamia cypselurus	33-15	Coral	23-5	*rubrum, Sargocentron*	14-11
gracilis	33-16	Dothead	26-2	Ruby Snapper	41-3
spilota	34-16	Flag-tailed	22-9	*ruconis, Secutor*	40-16
Rhina ancylostoma	3-14	Garish	26-3	*rueppelliae, Gymnothorax*	8-15
Rhinecanthus aculeatus	100-5	Hexagon	26-9	Ruffe, North-West	99-5
lunula	101-7	Large-spotted	26-10	*rufolineatus, Lutjanus*	42-5
rectangulus	100-6	Long-finned	23-13	*rufomaculatus, Parascolopsis*	50-2
verrucosus	101-6	Ocellated	23-4	Runner, Rainbow	39-2
Rhinoceros Leatherjacket	101-12	Peacock	22-4	*ruppelli, Apogon*	32-17
Rhiniodon typus	1-1	Radiant	23-15	*russelli, Decapterus*	38-12
Rhinobatus sp.	3-16	Rankin's	24-2	*Lutjanus*	42-12
Rhinomuraena quaesita	8-16	Red-flushed	23-1	*Pterois*	19-5
Rhinopias aphanes	18-13	Red-spotted	22-7	Russell's Mackerel Scad	38-12
Rhinoprenes pentanemus	54-9	Six-banded	24-3	Rust-banded Wrasse	79-5
rhinorhynchus, Plagiotremus	84-14	Six-blotch	26-5	Rusty Flathead	21-18
Rhynchostracion	103-5	Speckled-fin	25-10	Jobfish	43-1
Rhizoprionodon acutus	3-6	Spotfin	23-7	*rutidermatoides, Ophichthus*	9-10
rhodopterus, Apogon	33-8	Strawberry	26-4	*rutilans, Aphareus*	43-1
rhomboidea, Antigonia	16-6	Thinspine	26-1	*rutiodermatoides, Ophichthus*	9-10
Rhynchobatus djiddensis	3-15	Three-lined	23-9		
Rhynchostracion nasus	103-6	Tomato	22-8,		
rhinorhynchus	103-5		23-12	**S**	
Ribbon Eel	8-16	White-blotched	24-2	*Sacura parva*	28-3
Sweetlips	45-7	White-lined	22-1	Saddleback Anemonefish	63-8
Ribboned Pipefish	17-17	Woore's	24-6	Pigfish	70-6
richardsoni, Gymnothorax	8-11	Yellow-spotted	23-3	Saddled Butterflyfish	55-8
Pomachromis	62-12	Rockhampton Kingfish	69-10	Dwarf Monocle Bream	50-2
Richardsonichthys leucogaster	20-12	*rogaa, Aethaloperca*	23-1	Parrotfish	80-7
Richardson's Moray	8-11	*Rogadius asper*	21-20	Shrimp Goby	88-11
Reef Damsel	62-12	*rollandi, Chrysiptera*	62-5	Rainbowfish	75-9
rieffeli, Satyrichthys	21-11	Rolland's Demoiselle	62-5	Saddle-tailed Seaperch	42-4
Rifle Cardinalfish	36-15	*rosae, Pseudoplesiops*	28-17	*sageneus, Synodus*	11-13
Ringed Pipefish	18-6	Rose Devilfish	28-17	*sagittarius, Priacanthus*	31-6
Slender Wrasse	77-1	*roseigaster, Siphamia*	31-17	Sailfin Cardinalfish	31-12
Ring-eyed Hawkfish	68-7	*rostratus, Chelmon*	58-5	Catfish	11-7
Ringtail Surgeonfish	95-5	Rosy Dragonet	87-9	Dottyback	29-20
Ringtailed Unicornfish	93-3	Dwarf Monocle Bream	50-1	Perchlet	30-4
Ring-tailed Cardinalfish	33-12	Grubfish	83-5	Snapper	44-2
Surgeonfish	94-5	Snapper	41-6	Tang	94-14
Ringwrasse, Pastel	73-6	Threadfin-Bream	49-4	Sailfish, Indo-Pacific	96-5
Rippled Blenny	85-11	Rotund Blenny	84-12	*Salarias fasciatus*	85-15
River Shark	2-7	Rougefish, White-bellied	20-12	*patzneri*	86-20
rivoliana, Seriola	39-14	Rough Flutemouth	16-9	*ramosus*	85-16
rivulata, Canthigaster	105-3	Golden Toadfish	105-5	*segmentatus*	86-19
rivulatus, Epinephelus	23-14	Squirrelfish	14-8	*spaldingi*	85-17
Lutjanus	42-9	Rough-ear Scad	38-9	Salmon, Beach	53-7
Scarus	81-9	Rough-scaled Soldierfish	15-9	Cooktown	69-10
Roach	50-16	Round Skate	4-8	Striped	69-7
Robinson's Seabream	47-4	Roundbelly Cowfish	103-2	*saltator, Pomatomus*	40-5
Robust Bigeye	31-6	Roundface Firefish	19-7	*Samaris cristatus*	98-17
Garfish	13-5	Round-headed Blenny	84-13	*sammara, Neoniphon*	14-6
robusta, Sillago	30-15	Roundspot Surgeonfish	95-10	Samoan Cardinalfish	32-11
				Hardyhead	13-16

	Species No.		Species No.		Species No.
Samurai Squirrelfish	15-12	Scaly-head Threefin	87-3	*vosmeri*	50-9
Sand Bass	30-2	*scapularis, Blennodesmus*	68-12	*xenochrous*	50-11
Mullet	69-4	Halichoeres	76-15	*Scomberoides commersonnianus*	
sandaracinos, Amphiprion	64-4	*scapulostigma, Gnatholepis*	89-18		39-5
Sandbar Shark	2-14	Scarlet Seaperch	42-4	*lysan*	39-6
Sand-Diver, Elegant	91-20	Soldierfish	15-3	*tala*	39-4
Spotted	91-19	Scarlet-breasted Maori Wrasse	70-9	*tol*	39-3
Sandfish, Obtuse	68-9	*Scartelaos histiophorus*	92-3	*Scomberomorus commerson*	97-2
Sandpaper Squirrelfish	15-14	*Scarus altipinnis*	82-1	*munroi*	97-4
Velvetfish	21-1	*chameleon*	80-9	*queenslandicus*	97-5
sandracatus, Lepadichthys	12-7	*dimidiatus*	80-7	*semifasciatus*	97-3
Sandy Damsel	66-11	*flavipectoralis*	82-3	*scopas, Amanses*	101-9
Sangi Cardinalfish	35-11	*forsteni*	82-4	Zebrasoma	94-15
sangiensis, Apogon	35-11	*frenatus*	80-6	*Scorpaenodes guamensis*	20-6
sammara, Neoniphon	14-6	*ghobban*	80-8	*hirsutus*	18-16
Sarda orientalis	97-6	*globiceps*	81-2	*parvipinnis*	18-15
Sardine, Gold-stripe	6-17	*longipinnis*	82-6	*scaber*	20-9
Slender	6-8	*microhinos*	81-1	sp.	20-8
Smooth-belly	6-6	*niger*	82-7	*varipinnis*	20-7
Sardinella gibbosa	6-17	*prasiognathus*	81-4	*Scorpaenopsis diabolus*	20-11
Sargassum Fish	12-15	*psittacus*	81-5	*oxycephala*	18-17
Sargocentron caudimaculatum	15-10	*pyrrhurus*	82-8	*venosa*	20-10
cornutum	15-11	*quoyi*	82-10	Scorpioncod, Little	20-8
diadema	14-9	*rivulatus*	81-9	Scorpionfish, Butterfly	19-2
ittodai	15-12	*rubroviolaceus*	81-6	Cod	20-3
lepros	15-14	*schlegeli*	81-7	Deepsea	20-14
melanospilos	15-15	*sordidus*	81-8	False	22-2
punctatissimum	14-10	*spinus*	82-9	Guam	20-6
rubrum	14-11	*tricolor*	82-11	Hairy	18-16
spiniferum	14-12	Scat, Spotband	55-2	Leaf	20-13
tiere	14-13	Spotted	55-1	Mcadams's	18-14
tiereoides	15-13	Threadfin	54-9	Northern	20-5
violaceus	14-14	*Scatophagus argus*	55-1	Ornate	20-7
Satyrichthys rieffeli	21-11	*schismatorhynchus, Enchelycore*		Pitted	19-9
Saurida gracilis	11-16		8-3	Pygmy	20-9
tumbil	11-18	*schlegeli, Scarus*	81-7	Raggy	20-10
undosquamus	11-17	Schlegel's Parrotfish	81-7	Short-finned	18-15
Sawfish, Green	3-11	*schoenlenii, Choerodon*	74-8	Smallscale	18-17
Narrow	3-12	Schooling Bannerfish	56-13	Weedy	18-13
Wide	3-13	Goby	90-7	Yellow-spotted	18-18
scaber, Scorpaenodes	20-9	*schotaf, Plectorhinchus*	45-1	*scotochilopterus, Chromis*	65-9
Scad, Banded	38-6	*schultzi, Corythoichthys*	18-7	*scottorum, Cirrhilabrus*	72-7
Finny	38-15	Schultz's Pipefish	18-7	Scott's Wrasse	72-7
Indian	38-12	*schwartzi, Halichoeres*	76-11	Scribbled Angelfish	59-5
Long-bodied	38-13	Schwartz's Wrasse	76-11	Chisel-toothed Wrasse	70-2
Mackerel	38-11	*schwenkii, Pempheris*	53-11	Leatherjacket	102-2
Oxeye	39-7	Scissortail Sergeant	61-5	*scriptus, Aluterus*	102-2
Purse-eyed	39-8	*sclateri, Callogobius*	89-8	*scutatus, Centriscus*	17-1
Redtail	38-10	*scleratus, Lagocephalus*	105-8	Sea Mullet	69-5
Roughear	38-9	*sclerolepis, Arramphus*	13-2	Seabream, Blue-lined	47-4
Russell's Mackerel	38-12	*Scolopsis affinis*	51-7	Naked-headed	47-3
Small Mouth	37-4	*auratus*	51-8	Robinson's	47-4
Yellowtail	37-5	*bilineatus*	50-8	Swallowtail	47-2
Scaevius milii	50-12	*ciliatus*	51-9	Seahorse, Spiny	17-6
scalaris, Festucalex	17-13	*lineatus*	51-10	Spotted	17-4
Scalefin Anthias	27-4	*margaritifer*	51-11	Western Australian	17-5
Scalloped Hammerhead	3-7	*monogramma*	50-7	Seagrass Wrasse	79-17
Perchlet	30-5	*taeniopterus*	50-10	*sealei, Apogon*	36-7
Scaly Chromis	65-8	*temporalis*	51-12	Carcharhinus	2-15
Damsel	67-6	*trilineatus*	51-13	Seale's Cardinalfish	36-7

	Species No.		Species No.		Species No.
Seamoth, Short	16-14	Blacktail	63-2	Sliteye	3-2
Slender	16-13	Scissortail	61-5	Smalltooth Thresher	1-5
Seaperch, Bengal	43-3	Whitley's	63-3	Smooth Fanged	2-6
Bigeye	42-13	Yellow-tailed	63-4	Spot-tail	2-16
Black-banded	43-13	Sergeantfish	40-6	Tawny Nurse	1-3
Black-spot	42-8	Sergeant Major	61-3	Thickskin	2-14
Blue-banded	42-7	Blackspot	61-2	Tiger	3-1
Blue-striped	42-6	Narrow-banded	61-4	Western Angel	4-1
Checkered	42-3	*Seriola dumerili*	39-16	Whale	1-1
Crimson	42-1	*lalandi*	39-15	White Cheek	2-9
Dark-tailed	52-14	*rivoliana*	39-14	Whitetail	3-5
Ehrenberg's	43-6	*Seriolina nigrofasciata*	39-13	Winghead	3-8
Fingermark	43-9	*Serranocirrhitus latus*	29-1	Zebra	1-2
Five-lined	42-7	*serrulatus, Brachypterois*	19-7	Shark Bay Clingfish	12-7
Indian	43-11	*Setarches, guentheri*	20-14	Eel-Blenny	68-16
Indonesian	41-11	*setifer, Xiphasia*	84-17	Sharp-Headed Hawkfish	68-4
Lunartail	43-10	*setigerus, Trichonotus*	91-19	Sharp-nosed Duckbill	99-2
Maori	42-9	*Setipinna tenuifilis*	6-20	Wrasse	74-1
Moluccan	43-5	*setirostris, Thryssa*	6-21	Sharpnose Wrasse	79-12
Onespot	43-12	*setosus, Pentapodus*	51-4	Sharpsnout Snake Eel	10-1
Saddle-tailed	42-4	Seven-banded Cardinalfish	32-12	Sharptooth Jobfish	41-7
Scarlet	42-4	Wrasse	78-3	Short Seamoth	16-14
Striped	42-11	*sexfasciatus, Abudefduf*	61-5	Short-bodied Pipefish	17-10
Stripey	42-2	*Caranx*	38-8	Short Bill Spearfish	96-3
Timor	43-14	*Epinephelus*	24-3	Shortfin Mako	1-7
Two-spot Banded	43-4	*sexguttata, Valenciennea*	91-4	Short-finned Batfish	54-8
Yellow-lined	42-5	*sexlineatus, Grammistes*	28-5	Scorpionfish	18-15
Yellow-margined	43-7	*sexmaculata, Cephalopholis*	26-5	Waspfish	20-1
Seapike, Giant	69-11	*sexstriatus, Pomacanthus*	59-7	Short-headed Sabretooth Blenny	
Military	69-13	Shad, Gizzard	6-11		84-8
Striped	69-14	Shadow Goby	90-20	Short-nosed Pipefish	17-16
Searobin, Oriental	16-12	Shaggy Anglerfish	12-12	Shortsnout Spikefish	99-13
sebae, Lutjanus	42-10	Shark Bay Clingfish	12-7	Short-spined Porcupinefish	105-18
Sebastapistes cyanostigma	18-18	Shark Mackerel	97-13	Shortsnout Spikefish	99-13
Secutor insidiator	40-16	Ray	3-14	Short-tailed Pipefish	17-22
ruconis	40-16	Shark, Bignose	2-2	Shoulderspot Wrasse	73-12,
segmentatus, Salarias	86-19	Blackspot	2-15		77-5
Selar boops	39-7	Blacktip	2-10	Shrimpgoby, Flagtail	88-10
crumenthalmops	39-8	Blacktip Reef	2-11	Gold-streaked	88-15
Selaroides leptolepis	39-9	Blue	3-4	Masted	88-14
selas, Apogon	36-17	Bluntnose Sixgill	1-8	Periophthalma	88-6
Selenanthias analis	22-12	Bull	2-7	Randall's	88-7
selene, Chaetodon	57-8	Bullhead	1-6	Saddled	88-11
Selenotoca multifasciata	55-2	Dusky	2-13	Spotfin	88-4
semicinctus, Leiuranus	9-6	Epaulette	1-10	Steinitz's	88-5
Lethrinus	48-16	Ghost	5-11	Target	88-12
Lutjanus	43-13	Grey Nurse	1-4	Wide-barred	88-8
semicirculatis, Pomacanthus	59-10	Grey Reef	2-3	Yellow	88-9
semidoliatis, Priolepis	90-9	Inkytail	2-6	Shy Toby	104-10
semifasciatus, Scomberomorus	97-3	Lemon	3-3	Sicklefish	54-3
semilineatus, Apogon	32-6	Leopard	1-2	*Siderea picta*	7-14
semion, Chaetodon	57-4	Longnosed Grey	2-6	*thrysoidea*	7-15
semiornatus, Apogon	33-3	Milk	3-6	Sidespot Goatfish	52-9
semipunctuatus, Asterropteryx	89-3	Oceanic Whitetip	2-12	Sieve-patterned Moray	7-6
sephen, Dasyatis	5-2	Pig-eye	2-4	*Siganus argenteus*	93-14
septemfasciata, Thalassoma	78-3	River	2-7	*canaliculatus*	92-13
septemfasciatus, Abudefduf	61-1	Sand	2-14	*corallinus*	93-12
septemstriatus, Apogon	32-12	Sandbar	2-14	*fuscesens*	92-11
Sergeant, Banded	61-1	Silky	2-8	*guttatus*	93-15
Bengal	61-4	Silvertip	2-1	*lineatus*	92-12

	Species No.		Species No.		Species No.
javus	93-10	Sleeper, Small-eyed	92-5	Olive	9-10
puellus	93-13	Slender Armoured-Gurnard	21-10	One-banded	9-9
punctatus	92-10	Bullseye	53-8	Sharpsnout	10-1
spinus	93-6	Cardinalfish	33-16	Spotted	10-5
trispilos	92-14	Fusilier	46-13	Stargazer	9-3
vermiculatus	93-11	Grinner	11-16	Snapper, Black and White	43-15
virgatus	92-15	Leatherskin	39-3	Blue	44-3
vulpinus	93-16	Longtom	13-11	Deepsea	47-17
Single-spined Unicornfish	93-8	Pipefish	17-23	Long-spined	47-16
signatus, Parupeneus	52-5	Ponyfish	40-13	Midnight	43-16
Signigobius biocellatus	88-16	Sabre-tooth Blenny	84-1	North-West	47-8
sihama, Sillago	30-17	Sardine	6-8	Oblique-banded	44-7
Silhouettea hoesei	90-14	Seamoth	16-13	Pale	41-4
Silky Shark	2-8	Sprat	6-15	Rosy	41-6
Sillago analis	30-14	Suckerfish	40-7	Ruby	41-3
maculata burrus	30-16	Worm-eel	9-4	Sailfin	44-2
robusta	30-15	Yellow-tipped Threadfin		Tang's	41-5
sihama	30-17	Bream	49-8	Yellowtail Blue	44-4
vittata	30-18	Slendertail Moray	8-12	Snub-nosed Dart	39-11
Silty Wrasse	76-7	Slingjaw Wrasse	75-1	Garfish	13-2
Silver Batfish	53-6	Sliteye Shark	3-2	Soapfish, Arrowhead	29-2
Demoiselle	62-15	Smaller Salmon Catfish	11-2	Spotted	29-3
Spinefoot	93-14	Small-eyed Sleeper	92-5	Soela Wrasse	77-12
Toadfish	105-8	Small Mouth Scad	37-4	*soelae, Suezichthys*	77-12
Trevally	38-19	Smallmouth Squirrelfish	15-16	*solandri, Acanthocybium*	97-1
Tripodfish	99-9	Small-nosed Boxfish	103-6	*Canthigaster*	104-11
Silverbelly	50-15	Smallscale Scorpionfish	18-17	Solander's Toby	104-11
Silver-biddy, Banded	50-16	Small-scaled Groper	25-9	Sole, Dark-spotted	98-12
Common	50-15	Small-spoted Leatherjacket	101-11	Dark Thick-rayed	98-10
Long-finned	50-13	Smalltooth Emperor	48-8	Harrowed	98-15
Whipfin	50-14	Small Tooth Thresher Shark	1-5	Patterned Tongue	98-19
Silver-streaked Wrasse	77-11	Small-toothed Cod	23-11	Peacock	98-13
Silvertip Shark	2-1	Flounder	98-5	Spotted	98-16
sinensis, Bostrichthys	92-7	Jobfish	41-1	Tufted	98-11
Inimicus	19-16	Whiptail	51-2	Wicker-work	98-14
Single-striped Cardinalfish	35-5	*smithhursti, Leiognathus*	40-15	Soldierfish, Bigscale	15-1
Singular Bannerfish	56-11	Smithhurst's Ponyfish	40-15	Blackfin	14-1
sinensis, Inimicus	19-16	*smithi, Meiacanthus*	86-3	Crimson	14-5
Singapore Cardinalfish	34-3	Smith's Coral Goby	89-14	Doubletooth	14-2
singapurensis, Cheilodipterus	34-3	Fangblenny	86-3	Kuntee	14-3
singularius, Heniochus	56-11	*smithvanizi, Pseudanthias*	27-1	Lattice	15-4
Siokunichthys nigrolineatus	18-10	Smokey Chromis	64-10	Pale	14-4
Siphamia majimai	31-16	Smooth Flutemouth	16-8	Red	15-5
roseigaster	31-17	Golden Toadfish	105-6	Rough-scaled	15-9
Siphonfish, Pink-breasted	31-17	Hammerhead	3-10	Scarlet	15-3
Striped	31-16	Squirrelfish	15-6	Yellowfin	15-2
Sirembo imberis	12-20	Smooth-belly Sardine	6-6	*Solegnathus hardwickei*	17-19
sirm, Amblygaster	6-7	Smooth-fanged Shark	2-6	*Solenostomus cyanopterus*	17-3
Sixbanded Angelfish	59-7	Smooth-tailed Trevally	39-9	*paradoxus*	17-2
Parrotfish	80-6	Smudgespot Spinefoot	92-13	*solorensis, Halichoeres*	76-16
Wrasse	78-6	Snake-eel, Black-finned	9-11	*Ostracion*	104-12
Six-banded Rockcod	24-3	Black-striped	10-4	*solosus, Peristrominous*	20-3
Six-blotch Rockcod	26-5	Bonapart's	10-2	*sonnerati, Cephalopholis*	23-12
Six-lined Perch	28-5	Burrowing	9-13	*sordidus, Abudefduf*	61-2
Wrasse	77-6	Crocodile	10-3	*Paracaesio*	44-3
Sixspot Goby	91-4	Estuary	9-12	*Scarus*	81-8
Skate, Eyed	4-7	Flappy	9-8	*sorrah, Carcharhinus*	2-16
Western Round	4-8	Fringe-lipped	9-5	*Sorsogona tuberculata*	21-19
Skipjack Tuna	96-12	Harlequin	9-7	Southern Bluefin Tuna	96-10
Sleek Unicornfish	93-5	Marbled	9-2	Drummer	53-12

	Species No.		Species No.		Species No.
Tubelip	73-11	*spilota, Rhabdamia*	34-16	Cardinalfish	36-16
sp. *Alepes*	37-4	*spilurus, Smphorichthys*	44-2	Spot-tail Dartfish	91-14
Aptychotrema	4-2	Spindle-egg Cardinalfish	33-13	Shark	2-16
Callogobius	89-7	Spine-cheek Anemonefish	63-13	Spotted Archerfish	54-1
Coccotropus	21-3	Spinecheek, Barred-face	50-7	Armoured-Gurnard	21-11
Cypselurus	13-1	Bridled	50-8	Blenny	85-12
Dasyatis	4-11	Spinefoot, Black	92-11	Boxfish	103-4
Drombus	89-12	Blue-lined	93-13	Butterflyfish	57-2
Eviota	89-13	Doublebar	92-15	Cardinalfish	34-14
Gorgasia	10-12	Golden	93-15	Chisel-toothed Wrasse	70-1
Gymnocranius	48-4	Golden-lined	92-12	Croaker	53-1
Hydrolagus	5-11	Java	93-10	Dottyback	28-14
Irolita	4-8	Ocellated	93-12	Duckbill	99-1
Lethrinus	47-9	Silver	93-14	Eagle Ray	5-11
Nemipterus	49-12	Smudgespot	92-13	Eel-Blenny	68-15
Ogilbia	12-16	Spiny	93-6	Garden Eel	10-9
Opistognathus	83-9	Spotted	92-10	Golden Goatfish	52-7
Paraplotosus	11-7	Threespot	92-14	Hardyhead	13-14
Pentapodus	51-5	Vermiculate	93-11	Javelinfish	45-8
Pseudochromis	29-15	*spinidens, Calotomus*	80-3	Jewfish	53-1
Raja	4-7	*spinifer, Argyrops*	47-16	Moray	7-5
Rhinobatos	3-16	*Congrogadus*	68-13	Perchlet	22-14
Salarias	85-16	*spiniferum, Sargocentron*	14-12	Puffer	105-2
Scorpaenodes	20-8	*spinirostris, Halicampus*	17-16	Scat	55-1
Squatina	4-1	*spinosa, Onigocia*	21-15	Sand-Diver	91-19
Stephanolepis	102-14	*spinosissimus, Tylerius*	103-9	Seahorse	17-4
Trimma	90-12	*spinus, Scarus*	82-9	Shovelnose Ray	3-16
Uranoscopus	83-19	*Siganus*	93-6	Shrimpgoby	88-4
Urolophus	5-7	Spiny Chromis	63-1	Snake Eel	10-5
spadiceus, Lagocephalus	105-7	Eel-blenny	68-13	Soapfish	29-3
spaldingi, Salarias	85-17	Flathead	21-15	Sole	98-16
Spalding's Blenny	85-17	Seahorse	17-6	Spinefoot	92-10
Spangled Emperor	47-8	Spinefoot	93-6	Stinger	19-16
Spearfish, Shortbill	96-3	Squirrelfish	14-12	Spotted-chin Blenny	85-2
speciosus, Gnathodon	38-14	Spiny-cheeked Grunter	30-12	Spotted-gill Cardinalfish	36-8
Speckled Butterflyfish	55-7	Spiny-eyed Cardinalfish	32-13	Spotted-tail Anglerfish	12-14
Catshark	1-9	Spiny-headed Flounder	98-7	Wrasse	74-16
Garden Eel	10-12	Spinytooth Parrotfish	80-3	Sprat, Blue	6-16
Groper	25-7	*splendens, Pseudochromis*	29-10	Slender	6-15
Rainbowfish	75-5	Splendid Dottyback	29-10	*Spratelloides gracilis*	6-15
Squirrelfish	14-10	Garden Eel	10-8	*robustus*	6-16
Speckled-fin Rockcod	25-10	*splendidus, Synchiropus*	88-1	*springeri, Chrysiptera*	62-11
Spectacled Goby	90-5	Spindle-egg Cardinalfish	33-13	Springer's Demoiselle	62-11
Leatherjacket	102-4	Split-banded Cardinalfish	35-3	*Squamicreedia obtusata*	68-9
speculum, Chaetodon	56-3	Splitfin, Waite's	27-7	*squamipinnis, Pseudanthias*	27-4
Sphaeramia nematoptera	36-19	Splitlevel Pigfish	71-3	Squarespot Anthias	27-12
orbicularis	36-20	Spot-banded Butterflyfish	56-2	*Squatina* sp.	4-1
Sphyraena barracuda	69-12	Spotbase Burrfish	105-15	Squirrelfish, Black-finned	15-8
jello	69-11	Spotcheek Emperor	47-12	Blackspot	15-15
obtusata	69-14	Spotfin Lionfish	19-3	Bluestripe	14-13
qenie	69-13	Porcupinefish	105-16	Crowned	14-9
Sphyrna lewini	3-7	Rockcod	23-7	Deepwater	14-7
mokarran	3-9	Shrimpgoby	89-11	Pink	15-13
zygaena	3-10	Squirrelfish	14-6	Red	14-11
Spikefish, Longsnout	99-12	Waspfish	19-14	Rough	14-8
Shortsnout	99-13	Wrasse	72-12	Samurai	15-12
Trumpetsnout	99-14	Spothead Grubfish	83-3	Sandpaper	15-14
Spiky Bass	30-3	Spotless Cardinalfish	34-12	Smallmouth	15-16
spiloparaea, Cephalopholis	26-4	Firefish	19-5	Smooth	15-6
spilostylus, Chilomycterus	105-15	Spotnape Butterflyfish	57-9	Speckled	14-10

	Species No.		Species No.		Species No.
Spiny	14-12	Gross's	87-8	Surf Parrotfish	81-9
Spotfin	14-6	Japanese	87-7	Surge Demoiselle	62-9
Tailspot	15-10	Queensland	87-5	Surgeonfish, Blackstreak	95-9
Threespot	15-11	Stocky Anthias	27-14	Convict	94-8
Violet	14-14	*Stolephorus indicus*	6-19	Dark	95-12
Yellow-striped	15-7	Stonefish, Dampier	19-10	Dusky	94-7
Staghorn Damsel	61-6	Estuarine	19-18	Earbar	95-8
Stanulus talboti	84-6	False	20-11	Fowler's	95-14
starcki, Hoplolatilus	46-17	Reef	18-19	Mimic	95-2
Stargazer, Banded	83-14	Stout Longtom	13-13	Orange-socket	95-15
Double-banded	83-15	Whiting	30-15	Orange-spot	94-6
Kai	83-18	*Strabozebrias cancellatus*	98-15	Ornate	94-3
Marbled	83-16	*striata, Helcogramma*	87-2	Pale-lipped	95-11
White-spotted	83-19	Striated Blenny	85-9	Powderblue	95-1
Yellowtail	83-17	*striatula, Ilisha*	6-12	Ringtail	95-5
Stargazer Snake-eel	9-3	*striatus, Antennarius*	12-9	Ring-tailed	94-5
Starry Blenny	85-16	*Ctenochaetus*	94-9	Roundspot	95-10
Eel	7-1	*Entomacrodus*	85-9	Thompson's	95-7
Goby	89-3	*strigata, Valenciennea*	91-5	White-cheeked	94-2
Pufferfish	103-15	*strigatus, Aeoliscus*	18-1	Whitefin	95-6
Triggerfish	100-2	*Microcanthus*	53-15	White-spotted	95-4
Stars and Stripes Toadfish	103-13	*strigilifer, Pseudotriacanthus*	99-11	Yellowfin	95-13
Strawberry Rockcod	26-4	*strigilliceps, Cryptocentrus*	88-12	Yellowmask	94-1
steenei, Pseudochromis	29-11	*strigiventer, Stethojulis*	77-11	*surinamensis, Lobotes*	40-17
Steene's Dottyback	29-11	*strigosus, Ctenochaetus*	93-2	Swallowtail Seabream	47-2
Steephead Parrotfish	81-1	Striped Anglerfish	12-9	Swarthy Parrotfish	82-7
Stegastes fasciolatus	66-12	Anthias	27-15	Swarthy-headed Goatfish	52-2
lividus	66-13	Bullseye	53-11	Sweetlip Emperor	47-6
nigricans	66-15	Boxfish	104-2	Sweetlips, Brown	45-5
obreptus	66-14	Butterfish	55-2	Celebes	45-4
Stegostoma fasciatum	1-2	Cardinalfish	32-1	Diagonal-banded	44-13
steinitzi, Amblyeleotris	88-5	Catfish	11-5	Dotted	44-12
Steinitz's Shrimp Goby	88-5	Goatfish	52-13	Giant	44-11
stellatus, Abalistes	100-2	Goby	90-16	Gold-spotted	106-1
Arothron	103-15	Large-eye Bream	48-1	Many-lined	45-3
Stephanolepis sp.	102-14	Marlin	96-4	Many-spotted	45-2
Stethojulis bandanensis	77-10	Salmon	69-7	Minstrel	45-1
interrupta	79-3	Seaperch	42-11	Oriental	44-10
strigiventer	77-11	Seapike	69-14	Painted	45-6
trilineata	79-8	Siphonfish	31-16	Ribbon	45-7
stictus, Epinephelus	25-2	Sweetlips	44-8	Striped	44-8
stigma, Pomacentrus	67-11	Stripe-face Unicornfish	94-10	Yellowdot	44-9
stigmaticus, Cirripectes	86-6	Stripe-spot Goatfish	51-17	Swellshark, Reticulated	1-13
Stingaree, Blotched	4-4	Stripey	53-15	Swordfish	96-6
Brown	4-3	Stripey Seaperch	42-2	*Symphorichthys spilurus*	44-2
Patchwork	5-8	*Strongylura leiura*	13-11	*Symphorus nematophorus*	41-9
Stinger, Demon	19-15	Stout Whiting	30-15	*Synanceja horrida*	19-8
Spotted	19-16	Stout-body Chromis	65-6	*verrucosa*	18-19
Stingfish, Marbled	20-4	*subducens, Congrogadus*	68-14	*Synchiropus altivelis*	87-9
Plumb-striped	19-13	*subfasciatus, Gerres*	50-16	*morrisoni*	87-10
Stingray, Black	5-1	*subviridis, Liza*	69-2	*ocellatus*	88-2
Black-blotched	5-3	Suckerfish, Slender	40-7	*picturatus occidentalis*	87-12
Black-spotted	5-5	*Suezichthys soelae*	77-12	*picturatus picturatus*	87-12
Blue-spotted	4-11	*Suffamen bursa*	100-9	*rameus*	87-11
Blue-spotted Fantail	5-4	*chrysopterus*	100-8	*splendidus*	88-1
Brown	4-9	*frenatus*	100-10	*Syngnathoides biaculeatus*	17-21
Brown Reticulated	4-10	*sugillatum, Choerodon*	74-10	*Synodus hoshinonis*	11-12
Cowtail	5-2	*sulphureus, Upeneus*	52-11	*indicus*	11-8
Stinkfish, Australian	87-6	*sundaicus, Upeneus*	52-12	*jaculum*	11-9
Goodlad's	87-4	Sunrise Goatfish	52-11	*kainanus*	11-10

	Species No.		Species No.		Species No.
macrops	11-11	Tesselated Moray	7-7	Yellow-lipped	49-13
sageneus	11-13	*Tetrabrachium ocellatum*	12-8	Yellow-tipped	49-14
variegatus	11-14	*tetradactylum, Eleutheronema*	69-10	Three-banded Boarfish	68-1
		Tetrapturus angustirostris	96-3	Three by Two Garfish	13-5
T		*audax*	96-4	Three-barred Boarfish	68-1
tabl, Decapterus	38-9	*Tetrasomus concateanatus*	103-8	Puffer	105-1
taenianotus, Ablabys	19-11	*gibbosus*	103-7	Three-eyed Wrasse	76-6
taeniatus, Aspidontus	84-2	*thalassinus, Arius*	11-1	Threefin Velvetfish	21-4
Taenionotus triacanthus	20-13	*Entomacrodus*	85-7	Threefin, Black-throated	87-1
taeniophorus, Apogon	32-2	*Thalassoma amblycephalum*	78-2	Neon	87-2
Haliichthys	17-17	*hardwickei*	78-6	Scaly-head	87-3
taeniopterus, Scolopsis	50-10	*janseni*	78-7	Three-lined Grunter	30-12
Taeniura lymma	5-4	*lunare*	78-4	Monocle Bream	51-13
melanospila	5-3	*lutescens*	78-5	Rockcod	23-9
taeniurus, Novaculichthys	78-12	*purpureum*	78-1	Wrasse	79-8
Tailblotch Tubelip	73-8	*quinquevittatum*	79-9	Three-saddle Cardinalfish	32-10
Tailor	40-5	*septemfasciata*	78-3	Three-spined Frogfish	12-3
Tailspot Cardinalfish	36-1	*trilobatum*	79-10	Threespot Damsel	67-13
Lizardfish	11-9	*Thamnaconus modestoides*	102-13	Spinefoot	92-14
Squirrelfish	15-10	*thazard, Auxis*	97-10	Squirrelfish	15-11
tala, Scomberoides	39-4	Thermal Cardinalfish	36-18	Wrasse	75-6
talamparoides,Carangoides	37-13	*thermalis, Apogon*	36-18	Three-spot Angelfish	59-1
Talang Queenfish	49-5	*theraps, Terapon*	30-13	Cardinalfish	35-12
talboti, Chrysiptera	62-4	*thetidis, Dasyatus*	5-1	Dascyllus	64-13
Stanulus	84-6	Thicklip Trevally	37-14	Flounder	98-8
Talbot's Blenny	84-6	Thick-lipped Wrasse	75-12	Three-striped Fusilier	46-4
Demoiselle	62-4	Thickskin Shark	2-14	Whiptail	51-6
Tallegalane	69-4	Thin Velvetfish	21-2	Three-tooth Puffer	99-7
Tang, Blue-lined	94-15	Thinspine Rockcod	26-1	Thumbprint Emperor	48-15
Sailfin	94-14	*thompsoni, Acanthurus*	95-7	*thrysoides, Sidera*	7-15
tangaroae, Ctenogobiops	88-14	Thompson's Surgeonfish	95-7	*Thryssa hamiltoni*	6-21
Tang's Snapper	41-5	*thoracotaeniatus, Neoglyphidodon*		*Thunnus alalunga*	96-7
tanyactis, Parascolopsis	50-2		67-2	*albacares*	96-8
tapeinosoma, Plagiotremus	84-15	Thornback Cowfish	104-1	*maccoyii*	96-10
Pseudochromis	28-15	Threadfin Anthias	27-3	*obesus*	96-9
Target Shrimp Goby	88-12	Bigeye	31-5	*tonggol*	96-11
Tasselled Pipefish	17-15	Butterflyfish	55-5	*thynnoides, Naso*	93-8
Wobbegong	1-16	Cardinalfish	35-8	*tibicen, Centropyge*	59-4
taurus, Eugomphodus	1-4	Emperor	47-10	Tidepool Blenny	85-13
tauvina, Epinephelus	26-11	Leatherjacket	102-10	*tiere, Sargocentron*	14-13
Tawny Nurse Shark	1-3	Pearl-perch	30-8	*tiereoides, Sargocentron*	15-13
tayenus, Priacanthus	31-5	Scat	54-9	Tiger Mullet	68-1
taylori, Heteroconger	10-10	Wrasse	72-9	Moray	8-9
Taylor's Garden Eel	10-10	Threadfin, Black-finned	69-9	Pipefish	17-14
Teardrop Butterflyfish	56-1	Giant	69-10	Shark	3-1
Teira Batfish	54-6	Gunther's	69-8	*tigris, Filicampus*	17-14
teira, Platax	54-6	Northern	69-7	*tile, Pterocaesio*	46-6
Telkara Perchlet	30-6	Threadfin-Bream, Bali	49-1	Tilefish, Blue	46-17
temminckii, Cirrhilabrus	74-2	Doublewhip	49-10	Grey	46-16
Hypoatherina	13-16	Five-lined	49-3	Yellow-blotched	46-15
temporalis, Scolopsis	51-12	Japanese	49-7	*tille, Caranx*	38-5
tentaculata, Anaora	88-3	Notched	49-11	Tille Trevally	38-5
tenuifilis, Setipinna	6-20	Ornate	49-5	Timor Cardinalfish	32-18
Terapon jarbua	30-11	Pinkfin	49-12	Seaperch	43-14
puta	30-12	Purple	50-3	*timorensis, Apogon*	33-18
theraps	30-13	Rosy	49-4	*Epinephelus*	25-3
teres, Caesio	46-1	Slender Yellow-tipped	49-8	*Lutjanus*	43-14
Ternate Chromis	65-10	Twin-lined	49-6	Tiny Cardinalfish	35-19
ternatensis, Chromis	65-10	Yellowbelly	49-2	Toadfish, Black-spotted	103-16
tessellata, Pterocaesio	46-9	Yellow-cheeked	49-9	Brown-backed	105-7

		Species No.			Species No.			Species No.
	Darwin's	105-9			38-17	*Triodon macropterus*		99-7
	Hick's	105-10		Giant	38-1	*triostegus, Acanthurus*		94-8
	Milk-spotted	105-4		Golden	38-14	*trisignatus, Lophiocharon*		12-14
	Narrow-lined	103-11		Gold-spotted	37-9	*trophis, Epinephelus*		25-4
	Orange-spotted	105-12		Japanese	37-17	Triplespot Blenny		86-8
	Rough Golden	105-5		Lowly	38-1	Tripletail		40-17
	Silver	105-8		Malabar	37-16	Tripletail Maori Wrasse		70-12
	Smooth Golden	105-6		Onion	37-7	Tripodfish, Blacktail		99-8
	Stars and Stripes	103-13		Papuan	38-4		Blacktip	99-10
	Yellow-eyed	105-11		Port Hedland	37-12		Blotched	99-11
Toby, Ambon		104-5		Silver	38-19		Silver	99-9
	Bennett's	104-6		Smooth-tailed	39-9	*tripunctatus, Pomacentrus*		67-13
	Black-saddled	104-13		Thicklip	37-14	*trisignatus, Antennarius*		12-14
	Compressed	104-12		Tille	38-5	*trispeculare, Hemiscyllium*		1-9
	Lantern	104-7		Whitefin	37-10	*trispilos, Siganus*		92-14
	Leopard	104-9		White-tongued	37-13	*trispinosis, Batrachomoeus*		12-3
	Shy	104-10		Yellow-spotted	37-9	*Triso dermopterus*		24-6
	Solander's	104-11	*Triacanthodes ethiops*		99-13	*trivittatus, Pentapodus*		51-6
	Tyler's	104-8	*Triacanthus biaculeatus*		99-8	*Trixiphichthys weberi*		99-10
tol, Scomberoides		39-3		*nieuhofi*	99-9	Tropical Garfish		13-6
Tomato Anemonefish		63-11	*triacanthus, Taenionotus*		20-13	Trout Cod		26-7
	Rockcod	22-8,	*Triaenodon obesus*		3-5	Trout, Coronation		24-11
		23-12	Triangular Butterflyfish		58-1		Coral	24-7
tomentosus, Acreichthys		101-16	*Trichonotus elegans*		91-20	Trumpeter		30-10
tominiensis, Ctenochaetus		93-1		*setigerus*	91-19	Trumpeter, Whiting		30-16
Tomini Bristletooth		93-1	Tricolor Parrotfish		82-11		Yellowtail	30-9
Tommyfish		68-10	*tricolor, Scarus*		82-11	Trumpetsnout Spikefish		99-14
tonggol, Thunnus		96-11	*tridorsalis, Neoaploactis*		21-4	*truncatus, Apogon*		33-9
Toothpony		40-10	*trifascialis, Chaetodon*		56-5	Tubelip, Allen's		73-9
torquatus, Cymolutes		79-16	*trifasciatus, Chaetodon*		56-6		Southern	73-11
Torquigener hicksi		105-10	Triggerfish, Black		100-8		Tailblotch	73-8
	pallimaculatus	105-12		Blackpatch	101-6		Yellowback	73-10
	parcuspinus	105-11		Blue-finned	100-3		Yellowtail	73-5
toshi, Himantura		5-5		Brown	100-10	*tuberculata, Sorsogona*		21-19
Toxotes chatareus		54-1		Clown	101-2	*tuberosus, Naso*		94-4
	jaculatrix	54-2		Ebony	100-4	Tufted Sole		98-11
Trachinocephalus myops		11-15		Gilded	101-5	*tuka, Pseudanthias*		27-10
Trachinotus bailloni		39-10		Halfmoon	101-7	*tukula, Epinephelus*		24-5
	blochii	39-11		Lined	101-11	*tumbil, Saurida*		11-18
	botla	39-12		Pallid	100-9	*tumifrons, Dentex*		47-17
Trachyrhampus bicoarctatus		17-22		Pinktail	106-5	Tuna, Bigeye		96-9
	longirostris	17-23		Red-lined	100-1		Dogtooth	97-12
tragula, Upeneus		52-14		Redtooth	101-3		Mackerel	97-7
transluceus, Harpodon		11-19		Starry	100-2		Northern Bluefin	96-11
Trevally, Bassett-Hulls		40-3		Wedge-tailed	100-6		Skipjack	96-12
	Bigeye	38-8		White-barred	100-5		Southern Bluefin	96-10
	Black	38-2		White-spotted	101-4		Yellowfin	96-8
	Black-crested	40-1		Yellowmargin	101-1	Turretfish		103-8
	Bludger	37-15		Yellow-spotted	100-7		Black-blotched	103-7
	Blue	37-8	*trilineata, Pterocaesio*		46-4	Tuskfish, Anchor		71-12
	Bluefin	38-3		*Stethojulis*	79-8		Blackspot	74-8
	Blue-spotted	38-7	*trilineatus, Scolopsis*		51-13		Blue	74-3
	Brassy	38-4	*trilobatum, Thalassoma*		79-10		Bluespotted	74-9
	Bump-nosed	37-12	*trilobatus, Cheilinus*		70-12		Blue-toothed	78-9
	Cale Cale	40-2	*trimaculatus, Apogon*		35-12		Darkspot	74-11
	Club-nosed	37-6		*Apolemichthys*	59-1		Graphic	71-15
	Diamond	37-2		*Dascyllus*	64-13		Grass	71-13
	Epaulet	37-11		*Halichoeres*	75-6		Harlequin	71-14
	False Bluefin	37-14	*Trimma okinawae*		90-13		Purple	74-4
	Fringe-finned	37-3,		sp	90-12		Redstripe	74-7

	Species No.		Species No.		Species No.
Venus	71-16	Stripe-face	94-10	*varius, Gomphosus*	75-2
Wedge-tailed	74-10	Valming's	93-9	Heniochus	56-14
Zamboanga	74-12	*unicornis, Naso*	94-13	Veilfin, High-finned	16-7
Twin-lined Threadfin-Bream	49-6	*unifasciatus, Cheilinus*	70-11	*Velifer hypselopterus*	16-7
Twinspot Blenny	86-19	*unilineatus, Labrichthys*	77-4	*veliferum, Zebrasoma*	94-14
Cardinalfish	34-19	*unimaculata, Chrysiptera*	62-7	Velvetfish, Bearded	21-5
Chromis	65-6	*unimaculatus, Chaetodon*	56-1	Dark-finned	21-6
Flounder	98-3	*unitaeniata, Apogon*	35-5	Dusky	21-2
Goby	88-16	*Upeneus asymmetricus*	52-15	Sandpaper	21-1
Hawkfish	68-2	*moluccensis*	52-10	Thin	21-3
Twinspots Blenny	85-7	*sulphureus*	52-11	Threefin	21-4
twistii, Anampses	73-4	*sundaicus*	52-12	*venosa, Scorpaenopsis*	20-10
Two-banded Perch	28-1	*tragula*	52-14	*ventralis, Pseudanthias*	27-13
Two-eyed Cardinalfish	33-4	*vittatus*	52-13	*ventrifasciatus, Apogon*	35-14
Coralfish	58-9	*Uranoscopus bicinctus*	83-16	Venus Tuskfish	71-16
Two-spined Angelfish	58-13	*cognatus*	83-17	*venustus, Choerodon*	71-16
Twospot Anthias	27-16	*kaianus*	83-18	Vermicular Cod	24-10
Bristletooth	95-16	sp.	83-19	Vermiculate Spinefoot	93-11
Demoiselle	62-6	*Uraspis uraspis*	40-3	Vermiculated Angelfish	60-8
Goby	89-15	*uraspis, Uraspis*	40-3	*vermiculatus, Siganus*	93-11
Maori Wrasse	71-6	Urchin Clingfish	12-6	*verrucosa, Synanceja*	18-19
Pigfish	71-2	*urodeta, Cephalopholis*	22-9	*verrucosus, Rhinecanthus*	101-6
Two-spot Banded Seaperch	43-4	*Urolophus mitosis*	4-4	*versicolor, Minous*	19-13
Twostripe Goby	91-3	sp.	5-7	*vespa, Paracentropogon*	19-14
Twotone Dartfish	91-12	*westraliensis*	4-3	*vidua, Melichthys*	106-5
Wrasse	76-14	*Uropterygius marmoratus*	8-14	*violacea, Myripristis*	15-4
tyleri, Canthigaster	104-8			*violaceus, Sargocentron*	14-14
Tylerius spinosissimus	103-9	**V**		*violascens, Neopomacentrus*	62-18
Tyler's Toby	104-8	*vachelli, Ambassis*	30-6	Violet Demoiselle	62-18
Tylosurus crocodilus	13-12	Vagabond Butterflyfish	57-11	Squirrelfish	14-14
gavialoides	13-13	*vagabundus, Chaetodon*	57-11	Violet-lined Maori Wrasse	71-7
typus, Histiopterus	68-1	*vaigiensis, Abudefduf*	61-3	Parrotfish	81-2
Oxycirrhitus	67-18	Centrogenys	22-2	*virescens, Aprion*	41-2
Pristipomoides	41-7	Kyphosus	53-14	*virgatus, Nemipterus*	49-13
Rhiniodon	1-1	Leptoscarus	80-4	Siganus	92-15
Xiphocheilus	78-9	Liza	69-3	*viridescens, Balistoides*	100-3
		vaiuli, Pomacentrus	66-10	*viridis, Chromis*	64-8
U		*Valamugil buchanani*	69-6	Euleptorhamphus	13-8
uii, Carangoides	37-17	*Valenciennea alleni*	91-2	*vitta, Choerodon*	74-7
ulietensis, Chaetodon	56-4	*decora*	91-7	Lutjanus	42-11
Ulua aurochs	40-1	*heldingenii*	91-3	Pentapodus	50-6
mentalis	40-2	*immaculata*	91-8	*vittata, Myripristis*	15-5
undosquamus, Saurida	11-17	*longipinnis*	90-17	Sillago	30-18
undulata, Himantura	5-6	*muralis*	90-16	*vittatus, Upeneus*	52-13
undulatus, Balistapus	100-1	*parva*	91-6	*valmingii, Naso*	93-9
Cheilinus	71-11	*puellaris*	90-18	Vlaming's Unicornfish	93-9
Gymnothorax	7-12	*sexguttata*	91-4	*vogleri, Johnius*	53-5
undulostriatus, Epinephelus	25-11	*strigata*	91-5	Volitans Lionfish	19-6
unicolor, Apogon	33-6	*valentini, Canthigaster*	104-13	*volitans, Pegasus*	16-13
Gymnosarda	97-12	*Vanderhorstia ornatissimus*	90-19	Pterois	19-6
Unicorn Leatherjacket	102-1	*vanicolensis, Mulloidichthys*	51-15	*vosmeri, Scolopsis*	50-9
Unicornfish, Blunt	106-4	Variegated Cardinalfish	31-14	*vroliki, Centropyge*	60-5
Brown	94-13	Emperor	47-13	*vulpes, Albula*	6-2
Elongate	93-7	Lizardfish	11-14	*vulpinus, Bodianus*	70-8,
Humpback	93-4	Wrasse	72-13		106-3
Humphead	94-12	*variegatus, Fowleria*	31-14	Siganus	93-16
Longnosed	94-11	Lethrinus	47-13	Vulture Eel	9-15
Ringtailed	93-3	Synodus	11-14	*vulturis, Ichthyapus*	9-15
Single-spined	93-8	*Variola louti*	24-11		
Sleek	93-5	*varipinnis, Scorpaenodes*	20-7		

	Species No.
W	
Wahoo	97-1
waigiensis, Psammoperca	30-2
waitei, Luzonichthys	27-7
Waite's Splitfin	27-7
wardi, Orectolobus	1-15
Waspfish, Blackspot	19-12
Cockatoo	19-11
Long-finned	20-2
Short-finned	20-1
Spotfin	19-14
watanabei, Genicanthus	60-11
Watanabe's Angelfish	60-11
Wavy-lined Blenny	85-8
weberi, Chromis	64-12
Trixiphichthys	99-10
Weber's Chromis	64-12
Wedge-tailed Triggerfish	100-6
Tuskfish	74-10
Weed Blenny	84-8
Cardinalfish	31-13
Weedy Dragonet	88-3
Scorpionfish	18-13
Wrasse	77-5
westraliensis, Narcine	4-5
Urolophus	4-3
Western Australian Seahorse	17-5
Buffalo-Bream	53-13
Butterfish	50-6
Butterflyfish	55-4
Dragonet	87-12
Frogfish	12-2
Gregory	66-14
Round Skate	4-8
School Whiting	30-18
Yellowfin Bream	47-15
Wetmorella albofasciata	79-11
nigropinnata	79-12
Whale Shark	1-1
Whaler, Black	2-13
Bronze	2-5
wheeleri, Amblyeleotris	89-2
Plectranthias	22-14
Wheeler's Shrimp Goby	89-2
Whipray, Leopard	5-6
Whipfin Ponyfish	40-14
Silver-biddy	50-14
Whiptail, Butterfly	51-4
False	50-5
Paradise	51-3
Small-toothed	51-2
Three-striped	51-6
White-shouldered	51-1
Whiskered Goby	89-7
White Damsel	62-17
Pipefish	18-10
Ribbon Eel	8-17
Whiteband Maori Wrasse	70-11
White-banded Damsel	66-4
White-barred Bigeye	31-1

	Species No.
Triggerfish	100-5
White-bellied Rougefish	20-12
White-blotched Rockcod	24-2
Whitecap Goby	88-13
Whitecheek Shark	2-9
Monocle Bream	50-9
White-cheeked Surgeonfish	94-2
Whitefin Surgeonfish	95-6
Trevally	37-10
White-finger Anglerfish	12-10
whiteleggii, Cubiceps	99-6
Whitelegge's Eyebrowfish	99-6
White-lined Blenny	86-18
Rockcod	22-1
White-lipped Catfish	11-6
White-margined Moray	8-3
Whitepatch Damsel	62-13
Razorfish	79-14
White-saddled Pipefish	17-20
White-shouldered Whiptail	51-1
Whitespot Parrotfish	82-4
White-spotted Shovelnose Ray	3-15
Stargazer	83-19
Surgeonfish	95-4
Triggerfish	101-4
Whitestreak Monocle Bream	51-9
White-tail Angelfish	58-15
Whitetip Shark	3-5
White-tongued Trevally	37-13
Whiting, Golden-lined	30-14
Northern	30-17
Stout	30-15
Trumpeter	30-16
Western School	30-18
whitleyi, Abudefduf	63-3
Whitley's Sergeant	63-3
Wicker-work Sole	98-14
Wide Sawfish	3-13
Wideband Fusilier	46-7
Wide-barred Shrimp Goby	88-8
wilsoni, Pseudochromis	28-16
Winghead Shark	3-8
Wobbegong, Banded	1-14
Northern	1-15
Tasselled	1-16
Wolf Cardinalfish	31-9
Herring	6-5
Woore's Rockcod	24-6
Worm-eel, Slender	9-4
Wormfish, Curious	91-16
Onespot	91-15
Wrasse, Argus	76-12
Black Leopard	77-9
Black-striped	74-15
Blacktail	73-2
Blue and Yellow	70-3
Blue-headed	78-2
Blueside	72-1
Breastspot	76-8
Carpet	78-12

	Species No.
Celebes	71-8
Chain-lined	76-5
Chiseltooth	79-7
Choat's	73-13
Circle-cheek	76-10
Clubnosed	75-2
Connie's	72-15
Cryptic	79-4
Cutribbon	79-3
Diagonal-lined	79-11
Diana's	71-4
Diamond	75-8
Disappearing	72-17
Dotted	72-6
Double-headed Maori	71-11
Eight-lined	72-16
Exquisite	72-3
Feminine	73-1
Filamentous	72-10
Five-banded	75-11
Fivestripe	79-9
Flagfin	77-7
Fourspot	75-3
Golden	76-2
Goldstripe	76-4
Green	76-16
Green Moon	78-5
Green-spotted	76-1
Hump-headed	74-14
Jansen's	78-7
Japanese	79-1
Jordan's	74-6
Kuiter's	73-14
Ladder	79-10
Long Green	78-8
Lubbock's	72-4
Magenta-streaked	72-2
McCosker's	72-11
Midget	72-14
Moon	78-4
Nebulous	75-10
New Guinea	73-3
Ocellated	76-9
One-lined	77-4
Oriental	71-9
Ornamental	76-13
Ornate	77-8
Peacock	74-2
Philippines	79-6
Pointed-head Maori	71-10
Purple	75-7
Purple-lined	72-5
Randall's	72-8
Red and Green	78-1
Redhead	76-3
Red-lined	75-4
Redspot	77-10
Reticulated	73-15
Ringed Slender	77-1
Rust-banded	79-5

	Species No.		Species No.		Species No.
Scarlet-breasted Maori	70-9			Demoiselle	61-11
Schwartz's	76-11	**Y**		Fangblenny	86-1
Scott's	72-7	*yaeyamensis, Ecsenius*	85-6	Grunter	30 9
Scribbled Chisel-toothed	70-2	*yanashiroi, Pseudocoris*	79-1	Kingfish	39-15
Seagrass	79-17	Yellow Boxfish	103-3	Scad	37-5
Seven-banded	78-3	Cardinalfish	35-16	Stargazer	83-17
Sharpnose	79-12	Chromis	65-2	Trumpeter	30-9
Sharp-nosed	74-1	Devilfish	29-5	Tubelip	73-5
Shoulderspot	77-5	Emperor	28-1	Wrasse	70-4
Silver-streaked	77-11	Fangblenny	86-4	Yellow-tailed Butterflyfish	57-5
Silty	76-7	Sabretooth Blenny	84-15	Emperor	106-2
Six-banded	78-6	Shovelnose Ray	4-2	Perch	30-9
Six-lined	77-6	Shrimp Goby	88-9	Sergeant	63-4
Slingjaw	75-1	Yellow and Blueback Fusilier	46-1	Yellow-tipped Threadfin-Bream	
Soela	77-12	Yellow-Axil Chromis	65-12		49-14
Spotfin	72-12	Yellowback Fusilier	46-2	*Yirrkala lumbricoides*	9-14
Spotted Chisel-toothed	70-1	Tubelip	73-10	*Yongeichthys nebulosus*	90-20
Spotted-tail	74-16	Yellowbelly Cardinalfish	34-6		
Thick-lipped	75-12	Threadfin-Bream	49-2	**Z**	
Threadfin	72-9	Yellow-breasted Wrasse	73-4	*Zabidius novemaculeatus*	54-8
Three-eyed	76-6	Yellow-cheeked Threadfin-Bream		Zamboanga Tuskfish	74-12
Three-lined	75-6		49-9	*zamboangae, Choerodon*	74-12
Threespot	75-6	Yelowdot Sweetlips	44-9	*Zanclus cornutus*	92-9
Tripletail Maori	70-12	Yellow-dotted Butterflyfish	57-8	*zebra, Dendrochirus*	19-2
Twospot Maori	71-6	Maori Wrasse	70-10	Echidna	7-3
Twotone	76-14	Yellow-edged Moray	7-8	Heterodontus	1-6
Variegated	72-13	Yelloweye Blenny	86-16	Pterelotris	91-11
Violet-lined Maori	71-7	Leatherjacket	101-10	Zebra Eel	7-3
Whiteband Maori	70-11	Yellow-eyed Toadfish	105-11	Dartfish	91-11
Yellow-breasted	73-4	Damsel	62-8	Lionfish	19-2
Yellow-dotted Maori	70-10	Yellowfin Dottyback	28-16	Shark	1-2
Yellowtail	70-4	Goatfish	51-15	Wrasse	79-2
Zebra	79-2	Parrotfish	82-3	*Zebrasoma scopas*	94-15
Zigzag	76-15	Soldierfish	15-2	*veliferum*	94-14
Zoster	71-17	Surgeonfish	95-13	*Zebrius craticula*	98-14
		Tuna	96-8	*Zenarchopterus buffonis*	13-3
X		Yellowhead Dottyback	28-15	*Zenopsis nebulosus*	16-4
Xanthichthys auromarginatus	101-5	Yellow-lined Anthias	27-5	*Zeus faber*	16-5
lineopunctatus	100-11	Yellowlip Emperor	48-10	Zigzag Wrasse	76-15
xanthocheilus, Lethrinus	48-10	Yellow-lipped Threadfin-Bream		*zijsron, Pristis*	3-11
xanthochira, Chromis	65-12		49-13	*zonata, Pseudamia*	34-10
xanthometopon, Pomacanthus	60-12	Yellowmask Angelfish	60-12	*zonatus, Cheilodipterus*	34-6
xanthonota, Caaesio	46-2	Surgeonfish	94-1	*Pristipomoides*	44-7
Labropsis	73-10	Yellowmargin Triggerfish	101-1	*zonipectus, Gymnothorax*	7-13
xanthopterus, Acanthurus	95-13	Yellow-margined Seaperch	43-7	Zoster Wrasse	71-17
xanthura, Chromis	65-15	Yellowmouth Moray	8-8	*zosterophora, Archamia*	34-17
xanthurus, Chaetodon	57-5	Yellowsnout Large-eye Bream	48-6	*zosterophorus, Choerodon*	71-17
Diproctacanthus	73-5	Yellowspot Goby	89-18	*zygaena, Sphyrna*	3-10
Paracaesio	44-4	Yellow-spotted Emperor	48-11	*zysron, Nemipterus*	49-9
xenicus, Diplogrammus	87-13	Groper	25-3		
xenochrous, Scolopsis	50-11	Rockcod	23-3		
xenodontus, Phyllophichthus	9-8	Scorpionfish	18-18		
Xenojulis margaritaceous	77-13	Trevally	37-9		
Xiphasia setifer	84-17	Triggerfish	100-7		
Xiphias gladius	96-6	Yellowstripe Goatfish	51-14		
Xiphocheilus typus	78-9	Monocle Bream	51-8		
Xyrichtys aneitensis	79-14	Yellow-striped Goatfish	52-4		
dea	78-11	Squirrelfish	15-7		
pavo	78-10	Yellowtail Angelfish	59-6		
pentadactylus	79-15	Blue Snapper	44-4		